# Microbial Technology

SECOND EDITION/VOLUME II

Second Edition/Volume II

# Microbial Technology

Fermentation Technology

Edited by

**H. J. PEPPLER**
Universal Foods Corporation
Milwaukee, Wisconsin

**D. PERLMAN**
School of Pharmacy
University of Wisconsin
Madison, Wisconsin

**ACADEMIC PRESS**
**New York    San Francisco    London    1979**
A Subsidiary of Harcourt Brace Jovanovich, Publishers

ACADEMIC PRESS, INC.
111 Fifth Avenue, New York, New York 10003

*United Kingdom Edition published by*
ACADEMIC PRESS, INC. (LONDON) LTD.
24/28 Oval Road, London NW1 7DX

Library of Congress Cataloging in Publication Data

Peppler, Henry J
    Microbial technology.

    Vol. 2 edited by H. J. Peppler and D. Perlman.
    Includes bibliographies and index.
    CONTENTS: v. 1. Microbial processes.—v. 2.
Fermentation technology.
    1.  Industrial microbiology.  2.  Fermentation.
I.  Perlman, D.  II.  Title.  [DNLM: 1.  Bacteriologi-
cal technics.  2.  Fermentation.  QY100 M626]
QR53.P45  1979          660'.62          78–67883
ISBN  0–12–551502–2 (v. 2)

# Contents

v

**Chapter  8  Mushroom Fermentation**
Randolph T. Hatch and Stanley M. Finger

**Chapter  9  Inocula for Blue-Veined Cheeses and Blue
Cheese Flavor**
Gerard J. Moskowitz

**Chapter 10  Microorganisms for Waste Treatment**
Larry L. Gasner

**Chapter 11  Elementary Principles of Microbial Reaction
Engineering**
George T. Tsao

# List of Contributors

Numbers in parentheses indicate the pages on which the authors' contributions begin.

**Maynard A. Amerine** (131), Department of Viticulture and Enology, University of California, Davis, California 95616

**William B. Armiger** (375), BioChem Technology, Inc., Great Valley Corporate Center, Malvern, Pennsylvania 19355

**William H. Bartholomew** (463), Research Department, Stauffer Chemical Company, Westport, Connecticut 06880

**Paul A. Belter** (403), Fermentation Research & Development, Fine Chemical Division, The Upjohn Company, Kalamazoo, Michigan 49001

**W. T. Blevins** (303), Auburn University, Botany and Microbiology Department, Auburn, Alabama 36830

**L. T. Chang** (243), Bristol-Myers Company, Industrial Division, Syracuse, New York 13201

**Ichiro Chibata** (433), Research Laboratory of Applied Biochemistry, Tanabe Seiyaku Co., Ltd., 16-89 Kashima-3-chome, Yodogawa-ku, Osaka, Japan

**N. D. Davis** (303), Auburn University, Botany and Microbiology Department, Auburn, Alabama 36830

**Richard P. Elander** (243), Bristol-Myers Company, Industrial Division, Syracuse, New York 13201

**Stanley M. Finger** (179), Department of Chemical Engineering, University of Maryland, College Park, Maryland 20742

**Larry L. Gasner** (211), Department of Chemical Engineering, University of Maryland, College Park, Maryland 20742

**Randolph T. Hatch** (179), Department of Chemical Engineering, University of Maryland, College Park, Maryland 20742

**C. W. Hesseltine** (95), Northern Regional Research Center, Federal Research, Science and Education Administration, U.S. Department of Agriculture, Peoria, Illinois 61604

**N. J. Huige\*** (1), Jos. Schlitz Brewing Company, Milwaukee, Wisconsin 53212

**Arthur E. Humphrey** (375), School of Engineering, University of Pennsylvania, Philadelphia, Pennsylvania 19104

**W. F. Maisch** (79), Research Department, Hiram Walker & Sons, Inc., Peoria, Illinois 61601

**Irving Marcus** (497), Oblon, Fisher, Spivak, McClelland & Maier, Arlington, Virginia 22202

**Gerard J. Moskowitz** (201), Dairyland Food Laboratories, Inc., Waukesha, Wisconsin 53187

**G. B. Nickol** (155), Research Division, U.S. Industrial Chemicals Co., Cincinnati, Ohio 45237

**L. K. Nyiri†** (331), Department of Chemical Engineering, Lehigh University, Bethlehem, Pennsylvania 18015

**Norman F. Olson** (39), Department of Food Science, University of Wisconsin, Madison, Wisconsin 53706

**D. Perlman** (173), School of Pharmacy, University of Wisconsin, Madison, Wisconsin 53706

**A. J. Petricola** (79), Technical Division, Hiram Walker & Sons Limited, Walkerville, Ontario N8Y 4S5, Canada

**Herman J. Phaff** (131), Department of Food Science and Technology, University of California, Davis, California 95616

**Harold B. Reisman** (463), Stauffer Chemical Company, Westport, Connecticut 06880

**Tadashi Sato** (433), Research Laboratory of Applied Biochemistry, Tanabe Seiyaku Co., Ltd., 16-89 Kashima-3-chome, Yodogawa-ku Osaka, Japan

**M. Sobolov** (79), Technical Division, Hiram Walker & Sons Limited, Walkerville, Ontario N8Y 4S5, Canada

**L. P. Tannen\*\*** (331), New Brunswick Scientific Co., Edison, New Jersey

---

\* Present address: Miller Brewing Company, Milwaukee, Wisconsin 53208.
† Deceased.
\*\* Present address: The Great Atlantic and Pacific Tea Co., Inc., Manufacturing Group, Horseheads, New York, 14845.

**Tetsuya Tosa** (433), Research Laboratory of Applied Biochemistry, Tanabe Seiyaku Co., Ltd., 16-89 Kashima-3-chome, Yodogawa-ku, Osaka, Japan

**George T. Tsao** (223), School of Chemical Engineering, Purdue University, West Lafayette, Indiana 47907

**Hwa L. Wang** (95), Northern Regional Research Center, Federal Research, Science and Education Administration, U.S. Department of Agriculture, Peoria, Illinois 61604

**D. H. Westermann*** (1), Jos. Schlitz Brewing Company, Milwaukee, Wisconsin 53212

* Present address: Jayto, New Berlin, Wisconsin 53151.

# Preface

In the decade since the first edition of "Microbial Technology" appeared, applied microbiology has changed, expanded, and diversified. As new products were introduced in this period, and greater demand for some of the old ones developed, the total fermentation capacity increased at about the same rate noted in the previous 25 years. The number of fermentation products, it is estimated, has quadrupled, while the volume of products manufactured has increased tenfold. This growth has prompted publication of a second edition, completely revised and enlarged.

To accomplish a worldwide survey of industrial microbiology and to describe its contributions to agriculture, industry, medicine, and environmental control, the editors are indebted to 57 willing and expert contributors. Their comprehensive reviews of traditional fermentations and propagations, as well as newly developed microbe-dependent processes and products, are presented in a two-volume set.

Volume I, subtitled "Microbial Processes," describes the production and uses of economic bacteria, yeast, molds, and viruses, and reviews the technologies associated with products of microbial metabolism.

Volume II, subtitled "Fermentation Technology," deals principally with fermentations and modifications of plant and animal products for foods, beverages, and feeds, while reviewing salient aspects of microbial technology: general principles, culture selection, laboratory methods, instrumentation, computer control, product isolation, immobilized cell usage, economics, and microbial patents.

H. J. Peppler
D. Perlman

# Contents of Volume I

## MICROBIAL PROCESSES

# Chapter 1

# Beer Brewing

D. H. WESTERMANN
N. J. HUIGE

**1**

MICROBIAL TECHNOLOGY, 2nd ed., VOL. II
Copyright © 1979 by Academic Press, Inc.

## I. INTRODUCTION

### A. Historical Background

History indicates each society has developed an alcoholic beverage of some type based on the indigenous sources of starch or sugar. Wheat, corn, rye, millet, rice, oats, barley, potatoes, and other vegetables and fruits have been converted to potable alcoholic beverages. It is generally conceded that beer originated in the twin cradles of civilization, Mesopotamia and Egypt, about 6000 years ago.

According to Weeks (1949), the etymology of the word beer, as we know it today, indicates it originated from the Latin verb *bibere,* to drink. Similarly, the Spanish word for beer, *cerveza,* apparently originated from *cerevisia,* which combines the Latin *ceres,* goddess of grain, and *vis,* vigor.

It is not clear from historical records whether this vigor, which can be interpreted as the physiological effect of alcohol, or its preservative contribution to grains was the principal stimulus for long-term growing acceptance by societies. However, it is known that alcoholic beverages became closely associated with religious ritual during the infancy of early civilized societies.

The knowledge that a concoction of grains and water could ferment naturally to produce an alcoholic beverage followed the advance of agriculture across Europe with the Celts. The Teutons, who eventually settled in the Rhine area and became the Germanic Tribe, had followed the Celts in their westward migration. Some of the Teutons eventually settled in England by the fifth century and were followed by the Saxons. At least the possible sequence by which the brewing art could have spread across the continent and England appears historically logical.

The influence of the church on the art and science of brewing grew rapidly over a 1000-year period, reaching a climax in the sixteenth and seventeenth centuries. The influence of religious and civil law on brewing in Europe and England is a fascinating story, well documented by Arnold (1911).

Major brewing centers were eventually established in Pilsen, Czechoslovakia; Munich and Dortmund, Germany; Burton-on-Trent, England; and Dublin, Ireland. The local water supplies of each area gave rise to unique product characteristics associated only with that area.

Although the American Indians were already making a brew from maize, English-type beer was brought to America on the *Mayflower,* according to Weeks (1949). A journal of the *Mayflower* contained the following record: "We could not now take time for further search . . . our victuals being much spente, especially our beer."

English ale brewing traditions accompanied the colonists, and it was not until the 1840s that German lager beer became more widely accepted in the United States (Singruen, 1938).

English ales were fermented at ambient temperatures (21°–27°C) with top-fermenting yeast, while the German lager beers were fermented with bottom-fermenting yeast at the lowest temperatures which could be sustained year-round (10°–15°C). The lager beer was stored in cool caves for several months, resulting in improved physical stability (Baron, 1962). The superior keeping quality and preferred taste of lager beer rapidly led to its domination of the American market.

## B. Current Practice

Beer retains its unique taste and character because of historical acceptance coupled with legal restraints of composition. In the United States Federal Regulations, beer is defined as a malt beverage resulting from an alcoholic fermentation of the aqueous extract of malted barley with hops. It may include other sources of carbohydrates called adjuncts. Legally, malt liquor, ale, porter, stout, and sake are also malt beverages.

Current American beers are made from malted barley, with corn grits or syrup or rice grits as the adjunct carbohydrate source, and hops. Purified water, to which brewing salts are added, and a cultured yeast complete the five principal ingredients of brewing.

Barley which has been malted comprises the basic raw material for brewing. In the malting process, the barley kernel, separated from the stalk and chaff, is germinated under controlled temperature and humidity to generate enzyme systems and partially degrade endosperm starch and protein. Growth of the germinated kernel is stopped by drying of kilning in hot air, during which time some flavor and color components are also formed. The dried malt is milled before it is mashed with brewing water.

In the mashing process, milled malt and adjunct, such as corn grits or rice, are each made into a mash with brewing water, subjected to individual controlled time–temperature cycles, and combined for further enzymatic digestion and subsequent extraction of sugars and proteins.

The extract produced in the combined mashing process is separated from the nonsoluble husk and grain portions in a lauter tub or filter press. This extract is called wort.

Wort is boiled in the brew kettle, along with hops and syrup, if used as an adjunct. The dried hop cones used for brewing contain lupulin glands which are the source of bitter resins and essential oils that contribute some of the characteristic bouquet to beer.

The boiled wort is cooled, inoculated with a pure yeast culture, and fermented. Each brewery maintains a yeast strain considered to be unique to its own process and product character. The fermentable sugars are converted to ethyl alcohol and trace quantities of flavor ingredients, while the nonfermentable carbohydrates remain in the beer. Carbon dioxide from the fermentation is recovered, purified, and returned to the finished product.

The yeast generated during fermentation is separated from the beer by centrifugation or natural settling. After storage and aging for several weeks, recarbonation, and two or more filtrations, the finished product is packaged.

## II. RAW MATERIALS

### A. Barley

For centuries barley was the "staff of life" in Europe, as evidenced by barley kernels found in ancient ruins and tombs. Until the fifteenth century it was the principal ingredient of bread. However, because barley flour is not readily leavened, wheat, which makes a lighter bread, became the favorite of bakers.

Botanically, barley is a grass or member of the Graminae family. However, it has unique properties which make it ideally suited for the brewing of beer. The barleys used for brewing possess a very tough husk which is firmly cemented to the kernel. This husk provides protection for the kernel during handling and subsequent germination. Later in the brewing process the husks form a filter bed for the separation of extractable carbohydrates and protein from the mash and contribute a characteristic flavor. Barley can be readily malted, a process in which the events of plant reproduction are simulated and later stopped by the removal of moisture. During the malting process barley produces large amounts of amylases, proteases, and other enzymes which partially degrade starch, proteins, and some types of cellulose. The biochemistry of the malting process is described in Section III.

Two distinct barley types and many varieties of each type have been grown for brewing. The two-row and six-row designations of type describe the number of rows which form around the axis of the kernel head. In two-row barley, one kernel develops at each of two nodes, while three kernels develop at each node in six-row barley. Each type of barley has its own brewing characteristics. Historically, two-row barley produced a more mellow beer and contained less protein and enzymatic potential than six-row barley. This resulted largely from the greater plumpness of the two-row kernel (i.e., less surface to volume ratio). Agronomic crossbreeding during the last decade has tended to minimize this difference.

To produce typical light American beers, barleys must be used which provide ample enzymatic activity to hydrolyze adjuncts. The six-row barley varieties, Larker, Dickson, and Conquest, are typical of those grown in North and South Dakota and Minnesota. These varieties provide about 80% of the barley for the American brewing industry. They have high enzymatic potential and protein contents of up to 13.5%. While higher enzymatic potential and protein content can be generated, brewers usually avoid such barley because of product instability. A western six-row barley with less desirable brewing characteristics is also grown primarily in California.

The two-row barleys, known for their contribution of mellowness and flavor, include varieties such as Piroline, Betzes, Klages, Firlbecks III, Vanguard, Hannchen, and Shabet. They are grown in the western states of Washington, Oregon, Idaho, and Montana.

The proximate composition of the three major barley types, described by Winton and Winton (1932), is shown in Table I, and the analysis of malts from these barleys in Table II.

## B. Adjuncts

Approximately 70% of the malted barley kernel is potentially available as soluble carbohydrate. It is this carbohydrate, principally starch, which is converted to fermentable sugar and nonfermentable dextrins. The fermentable sugars are ultimately converted primarily to ethyl alcohol. The maximum alcohol and remaining solids content of an all-malt beer is therefore established by the milled malt solids concentration which can conveniently be handled in an all-malt mash. The ratio of protein to carbohydrate and amount of alcohol in the finished beer is primarily a fundamental characteristic of the malted barley.

All-malt beers historically were more satiating and had poor physical

**TABLE I.** Proximate Composition of Three Barley Types

| Analysis | Midwestern six-row | California six-row | Western two-row |
|---|---|---|---|
| Kernel Weight (mg) | 36 | 44 | 40 |
| Husk (%) | 12 | 14 | 10 |
| Protein (%) | 12 | 11 | 10 |
| Fat (%) | 2 | 2 | 2 |
| Starch (%) | 58 | 58 | 60 |
| Fiber (%) | 5.7 | 6.6 | 5.2 |
| Ash (%) | 2.7 | 3.0 | 2.5 |
| Enzyme potential after malting | High | Low | Medium |

**TABLE II.**  Typical Analyses of Malts from Three Barley Types

| Malt analysis[a] | Midwestern six-row | California six-row | Western two-row |
|---|---|---|---|
| Kernel weight, dry basis (mg) | 30.0 | 39.0 | 37.0 |
| Growth; length of acrospire | | | |
| 0–¼ (%) | 0 | 1 | 1 |
| ¼–½ (%) | 1 | 5 | 1 |
| ½–¾ (%) | 5 | 9 | 6 |
| ¾–1 (%) | 92 | 85 | 91 |
| Overgrown (%) | 2 | 0 | 1 |
| Assortment | | | |
| On 7/64 screen (%) | 32 | 68 | 85 |
| On 6/64 screen (%) | 53 | 26 | 10 |
| On 5/64 screen (%) | 14 | 6 | 1 |
| Through screen (%) | 1 | 6 | 1 |
| Moisture (%) | 4.2 | 4.4 | 4.2 |
| Extract, fine grind, dry basis (%) | 76.8 | 77.0 | 80.5 |
| Extract, coarse grind, dry basis (%) | 74.8 | 75.0 | 79.0 |
| Difference (%) | 2.0 | 2.0 | 1.5 |
| Wort color, ½ inch cell, °Lovibond | 1.6 | 1.4 | 1.2 |
| Protein (N × 6.25), dry basis (%) | 12.3 | 11.0 | 10.5 |
| Soluble protein as percentage of total | 40.0 | 35.0 | 38.0 |
| Diastatic power, dry basis, °L | 135 | 65 | 90 |
| $\alpha$-Amylase, dry basis, 20° units | 38 | 25 | 30 |

[a] From American Society of Brewing Chemists (1976).

stability due to their high concentration of soluble protein. American malts had three properties which were different from European malts. Protein content and enzymatic activity were substantially higher, and the husk was thicker. Early American brewers recognized that these properties could be exploited to use other sources of starch as a supplemental brewing material. The higher enzymatic activity would allow hydrolysis of gelatinized starch from other cereal grains, and the heavy husk would provide an efficient filter bed for extraction of solubles from mashes containing adjuncts. The limited residual proteins from these supplemental materials, such as corn and rice, were not readily hydrolyzed by malt proteases. Therefore, beers produced with these supplemental adjuncts had lower protein content for a given alcohol content, resulting in better physical stability with a lighter character.

With the exception of specialty products, all American brewers use an adjunct for brewing, as described by Matz (1970). While wheat, milo, and unmalted barley have been used, corn grits, corn syrup, and rice grits are dominant. Of the 155,000,000 barrels of beer currently produced, two-thirds are brewed with corn grits or corn syrup adjunct.

When corn grits, refined grits, flakes, or rice are used by the brewer, they are pregelatinized by boiling in a "cooker" before adding to the malt mash. Cooker mashes generally contain about 10% malt to provide the enzymes which partially hydrolyze the gelatinized starch to reduce viscosity.

Corn syrups produced for the brewing industry by the corn wet milling industry are made from acid and enzymatic hydrolysis of starch slurries. Dependent upon the method of hydrolysis, it is possible to produce syrups with 0–100% fermentability. Typical analyses of brewing adjuncts are shown in Table III.

Brewing technologists generally acknowledge that to produce presently accepted beers, yeast requires 120–140 mg/liter of free amino nitrogen for proper nutrition. Since the nitrogen is available only from malt, this limits the malt to adjunct ratio theoretically to approximately 50/50 for lager beers. American industry beers vary from 100% malt for the specialty products to about 60/40, malt/adjuncts, for lager beers.

## C. Hops

Since medieval times herbs have been used to flavor beers, but only hops are used commercially today. The characteristic aroma and bitterness imparted to beer by the oils and resins of hops make beer and similar malt beverages quite unique.

According to Arnold (1911), records exist indicating hops were grown in the seventh or eighth century, but the first recorded use in brewing was made by a nun, St. Hildegard (1098–1197), in a convent at Rupertsberg, Bingen-on-Rhine. The record states: "The hop is of a heating and drying nature, but does contain some little moisture; is, however, of slight benefit to man, since it promotes melancholy and creates in man a sad mood, and also it affects his bowels unpleasantly by reason of its heating properties. Its bitterness, though, when added to beverages, prevents in the latter putrification, and gives to them a longer durability." (Sic)

While the preservative value of hop resins is limited, it is known to contribute some microbiological stability during beer processing. Whether this attribute or its uniquely acceptable flavor contribution stimulated its use may never be put in proper perspective. Both attributes were important.

About 20% of the world's hops are grown in the United States, in the states of Washington, Oregon, California, and Idaho. The Old World famous centers of hop growing still prevail in Kent, England; Saaz, Bohemia; and Hallertau, Bavaria.

The natural family of Cannabinaceae consists of two genera, *Humulus* and *Cannabis*. *Cannabis* is represented by *C. sativa,* which includes Indian Hemp, marihuana, and hashish, while *Humulus* consists of *H. lupulus* and *H. japonicus*. The *H. lupulus* provides resins for brewing,

**TABLE III.** Typical Analyses of Brewing Adjuncts

| Composition | Amount |
| --- | --- |
| *Cereal Adjuncts* | |
|   Corn Grits | |
|     Moisture (%) | 10.9 |
|     Extract, dry basis (%) | 91.4 |
|     Oil, dry basis (%) | 0.76 |
|   Corn Flakes | |
|     Moisture (%) | 9.0 |
|     Extract, dry basis (%) | 92.4 |
|     Oil, dry basis (%) | 0.50 |
|   Refined Grits | |
|     Moisture (%) | 9.6 |
|     Extract, dry basis (%) | 103.3 |
|     Oil, dry basis (%) | 0.03 |
|   Rice | |
|     Moisture (%) | 12.0 |
|     Extract, dry basis (%) | 93.0 |
|     Oil, dry basis (%) | 0.86 |
| *Liquid Adjuncts* | |
|   Corn Syrup | |
|     Extract, as is (%) | 82.0 |
|     Extract, dry basis (%) | 100.0 |
|     Fermentable extract, as is (%) | 60.2 |
|     Reducing sugars (as dextrose), as is (%) | 50.5 |
|     Ash, as is (%) | 0.19 |
|     pH (10% solution) | 4.95 |

whereas the *H. japonicus* is an ornamental plant devoid of brewing value. It is the dried fruit or flower of the female hop plant which is used for brewing.

The hop plant possesses an extensive root system, which each year sends out a large number of shoots which grow very rapidly to a length of 20–25 ft. These climbing vines are supported on a wire network, and in late August or early September the flowers or cones of these female plants are picked, dried to 9–10% moisture, and baled. Extensive effort is taken to eliminate or minimize the growth of male plants to prevent pollination and seed production in the female flowers. The typical composition of a dried hop cone is given in Table IV, from Hough *et al.* (1971).

The structure of the hop cone is shown in Fig. 1. The cone consists of bracts and seed-bearing bracteoles which are attached to a supporting structure or strig. Lupulin glands, small resinous beads which contain the materials of brewing value, develop along with seeds at the base of the bracteole as the hop cone ripens. The lupulin glands contain the

**TABLE IV.** Proximate Composition of Hops[a]

| Component | Percent of total weight |
|---|---|
| Water | 10.0 |
| Total resins | 15.0 |
| Essential oils | 0.5 |
| Tannins | 4.0 |
| Monosaccharides | 2.0 |
| Pectin | 2.0 |
| Amino acids | 0.1 |
| Proteins (N × 6.25) | 15.0 |
| Lipids and Wax | 3.0 |
| Ash | 8.0 |
| Cellulose, lignin, etc. | 40.4 |
| Total | 100.0 |

[a] From Hough et al. (1971).

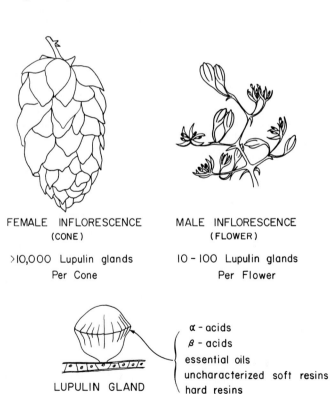

FEMALE INFLORESCENCE
(CONE)

>10,000 Lupulin glands
Per Cone

MALE INFLORESCENCE
(FLOWER)

10 - 100 Lupulin glands
Per Flower

LUPULIN GLAND
⊢0.1 mm⟶

$\alpha$ - acids
$\beta$ - acids
essential oils
uncharacterized soft resins
hard resins

**FIGURE 1.** Hop cone, male flower, and lupulin gland.

resins and essential oils which characterize the hop variety and determine its value in brewing.

The hop resins are broadly classified by their solubility, as indicated
by Findlay (1971), and the specific compounds which comprise the
approximate 15% total resins are shown by Burgess (1964) in Table V.
The major constituents of hop oil are the hydrocarbons myrcene,
farnesene, humulene, and caryophyllene, which account for 50–80% of
the oil and the 80 or more oxygenated compounds which are currently
identified.

The lupulin of hop varieties differs largely in its volatile oil composition
and percentage of cohumulone in the $\alpha$-acids. The European hop varieties
Hallertau and Saaz contain about 20% cohumulone in $\alpha$-acids, while the
American varieties Bullion and Brewer Gold contain about 40%. These
hops also contain more humulene and less myrcene than American
hops.

When the hop cones are added to the brew kettle the water-soluble
resins and oils are dissolved in the wort. The $\alpha$-acids are isomerized,
providing 85–90% of the characteristic hop bitterness. Some additional
bitterness is derived from oxidation products of $\alpha$-acids, with a very
minor contribution from oxidized $\beta$-acids. Methods of analyzing hops
and beer for $\alpha$-acids and iso-$\alpha$-acids, respectively, are given in "The
Methods of Analysis" (American Society of Brewing Chemists, 1976). An
excellent review of hops and hop chemistry is given by Hough et al.
(1971). The iso-$\alpha$-acids not only contribute the principal bitterness to
beer but are a major reactant, along with protein, for the formation of
stable foam with good adhesion.

The yield of iso-$\alpha$-acids in beer from $\alpha$-acids in hops added to the

**TABLE V.**  Proximate Composition of Hop Resins[a]

|                                           | Percent |
|-------------------------------------------|---------|
| Total hop resins                          | 15.0    |
| Soft resins                               | 11.5    |
| $\alpha$-Acids                            | 5.0     |
| Humulone                                  | 3.0     |
| Cohumulone                                | 1.5     |
| Ad-, pre-, posthumulone                   | 0.5     |
| $\beta$-acids                             | 3.0     |
| Lupulone                                  | 1.5     |
| Colupulone                                | 1.2     |
| Ad-, pre-, postlupulone                   | 0.3     |
| Uncharacterized resins                    | 3.5     |
| Hard resins (humulinone, hulupones)       | 3.5     |

[a] From Burgess (1964).

brew kettle is about 25–30%, so the financial incentive for improvement is substantial. One simple method of increasing yield is to grind the hop cones, thus increasing surface exposure of fractured lupulin glands to wort in the brew kettle. This increases yield to 35–40%. Solvent extracts of hops are also made with hexane or dichloromethane. These extracts added to the brew kettle give yields of about 40–45%.

Some commercially available extracts contain isomerized $\alpha$-acids which have been prepared in aqueous systems with sodium or potassium hydroxide, calcium carbonate, or magnesium carbonate. Isomerization rates are very rapid at higher pH compared to the pH of 5.0–5.4 which prevails in the brew kettle.

The yield of purified potassium salts of $\alpha$-acids properly added to beer after fermentation and filtration, as described by Westermann (1976), reaches 95+%. These salt extracts are void of the essential oils and $\beta$-resins normally contained in whole hops or whole hop extracts. The residual portion of extract from which the purified salts are made is added to the kettle to retain the contribution of remaining hop ingredients to beer character.

## D. Water

The chemical composition and concentration of salts in brewing water has a profound effect on the brewing process and resulting beer characteristics. Two ions, calcium and bicarbonate, are critical because of their impact on pH during the brew house operations. The pH of the water and resulting malt extract governs the activity of numerous enzyme systems of the malt, the malt yield, the solution of tannins and other coloring ingredients, the precipitation of proteins in the brew kettle, the rate of browning reactions, isomerization rate of humulone, and the physical and foam stability of the beer. Indeed, brewing water composition is of vital importance. Siebel (1950) has described the composition of water as suitable for brewing lager beer as shown in the tabulation below.

| Composition | Amount |
|---|---|
| Sodium carbonate | None |
| Calcium and magnesium carbonate | Up to 50 ppm |
| Sodium and magnesium sulfate | Up to 50 ppm |
| Calcium sulfate | 240 to 400 ppm |
| Sodium chloride | Up to 100 ppm |
| Calcium and magnesium chloride | Not over 75 ppm |
| Iron, as $Fe^{2+}$ | Less than 0.2 ppm |
| Silica, as $SiO_2$ | Less than 10 ppm |

The composition of brewing waters of the world famous brewing centers has been compiled by Scholefield (1956), as shown in Table VI.

The high permanent hardness of water used for Burton bitter ales should be noted. The sweeter, darker beers of Dublin, London, and Munich contain less calcium sulfate and more calcium carbonate. Pilsen pale lager is brewed with very soft water.

An ideal water for American lager beers would contain a hardness level midway between Dortmund and Pilsen water, i.e., about 100–125 ppm of $Ca^{2+}$ with generally less than 20 ppm of bicarbonate hardness. At these levels the mash would have a pH of less than 5.6, minimizing the extraction of silicates, tannins, and other harshly flavored materials from the malt. Similarly, the extract in the brew kettle (wort) would ultimately have a pH of 5.1–5.2, producing a beer with potentially good foam and physical stability.

Since waters theoretically ideal for brewing lager beer are not naturally available, two corrective actions can be taken when appropriate. When temporary hardness (bicarbonate) is too high, a mineral acid is added to the water, followed by heating and gas stripping to remove $CO_2$ from the decomposition of carbonic acid. In addition, gypsum ($CaSO_4$) is added to grains sparge water and to the brew kettle to raise the $Ca^{2+}$ content of the resultant wort.

The desired pH in both mashing and the brew kettle result from the following reactions. The carbonate ion, bicarbonate ion, and $CO_2$, or carbonic acid, exist in equilibrium, depending upon pH of the water. At high pH the carbonate ion prevails, at neutral pH the bicarbonate, and at low pH, dissolved $CO_2$ exists. When heating water which contains the bicarbonate ion, an undesirable rise in pH occurs as follows:

$$Ca(HCO_3)_2 \rightleftharpoons Ca^{2+} + 2\ HCO_3^-$$

$$HCO_3^- + H_2O \rightleftharpoons H_2CO_3 + OH^-$$

$$H_2CO_3 \rightleftharpoons H_2O + CO_2 \uparrow$$

Therefore, as previously mentioned, bicarbonate ion concentration is usually reduced before water is used for brewing.

The calcium ion from gypsum ($CaSO_4$) is used to lower the pH in mashing and the brew kettle due to its reaction with secondary phosphates, phytate, and some proteins, peptides, and nucleic acids.

$$3\ Ca^{2+} + 2\ HPO_4^{2-} \overset{\Delta}{\rightleftharpoons} 2\ H^+ + Ca_3(PO_4)_2 \downarrow$$

A similar reaction occurs with phytic acid. The calcium salts tend to precipitate on boiling, exceeding the concentration allowed by virtue of their solubility product constant. If an increase in sweetness or palate

**TABLE VI.** Ionic Concentrations of Typical Brewing Liquors[a,b]

| Ions | Burton-on-Trent | Dortmund | Edinburgh (sandstone) | London (off chalk) domestic | Munich (Dublin-type) | London (deep-well) | Pilsen |
|---|---|---|---|---|---|---|---|
| $Na^+$ | 1.3 | 3.0 | 4.0 | 1.1 | 0.1 | 4.3 | 0.1 |
| $Mg^{2+}$ | 5.2 | 1.9 | 3.0 | 0.3 | 1.6 | 1.6 | 0.1 |
| $Ca^{2+}$ | 13.4 | 13.0 | 7.0 | 4.5 | 4.0 | 2.6 | 0.4 |
| $NO_3^-$ | 0.5 | — | 0.5 | 0.1 | 0.1 | — | — |
| $Cl^-$ | 1.0 | 3.0 | 1.7 | 0.5 | 0.1 | 1.7 | 0.1 |
| $SO_4^{2-}$ | 13.7 | 5.9 | 4.8 | 1.2 | 0.1 | 1.6 | 0.1 |
| $CO_3^{2-}$ | 4.7 | 9.0 | 7.0 | 4.1 | 5.5 | 5.2 | 0.3 |
| $(Ca^{2+} + Mg^{2+})/$ $CO_3^{2-}$ | 100/25 | 100/60 | 100/70 | 100/86 | 100/77 | 100/24 | 100/71 |
| Total salts (ppm) | 1226 | 1011 | 800 | 319 | 273 | 463 | 30.8 |

[a] Figures rounded to one decimal place. Concentration of ions in millivals (N/1000).
[b] From Scholefield (1956).

fullness is desired, $CaCl_2$ can be used to obtain the apparent results from the chloride ion. An interesting review of ion impact on product properties and processes is given by Comrie (1967).

## E. Yeasts

Yeasts are unicellular fungi which are classified by their spore-forming characteristics into one of three out of four fungi groups: Ascomycetes, Basidiomycetes, or Fungi Imperfecti. Of the approximately 350 species of 39 genera of yeasts in these three groups, only two species of the genus *Saccharomyces* are important in brewing. These are *S. cerevisiae* and *S. carlsbergensis*. "There is no clear indication from yeast ecology how *S. carlsbergensis* and *S. cerevisiae* came to be universally employed for beer production" (Hough *et al.*, 1971). *Saccharomyces cerevisiae* is traditionally used in baking, distilling, brewing of sake and ale, production of industrial alcohol, and for animal feed.

According to Lodder and Kreger-Van Rij (1952), Stelling-Dekker established in 1931 that top-fermenting yeasts used for the production of ales and the bottom-fermenting yeasts used for the production of lager beer were actually different. She classified those ale yeasts which rose to the top of the fermenter at the completion of fermentation as *S. cerevisiae*. Those yeasts, traditionally used for lager, which agglomerated and settled to the bottom at the completion of fermentation were classified as *S. carlsbergensis* (now *S. uvarum*). The continued application of these species to their respective brewing process conditions may have enhanced divergence of their physiological and morphological characteristics.

In order to assure continued use of the desired yeast species and strain, it is necessary to have clear criteria for identification. This is accomplished through a wide variety of measurements and procedures, none of which is solely adequate. Cell shape, growth on selected solid media, spore formation, and colony geometry are the principal morphological criteria. The cell's growth kinetics, carbohydrate and nitrogen assimilation characteristics, and nature of metabolite formation aid in characterizing a given strain. Identification of cell wall components and immunological characteristics are also used. These scientific assessments, along with yeast performance in the brewery, are used to evaluate the yeast strain.

Yeast is generally added to the fermentor at a level of 10–15 million cells per milliliter of wort, or a range of 1.3–2 lb of yeast slurry per barrel of wort. During fermentation, cells multiply about four- to fivefold, depending upon initial wort aeration and wort composition. The size and density of the yeast cell varies according to the strain and the process time at which measurement is made. During the initial portion of a fermentation cycle, when glycogen reserves are being built up after budding, the cell has a greater mass and the ratio of carbohydrate to protein is high. At the end of the fermentation cycle, when glycogen stores are depleted and excretion occurs, the mass is decreased. The cell mass fluctuates through a sine curve cycle during the course of fermentation, but protein mass stays fairly constant. The percentage of protein (Kjeldahl N × 6.25) varies from 35 to 50%, whereas cells per gram of yeast solids ranges from 7 to 16 billion. Yeast cells range from 4 to 14 $\mu$m in length and from 3 to 10 $\mu$m in width.

At the end of fermentation, yeast cells flocculate or form aggregate clumps. This physical phenomenon of producing clumps lighter than or heavier than the beer is a principal characteristic which distinguishes the ale or top-fermenting yeast from the lager or bottom-fermenting yeast. The flocculation tendency can be measured by a standard test (Burns test), but quantitative data are highly speculative. The lager yeasts which settle are classed as flocculating (rapid settling) or powdery (slow settling). The rate of flocculation is important since it drastically affects metabolism of some important secondary yeast by-products and the efficiency of filtration following fermentation due to the blinding effect of yeast on filters.

The solids content of the yeast slurry recovered from the fermentor is totally dependent upon the mechanism of recovery. Well-settled yeast slurry will contain 15–17% solids, whereas centrifuged yeast contains 18–26% solids, depending upon machine type and process conditions. Yeast pressed at 40–50 psi will contain about 25–28% solids.

The yeast harvested from the fermentor can be added at once to another batch of wort, but it is often washed prior to being recycled. Washing frees the yeast of mechanical impurities, such as coagulated protein, hop resins, and some dead yeast cells. Treatment with ammonium persulfate adjusted to pH 2.2 with phosphoric acid has been shown by Bruch *et al.* (1964) to eliminate most bacterial contamination in the harvested yeast.

Laufer and Schwarz (1936) have described apparatus for growing a pure culture yeast in the plant under sterile conditions to obtain sufficient pitching yeast for a full-size brew or a portion of it. It consists of a series of two or three vessels, each about ten times larger than the preceding vessel. After 1–2 days of growth, the culture is transferred to the next largest vessel containing cold, sterilized wort. Rather than start a new laboratory culture each time, a small portion of the contents of the second vessel is sometimes transferred back to the starter vessel.

The amount of fresh culture yeast grown is dependent upon the viability of the yeast after multiple fermentations, the amount of contamination of the recovered yeast, and the method of recovery. In commercial brewing practice, new yeast tends to decrease in viability or becomes contaminated with continued recycle; therefore, recycled yeast is replaced with a new culture after from 5 to as many as 100 fermentation cycles.

### III. MALTING PROCESS

Malting is the process in which grains are intentionally germinated and subsequently dried by the application of heat. Although the term "malt" is commonly used for barley malt only, wheat and rye are also commercially malted in small quantities in the United States. Reviews of the malting process and design of malting plants are given by Macey (1963), Kellett (1965), and Bradee and Westermann (1975).

During the malting process the enzyme systems necessary for reproduction of the barley plant are fully developed and proteins and carbohydrates of the barley kernel are modified or made soluble for subsequent hydrolysis in the mash. The resulting carbohydrates furnish the energy source for the production of alcohol, while the amino acids and other trace materials provide the nutrition for yeast growth. During the kilning process, color and flavor ingredients are developed, and the moisture content is lowered to allow storage of the malted barley.

Before malting, barley is stored several weeks to overcome dormancy. During storage it is kept at 10–12% moisture and is frequently transferred to maintain low temperatures and break the dormancy. The moisture content must be kept low to protect the barley from mold growth.

The malting process can be divided into five basic steps listed below.

1. Barley cleaning and grading.
2. Steeping in water to increase moisture content.
3. Germination of moist grain under controlled conditions.
4. Kilning to arrest growth, develop color and flavor, and reduce moisture to a suitable storable level.
5. Rootlet removal.

Figure 2 is a schematic flow diagram of the malting process, including a material balance.

### A. Cleaning and Grading

Barley received by the malster is cleaned to remove weed seeds, other grains, broken kernels, chaff, sand, dust, and foreign objects such as iron. Iron particles are removed by an electromagnet. Small particles,

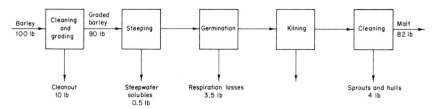

**FIGURE 2.** Schematic flow diagram of the malting process; material balance on a bone-dry basis.

such as sand and seeds, are usually removed through a fine vibrating screen. Broken kernels and other types of grains are often removed in inclined rotating cylinders with inner periphery pockets that will carry the smaller grains upward, subsequently emptying into a trough for removal.

Size separation according to kernel width is accomplished by passing the barley over three width-separation graders, producing A, B, and C grades. Undersize kernels passing through the C grader are collected and sold as feed. The small C-grade barley is used in the production of distillers malt. Brewers malt is generally produced from the larger A and B grades.

## B. Steeping

Graded barley is conveyed to cold water-filled (10°–15°C) steep tanks by pneumatic, bucket, belt, or screw conveyors. The first stage of the steeping process is used to wash the barley and to remove dirt, loose hulls, and other foreign material by overflowing the tank with water. Air under pressure is injected to mix the barley and water and initiate respiration.

During steeping the grain absorbs moisture and dormant embryos become activated. As soon as the kernel absorbs moisture, the respiration rate of the embryo increases rapidly. Aeration may be intermittent or continuous.

Steep water is changed every 10–15 hours. Steeping is continued until the moisture content of the grain reaches 44–45%, which may take from 48 to 60 hours. A total of 1.5–2 lb of water per pound of barley is generally used.

## C. Germination

During the germination period the endosperm serves as a nutrient source for the embryo growth. The embryo synthesizes gibberellic acid, which diffuses into the aleurone layer as described by MacLeod (1976). Various hydrolytic enzymes are synthesized by the aleurone cells. These enzymes diffuse into the endosperm and break down hemicellulose, the protein and peptides, lipids, organophosphates, and some of the starch. Barley endosperm cell walls consist of about 75% $\beta$-glucan, which must be enzymatically hydrolyzed before the cell contents can be modified by proteolytic enzymes, phosphatases, and amylases. $\alpha$-Amylase is formed during germination. $\beta$-Amylase is present in barley but is bound to albumin proteins and is released during germination by the action of proteolytic enzymes. During malting, only 5–10% of the starch is hydrolyzed to various degrees.

Modification of the endosperm starts near the embryo and progressively extends toward the distal end of the kernel. Daily samples are taken to test the progress of the modification, which is usually complete after about 5 days.

Relatively low temperatures, from 15° to 21°C, depending on the type of malt, must be maintained during the entire germination time to reduce respiration and rootlet losses and still bring about the desired physical and chemical changes. Fresh air is necessary to supply oxygen for respiration, prevent the accumulation of carbon dioxide, and control temperature in the grain bed by removing heat formed during respiration. The moisture content during germination is kept constant by humidifying the fresh incoming air or by water sprayers.

Germination is currently carried out in either drum or compartment systems. Drum systems have a capacity of 600–700 bushel of barley per drum. Drums are equipped with a device for distributing conditioned air and revolve slowly to prevent matting of rootlets.

In compartment malting, the steeped barley is put in long, rectangular, concrete open-top boxes equipped with perforated floor plates to permit flow of conditioned air. The grain bed, about 18–30 inches deep, is slowly turned over by mechanical screws that move back and forth through the compartment.

Drum systems require a higher initial investment and have higher operating and maintenance costs than the compartment system.

## D. Kilning

After the desired degree of modification has been attained, the so-called "green" malt is transported to the kiln. The kilning process is carried out in two phases: (1) drying and (2) heating. During the drying phase, the moisture content is reduced from 45 to about 6% to terminate respiration, arrest biological activity, prevent bacterial spoilage, and preserve the enzymes and substrates which will be required in the subsequent mashing process. Additionally, during the first part of the drying phase, some of the characteristic flavor and aroma components are synthesized in a series of complex biochemical and chemical reactions. Amino acids are formed during further breakdown of proteins and react with sugars to form the colored and flavored melanoidins.

The enzymatic activity and flavor characteristics of the malt are markedly affected by the moisture content of the malt and the kiln air temperature applied at a given moisture content. The moisture content of regular brewers malt is reduced to about 6% with lower temperature air before the final kilning temperature is raised to about 82°C, reducing moisture to about 4.5%. Conversely, a crystal or caramel malt, from which very dark beer is made, is rewetted after light kilning and held at 66°–72°C for several hours without ventilation. It is kilned off up to 121°C and contains about 3% moisture. It is apparent that a wide variety of moisture contents and temperatures can be used to produce specialty malts.

Some kilns consist of an oven below two or three perforated floors, through which hot air is passed. Green malt is first deposited on the top floor for the initial drying process. After 12–24 or more hours, depending on the number of decks and type of malt, the malt is dropped to a lower floor and dried at the desired temperature to final moisture content. Each floor is equipped with turning machines.

Some malsters introduce sulfur dioxide into the air stream, which passes through the malt for a short period each day. This sulfur dioxide serves as a bactericide and fungicide for surface organisms, bleaching the malt, and lowering mash pH. This procedure has been minimized for ecological reasons.

The rootlets formed on the malt become dry and brittle during kilning and are removed in a scalper provided with a dust removal system. The rootlets are accumulated and sold as a valuable food and feed product.

The finished malt is transferred to storage, where it is held until scheduled for brewing.

## IV. BREWING PROCESS

After storage, malt is elevated to the mill house, where cleaning and milling take place. In the commonly used batch brewing process (see Fig. 3), a batch of milled malt is weighed and mixed with water in the mash tub. A portion of the malt is mixed with adjunct and water in the cooker. After liquefaction the cooker mash is added to the main mash. After the enzymatic conversion of starch and proteins, wort is separated and extracted from the spent grains. The strained wort is then boiled with hops or hop extract. During boiling a precipitate is formed. It is

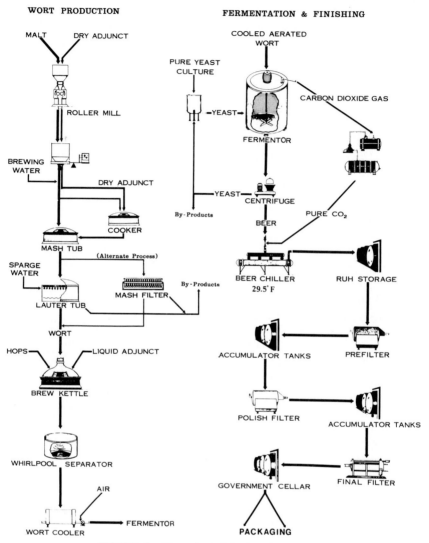

**FIGURE 3.** Flow chart of the brewing process.

removed after the wort is transferred to the hot wort settling tank. During subsequent transfer to the fermentor, wort is cooled and aerated. Each of these process steps is described in detail below.

## A. Malt Preparation, Milling, and Weighing

Malt blended to proper specification is stored and kept at 4–5% moisture in silos of capacities up to 800,000 lb of malt. Elevation of malt to the mill house is done pneumatically or by bucket elevator. A considerable amount of abrasion may take place during pneumatic conveying. Typically, the amount of loose husks and fines after pneumatic elevation is 3–5%, while after bucket elevation, it is 1–2%.

Prior to milling, the malt is passed over a magnetic separator and sometimes through an aspirator. In the aspirator, loose husks and dust are collected to bypass the mills.

The purpose of milling is to provide ground malt with a particle size distribution that will give the brewer the optimum extract yield after mashing and wort separation within an allotted time. Generally, the degree of conversion and extraction increases with decreasing particle size, whereas the rate of wort separation decreases with decreasing particle size, as reported by Huige and Westermann (1975). Every brewer establishes his preferred optimum balance between yield and rate.

In most breweries, malt is milled in roller mills consisting of three pairs of rollers which operate using both direct pressure and shear. Double screens are located between the pairs of rollers to separate the various particles when they have attained proper size. Large mills have capacities of up to 13,000 lb of malt per hour. The three sets of rollers are generally adjusted to progressively decreasing gaps of about 0.040–0.030 inches.

The endosperm is typically crushed to produce fine grits rather than flour, though some flour production is inevitable. Schöffel (1972) indicated that the optimum particle size not only depends on the gap between the rollers, but also on the type, quality, and moisture content of the malt and the type of wort separation equipment that is used. Generally, lauter tubs and Strainmasters require a smaller fraction of flour than mash filters or continuous lautering devices, such as centrifuges.

To produce less flour and a faster wort separation, it is advantageous to minimize husk fracture. This can be accomplished by soaking the malt in cold or warm water or treating with steam for a period of up to 30 minutes. The moisture content of the malt increases up to 35%. Special mills are used for this so-called wet milling process, as described by Schauss (1964) and Narziss (1974, 1976, 1977).

## B. Adjunct Cooker

Since cereal adjuncts have not gone through a malting process, their cell walls have not been modified. They contain starch and proteins that are difficult to solubilize because of their high molecular weight. Before these adjuncts are added to the main mash, therefore, they are milled, mixed with water, and brought to a boil in the steam-heated adjunct cooker to bring about gelatinization of starch. A small percentage of malt (approximately 10%) is added to aid liquefaction and decrease viscosity. The ratio of water to adjunct is about 2.5–3.5 lb/lb, and the total time to heat and boil 1–1½ hours. The cooker is agitated to increase heat transfer rates and to prevent burn-on.

## C. Mashing

The purpose of mashing is to extract from the grains the available soluble substances (about 17% of malt) and to convert additional insoluble solids through controlled enzymatic reactions (additional 58% of malt). Yields for adjuncts such as corn grits may be as high as 87%. Because temperature and pH affect enzymatic activity, they must be optimized and controlled during mashing.

The pH of the mash is controlled at 5.2–5.5 by the addition of lactic acid and/or mineral acids and calcium sulfate. Calcium sulfate reacts with phosphates in the malt to provide buffer action, as described in the section on brewing water.

Many different time–temperature cycles may be used during mashing, depending on the raw materials and the type of wort desired. Temperatures are increased by heating with steam or by adding mash from other vessels, such as the cooker. Between temperature rises the mash is held for certain rest periods at temperatures that are optimum for certain enzymatic reactions. During mashing-in (mixing with water) and during heating periods, the mash is slowly agitated.

In a typical American mash schedule, known as the "double mash method," ground malt is mashed with water at a temperature of 35°C. During the so-called "protein rest" that follows, the temperature is kept constant for about 1 hour. The completed cooker mash is then added to raise the temperature of the combined mashes to 67°–68°C and begin starch hydrolysis. A rapid temperature rise (10 minutes) will increase the wort dextrin content, while a slow rise (30 minutes) will increase wort fermentable sugars. The mash temperature is kept constant for 20–30 minutes to allow completion of starch hydrolysis. After the absence of starch is confirmed by an iodine test, the mash temperature is increased to the "mashing-off" temperature of 75°–80°C, completing the mashing process.

The biochemical reactions occurring during mashing are an extension of those started during malting. Most of the soluble nitrogenous compounds of wort are formed during malting. During the protein rest when the temperature is close to the optimum for most proteolytic enzymes, a further breakdown of proteinaceous material takes place. Amino acids, peptides, peptones, and ammonia are formed. However, proteolysis is far from complete, and soluble higher molecular weight proteins are also extracted into the wort. Some of these high molecular weight proteins are subsequently precipitated during kettle boil. The final molecular weight distribution of proteins determines palate fullness, physical stability, and foam properties of the beer. Excessive hydrolysis of proteins may result in a beer with little foam stability, whereas too little hydrolysis might give rise to a cloudy, unstable beer.

The most important enzymatic reactions involve the degradation of starch to the fermentable sugars glucose, maltose, and maltotriose, and to nonfermentable higher molecular weight dextrins. The ratio of fermentable sugars to dextrins can be controlled by proper adjustment of the temperature–time cycle.

Malt starch consists of 20–25% amylose, a linear polymer composed of 20–25 glucose units joined by 1—4 chemical bonds. Amylose is hydrolyzed by $\alpha$-amylase to linear dextrins containing about six glucose units, which are hydrolyzed further to maltose by the action of $\beta$-amylase. The other 75–80% of starch is amylopectin, a branched polymer composed of several thousand glucose units joined by both 1—4 and 1—6 chemical bonds. $\alpha$-Amylase catalyzes the liquefaction of amylopectin by attacking the 1—4 bonds and bypassing the 1—6 bonds, yielding branched dextrins of lower molecular weight. This results in a rapid viscosity decrease. $\beta$-Amylase partially hydrolyzes the branched dextrins by attacking the 1—4 bonds and removing maltose from the linear portion of the chain.

The enzyme phytase is active during the "protein rest" period. This enzyme hydrolyzes phytin to inorganic phosphates, which act as buffers during mashing and fermentation, and inositol, an essential growth factor for yeast nutrition.

### D. Wort Separation

After the mash cycle, malt husks and other insoluble material must be separated from the dissolved solids solution, wort, before it flows into the brew kettle. The process consists of two steps: (1) the principal separation of wort from insoluble solids by filtration, and (2) washing of the remaining insoluble solids bed to displace residual wort between the particles and to extract dissolved solids from within the particles. Two types of commercial wort separators include: (a) equipment in which the particles themselves form the sole filter medium. The lauter tub, in which wort is withdrawn from a slotted false bottom vessel, and the Nooter Strainmaster, in which wort is withdrawn from a plurality of horizontal strainer pipes at different heights of the bed, are widely used. (b) Equipment in which a filter screen or filter cloth is used as the main separating medium. Examples of this type are the plate-and-frame filter press and continuous lautering devices, such as a screening centrifuge.

A lauter tub is a cylindrical tank with a height to diameter ratio of about 1:5 and a maximum diameter of approximately 32 ft. The false bottom consists of plates with 0.020-inch-wide slots which provide 8–10% of open area and form a support for the filter bed of undissolved grain particles. The lauter tub is provided with vertical knives mounted on an arm that rotates around a central shaft. The knives, which can be lowered to cut into the bed while rotating, are used to prevent the bed from getting packed too tightly. After the mash has been pumped to the lauter tub, the particles are allowed to settle for 10–30 minutes. The first wort that is withdrawn is cloudy and is recycled until the desired clarity is obtained. After most of the wort has run off to the brew kettle, sparging is started with water of about 76°C, adjusted to the desired pH. Sparging is usually continued until the dissolved solids content of the withdrawn liquid drops below 1%. A typical analysis of wort brewed with corn grits is shown in Table VII.

The rate of wort withdrawal is generally about 0.25 gal/minute/ft$^2$ of cross-sectional area and may be as high as 0.04 gal/minute/ft$^2$. At a high rate of wort withdrawal, care has to be taken that the pressure differential across the bed is kept below 10 inches of water to prevent the grain bed from becoming too compressed. Also, upon increasing the runoff rate, the contact time between sparge water and grains to be extracted is reduced, thereby decreasing the degree of extraction of dissolved solids. After lautering is completed, the so-called spent grains are removed from the lauter tub by the action of the knives which are positioned to sweep out the bed. Spent grains are sold as cattle feed, either wet or dry. Before drying, the moisture of spent grains is usually reduced from about 80% to 60–72% in grain presses or centrifuges. The liquid containing 2–5% dissolved solids is sometimes returned to the cooker to be used as part of the cooker water, as reported by Coors and Jangaard (1975).

**TABLE VII.**  Typical Wort Analysis

| | |
|---|---|
| Wort color (°Lov. 52–1 inch cell) | 6.0 |
| Specific gravity (20°C/20°C) | 1.04545 |
| Extract (°Plato) | 11.3 |
| Limit attenuation (°Plato) | 2.5 |
| Fermentable extract (%) | 63.0 |
| Wort protein (N × 6.25) (%) | 0.41 |
| Soluble nitrogen (gm/100 gm) | 0.065 |
| Wort pH | 5.35 |
| Titr. acidity; ml N/10 alkali/100 gm | 9.0 |
| Calcium (ppm) | 72 |
| Iodine reaction | Negative |

## E. Brew Kettle Boil

After clarification the wort is heated in the brew kettle and boiled for 1–1½ hours, during which 5–8% of the water evaporates. During the brew kettle boil, hops or hop extracts are added all at once or in increments. When liquid adjunct, i.e., corn syrup, is used, it is generally added at the start of boil to maximize brewing reactions.

Historically, brew kettles have been constructed of copper because of material availability and ease of construction of the peculiar shape of the dome and draft stack. Some brewers believe that copper reacts with some sulfur-containing compounds that could adversely affect beer quality. All modern brew kettles are made with stainless steel. Kettles are steam heated by a bottom jacket and/or a central heat transfer surface that acts as a percolator. Heating can also be accomplished in outside heat exchangers, a technique more frequently practiced by European brewers. The objectives of the brew kettle boil are discussed below.

### 1. Wort Sterilization

Since wort is an ideal growth medium for microorganisms, sterilization is required before wort enters the fermentors. Hops that are added to the kettle also aid in preservation.

### 2. Enzyme Inactivation

Boiling will destroy any enzyme system that has not been completely inactivated at the end of the mash cycle. Inactivation is necessary to fix the ratio of fermentable sugars to dextrin to control the final alcohol content.

### 3. Protein Precipitation

High molecular weight proteins are precipitated during kettle boil, forming large flocs called "hot break" or trub. This precipitation is essential to enhance the colloidal stability of the beer. Precipitation occurs by (a) heat denaturation of those proteins which have an isoelectric point of the wort (pH 5.1–5.2), and (b) complexing with tannins and their oxidation products from malt husk or hops.

The pH of wort, which drops by 0.3–0.5 units during kettle boil, must be controlled to provide sufficient precipitation of "undesirable proteins" for a colloidally stable beer but no precipitation of "desirable proteins" which would result in a beer with poor foam quality.

### 4. Hops Extraction and Isomerization

During kettle boil, water-soluble resins, oils, tannins, amino acids, and monosaccharides are extracted from hops. Tannins aid in the precipitation of proteins.

Some of the about 100 hop oil constituents were originally thought to directly contribute to beer hop aroma, but since none of those components can be identified in finished beer, it is now believed that hop oil degradation products impart the hoppy aroma, according to Likens and Nickerson (1964).

The soft resins that contribute to beer bitterness in varying degrees are $\alpha$- and $\beta$-acids and their isomerization, oxidation, and degradation products. The main contributors to bitterness are isomerized $\alpha$-acids. The rate of isomerization increases with increasing pH. At a kettle pH of about 5.2, it takes about $1\frac{1}{2}$ hours for 90% of the important $\alpha$-acid humulone to isomerize to isohumulone. Some isomerization may take place in the hot wort settling tank. Only about 25–30% of $\alpha$-acids appear as isomerized products in beer due to incomplete extraction, incomplete isomerization, absorption to colloidal particles, and further degradation. For better control and utilization, preisomerized hop extracts are sometimes added after kettle boil, as noted by Westermann (1976) in a United States patent.

### 5. Distillation of Volatiles

During kettle boil, about 90% of the essential hop oils are removed by steam distillation, along with some fatty acids and low molecular weight degradation products.

### 6. Color Development

About two-thirds of the final wort color is produced during kettle boil. One-third is developed during malting, as reported by Nakayama (1962). Color and flavor development take place by the following types of reactions: (a) oxidation of phenolic compounds; (b) Maillard reaction between reducing sugars and amino acids; (c) caramelization of polyhydroxycarbonyl compounds.

## F. Trub Separation, Wort Cooling, and Aeration

The protein precipitate formed during kettle boil and the spent hops material have to be separated from wort prior to fermentation. This is usually done in a Whirlpool separator or a combination of settling tank and filter or centrifuge. If whole hops are used, a hop strainer is often used to remove most of the hop solids prior to trub separation.

Trub consists of 50–60% protein, 16–20% hop resins, 20–30% polyphenols, and 2–3% ash by weight (Hough et al., 1971). The particle size is 30–80 $\mu$m. To aid separation, some brewers add materials to the kettle that enhance floc formation, such as Irish moss or bentonite.

During the last 10 years the Whirlpool separator has become popular.

In this sedimentation tank, which has a height to diameter ratio of about 1:4, wort is fed tangentially, thereby creating a whirlpool pattern. Due to the centripetal force, the wort moves down the sides and up the middle. Particles with a settling velocity greater than the upward liquid velocity will settle out in the middle, forming a trub cone. Wort is drawn off from the outer periphery of the bottom. The trub solids are flushed to the sewer, are recovered as by-products, or are recycled back into the process.

After separation of the trub, wort is cooled in a plate heat exchanger to the starting temperature required for fermentation. As the temperature reaches about 48°C, an additional protein precipitate starts forming (particles of <5 $\mu$m) known as the "cold break." This cold break is removed by some brewers, who feel that it alters yeast metabolism. Separation is done by cold wort diatomaceous earth filtration or in a flotation tank.

Wort flowing to the fermentor is aerated to bring the oxygen concentration up to 8–10 ppm. Oxygen is necessary in the initial growth phase of fermentation.

## G. Fermentation

### 1. Fermentation Process

After cooling and aeration, wort is inoculated with yeast, usually obtained from a previous fermentation. about 1–2 lb of yeast slurry per barrel of wort is used, equivalent to 7–15 million yeast cells per milliliter of wort. At higher yeast concentrations the start of active fermentation is faster and the fermentation rate increased. Wort is pumped directly to the fermentors or to starting tanks where fermentation begins and dead yeast cells, hop resins, and part of the cold break are separated.

Changes taking place during a typical lager fermentation are graphically depicted on Fig. 4. A lag period of about 12 hours is necessary for yeast to acclimatize to its new environmental conditions. After this time, yeast multiplication starts. The dissolved oxygen originally present in wort is sufficient to give a four- to fivefold increase in yeast cell population in 3–4 days. The fermentable carbohydrates furnish energy and carbon to produce yeast metabolites of ethanol, carbon dioxide, and other minor by-products, resulting in a decrease in specific gravity. Since the reaction is exothermic, 24 kcal are liberated per mole of glucose, the temperature rises. When a desired maximum temperature (in the example of Fig. 4, 13°C) is reached, it is kept constant by cooling the wort.

Within 12–26 hours after pitching, the wort becomes saturated with

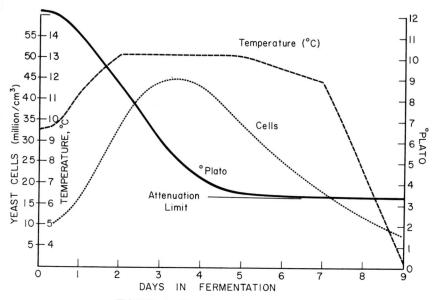

**FIGURE 4.** Brewery batch fermentation.

carbon dioxide and small bubbles appear on the surface, forming a creamy head called kraeusen. As the bubbles rise to the surface, they expand, thereby agitating the wort and causing the yeast to be suspended. Agitation and suspension of wort affect the rate of various reactions that take place during fermentation. Agitation is sometimes aided by providing mechanical agitators. Also, the use of large vertical fermentors promotes agitation.

As fermentation proceeds the foam head becomes thicker and yellowish due to precipitation of malt, hops substances, and dead yeast cells. The foam also contains brown spots from oxidized hop resins.

After 4–5 days the rate of $CO_2$ production decreases, the foam head begins to collapse, and some of the yeast cells start settling out. In the traditional process, cooling is further intensified.

The so-called primary fermentation is complete after 4–5 days, but virtually complete utilization of fermentables may take as long as 12 days. At this point all fermentable sugars, except for a small portion of the maltotriose, are fermented and the yeast has settled.

The flocculation tendency of the yeast is an important parameter. If a yeast flocculates too quickly, fermentation may not have proceeded far enough; whereas a yeast that flocculates with difficulty and stays in suspension too long may partly autolyze, imparting an undesirable yeasty flavor to the beer. The degree of flocculation of a yeast strain, although a genetic characteristic, can vary widely due to mutation within that strain and the concentration of divalent ions, such as calcium, outside the cell. Stewart (1975) concluded, "The contribution of calcium to flocculation is not the absolute amount of this ion absorbed by the yeast cell wall, but rather the stereo-specific manner by which it is bound. . . ."

After fermentation, yeast is removed by decantation, slurry centrifugation, or a combination of the two.

## 2. Fermentation Biochemistry

Yeast, specifically *S. uvarum,* requires fermentable carbohydrates for adequate growth and maintenance and also growth factors, amino acids, vitamins, and essential minerals for proper nutrition. In a typical lager wort, the proportion of solids in the substrate is given in the tabulation below.

| | |
|---|---|
| Monosaccharides (principally glucose) | 8–10% |
| Disaccharides (principally maltose) | 50–55% |
| Trisaccharides (maltotriose) | 10–12% |
| Higher dextrins | 20–25% |
| Proteins (peptones, peptides, polypeptides, amino acids) | 4–6% |
| Minerals (calcium, potassium, zinc, magnesium, etc.) | 2% |
| Growth factors (B complex vitamins) | Traces |

The addition of adjuncts, particularly syrup, can change the amounts of glucose, maltose, and maltotriose in the wort.

Yeast is facultative, i.e., it can grow aerobically as well as anaerobically. If sufficient oxygen is present, all fermentable carbohydrates are oxidized to water and carbon dioxide, and about ten times the amount of heat is liberated as during anaerobic respiration. The energy not dissipated as heat is used for cell growth and multiplication. During fermentation of wort, anaerobic conditions prevail, although some oxgen is initially required for cell mass production. Fermentable carbohydrates are transported into the cell, converted to glucose, and metabolized to produce principally ethanol and carbon dioxide.

The fermentation proceeds primarily via the glycolytic or Embden–Meyerhof-Parnas metabolic pathway. In the first step in this pathway, glucose is phosphorylated and isomerized to fructose phosphate, which is phosphorylated again and cleaved into two phosphorylated compounds. Several oxidation–dephosphorylation–reduction reactions take place to yield ethanol and carbon dioxide.

The initial amount of glucose in the wort (*Ig*) is an important factor in the carbohydrate assimilation kinetics, as shown by Westermann (1976). For glucose assimilation, which occurs first, $lg$ is the rate-limiting factor at 5–6 gm/100 gm of wort. Above this level the yeast concentration is limiting. Sucrose is hydrolyzed early by invertase, which is exogenous to the yeast cell. Maltose is first transported into the cell by a permease system and is subsequently hydrolyzed to glucose by maltase. Since glucose inhibits the activation of the permease system, a higher initial glucose concentration increases the lag time for initial maltose uptake and decreases the maltose assimilation rate. The effect of glucose on maltotriose is quite similar, although maltotriose assimilation occurs at a much slower rate. In most American beers about 10–20% of the malto-triose remains unfermented. Maltotetraose and dextrins are not fermented.

Amino acids, necessary for yeast nutrition, are assimilated in the cell at various rates, as described by Jones and Pierce (1969). Purines and pyrimidines necessary for nucleus formation can be synthesized by the yeast cell but are usually present in wort.

Many of the desirable and undesirable aroma and flavor compounds of beer are products of amino acid assimilation. Yeast strain, amino acid profile and concentration in wort, and pH are all important factors.

Higher alcohols, such as propanol and isobutanol, are generated from amino acids, but also from keto- acids derived from carbohydrate metabolism. The concentrations of individual higher alcohols in finished beer range from several to 60 ppm.

Aldehydes are precursors of alcohols. The concentration of acetaldehyde, the precursor of ethanol, can be as high as 10 ppm. Organic acids, such as acetic, lactic, pyruvic, citric and malic acids, are all derived from the tricarboxylic acid cycle. Fatty acids ($C_3$–$C_{22}$) are extracted from grains or are synthesized by yeast. The unsaturated fatty acids from wort may partially supplant oxygen required for yeast viability, as shown by Thompson and Ralph (1967).

Ester formation occurs largely during the earlier stages of active yeast growth from the condensation of alcohols and an enzyme intermediate. Little ester formation occurs due to condensation of alcohols and organic acids during aging.

Hydrogen sulfide is formed during the enzymatic breakdown of cysteine or methionine by yeast. Most of the hydrogen sulfide is purged out with $CO_2$.

Diacetyl (2,3-butanedione) and 2,3-pentanedione, collectively called vicinal diketones, impart a buttery flavor which is generally considered undesirable in American beers. Since the threshold level for detection of diacetyl by the consumer is about 0.1 ppm, substantial effort is made to control its concentration below this level. $\alpha$-Acetolactic acid, the precursor of diacetyl, is formed during yeast cell metabolism from pyruvic acid. Diacetyl is formed by the oxidative decarboxylation of $\alpha$-acetolactic acid. The decarboxylation rate increases with increasing temperature and decreasing pH and is independent of yeast cell concentration. Once diacetyl is formed, it can be assimilated by yeast to produce acetoin or 2,3-butanediol, which are not readily detected in beer. The assimilation of diacetyl occurs principally during secondary fermentation or aging. The assimilation rate of diacetyl is strongly yeast cell concentration dependent, and continued agitation during secondary fermentation to keep yeast suspended has been shown to be beneficial by Westermann (1976).

## H. Lagering, Aging, and Finishing

The primary objectives of these final steps of the brewing process are: (1) flavor maturation; (2) physical stabilization; (3) clarification; and (4) adjustment of carbon dioxide level.

Ruh storage, or lagering, and aging are terms referring to different time periods for maturing beer. In the traditional ruh storage or lagering process, which is accomplished in closed vessels, maturation of the beer involves a secondary fermentation to assimilate the small amount of residual maltotriose and to complete diacetyl assimilation. This is accomplished by the action of 6–10 million cells per milliliter or residual yeast transferred from the primary fermentor. During the 2- to 6-week process the temperature may start at 3°–6°C and gradually be decreased to −1°C.

In the more recent aging process, completely fermented beer is matured at temperatures of about −1°C in the presence of 1 million or less yeast cells per milliliter.

### 1. Saturation with $CO_2$

Beer can be saturated with $CO_2$ by one or more of the following methods:

1. Natural buildup of $CO_2$ during secondary fermentation or kraeusening by applying counter pressure. In the kraeusening process, carbon dioxide is produced in an after-fermentation. Usually 10–20% of wort in the active, or so-called kraeusen, stage of fermentation is added to the partially fermented beer to bring about secondary fermentation.

2. Sparging with $CO_2$ in storage tanks.

3. Sparging with $CO_2$ in line. The carbon dioxide that is used during sparging is usually $CO_2$ collected during primary fermentation and

purified by distillation or by water scrubbing, permanganate scrubbing, and carbon absorption.

## 2. Flavor Maturation

Diacetyl, hydrogen sulfide, mercaptans, and dimethyl sulfide produced during fermentation are decreased during secondary fermentation and aging. The removal of diacetyl is yeast cell dependent, and higher temperatures and high cell concentration are favored. Hydrogen sulfide and undesirable acetaldehyde can be removed by purging with $CO_2$.

Purging with $CO_2$ is also important to minimize oxygen concentration. Low oxygen levels will minimize staling reactions in packaged beer and reduce the formation of haze precursors.

The concentrations of some esters, such as ethyl acetate and isoamyl acetate, that are responsible for the fruity background flavor in beer increase slightly during lagering, as Hashimoto and Kuroiwa (1972) have reported.

## 3. Beer and Foam Stabilization

Three types of stability are required for a good beer: (1) microbiological; (2) physical, and (3) foam.

Microbiological stability is obtained by pasteurization (to be discussed later) or by sterile filtration of beer through very fine sheet or membrane filters. The filters which are used for beer have a pore size of 0.45–0.8 $\mu$m, become clogged very easily, and require adequate prefiltration. The technique was popular for the production of "draft beer" in packages in the 1960s. It has only very limited use today.

Wainwright (1974) describes two types of haze. Chill haze (0.1–2 $\mu$m) is defined as a haze that forms at 0°C and redissolves at 20°C, and permanent haze (1–10 $\mu$m) is defined as that remaining above 20°C. Haze is composed of polypeptides, polyphenols, and carbohydrates in various ratios, with a small percentage of ash. Stable linkages between polypeptides and polyphenols can be formed after oxidation of the latter to quinones, according to Bishop (1975). Methods used to prevent or delay haze formation are: (1) hydrolysis of polypeptides by papain to low molecular weight components that cannot form hazes, (2) absorption of polypeptides by silica gel or bentonite, (3) precipitation of polypeptides by tannic acid, (4) absorption of polyphenols by PVPP or Nylon 66.

Foam stability is improved with increased viscosity and increased surface tension and elasticity. Foam-promoting substances are proteinaceous-type materials, surface-active hop components, and gums that increase beer viscosity (Runkel, 1976). Glycoproteins increase foam stability considerably. Foam adhesion or cling is promoted by iso-$\alpha$-acids from hops.

### 4. Clarification

After lagering or aging, beer still contains yeast cells, proteinaceous-type precipitates, microorganisms, and colloidal matter. A first crude clarification which sometimes takes place prior to aging is usually carried out by centrifugal separation. Batch or continuous centrifuges are used with capacities of up to 250 barrels of beer per hour.

Clarification by filtration is carried out in sheet filters, diatomaceous earth filters, or pulp filters.

Pulp filters use a medium which contains cotton or wood pulp fiber made into a mat several inches thick. The filter mat is repulped, washed, sterilized, and reformed frequently. Because of their small capacity and high labor cost, pulp filters are rarely used in the United States.

Sheet filters consist of typical plate and frame filters containing filter sheets of cellulose, with or without a small percentage of asbestos. The sheets clog easily if beer has not been prefiltered. Depending on the density of the filter sheets, a high degree of beer clarity and sterility can be achieved.

Diatomaceous earth leaf filters are the main type of filter used for prefiltration. Diatomaceous earth, the calcined skeletons of the single-cell diatoms, is an excellent filter aid because of its high porosity. Diatomaceous earth is available in different grades. The grade that is selected depends on desired clarity of the filtered beer and the size and nature of the particles to be removed from the beer. At the start of the filtration cycle, the filter support, a mesh screen or a filter pad, is precoated with a thin layer (several millimeters) of filter aid. After 10–15 minutes of recycling, the main filtration starts. Filter aid is injected in beer that is going to the filters to build up a porous filter bed. After the pressure drop exceeds approximately 50 psi, the cycle is ended and the filter cake is removed manually or automatically. Generally about 25–40 lb of diatomaceous earth per 100 barrels of beer is used. Flow rates are in the order of 0.4–0.5 gal/minute/ft² of filter area. Some of the larger filters of approximately 1000 ft² can handle 900 barrels of beer per hour.

### V. PACKAGING

The brewing industry has historically been the pacesetter in high-speed, high-efficiency packaging. Modern 12-oz can and bottle lines produce 1500 cans per minute and 900 bottles per minute. Today, 87.6% of all beer sold in the United States is packaged in bottles or cans. The

26.0 billion steel and aluminum cans used for beer account for over 25% of the can industry's total production. The 11 billion returnable and nonreturnable bottles account for 39% of packaged sales. Cans account for over 60%. The ratio of nonreturnable to returnable bottles shipped is 18.8 : 1.

In 1934, 2 years after the repeal of prohibition, 75% of the beer sold was draught. By 1937, 44% was sold in bottles, all returnable. In 1935 the first cone top can was introduced and found acceptance because it could be filled and crowned on bottling equipment. Its relative bulk and higher cost contributed to its rapid replacement by the flat top steel can.

Can competition forced the glass industry to develop the "one-way" ($7\frac{1}{2}$-oz glass) and the throwaway bottles ($6\frac{1}{2}$-oz glass) during World War II. In 1959 the "glass can" was developed. This short-necked stubby bottle concept prevails today for disposable bottles.

The can industry introduced the aluminum "soft-top can" for easy opening, and in 1962 introduced the "pop-top" lid. The bottle industry countered with the "twist-off" crown in 1963 to retain market share.

By 1964 aluminum cans became established, and the competitive battle for container market share continued. "Tin-free steel," welded side seams, cemented side seams, and drawn and iron aluminum and steel cans reflected technological improvements fostered by intense competition.

Imprudent disposal of the tabs from "pop-top" lids fostered the introduction of various ecology lids on which the opening device is an integral part of the lid.

Plastic beer bottles, although evaluated by some major brewers, are not yet a reality.

Returnable bottles are washed in a compartmented bottle washer in which the bottle, carried in a multipocket conveyor, is subjected to caustic solutions and freshwater rinses. Bottles and cans are filled on rotary continuous operating fillers containing up to 140 valves with maximum speeds of 1500–1600 containers per minute.

Both bottles and cans are pasteurized while being conveyed through a multicompartment pasteurizer in which the package is subjected to sprayed water at temperatures up to 71°C. The three principal organisms to be inhibited are wild yeast (*Saccharomyces diastaticus*), pediococci (*Pediococcus cerevisiae*), and long rods (*Lactobacillus brevis*). The pasteurization effect required is comparable to about 6–8 minutes at 60°C. Following pasteurization, bottles are labeled, and both bottles and cans are automatically placed in a proliferation of secondary packages. Unpasteurized draught beer, principally in half and quarter barrels, accounts for about 12% of the beer sales.

**TABLE VIII.** Beer Production and Per Capita Consumption

| Country | Production in barrels | Per capita consumption in gallons |
|---|---|---|
| West Germany | 79,628,000 | 39.7 |
| England | 55,056,000 | 30.4 |
| Russia | 51,131,000 | 6.2 |
| Japan | 33,482,000 | 9.6 |
| Czechoslovakia | 19,066,000 | 40.3 |
| France | 19,017,000 | 11.2 |
| Canada | 18,117,000 | 24.8 |
| East Germany | 17,214,000 | 31.0 |
| Mexico | 16,509,000 | 8.7 |
| Australia | 16,481,000 | 37.8 |
| Brazil | 14,775,000 | 4.3 |
| Spain | 14,162,000 | 12.4 |
| Belgium | 11,902,000 | 37.8 |
| Poland | 10,959,000 | 10.2 |
| Netherlands | 10,596,000 | 24.2 |
| Denmark | 7,567,000 | 45.9 |
| Yugoslavia | 7,224,000 | 10.5 |
| Colombia | 6,580,000 | 8.4 |
| Austria | 6,478,000 | 26.7 |
| Romania | 6,348,000 | 9.3 |

## VI. AMERICAN INDUSTRY STATISTICS*

In 1975, the 102 breweries in the United States produced a total of 160,575,908 barrels (31 gal/barrel) and sold 148,632,852 barrels, resulting in a per capita consumption of 21.6 gal. Although many other countries produce less, their per capita consumption is substantially higher, as shown in Table VIII.

The Unites States brewing industry produces retail product sales in excess of $15 billion per year and employs approximately 51,000 persons. Over $850 million worth of agricultural products are used, including 4.2 billion lb of malt, made from 125 million bu of barley, valued at $610 million; other grains, chiefly corn and rice, valued at $188 million; and $49 million worth of hops.

The direct salaries, excluding fringe benefits, amounted to $860 million in 1975.

Close to $2.5 billion is spent annually on packaging and containers.

* The statistics provided in this section have been taken from the "Brewers Almanac."

The industry annually spends $80 million on fuel power and water, $360 million on transportation, $600 million on equipment and improvements, and $230 million on advertising.

The present Federal Excise Tax on beer is $9 per barrel, set in 1951. In 1975, the United States Government received $1,308,673,000 in excise taxes on malt beverages.

With the state taxes levied in addition to the $9 per barrel Federal tax, each barrel of beer sold earns an average of $13 in taxes. The state taxes range from $0.62 per barrel in Wyoming to $23.81 per barrel in South Carolina. The 1976 "Brewers Almanac" states, "By contrast, the brewing industry's average profit per barrel after taxes has been estimated at about a dollar and one-half."

## REFERENCES

American Society of Brewing Chemists (1976). "Methods of Analysis," 7th ed., Am. Soc. Brew. Chem., St. Paul, Minnesota.

Arnold, J. P. (1911). "Origin and History of Brewing." Wahl-Henius Inst. Technol., Chicago, Illinois.

Baron, S. (1962). "Brewed in America—A History of Beer and Ale in the United States." Little, Brown, Boston, Massachusetts.

Bishop, L. R. (1975). *J. Inst. Brew.* **81,** 444–449.

Bradee, L. H., and Westermann, D. H. (1975). Internal communication, p. 37. Jos. Schlitz Brewing Co., Milwaukee, Wisconsin.

Brewers Almanac (1976). United States Brewers Association, Washington, D.C.

Bruch, C. W., Hoffman, A., Gosine, R. M., and Brenner, M. W. (1964). *J. Inst. Brew.,* **70,** 242.

Burgess, A. H. (1964). "Hops—Botany, Cultivation and Utilization." Wiley (Interscience), New York.

Comrie, A. A. D. (1967) *J. Inst. Brew.* **73,** 335–341.

Coors, J. H., and Jangaard, N. O. (1975). *Proc. Eur. Brew. Conv.,* p. 311. Elsevier Scientific Pub Co., Amsterdam.

Findlay, W. P. K., ed. (1971). "Modern Brewing Technology." Macmillan, New York.

Hashimoto, N., and Kuroiwa, Y. (1972). *Brew. Dig.* **47,** 64–71.

Hough, J. S., Briggs, D. E., and Stevens, R. (1971). "Malting and Brewing Science." Chapman & Hall, London.

Huige, N. J., and Westermann, D. H. (1975). *MBAA Tech. Q.* **12,** 31–40.

Jones, M., and Pierce, J. S. (1969). *Eur. Brew. Conv. Proc. Congr.* p. 315.

Kellett, O. S. (1965). *MBAA Tech. Q.* **2,** 69.

Laufer, S., and Schwarz, R. (1936). "Yeast Fermentation and Pure Culture Systems." Schwarz Lab. Int., Inc., Mt. Vernon, New York.

Likens, S. T., and Nickerson, G. B. (1964). *Annu. Proc. Am. Soc. Brew. Chem.* **4,** 5.

Lodder, J., and Kreger-Van Rij, N. J. W. (1952). "The Yeasts—A Taxonomic Study." Wiley (Interscience), New York.

Macey, A. (1963). *Brew. Guardian* **92,** 41.

MacLeod, A. (1976). *MBAA Tech. Q.* **13,** 193–198.

Matz, S. A. (1970). "Cereal Technology." Avi Publ. Co., Westport, Connecticut.

Nakayama, T, O. M. (1962). *Annu. Proc. Am. Soc. Brew. Chem.* 137–139.

Narziss, L. (1974). *Brauwelt* **114,** 851–892.
Narziss, L. (1976). *Brauwelt* **116,** 1736–1740.
Narziss, L. (1977). *Brauwelt* **117,** 50–53.
Runkel, U. D. (1976). *Monatschr. Brau.* **29,** 248–260.
Schauss, O. O. (1964). *MBAA Tech. Q.* **1,** 112–119.
Schoffel, F. (1972). *Brauwissenschaft* **25,** 301–312.
Scholefield, A. J. B. (1956). "The Treatment of Brewing Water." Published privately, Rockcliff, Liverpool.
Siebel, F. P., Jr. (1950). *Brew. Dig.* **25,** 45.
Singruen, E. (1938). *Mod. Brew.* **19,** 40.
Stewart, G. G. (1975). *Brew. Dig.* **50,** 42–56.
Thompson, C. C., and Ralph, D. J. (1967). *Eur. Brew. Conv., Proc. Congr.* p. 177, Elsevier Scientific Pub. Co., Amsterdam.
Wainwright, T. (1974). *Brew. Dig.* **49,** 38–48.
Weeks, M., Jr. (1949). "Beer and Brewing in America." U.S. Brew. Found., New York (Designed and produced by Appleton, Parsons, and Co., New York).
Westermann, D. H. (1976). *Abstr. Pap., 171st Centen. Meet., Am. Chem. Soc., 1976.*
Westermann, D. H., Chicoye, E., and Hoffman, D. R. (1976). U.S. Patent 3,965,188.
Winton, A. L., and Winton, K. B. (1932). "The Structure and Composition of Foods," Vol. I. Wiley, New York.

# Chapter 2

# Cheese

## NORMAN F. OLSON

<table>
<tr><td>I.</td><td colspan="2">Introduction</td><td>40</td></tr>
<tr><td></td><td>A.</td><td>History</td><td>40</td></tr>
<tr><td></td><td>B.</td><td>Classification of Cheese</td><td>40</td></tr>
<tr><td>II.</td><td colspan="2">Fundamental Processes in Cheese Manufacture</td><td>41</td></tr>
<tr><td></td><td>A.</td><td>Control of Milk Properties</td><td>41</td></tr>
<tr><td></td><td>B.</td><td>Coagulation of Milk</td><td>46</td></tr>
<tr><td></td><td>C.</td><td>Separation of Curd and Whey</td><td>49</td></tr>
<tr><td></td><td>D.</td><td>Cheese Ripening</td><td>52</td></tr>
<tr><td></td><td>E.</td><td>Cheese Packaging</td><td>53</td></tr>
<tr><td>III.</td><td colspan="2">Cheese Varieties in Which Milk Is Clotted by Acid</td><td>54</td></tr>
<tr><td></td><td>A.</td><td>Cottage Cheese</td><td>54</td></tr>
<tr><td></td><td>B.</td><td>Cream and Neufchatel Cheeses</td><td>56</td></tr>
<tr><td></td><td>C.</td><td>Baker's Cheese</td><td>56</td></tr>
<tr><td>IV.</td><td colspan="2">Cheese Varieties in Which Milk Is Clotted by Proteases</td><td>57</td></tr>
<tr><td></td><td>A.</td><td>Cheddar Cheese</td><td>57</td></tr>
<tr><td></td><td>B.</td><td>Colby and Stirred Curd (Granular) Cheeses</td><td>61</td></tr>
<tr><td></td><td>C.</td><td>Surface-Ripened Cheeses</td><td>62</td></tr>
<tr><td></td><td>D.</td><td>Other Semisoft Cheeses</td><td>64</td></tr>
<tr><td></td><td>E.</td><td>Swiss Cheese</td><td>65</td></tr>
<tr><td></td><td>F.</td><td>Italian-Type Cheeses</td><td>68</td></tr>
<tr><td></td><td>G.</td><td>Mold-Ripened Cheeses</td><td>71</td></tr>
<tr><td>V.</td><td colspan="2">Process Cheese</td><td>73</td></tr>
<tr><td></td><td>A.</td><td>Manufacture</td><td>74</td></tr>
<tr><td></td><td>B.</td><td>Packaging and Mechanization</td><td>74</td></tr>
<tr><td></td><td>C.</td><td>Cold-Pack Cheese</td><td>74</td></tr>
<tr><td></td><td colspan="2">References</td><td>75</td></tr>
</table>

MICROBIAL TECHNOLOGY, 2nd ed., VOL. II

## I. INTRODUCTION

### A. History

The exact origin of cheese is difficult to determine since it undoubt-edly preceded use of written records. Cheese of some sort would have been made when milk was allowed to stand undisturbed until it coagu-lated. Use of animal stomachs to carry milk may have provided humans with the first cheese curd which was formed by the action of enzymes in the stomach walls and by metabolism of microbial contaminants. With a little ingenuity, humans learned that stomach linings and various plant extracts could curdle milk if they were soaked in or added to milk.

Early records indicate that the art of cheese making spread from the Middle East to the northern Mediterranean area and northern Europe. Romans introduced cheese making into England. The art of making many of the important present-day European types of cheese was pre-served and at times jealously guarded in monasteries during the Middle Ages.

The taste for and techniques of making cheese were carried by English colonists to the United States, Australia, and New Zealand. Cheese making was a farm industry in England and America until the middle of the nineteenth century. Housewives converted surplus milk, from their farm and possibly neighboring farms, into cheese for family consumption. Centralization of cheese operations into manufacturing plants was inevi-table so that this system eventually replaced most of the farm-based manufacturing.

### B. Classification of Cheese

The long history of cheese, the accidental modification of the cheese-making processes, its development in diverse areas during an era of limited communications, and global production has resulted in an enor-mous number of cheese varieties. Over 400 varieties have been des-cribed and more than 800 indexed by Walter and Hargroves (1972). Attempts have been made to categorize these varieties. Classification by countries in which varieties originated is satisfactory from a merchandis-ing standpoint but is unsatisfactory since the method does not charac-terize the variety. In addition, varieties may be essentially the same and differ only in their national name. Cheese varieties can be grouped into the following eighteen types based on differences in manufacturing: brick, Camembert, Cheddar, cottage, cream, Edam, Gouda, hand, Limburger, Neufchatel, Parmesan, Provolone, Romano, Roquefort, Sapsago, Swiss, Trappist, and whey cheeses (Walter and Hargrove, 1972). This system is logical and informative but is incomplete since two varieties may be made by the same process but differ markedly from the mode of curing. A third system of classifying differentiates on the basis of techniques of curing and firmness of the cheese varieties (Walter and Hargrove, 1972).

## II. FUNDAMENTAL PROCESSES IN CHEESE MANUFACTURE

The manufacture of cheese can be divided into five general phases: (1) control of properties of milk, (2) coagulation of milk, (3) separation of the solid curd from fluid whey, (4) forming curd into its final physical form, and (5) controlled storage of cheese to maintain the flavor of curd or to develop desired changes in flavor and rheological characteristics of cheese before consumption. Variations in each of these phrases result in cheese with different characteristics and explain the diversity in types of cheese.

### A. Control of Milk Properties

Some of the most important factors in controlling the characteristics and quality of cheese are the biological, chemical, and physical properties of milk. The type of animal used to produce milk affects all three properties and is the basis for differentiating certain cheese varieties. Cheese usually is made from bovine milk but some varieties are made from milk of sheep, goats, water buffalo, mares, and camels.

#### 1. Biological Properties

**a. Adventitious Microorganisms.** The biological properties of milk can determine the type of cheese made from that milk but can have a greater influence on quality of any given type. Milk with relatively low numbers of microorganisms is preferred for cheese making, even though certain contaminants can contribute to flavor development in cheese (Mocquot, 1971). Problems with excessive gas production, development of undesirable flavors, and potential contamination with pathogenic microorganisms dictate against use of milk with a high microbial population.

Numbers and types of contaminants in milk depend upon health of the cows and sanitary practices during the production and processing of milk (Foster et al., 1957). Numbers of psychrotrophic bacteria have assumed greater importance because of their ability to grow in refrigerated milk, especially at temperatures exceeding 7°C, and their lipolytic and proteolytic activities. Growth and activities of these species reduce yield of cottage cheese and cause spoilage during storage (Cousin and Marth, 1977a). Psychrotrophs can hydrolyze casein in milk, resulting in greater losses of nitrogenous materials in whey, and decrease yield of other cheese varieties (Cousin and Marth, 1977b; O'Leary et al., 1977; Nelson and Marshall, 1977).

**b. Enzymes in Milk.**  In addition to microorganisms, milk contains enzymes that are elaborated by the mammary tissue or by microorganisms contaminating milk. The principal, naturally occurring enzymes in milk are catalase, peroxidase, xanthine oxidase, alkaline and acid phosphatases, amylases, proteases, lipases, and aldolase (Jenness and Patton, 1959). Several of these enzymes have direct or indirect effect on cheese making. Catalase has no direct effect but is an indicator of mastitic milk, which retards normal acid production and may affect rheological properties of cheese curd. Lactoperoxidase is inhibitory to certain lactic acid bacteria used in cheese making (Reiter, 1973). Alkaline phosphatase serves as an index of adequate pasteurization since it is destroyed during normal pasteurization of milk. Lipases and proteases hydrolyze casein and milk fat, affecting flavor and rheological properties of cheese. It is doubtful that hydrolytic enzymes naturally occurring in milk contribute greatly to ripening of cheese varieties such as Cheddar since only a small fraction of the total activity has an optimum in the pH range of cheese while it is curing (McGugan, 1975). Proteolytic and lipolytic activities of added and adventitious microorganisms contribute significantly to the characteristics of many varieties of cheese, especially the mold-ripened and surface-ripened cheeses (Mocquot, 1971). Esterases isolated from the oral glands of calves, kids, and lambs produce the desired lipolysis and flavor in Italian varieties of cheese such as Romano and Provolone (Nelson et al., 1977).

**c. Control of Biological Properties.**  The cheese maker can exert some control over biological properties of milk by certain treatments and additives. The most common is heat treatment, including pasteurization (Wilster, 1969). To be labeled as pasteurized-milk cheese, cheese must not contain levels of alkaline phosphatase greater than those specified in federal and state regulations (U.S. Department of Health, Education and Welfare, 1974). Certain varieties of cheese, such as cottage and cream, must be made from pasteurized milk. Other varieties need not be pasteurized but must be stored for at least 60 days at temperatures of 2°C or higher if they are made from nonpasteurized milk. Pasteurization treatments of 72°C for 15 seconds and 63°C for 30 minutes or equivalent treatments have been selected to ensure destruction of pathogenic microorganisms which might contaminate milk. These treatments are used also to reduce numbers of microorganisms in milk and to give greater control over fermentations during manufacturing and curing. However, such treatments reduce the rate of flavor development in cheese since microorganisms involved in ripening are destroyed. Subpasteurization treatments are used as a compromise to attain faster flavor development but still control undesirable fermentations. Hydrogen peroxide treatment is used also to reduce numbers of microorganisms in milk (Kosikowski, 1977). It is not a substitute for pasteurization; the resulting cheese must be stored as is that made from unpasteurized milk.

### 2. Physical Properties

Physical properties of milk may be modified by homogenization, filtration and clarification, and vacuum treatments (Bradley, 1969; Harper and Hall, 1976). Milk fat globules are reduced in size (diameter) from about 4 $\mu$m to 1 $\mu$m by homogenization by forcing milk through a small orifice or specially designed wire mesh. The treatment results in a weaker curd formed by milk-clotting enzymes and decreased whey syneresis from curd but reduced fat losses during cheese making and less fat leakage from cheese (Ernstrom and Wong, 1974). Homogenization is used in the manufacture of cream cheese, blue cheese, certain soft types, and process cheese spreads. In the manufacture of cream cheese, cream is homogenized prior to cheese making and the curd may be homogenized if the "hot-pack" method is used. Use of this treatment on the finished cream cheese or other soft types and cheese spreads increases smoothness of the product. Decreasing the size of fat globules, with concomitant increase in surface area, accelerates lipolysis and flavor development in blue cheese. Even though fat losses are reduced by homogenization, problems of reduced curd strength and whey syneresis and development of hydrolytic rancidity have prevented application of the treatment to most varieties of cheese. Mild homogenizing treatments of milk have been used experimentally in making Cheddar cheese and in an accelerated direct-acidification procedure for cheese making (Peters, 1964; Quarne et al., 1968).

Filtration and clarification affect physical properties of milk by removing extraneous material from milk. Clarification can be more efficient than filtration, removing materials such as leukocytes in addition to particulate material. Milk is clarified by flowing through a revolving bowl similar in construction to a milk separator (Wilster, 1969; Harper and Hall, 1976). All substances which are more dense than normal milk are sedimented out by the centrifugal force. This process is used widely in making Swiss cheese; more uniform eye formation results from removal of extraneous matter which can act as foci for developing holes or eyes.

Vacuum pasteurization of milk removes gases and volatile odors from milk (Wilster, 1969). This can be beneficial in areas plagued with weed or similar flavors in milk. The removal of some water vapor from milk reduces milk volume and can have economic advantages by increasing milk throughput in a large plant.

### 3. Chemical Properties

**a. Treatments of Milk.** Altering relative amounts of the principal milk constituents is the most universal method of controlling chemical properties or composition of milk. Removal of fat from milk by centrifugation using a "separator" is a practical method of adjusting composition (Bradley, 1969). Cream cheese is made from cream having a fat content higher than normal milk; cottage cheese is made from skim milk.

Chemical properties of milk can be affected by chemical additives permitted under regulations promulgated by the U.S. Food and Drug Administration (1974). Food-grade colors are used to enhance color of certain cheese varieties. They do not affect flavor or rheological properties of cheese but their use is dictated by tradition and personal or regional preferences of consumers. Several types of colors are certified under federal regulations but annatto, a carotenoid pigment extracted from the seeds of the tree *Bixa orellana,* is used most widely. Conversely, intensity of natural milk pigments is reduced in certain cheeses (blue, Swiss, and Italian varieties). This is done by masking the color with harmless artificial blue or green coloring or by bleaching milk with benzoyl peroxide. That portion of vitamin A potency of milk destroyed by bleaching must be restored.

Calcium chloride may be added at levels not exceeding 0.02% of milk weight to aid in rennet coagulation, especially milk heated to higher than normal pasteurization temperatures. Food-grade acids can be used to aid in curd formation in certain types of cheese, such as Ricotta, and as acidulants in process cheese products. Accelerated, continuous procedures have been suggested using direct acidification of milk (Olson, 1975). Hydrogen peroxide may be used, at levels not to exceed 0.05%, to treat milk prior to the manufacture of Cheddar, Colby, and Swiss cheeses. Residual hydrogen peroxide must be destroyed by catalase before cheese making.

**b. Yield of Cheese.** When milk is converted into cheese, milk constituents are partitioned between curd and whey as illustrated in Table I (see Section II, C). The principal milk constituents contributing to yield of cheese are milk fat and casein. Based upon a large number of analyses, a formula was developed which related yield of Cheddar cheese to casein and fat contents of milk (Van Slyke and Price, 1952).

$$\text{Yield} = \frac{(0.93F + C - 0.1)\,(1.09)}{1.00 - W}$$

where yield = kilograms of cheese per 100 kg of milk
   $F$ = kilograms of fat per 100 kg of milk
   $C$ = kilograms of casein per 100 kg of milk
   $W$ = kilograms of water per kilogram of cheese

It assumes that 7% of fat and 0.1 kg of casein are lost in whey, pressings, curd fines, etc. Higher losses with some manufacturing methods necessitate changes in both factors for fat and casein. The factor of 1.09 corrects for solids in cheese other than fat and casein.

 D. B. Emmons (personal communication, 1977) adapted the above formula to cheeses of varying fat contents. In the formula below, it is assumed the casein constitutes 78% of total milk protein and that 3.89% of casein is lost in whey.

$$\text{Yield} = 0.93F + \frac{0.907P}{1.00 - MNFP}$$

$$= 0.93F + \frac{1.163C}{1.00 - MNFP}$$

where yield = kilograms of fat-reduced cheese per 100 kg
      of fat-reduced milk
    $F$ = kilograms of fat per 100 kg of fat-reduced milk
    $P$ = kilograms of protein per 100 kg of fat-reduced milk
    $C$ = kilograms of casein per 100 kg of fat-reduced milk
  $MNFP$ = kilograms of water per kilogram of nonfat portion
       of cheese

In addition to milk composition, yield is affected by manufacturing practices which affect retention of constituents (Olson, 1977).

**TABLE I.** Partition of Milk Solids between Cheese and Whey during Manufacture of Cheddar Cheese Containing 37% Moisture[a]

| Constituent | In cheese[b] | In whey[b] |
|---|---|---|
| Fat | 3.7 | 0.3 |
| Casein | 2.4 | 0.1 |
| Lactose | 0.2 | 4.9 |
| Whey proteins and salts | 0.4 | 1.0 |
| Water | 3.9 | 83.1 |
| Total | 10.6 | 89.4 |

 [a] Adapted from Van Slyke and Price (1952).
 [b] Kilograms of milk constituents from 100 kg of milk containing 4.0% fat.

## B. Coagulation of Milk

In the manufacture of almost all varieties of cheese, milk proteins are coagulated to form a continuous solid curd in which milk fat globules, water, and water-soluble materials are entrapped. The casein fraction of milk normally forms this curd but in a few varieties of cheese, whey proteins are coprecipitated with the casein fraction. Coagulation may be induced (1) by adjusting the pH of milk to the isoelectric point of casein, (2) by action of milk-clotting enzymes, (3) by application of heat, or (4) by addition of salts. The first two systems are commonly used alone or in combination to form cheese curd. Application of heat is used for only a small number of cheese varieties and salt addition is rarely used because of effects on flavor.

### 1. Lactic Starter Cultures

Acidification of milk is usually accomplished by fermentation of lactose to lactic acid by selected cultures of bacteria. *Streptococcus cremoris* and *Streptococcus lactis* are used in cheese varieties heated to 40°C or less during the acid fermentation. High-temperature homolactic bacteria, such as *Streptococcus thermophilus, Lactobacillus bulgaricus,* and *Lactobacillus helveticus,* are used in manufacture of cheese heated to higher temperatures (Ernstrom and Wong, 1974).

Lactic acid-producing bacteria are usually selected, propagated, and distributed in the United States by industrial firms. Research on physiology, propagation methods, and preservation of lactic cultures and improvements in technology of preparation, distribution, and use of these cultures have significantly improved their activity and reliability.

Commercial mesophilic cultures used for such cheese varieties as Cheddar and cottage may be single strains of *S. cremoris* or *S. lactis* or combinations of strains of these two species, or combinations of these two species with low numbers of *Leuconostoc* species or *Streptococcus diacetilactis.* Strains are selected for acid-producing activity, bacteriophage sensitivity, growth compatibility with other strains, lack of gas production, and effect on flavor of cheese. Lactic cultures are supplied to cheese plants in the lyophilized or frozen state, with the latter being more prevalent in North America. Freezing, preservation, and distribution is accomplished by use of liquid nitrogen or dry ice. Traditional lyophilized and frozen cultures are propagated several times to attain the volume of culture to inoculate milk for cheese making. Concentrates of lactic cultures have been prepared by centrifuging cells out of a growth medium and resuspending in menstrua to give a culture "paste" with approximately 100 times more cells than a normal culture (Gilliland and Speck, 1974; Reinbold, 1973). The concentrated cultures have been inoculated directly into bulk cultures for cottage and Cheddar cheese manufacture and more recently inoculated directly into milk in the cheese vat (Gilliland, 1977; Volume I, Chapter 3).

One of the principal causes of lack of acid production during cheese making is destruction of lactic cultures by the lytic action of bacteriophage (Sandine, 1977). Effects of bacteriophage can be limited by using properly designed equipment and facilities, asepsis in culture handling, sanitation, proper handling of whey, use of bacteriophage inhibitory media, and a culture rotation program (Collins, 1977: Lawrence *et al.*, 1976). The latter technique is effective if the cultures used in sequence are not attacked by the same bacteriophage. Collins (1952) felt that attack by bacteriophage would not be noticed during cheese manufacture if at least 50% of the strains were resistant to the infecting phage strains.

Lactic acid-producing cultures have significant effects on the curd-making and cheese-curing phases of cheese manufacture (Lawrence *et al.*, 1976; Ernstrom and Wong, 1974). Acid production effects coagulation of milk at the isoelectric point of casein (pH 4.6) in the manufacture of such varieties as cottage cheese and accelerates enzymatic coagulation of milk in other cheese varieties. Further acid production and reduction in pH of rennet curd during cheese making aids in whey syneresis from curd, controls undesirable microorganisms, and increases the plasticity of curd (Kosikowski, 1977; Ernstrom and Wong, 1974). Acids (lactic, acetic, and propionic), ketones and aldehydes or their precursors, and esters produced in total or in part by lactic cultures contribute directly to the flavor of cured cheese (McGugan, 1975; Ernstrom and Wong, 1974; Lawrence *et al.*, 1976). Indirect effects on flavor and appearance of cured cheese result from control of the growth and metabolism of adventitious bacteria by reduction of pH of curd. Certain strains of lactic streptococci are undesirable for cheese making because of production of excess gas, antibiotics against other strains of lactic bacteria, bitterness, and fruity flavors (Lawrence *et al.*, 1976).

### 2. Milk-Clotting Enzymes

Enzymes used to clot milk are classified as acid proteases (Ernstrom and Wong, 1974). Requirements considered important for milk-clotting enzymes are: clotting and proteolytic activities, proteolytic specificity, properties of curd, presence of contaminating enzymes, and heterogeneity of proteinases (Green, 1977). Milk-clotting enzymes traditionally have been selected on the basis of their high ratio of milk-clotting to proteolytic activities. Low proteolytic activity is essential to avoid excess protein degradation in cheese or in curd during manufacturing which results in reduced yields of cheese. Proteolytic specificity refers to relative rates of activity on $\alpha_s$- and $\beta$-caseins with the resulting release of bitter peptides. Commercial enzyme preparations are tested for this tendency before wide-scale commercial use. Rates of increase in curd firmness vary with different enzymes but cutting the curd at the same firmness apparently negates these differences and any effect on properties of curd during subsequent manufacturing (Ramet et al., 1969; Richardson et al., 1971). Some preparations of microbial milk-clotting enzymes reportedly contained contaminating lipases and proteases (Green, 1977). Purification procedures were instituted to eliminate these contaminants in commercial enzymes. Commercial milk-clotting enzyme preparations may contain several proteases. Differing responses of these enzymes to pH, temperature, and salts must be taken into consideration by the cheese manufacturer. Adjustments in the manufacturing schedule, such as the time of cutting the curd, may have to be made.

A number of different enzyme preparations are used commercially to clot milk: rennet extract, porcine pepsin, and proteases from selected microorganisms. The most widely used microbial proteases are derived from *Mucor miehei, Mucor pusillus,* and *Endothia parasitica* (Ernstrom and Wong, 1974; Green, 1977). Rennet is the partially purified enzyme extract from the fourth stomach of calves. The principal milk-clotting enzyme in the extract is chymosin (rennin). A second enzyme, bovine pepsin, contributes a small portion of the total activity of extracts from young, suckling calves. The relative concentration of pepsin increases as the calf ages. A mixture of rennet and porcine pepsin called 50:50 is used to manufacture cheese which is not ripened extensively.

The action of milk-clotting enzymes on milk can be divided into three phases: the primary or enzymatic, secondary or clotting, and tertiary or proteolytic phase (Ernstrom and Wong, 1974; Green, 1977). The total effect of the primary and secondary phases is the conversion of one of the milk protein fractions, casein, from a colloidal suspension to a fibrous network. The casein fraction in milk is heterogeneous; its main components are designated as $\alpha_s$-, $\beta$-, $\gamma$-, and $\kappa$-caseins. These components interact in milk to form a stable suspension of spherical particles called micelles; stability of the suspension in the presence of calcium depends on $\kappa$-casein. Milk-clotting enzymes attack $\kappa$-casein causing a release of macropeptides from this fraction; loss of the peptides destroys the stabilizing effect of $\kappa$-casein and sensitizes micelles to calcium. Coagulation of casein in the presence of calcium constitutes the second phase of enzyme action. The first two phases of milk clotting are accelerated by lowering the pH and by increasing temperatures up to 45°C. The rate of the secondary phase is more temperature-dependent than the primary phase so that coagulation of milk can be delayed at low temperatures for a considerable time after the primary phase is completed.

Instantaneous coagulation occurs upon warming the milk. This characteristic is used in some continuous cheese-making systems in which milk is cold-renneted and then coagulated continuously in special devices (Kosikowski, 1975).

Proteolytic action of milk proteases, lactic cultures, and adventitious bacteria have complicated investigation of the tertiary phase of chymosin action in cheese. Studies on proteolytic action of chymosin on purified casein suggest that chymosin hydrolyzes protein in cheese to polypeptides (Green, 1977; Lawrence et al., 1976; O'Keefe et al., 1976). Lactic cultures and adventitious microorganisms further degrade polypeptides to lower molecular weight peptides and amino acids.

## C. Separation of Curd and Whey

Subsequent to coagulation, partition of the solid curd and whey (water plus water-soluble compounds) occurs. This is accelerated in the cheese-making process by heating, a decrease in pH, and physical disruption and manipulation of the curd. In the procedures for most varieties of cheese, the milk is coagulated into a contiguous curd mass in large vats (Fig. 1). Special curd knives, steel frames strung with wire, are used to cut the curd into particles of the desired size. These may be small irregularly shaped particles resulting from extensive cutting or discrete cubes from more precise cutting procedures. Cutting curd greatly accelerates initial syneresis of whey by increasing the surface to volume ratio of curd. Hence, rates of syneresis are inversely related to curd size.

Production of acid by lactic starter cultures lowers the pH and water-holding capacity of curd. Temperature and pH probably are the most important factors in controlling whey syneresis or moisture expulsion from curd during manufacture and in establishing moisture levels for a given variety of cheese (Patel et al., 1972). Stirring curd in whey also accelerates whey syneresis, presumably from physical stresses on curd (Lawrence, 1959). The curd is stirred in whey until the pH has decreased sufficiently and an optimum amount of whey expulsion has occurred. Then the curd is physically separated from whey by various means which differ between cheese varieties. Whey can be removed from curd by settling the denser curd and allowing whey to flow from the curd or by placing curd and whey in perforated forms (hoops) in which curd is trapped and whey escapes. The first method is used in manufacture of Cheddar and cottage cheeses; the second in brick and blue cheeses. Curd can be removed from whey by gathering it in a cloth and lifting the cloth and curd out of the whey and into a drainage form as was done in the traditional method for Swiss cheese.

**FIGURE 1.** Horizontal vats used in cheese making. Revolving agitators shown in the center of vats traverse the length of vats and stir the curd and whey.

The separation of curd and whey results in a partition of milk solids in the two phases as shown in Table I. Approximately 50% of the milk solids are lost in the whey in the manufacture of Cheddar cheese. This percentage loss would be higher for cheese made from milk with a lower fat content since most (90%) of the milk fat is recovered in curd. This value varies between cheese varieties and depends upon treatment of curd during manufacturing. Extensive cutting of curd, exposure to high temperatures, and excessive physical stress on curd increases fat losses. Loss of casein in whey usually is low but increases with use of inferior quality milk and poor manufacturing techniques. Whey proteins, as indicated in Table I, partition primarily into whey because of their high solubility under conditions of normal cheese manufacture. Levels retained in curd depend on amounts of whey retained in the curd; more are retained in high-moisture cheese. Exposing milk to high temperatures before or during coagulation causes an interaction of whey proteins with casein and their precipitation with curd (Ernstrom and Wong, 1974). Such treatments are used in making Ricotta and certain Norwegian and South American varieties of cheese. It has been suggested for cottage cheese but not adopted widely because of problems with curd formation and firmness (Emmons and Tuckey, 1967).

Water-soluble salts will partition in the same manner as whey proteins. A portion of the calcium is bound to casein or is associated with casein as colloidal calcium phosphate. Retention of calcium with curd varies inversely with the pH of curd at the time of whey drainage; cottage cheese has low levels, whereas Cheddar and Swiss have higher levels. The milk sugar, lactose, is the greatest contributor to solids loss in cheese making and constitutes 70% of the whey solids. A few cheese varieties made from whey and whey–milk mixtures contain high levels of lactose but are not normally cured after manufacture.

Most (95%) bacteria present in milk are trapped in the curd (Ernstrom and Wong, 1974). Consequently, lactic acid fermentation occurs primarily within curd particles and the lactic acid diffuses out into whey. The environment for bacteria within cheese curd during cheese making has not been completely characterized but is dynamic. Acid is being produced within the curd, causing a decrease in pH of the serum in curd and an increase in soluble calcium salts in the serum and whey. Serum or whey is being expelled, carrying with it lactose, the substrate for lactic acid fermentation. However, diffusion is not unidirectional since Czulak *et al.* (1969) found that lactose diffused back into Cheddar cheese curd from whey at about the same rate as it was fermented in curd. The equilibration was not rapid enough in cheese in which lactic acid fermentation is abnormally fast. Consequently, composition of curd at the end of whey expulsion depended upon the lactic fermentation and the rate and time for diffusion of constituents in and out of curd.

The low molecular weight of milk-clotting enzymes precludes their entrapment in curd. Retention of rennet in Cheddar cheese is low, with only 6% of the original activity evident in curd after pressing (Holmes *et al.*, 1977). Approximately 35% of the activity is destroyed during the manufacturing process up to the point of whey drainage; the remainder is partitioned in whey. No activity of proteases from *M. miehei* and *M. pusillus* was lost during cheese making with 2–3% remaining in the cheese and the rest in the whey.

After physical separation of curd and whey, the curd is subjected to a series of treatments to allow acid production and attainment of optimum pH levels, to induce sufficient moisture loss from curd, to apply salt to the curd, and to fuse the curd into its final shape. The techniques used in these steps vary between varieties and types of cheese and influence characteristics of the varieties. Certain cheese varieties are distinguished by shape alone. The Italian varieties, Provolone and salami, may be made from the same curd but molded into different shapes.

## D. Cheese Ripening

The transformation of the elastic or chalky cheese curd with an acid flavor to a ductile, full-flavored cheese is accomplished by controlled storage called ripening. Agents responsible for these changes are milk enzymes, adventitious microorganisms, and added microbial cultures (Ernstrom and Wong, 1974). The changes which occur are extremely complex in almost all cheese varieties. They involve hydrolysis of a portion of the fat and protein with subsequent decarboxylation, deamination, and dehydrogenation of some of the hydrolysis products. Cheese flavor has been attributed to carbonyls, nitrogenous compounds, fatty acids, sulfur compounds, and miscellaneous minor components.

In certain cheese varieties, flavor originates predominately from a specific group of microorganisms and enzymes. Propionibacteria contribute significantly to flavor and eye formation in Swiss cheese; pregastric esterases are vital for the "piccante" flavor of Romano and Provolone cheeses; *Penicillium roqueforti* and *Penicillium camemberti* produce the typical flavors and appearance of blue and Camembert cheeses. Progressive growth of yeast, micrococci, and *Brevibacterium linens* impart characteristic flavors to smear-ripened cheese such as Limburger. Other varieties, such as Cheddar and Gouda, do not depend on a specific class of compounds for their typical flavor (Ernstrom and Wong, 1974). A balance of various components is essential for the overall flavor of these varieties. In such a complex system, addition of single bacterial species has not consistently accelerated the development of typical Cheddar flavor.

Fermentation patterns in most varieties of cheese have not been defined sufficiently to allow direct control of cheese ripening. Indirect control of ripening is attained by establishing proper conditions in cheese and in the external environment. Low numbers of microorganisms in milk make such control of cheese quality more effective. Fermentations are regulated by pH, moisture, salt, osmotic pressure, oxidation–reduction potential, and depletion of reducing sugars in cheese. Temperature affects biological changes occurring throughout cheese; relative humidity influences the type and extent of microbial growth on the cheese surface.

Improvements in packaging and in ripening rooms have contributed significantly to standardization of cheese quality. Development of mechanical refrigeration for temperature control obviated many quality defects associated with high ripening temperatures. Even though ripening temperatures are fairly low there is considerable variation between varieties of cheese; Cheddar cheese is ripened at 2°–10°C but certain firm-bodied Italian varieties may be ripened at higher temperatures.

Swiss cheese is held initially at 21°C to induce eye formation and then ripened at lower temperatures. Surface-ripened cheeses (brick, Limburger, and Bel Paese) are held at 13°–15°C and under high relative humidities to allow surface microbial growth before final storage at lower temperatures.

### E. Cheese Packaging

Cheese is normally covered with a protective coating at some stage of the ripening process. Coatings vary from vegetable oils to special plastic films. The shape and characteristics of cheese and changes to be induced in cheese dictate the type of packaging material. Cheese which is to be slightly dehydrated, such as Parmesan and Romano, may be rubbed with a vegetable oil to retard surface microbial growth but allow loss of moisture from the cheese. Waxes have been formulated for cheese packaging which cling to cheese to minimize surface growth and have low water-transmission rates. It is necessary to dry the surface of cheese slightly before waxing; this causes some loss of moisture and yield and necessitates some special handling. Waxes are especially useful for cheese varieties with unusual shapes where application of molten wax is a simple method of packaging. Waxes are also used in combination with plastic films.

The most significant change in packaging of cheese prior to ripening was the development of the rindless block style of cheese. Most Cheddar and Swiss cheese manufactured in the United States is made in this form. The cheese is pressed in rectangular hoops or forms and wrapped in a plastic film. The films are usually laminants of different plastic films and waxes with specific properties (National Research Council, 1958). This system of packaging has several advantages. It is not necessary to dry the cheese surface as thoroughly as is done in waxing of cheese, which eliminates yield loss. Ripening of film-wrapped cheese can be done at lower humidity levels without excessive moisture losses from the cheese.

Cheddar and Cheddar-type cheeses are pressed into larger units such as large cylinders (barrel cheese) and blocks. The barrel style of Cheddar cheese is made by placing 215 kg of curd, after sufficient whey drainage and salting, in a cylindrical barrel. A uniform fusion of the curd, without external pressure, is attained by inverting the barrel at appropriate intervals; or pressure may be applied. Vacuum treatments can be used also to improve fusion of this style and "block" Cheddar cheese (Price and Olson, 1965; Robertson, 1966). The trend to pressing cheese curd into large forms with the aid of vacuum undoubtedly will continue.

## III. CHEESE VARIETIES IN WHICH MILK IS CLOTTED BY ACID

### A. Cottage Cheese

#### 1. Curd Formation

Cottage cheese curd is formed by acid coagulation of skim milk (Emmons and Tuckey, 1967). Traditionally, acid production results from fermentation by added lactic starter cultures containing *S. cremoris, S. lactis*, and small amounts of *Leuconostoc cremoris*. Lactic fermentation in skim milk may be carried out at a lower temperature, 22°C, for a longer time (16 hours), using a low inoculum (0.5–1.0%) of a normal lactic culture or at a higher temperature of 32°C for 4–5 hours using 4–5% culture. Intermediate conditions may be used to fit manufacturing schedules. Reduction of pH causes aggregation of one fraction of milk proteins, casein, to form a continuous gel in the cheese vat. When the pH has decreased to 4.6–4.7 and the proper firmness of the gel has been attained, the gel is cut into cubes with curd knives. The size of cubes may be 0.63 cm to make "small-curd" or 1.3–1.9 cm to make "large-curd" cottage cheese.

#### 2. Cooking the Curd

Whey expulsion from curd begins immediately after cutting and is accelerated by stirring and heating the curd–whey slurry. Heating or "cooking" is done slowly to avoid case-hardening, formation of an excessively impervious skin on the curd, which retards further whey syneresis. The final cooking temperature will approximate 55°C. Growth and metabolism of the lactic culture diminishes and ceases when the temperature is raised beyond 40.6°–43°C (Emmons and Tuckey, 1967). The combination of low pH and high cooking temperature destroys virtually all contaminating psychrotrophic bacteria and lactic bacteria in the curd and whey (Collins, 1961). Consequently, microbial spoilage of packaged cottage cheese results from postcooking contamination.

#### 3. Acidity Effects

The pH attained at cutting and the minimum pH reached when cooking temperature exceeds 43°C have significant effects on characteristics of the finished cheese. If the pH drops below the isoelectric point (pH 4.6) of casein, the curd will retain more water and the cheese may have a soft, pasty body. Excess acidity may also produce a mealy body if the curd is firm. A high pH may produce curd that is too firm and rubbery and may become translucent. Susceptibility to spoilage is greater in a low-acid curd with spoilage manifesting itself as a slimy, gelatinous surface growth, discoloration, and fermented, fruity, and unclean flavors.

### 4. Whey Draining and Curd Washing

The curd and whey are stirred, after attaining the desired cooking temperature, until the curd has become adequately firm. The curd particles are allowed to settle but not fuse while most of the whey is drained. The curd is then subjected to a series of washings with progressively colder water. This cools the curd, reduces acid flavor, and raises the pH of curd slightly.

Microbial contamination of cheese occurs most frequently from contact with water. This can be minimized by treating water to adjust available chlorine levels to 5–10 ppm. Acidification of water prior to chlorination increases effectiveness of the sanitizer and prevents curd defects resulting from excessively high pH.

### 5. Creaming the Curd

A cream mixture is added to the washed and drained curd to yield a finished creamed cottage cheese. The minimum fat content of creamed cottage cheese in the United States is 4% but "low-fat" cottage cheese is also produced. When cream is mixed with curd, a portion of the serum phase of cream is absorbed by curd, leaving a desirable, viscous coating of cream on the cheese. Insufficient absorption, excess cream, or low viscosity of cream results in "free cream" which flows from the curd.

Flavor intensity of creamed cottage cheese has been increased through citrate fermentation by heterofermentative lactic streptococci. A widely used method involves growth of *L. cremoris* in skim milk, acidfying to pH 4.3 with sterile citric acid, and further incubation to develop flavor (Mather and Babel, 1959). The cultured skim milk containing desired concentrations of diacetyl and acetoin is added to the cream dressing. A special strain of *S. diacetilactis* has been used also to increase flavor intensity (Lundstedt and Fogg, 1962).

### 6. Mechanization

Alternative and mechanized systems for cottage cheese manufacturing have been used experimentally or on a commercial basis (Ernstrom and Kale, 1975; Kosikowski, 1977). Replacement of biological production of acid with direct acidification methods has been done on laboratory, pilot-plant, and commercial scales. A process of partial acidification of skim milk with food-grade acids and final acidification with D-glucono-δ-lactone is used commercially (Ernstrom and Kale, 1975). This system is feasible because hydrolysis of lactone to gluconic acid is slow enough to allow mixing with warm milk and clotting of the milk in traditional vats. A continuous system of heating acidified skim milk in a tubular heat exchanger to cause clotting and equipment to process the stream of curd has been evaluated on a pilot-plant basis (Ernstrom and Kale, 1975).

An enclosed system of washing, cooling, draining, creaming, and packaging cottage cheese is used commercially (Ernstrom and Kale, 1975). Curd and a portion of the whey, at the end of heating (cooking) is pumped to closed, cylindrical tower-vats where it is drained and washed before flowing through the rest of the system.

## B. Cream and Neufchatel Cheeses

Cream cheese is a soft, fresh variety made from cream which is standardized to yield a cheese containing at least 33% fat and not more than 55% moisture (U.S. Food and Drug Administration, 1974). Cream cheese curd is formed by acid coagulation, as is cottage cheese, but is thoroughly disrupted by agitation instead of being cut into cubes (Emmons and Tuckey, 1967; Kosikowski, 1977). The same species of lactic bacteria are used for starter as those used for cottage cheese.

Curd and whey are separated by differential centrifugation in special separators or, to a limited extent, by drainage in cloth bags and filter presses. The curd can be packaged and marketed after whey removal or it can be processed by the "hot-pack" method. This involves heating, homogenizing, and packaging hot curd. Cheese made by this process has better storage stability than cheese packaged immediately after whey drainage.

Neufchatel cheese is similar to cream cheese but differs in composition; the minimum milk fat content according to United States standards is 20% and the maximum moisture content is 65% (Emmons and Tuckey, 1967; Kosikowski, 1977). Manfacturing methods are similar to cream cheese with the "hot-pack" method being most prevalent.

## C. Baker's Cheese

Skim milk is clotted by lactic fermentation using S. cremoris type culture and small amounts of rennet (4 ml/454 kg of skim milk) and procedures similar to those for cottage cheese (Emmons and Tuckey, 1967; Kosikowski, 1977). The curd is disrupted in a manner similar to that used for cream cheese. Traditionally, the curd was drained in filter bags under pressure but centrifugal separation has been used in some operations.

## IV. CHEESE VARIETIES IN WHICH MILK IS CLOTTED BY PROTEASES

### A. Cheddar Cheese

Cheddar cheese is named after the village of Cheddar in Somerset-shire, England. It has evolved into the most important single variety of cheese in the world. In the United States, Cheddar plus its modifications, Colby, granular Cheddar, washed curd, Jack, and Monterey are referred to as American cheese. The manufacturing techniques for Cheddar cheese have undergone and are still undergoing modifications (Robertson, 1966; Olson, 1971, 1975). Cheddar cheese was first made by a "stirred curd" procedure, described in Section IV, B, but evolved into the typical Cheddar procedure outlined in Table II.

### 1. Manufacturing Procedures

This variety is made from raw, pasteurized, or partially pasteurized whole milk (Kosikowski, 1977; Price et al., 1971; Wilson and Reinbold, 1965; Wilster, 1969). In the United States, the fat content of milk can be standardized so that the finished product contains a minimum of 50% fat in the dry matter. Lactic acid-producing cultures normally used contain S. lactis and S. cremoris, the latter species being preferred. The rate of growth and acid production by the culture must be controlled and corre-lated with rate of whey syneresis and time schedule of manufacturing. This is accomplished in modern plants by using cultures of known activity, accurately measuring the desired amount of culture into cheese milk, and standardizing manufacturing times and temperatures.

Milk is clotted with suitable proteases and it is cut into 0.63-cm cubes with curd knives. Syneresis of whey occurs rapidly after cutting and is accelerated by stirring, heating, and acid development. Maximum cook-ing temperatures, 40°–41°C, are fixed by effects on metabolism of lactic streptococci. Collins (1977) found that holding strains of S. cremoris at 38.9°C, to simulate conditions of cooking Cheddar cheese, arrested growth of most strains but harmed only two out of nine test strains if the holding time was 45 minutes. A third strain was adversely affected with a 90-minute holding period. Growth rates differed between strains at tem-perature simulating those of Cheddar cheese curd after cooking and after whey drainage. The above differences in temperature responses of strains could affect rate of acid production in later stages of cheese making.

After the curd has reached proper pH and firmness, it is allowed to settle in whey to form a cohesive block at the bottom of the vat. This aids in producing the smooth, closely knit appearance of Cheddar cheese.

**TABLE II.** Procedure of Manufacturing Cheddar Cheese Containing 37–38% Moisture from 1000 kg of Milk[a]

| Steps in making | Time of step | Temperature (°C) | Acidity measurements | | Comments |
|---|---|---|---|---|---|
| | | | % | pH | |
| Add starter | 8:15 | 31 | 0.16 | 6.56 | 7 kg |
| Add color | 8:45 | 31 | 0.16 | — | 65 ml |
| Add rennet | 9:00 | 31 | 0.17 | 6.50 | 195 ml |
| Coagulation | 9:12 | 31 | — | — | |
| Cut curd | 9:30 | 31 | 0.10 | — | 0.64-cm knives |
| Steam on | 9:45 | 31 | 0.10 | — | |
| Steam off | 10:15 | 39 | 0.11 | 6.4 | Stir slowly |
| Drain portion of whey | 10:30–10:45 | — | — | — | |
| Drain whey | 11:00 | 39 | 0.13 | 6.2 | |
| End dipping | 11:30 | | 0.15 | 6.0 | 45-cm trench |
| Pack | 11:45 | | 0.17 | 5.9 | Blocks 17-cm wide |
| Pile 2 high | 12:30 | | 0.25 | 5.7 | Cut blocks in half |
| Pile 3 high | 1:00 | | 0.32 | 5.5 | |
| Mill | 1:30 | | 0.45 | 5.4–5.5 | |
| Salt | 1:50 | | — | — | 2.7 kg |
| Hoop | 2:30 | | — | — | All salt dissolved |
| Prepress | 2:50 | | — | — | Full pressure in 15 min |
| Dress | 3:20 | | — | — | |
| Press | 3:30 | | Press for 5–20 hours at full continuous pressure | | |

[a] Adapted from Price et al. (1971).

Stirring curd after whey removal maintains the granular form, thereby allowing more openings in the finished cheese. After whey drainage, the large mass of curd is cut into smaller blocks which are turned and piled in a process called "cheddaring." The blocks of curd exhibit plastic flow and deform under the influence of increased acidity and physical stress. Individual curd particles in the blocks change from a granular form to contiguous flat platelets, producing a fibrous network of protein which imparts some of the typical elasticity and plasticity to Cheddar cheese. Cheddaring is continued until optimum acidity and firmness levels are reached. Mechanized systems of cheddaring cheese have been developed; one example is illustrated in Fig. 2. After cheddaring, the blocks are cut or "milled" into small pieces with special cutting devices called "curd mills." Subdividing the blocks cools the curd to avoid excessive fat losses during subsequent pressing and facilitates mixing and uniform salting. The salted curd is put into forms (hoops) and pressed into the desired shape by mechanical pressure.

**FIGURE 2.** Mechanized cheddaring system. Whey is drained and curd is fused and flows to obtain the "cheddared" structure.

### 2. Ripening

Cheese is usually packaged in various types of plastic films which are sealed to act as oxygen and moisture barriers. The cheese is placed in storage at selected temperatures ranging between 2° and 7°C to undergo ripening. This is a series of complicated and interrelated reactions involving metabolism of carbohydrates, protein, and fat. Simple sugars in cheese are fermented by the lactic bacteria to reduce the pH to 5.0–5.2 during the first few days after manufacturing (Brown and Price, 1934). This minimum pH value depends upon the pH attained during manufacturing, the amount of moisture or serum, and consequently the concentration of sugars in curd, buffering capacity of curd, and any inhibition of culture (Breene *et al.*, 1964). These interrelationships are taken into consideration when Cheddar cheese is made with varying moisture contents. A lower pH will have to be attained at milling when "low-moisture" Cheddar is manufactured since subsequent decrease in pH will be less because of low amounts of sugar to be fermented and the slightly higher buffering capacity of cheese. High-moisture cheese will be milled at a higher pH value since this cheese will have a higher concentration of fermentable sugar. Consequently, the drop in pH will be greater during the first few days of ripening.

Initial proteolysis of Cheddar cheese probably results from action of milk-clotting enzyme (Ernstrom and Wong, 1974; Lawrence et al., 1976; McGugan, 1975). The rubbery, elastic structure of fresh curd is converted to a smooth consistency by hydrolysis of casein in the cheese. This change occurs within the first 1–2 months of ripening. Proteolysis also contributes to flavor by releasing amino acids which impart a background flavor and are precursors for other flavor compounds (McGugan, 1975).

The flavor defect, bitterness, results from proteolysis, but there is disagreement as to whether milk-clotting enzymes or certain lactic streptococcal strains are causative agents (Lawrence et al., 1976). It has been demonstrated that cheese made by chemical acidification without lactic bacteria can become bitter, presumably from the action of milk-clotting enzymes (Micketts and Olson, 1974; Perry and McGillivray, 1964). However, conditions of manufacture and composition of this cheese did not duplicate that of Cheddar cheese. A direct role of lactic bacteria in producing bitterness was suggested by use of bacteriophage to restrict growth of S. cremoris HP but not acid production near the end of manufacture (Lowrie, 1977). Cheese made under these conditions was very slightly bitter after 6 months of aging, whereas control cheese made with the same strain was definitely bitter.

Conditions of manufacture and cheese composition may also affect development of bitterness. Fox and Walley (1971) found that 5–10% NaCl significantly reduced hydrolysis of one casein fraction, $\beta$-casein, by chymosin but had little effect on hydrolysis of $\alpha_s$-casein. It was suggested that NaCl in cheese might control bitterness since $\beta$-casein hydrolysates are bitter whereas $\alpha_s$-casein hydrolysates are not.

### 3. Microbiology of Ripening

The relative importance of various groups of bacteria in development of flavor of Cheddar cheese has been investigated but has not been resolved. Reiter (1973) summarized studies conducted in England and Canada which indicated a marked increase in flavor intensity of cheese made with lactic starter culture compared to that made by chemical acidification using glucono-δ-lactone. However, the greatest flavor intensity was obtained when a mixed bacterial flora, isolated from fresh cheese curd, was used with the lactic starter. The importance of nonlactic flora was questioned in other studies. Lawrence *et al.* (1976) pointed out that choice of appropriate strains of lactic bacteria is important in attaining flavor intensity. Results of several studies suggested that added nonstarter bacteria produced off-flavors or had no effect on flavor (Lawrence *et al.*, 1976).

### 4. Cheddar Flavor

The complex flavor profile of Cheddar cheese has complicated identification of important flavor components and mechanisms of forming them. Inability to find a compound which imparts cheese flavor prompted the "component balance theory" of cheese flavor (Mulder, 1952; Kosikowski and Mocquot, 1958). As the name indicates, flavor depends upon a proper blend of components, none of which completely possesses the unique cheese flavor.

Factors affecting flavor development and compounds contributing to Cheddar cheese flavor were discussed in a review by McGugan (1975). The role of milk fat was thought to be direct as a source of free fatty acids or indirect as a solvent for flavor components. Sulfur-bearing amino acids provide sulfur moieties which are converted to compounds implicated in cheese flavor.

Several compounds appear to be especially important in Cheddar cheese flavor: methanethiol, free fatty acids, acetate, diacetyl and methyl ketones. Many other compounds identified in Cheddar cheese may contribute to flavor. These include aldehydes, $H_2S$, dimethyl sulfide, lactones, keto acids, phenolics, and pyrazines. However, the importance of various components has been disputed. The proper ratio of free fatty acids to acetic acid was felt to be essential by some workers but other research suggested that free fatty acids did not play an important role. McGugan (1975) suggested that part of the dilemma may result from differences in perception and preferences of typical Cheddar cheese. Diversity between countries and effects of different ages of cheese probably complicate a concise definition of Cheddar cheese flavor.

### B. Colby and Stirred-Curd (Granular) Cheeses

Colby cheese is the most popular modification of Cheddar in the United States and is made like Cheddar cheese up to whey drainage (Kosikowski, 1977; Wilson and Reinbold, 1965; Wilster, 1969). Instead of fusing the curd under whey, a portion of the whey is drained and cold water is added to reduce the temperature of curd and whey to approximately 31°C. The curd and whey–water mixture are stirred and drained after 10–20 minutes. The granular curd is salted, hooped, and pressed by the methods used for Cheddar.

The water treatment reduces curd temperature, resulting in more openings in the finished cheese, and leaches out a portion of the acids, salts, and lactose to give a mild flavored cheese with a shorter shelf-life than Cheddar. Lower water temperatures produce higher moisture levels in cheese, the legal maximum being 40%. Water treatment of curd is used also in the manufacture of Monterey and washed-curd cheese. Curd-washing treatments are essential for these high-moisture cheeses to reduce the lactose content and thereby control the pH of the finished cheese (Price and Buyens, 1967).

The granular or stirred-curd modification of Cheddar is made like Colby but the curd is not washed with cold water; cooking temperatures are similar to those used for Cheddar cheese (Price et al., 1971; Wilson and Reinbold, 1965; Wilster, 1969). Its composition is similar to Cheddar but its appearance or openness resembles Colby. The mechanical openings can be eliminated with vacuum treatments (Price and Olson, 1965).

## C. Surface-Ripened Cheeses

### 1. General Description

These are soft and semisoft varieties which are ripened under conditions to induce a progression of microbial growth on the cheese surface (Kosikowski, 1977; Olson, 1969). Examples are brick from the United States, Limburger from Belgium, Port du Salut from France, and Bel Paese from Italy. All use the same basic procedure of coagulation of milk with milk-clotting enzymes, drainage of curd in perforated hoops, brine-salting or application of dry salt to cheese, and initial ripening at 12°–18°C under high relative humidities. Differences in varieties result from the shape of cheese and the amount and, to a slight extent, type of surface growth. Microorganisms on the surface usually are not added as pure cultures but develop under ripening conditions and from carryover contamination in the ripening room.

### 2. Brick Cheese

Two methods have been used to manufacture brick cheese (Olson, 1969). In one method, S. cremoris-type lactic culture and washing of curd with water are unique features (Price and Buyens, 1969). A "whey" culture is used and the curd is not washed in the second method (Olson, 1969). The culture for the latter method is made by heat-treating whey from previous day and incubating it to allow growth of lactic acid-producing bacteria surviving the heat treatment. The "washed-curd" method is the most widely used procedure in the United States.

The basic manufacturing procedure for "washed-curd" brick cheese to the point of whey drainage is similar to that of Cheddar cheese with the exception that smaller amounts of lactic starter are used, ripening or incubation of milk before addition of milk-clotting enzymes is eliminated, very little acid is produced prior to whey drainage, and lower cooking temperatures are generally used. When the desired cooking temperature is reached and the curd has attained the proper firmness, it is allowed to settle but not fuse and the whey is drained to within 2.5 cm of the curd surface. Water at curd temperature and equivalent to 50% of the original milk volume is added and the curd slurry is stirred for 15–20 minutes. The liquid is again drained down to 2.5 cm above the curd level. The curd and whey slurry is pumped over to perforated draining hoops, where the curd fuses into a cohesive mass. The fused cheese mass contains a number of mechanical openings, presumably from the high pH of curd, which reduces its ductility, and because of the rapid drainage of whey before the curd mass completely fuses. The blocks of curd are turned routinely to obtain smooth surfaces and are allowed to drain for approximately 6 hours before brine-salting. The pH of cheese decreases from 6.4 to about 5.6 during this period. Salting is accomplished by immersion of cheese blocks in 23% NaCl brine for 18–24 hours.

The salted cheese is held at 16°C and at a high relative humidity (ca. 95%) to allow microbial growth on its surface. The growth is a progression of yeasts, micrococci, and *Brevibacterium linens* (Langhus et al., 1945; Lubert and Frazier, 1955). Like many food fermentations, the type and extent of growth is not controlled by inoculation but by conditions such as high NaCl concentration on the cheese surface, low pH of 5.0–5.2, reduced water activity of curd, and high relative humidity in the room. Yeasts metabolize lactic acid in the surface layer of curd, raise the pH, and synthesize vitamins to stimulate subsequent growth of the micrococci and *B. linens* (Olson, 1969).

The microbial flora produces flavor components on the cheese surface which diffuse into the cheese. Hydrogen sulfide and methyl mercaptan released from sulfur-bearing amino acids are important flavor components in surface-ripened cheese and presumably are important in brick cheese. Slight lipolysis is necessary for acceptable flavor development in surface-ripened varieties (Olson, 1969).

The extent of ripening to develop surface growth depends upon the desired flavor intensity. A mellow to moderately intense flavor develops in curing periods of 4–10 days. It is necessary to maintain moist cheese surfaces to favor yeast and bacterial growth and limit mold growth. After sufficient surface growth has developed, it is washed off and the cheese is covered with a wax coating or packaged in plastic film. The cheese then is ripened at 2°–8°C for 1–3 months. During this time, the initial elastic structure of cheese is converted by proteolysis to a smooth consistency. This ripening period also allows flavor components to diffuse in from the surface.

### 3. Other Surface-Ripened Varieties

Varieties such as Limburger, Port du Salut, Bel Paese, and Tilsit are manufactured and ripened in a manner that is essentially the same as surface-ripened brick cheese (Kosikowski, 1977; Olson, 1969). Acidity control of cheese is accomplished by washing curd and limiting acid production before whey drainage. Whey syneresis is accelerated in manufacture of Port du Salut by cutting curd into smaller cubes and using higher cooking temperatures as compared to brick cheese. High-temperature lactic cultures composed of S. thermophilus and Lactobacillus species, used for some surface-ripened varieties, produce slightly different flavor profiles as compared to S. cremoris cultures. Streptococcus diacetilactis has been isolated from St. Paulin cheese and presumably is responsible for eye formation and diacetyl which would contribute to flavor.

Higher than normal clotting temperatures (40°–45°C) are partially responsible for the smooth, elastic body of Bel Paese, an Italian soft to semisoft cheese. Lower temperatures during development of surface growth on this variety results in greater dominance of yeast. Temperatures and relative humidities recommended for ripening Port du Salut vary widely. Lower temperatures favor growth of yeasts over B. linens; lower humidities allow growth of mold such as Geotrichium species. Consequently, flavor characteristics of this and similar varieties can be modified significantly by ripening conditions.

### D. Other Semisoft Cheeses

A great number of varieties such as Edam, Gouda, Monterey, Muenster (in the United States), and Danish-type cheeses are included in this class. The type of lactic culture and methods of milk clotting, cutting, and cooking curd are similar to the Cheddar schedule shown in Table II (Kosikowski, 1977; Olson, 1969). Higher moisture contents in these varieties necessitate lower cooking temperatures, minimal acid production prior to whey drainage, and washing of curd with water to control pH of cheese. Washing treatments remove lactose and thereby limit the amount of lactic acid produced after whey drainage. It may not be necessary to wash curd of varieties such as Edam and Gouda if sufficient whey (lactose) is removed from curd by syneresis and if the pH of curd is high enough at whey drainage. With the proper balance of lactose retained and desired pH of curd, the drop of pH will be optimal (see Section IV, A, 2).

Depending upon the variety, curd handling and pressing can be done by procedures similar to those for Cheddar or brick cheese. These processes have been mechanized, especially for those varieties in which individual pieces of cheese are small (Olson, 1975). The size and shape of cheese varies widely between and within each variety (Olson, 1969).

Most varieties in this class are salted by immersion in NaCl brine. Cheese is waxed, covered with a "plastic" solution, or wrapped in heat-shrinkable plastic film with application of a vacuum treatment. Ripening of cheese is done at 2°–10°C for a few months to develop desired flavor intensity. The "component balance theory" can be applied to flavor development in most of these varieties. Metabolism of S. diacetilactis may produce "eyes" in Edam and Gouda cheeses and some Danish varieties.

## E. Swiss Cheese

### 1. Manufacturing Methods

The basic steps in manufacturing (milk clotting, cutting curd, cooking curd and whey, and separation of curd and whey) are similar to those for Cheddar cheese. A cylindrical, copper kettle was the traditional vat but the horizontal or vertical tower-vats are used in many operations (Reinbold, 1972). The fat content of milk is usually standardized to a slightly lower level as compared to that for Cheddar cheese.

High-temperature lactic cultures, S. thermophilus and Lactobacillus species, are necessary for Swiss cheese because of the high cooking temperatures (50°–53°C) attained during manufacturing. Rennet extract was the traditional milk-clotting enzyme but a protease preparation from E. parasitica was found to be very suitable for manufacture of Swiss cheese because of its low heat stability (Ernstrom and Wong, 1974). High cooking temperatures and high pH values of curd during cooking result in very low residual enzyme levels in cheese. This minimizes proteolytic attack on the cheese structure which is beneficial for good "eye" formation.

The curd formed by milk-clotting enzymes is cut into small particles with wire knives (harps) by hand or more recently by mechanically operated knives (Reinbold, 1972). The small curd particles result in rapid whey syneresis which compensates for decreased whey expulsion because of low acid development during initial stages of manufacture. The small curd particles also form a tightly knit structure in the finished cheese which produces more uniform "eye" formation.

Cooking temperatures of 50°–53°C are commonly used to attain the proper moisture level, firmness, and plasticity of curd. In the traditional method, the warm curd was dipped from the kettle with a cloth and transferred to a wooden circular hoop if the curd was to be pressed into a traditional Swiss wheel (Kosikowski, 1977). This form has been replaced with a rectangular hoop to yield a rindless block style (Reinbold, 1972). More highly mechanized manufacturing systems consist of the horizontal or tower-vats in which curd is formed, cut, and cooked. After the curd has attained optimum firmness, it flows by gravity or is pumped with some whey to special draining forms. The curd settles and fuses, under applied pressure, into dense blocks while the whey drains from the perforated forms. The method of pressing and condition of curd are critical for proper eye formation in the finished cheese.

The pressed blocks or wheels are salted by immersion in a NaCl brine and the surfaces of blocks are dried before packaging in plastic film. The packaged cheese is ripened through the series of stages described in Section IV, D, 3.

### 2. Lactic Bacteria

Lactic acid bacteria in Swiss cheese serve the typical functions of production of lactic acid, control of undesirable fermentations, and contributing to flavor development. Presence of both S. thermophilus and high-temperature lactobacilli appears to be essential to obtain a satisfactory cheese (Langsrud and Reinbold, 1973a). Only one-half as much acid was produced in cheese made with S. thermophilus, whereas acidity levels were higher in cheese made with L. helveticus, but the maximum level was reached fastest in cheese containing both species. The complexity of symbiosis between high-temperature lactic bacteria necessitates careful selection of compatible strains to obtain predictable growth rates and acid production during cheese manufacturing.

The two genera of lactic bacteria exhibit different growth patterns during cheese manufacturing. Limited growth of S. thermophilus occurs during initial stages of manufacture up to the stage when the cheese is being pressed (Frazier et al., 1934, 1935). Rapid growth of this species occurs during the first 4–8 hours after whey drainage and then numbers decline slowly. Numbers of lactobacilli and propionibacteria may decline during initial stages of manufacturing but begin to grow midway through pressing. The high metabolic activity, with concomitant acid production by lactic bacteria and high temperature of curd, suppresses growth of undesirable microorganisms during pressing.

Limited production of acid occurs during the early stages of manufacture; the pH of curd at the time of draining whey (dipping) is 6.3–6.5. The pH of curd should not be below 5.3 at the end of pressing since excessive acidity inhibits growth of propionibacteria and eye formation (Langsrud and Reinbold, 1973a). Also, cheese structure may lack ductility resulting in abnormal eye formation or cracks in cheese. High pH values can cause soft cheese structure, excessive eye formation, and possibly abnormal fermentations during ripening.

### 3. Ripening

Ripening of Swiss occurs in three phases. After salting and packaging, the cheese may be stored at 7°–13°C for up to 10 days to allow some diffusion of NaCl into the cheese block. The blocks of cheese are held at 21°–25°C for 2–7 weeks to induce growth and gas (carbon dioxide) production by *Propionibacterium* species to form the characteristic eyes (Reinbold, 1972). Since these bacteria are dispersed throughout the curd, gas must diffuse through the curd matrix and collect at points of weakness. Sufficient gas pressure develops to stretch the protein structure surrounding weak focal points and form smooth, round eyes. Thus the number, size, and uniformity of eyes depends on rate of gas production, presence of contaminating gas-producing bacteria, number of weak points in curd, and plasticity of curd (Langler and Reinbold, 1974; Reinbold, 1972). Lactic bacteria continue to function during ripening in symbiosis with other bacteria. They ferment most of the simple sugars in curd to lactic acid via the homofermentative Embden–Meyerhof–Parnas (EMP) pathway but use a heterofermentative pathway to a slight extent since small amounts of acetic acid and $CO_2$ are produced (Langsrud and Reinbold, 1973b). Growth of propionibacteria is stimulated by $CO_2$ and by the reduced oxidation–reduction potential resulting from metabolism by the lactic starter.

Numbers of lactic bacteria decrease during the "warm-room treatment" phases of ripening presumably because of lack of fermentable sugars as energy sources (Langsrud and Reinbold, 1973b). Numbers of propionibacteria increase 3- to 980-fold over levels in curd at pressing. The population attained is correlated directly with pH values of cheese. Although propionic and acetic acids may be produced by metabolism of propionibacteria on several substrates, lactate is the principal substrate in cheese (Hettinga and Reinbold, 1972).

The component balance theory of cheese flavor is considered applicable to Swiss cheese (Langsrud and Reinbold, 1973b). However, high-quality cheese contains relatively large amounts of acetic and propionic acids, very little or no butyric acid, and small amounts of higher fatty acids. The level of acetaldehyde in Swiss cheese exceeds its flavor threshold and is thought to be an important flavor component. Amino acids capable of imparting a sweet taste are considered important for this flavor note in Swiss cheese.

## F. Italian-Type Cheeses

Commercially important varieties of Italian-type cheeses can be classified as hard (grating) and "pasta filata" varieties. Parmesan and Romano cheese are examples of grating cheeses. "Pasta filata" means filamentous curd; Provolone and Mczzarella are examples of this type.

### 1. Parmesan Cheese

Parmesan is typical of several very hard Italian cheese varieties which are commonly grated for use in foods. The fat content of milk for this variety usually is reduced since U.S. Standards of Identity require only 32% of the cheese solids must be milk fat. Manufacturing methods are similar to Swiss up to the point of whey drainage and the same species of lactic bacteria are used but no propionibacteria are added (Kosikowski, 1977; Reinbold, 1963). The ratio of *Lactobacillus* species to *S. thermophilus* in the lactic starter is about 1:1, which is higher than that used in starters for Swiss cheese. Very small amounts (about 5 gm/454 kg milk) of lipases (pregastric esterases) may be added to milk to produce the desired flavor intensity in the ripened cheese as described in Section IV, F, 5. Calf-derived enzymes are usually used for Parmesan cheese since the desired flavor is more subtle.

Deviation from procedures for Swiss occurs at whey drainage since the curd is not fused under whey into a dense block. It is either stirred to prevent fusion of curd particles or may be matted into a porous mass. After the whey has drained, the curd is placed in cylindrical forms and pressed. The wheels of cheese are brine-salted for approximately 14 days depending upon the size of cheese. The long salting time is required because of the low moisture content of the cheese which retards salt diffusion. The cheese is allowed to dry slowly during initial stages of ripening so that the final moisture content is 32% or less. The minimum curing time for this variety in the United States is 10 months; curing for 18–24 months produces a more intense sweet, nutty flavor.

### 2. Romano Cheese

Romano cheese is made in essentially the same way as Parmesan cheese but has higher moisture and fat contents (Kosikowski, 1977; Reinbold, 1963). Lipases derived from kid goats or mixtures of kid and lamb enzymes are commonly used to produce fairly intense flavors (see Section IV, F, 5). Enzymes are added in greater amounts to cheese milk as compared to Parmesan, resulting in higher levels of free fatty acids in cheese. After whey drainage, the curd is pressed into cylindrical forms, brine-salted for 1–2 days, and salted for several weeks by application of salt to the cheese surface. This variety usually is not ripened as long as Parmesan and can be used as a table cheese after 5–8 months or for grating purposes with longer ripening.

### 3. Other Hard Italian Cheeses

There are many other Italian-type cheeses that are made by a process similar to Parmesan but differ in composition, flavor, regional name, and the breed of animal used for milk source. Asiago may be less firm, may possess a more delicate nutty flavor, and is used as table cheese. Inclusion of the term Pecorino in a varietal name indicates that it was made from goat's milk. The terms Parmigiano, Sardo, and Reggiano indicate regions of Italy in which these particular types are made or originated. Fontina is a "table cheese" which can be cut or sliced but is used as a grating cheese when aged. It has eyes similar to Gouda cheese and may have a surface flora of *B. linens* (Kosikowski, 1977).

### 4. Provolone Cheese

This variety is one of several pasta filata types of cheese. The name "pasta filata" can be translated as filamentous dough or curd, which describes the structure of the cheese produced by stretching the curd during manufacture. Manufacturing procedures for Provolone are similar to Cheddar up to the milling stage except for use of high-temperature lactic cultures and higher cooking temperatures (Kosikowski, 1977; Reinbold, 1963). The ratio of streptococci to lactobacilli is about 1:1; lactobacilli are important to attain the low pH at later stages of manufacturing.

When the "cheddared" curd reaches a pH of 5.2–5.3, it is warmed to 50°–55°C. This usually is done in a mechanical system using hot water or steam as the heating medium. The warm curd is mixed mechanically until its texture resembles taffy candy. The filamentous curd is then manually or mechanically molded into its final shape (Kosikowski, 1977). Smaller, cylindrical pieces are made by extruding warm curd into forms; larger pieces are formed manually. The final shape is used as a basis of naming many pasta filata cheeses, including the cylindrical salami, pear-shaped Provolone, large round Mandarini Provolone, and small Provolone styles, Provoletti and Bocchini.

Provolone styles are salted by immersion in salt brine; the duration of salting varies indirectly with the surface to volume ratio of cheese pieces. A smoke flavor can be imparted to cheese by exposure to smoke or by applying a liquid concentrate of wood smoke.

### 5. Flavor Development

Lipolysis, which is important in flavor development in Italian-type cheeses, results to a great extent from action of lipolytic enzymes in rennet paste or by pregastric esterases (Neelakantan *et al.*, 1971; Nelson *et al.*, 1977). Pregastric esterase preparations are liquid extracts of selected oral tissue of young calves, kid goats, or lambs. Rennet pastes are prepared from the abomasum of the above species which are slaughtered immediately after suckling. The engorged stomachs are processed into a product comprised of stomach tissue and milk-curd contents. Lipolytic activity of the paste arises from lipases (pregastric esterases) which were released into milk by the oral tissues. Esterases from *Mucor miehei* were used experimentally for Fontina and Romano cheeses (Peppler *et al.*, 1976). Fungal esterases produced equivalent flavor intensities as pregastric esterases in Fontina cheese when both preparations were standardized to equivalent lipolytic activities. Five times more fungal esterases as compared to pregastric esterases were required to attain desired flavor intensities in Romano cheese.

### 6. Mozzarella Cheese

This variety is a soft, usually unripened, pasta filata-type cheese (Kosikowski, 1977; Reinbold, 1963). It is manufactured by the same basic procedures and with the same type of lactic cultures but cooking temperatures may be lower since moisture content is generally higher than in Provolone cheese. Standards of Identity in the United States specify four types of Mozzarella cheese which differ in moisture and fat contents: Mozzarella, part-skim Mozzarella, low-moisture Mozzarella, and part-skim, low-moisture Mozzarella (U.S. Food and Drug Administration, 1974).

Mechanical systems for stretching and molding curd are universally used (Olson, 1975). Size of the finished cheese varies but a 2.3-kg rectangular loaf is dominant. Brine-salting is accomplished by floating cheese loaves in NaCl brine or by immersing mechanized trolleys holding cheese loaves. Cheese is packaged in heat-shrinkable bags or in pouches; air is removed by vacuum prior to sealing. Storage of cheese is sufficient to obtain adequate proteolysis of the protein structure to attain desired physical characteristics; this may require 7–30 days. Most Mozzarella is used as a food ingredient which makes sliceability, meltability, cohesiveness, and optimum firmness important physical characteristics.

## G. Mold-Ripened Cheeses

### 1. Blue-Veined Cheeses

These types of cheese have been characterized as being unique because an almost pure culture of *Penicillium roqueforti* grows within the cheese (Morris, 1969). Growth of mold in cheese is facilitated by control of curd structure, pH, NaCl concentration, and the $O_2$–$CO_2$ interchange in cheese. It is critical to handle curd properly during manufacturing to obtain an open texture within the cheese so that mold can grow throughout the openings and produce the "blue veining" which is typical of these varieties.

Several varieties of cheese fall within the blue-veined class, of which blue (Bleu), Roquefort, Gorgonzola, and Stilton are examples. Manufacture of blue cheese will be described in this section; principles of manufacturing for this variety apply to other blue-veined cheeses.

**a. Milk Treatment and Clotting.**   Milk is prepared by blending pasteurized or heat-treated skim milk with homogenized cream containing about 12% fat. Homogenization pressures of 35 kg/cm² for the first stage and 70 kg/cm² for the second have been suggested (Kosikowski, 1977). The cream may be bleached with benzoyl peroxide, or chlorophyll may be added to produce the whiteness desired for this variety.

Lactic culture (*S. cremoris* and *S. lactis*) is added and the milk is ripened at 30°–31°C for up to 1 hour. *Penicillium roqueforti* mold powder is inoculated into the milk to obtain a uniform distribution in curd. The powder may be sprinkled on curd at hooping but this is less convenient.

Milk-clotting enzyme is added at the rate of 165–200 ml per 1000 kg of milk. The firmness of curd at cutting is equal to or greater than that of Cheddar cheese. This aids in maintaining the cube shape of the curd and increases openness of the finished cheese. Cutting curd in large 1.5-cm (⅝-inch) cubes also increases openness.

**b. Curd Handling.**   Stirring of curd and whey commences after sufficient whey is expelled to act as a lubricant between cubes of curd. Care must be exercised in stirring to maintain the large, cube shape which is important in obtaining open texture in cheese. The temperature of curd and whey may be raised slightly to adjust the moisture content of the finished cheese.

When the desired whey expulsion and curd firmness is attained, the curd is settled and a portion of the whey is drained off. The curd and remaining whey are transferred manually to draining forms (hoops) with perforated scoops or by special curd pumps; a higher proportion of curd to whey is required for pumping. Distortion (flattening) and breakage of curd during whey drainage and curd transfer must be avoided to obtain open texture in the finished cheese. The curd settles and fuses while whey drains through the cylindrical, perforated hoops. Turning hoops during initial stages of draining produces smooth outside surfaces. Drainage in hoops continues overnight.

The wheels of cheese are salted by daily application of crystalline NaCl over a 5-day period or by immersing wheels in a 23% NaCl brine for 24–48 hours and then three daily applications of crystalline salt. The concentration of NaCl in the final cheese is about 4% or 8–9% in the serum phase. The salt content is much higher initially on the cheese surface which controls the extent and type of microbial growth.

One day after salting, the wheels of cheese are pierced with a gang of needles. This allows subsequent escape of $CO_2$ and entrance of $O_2$ which is essential for mold growth and metabolism. The cheese is stored at 10°–13°C for 2–3 months to induce mold growth; subsequent ripening is carried out at lower temperatures of 2°–5°C.

Lactic fermentation continues during draining so that the minimum pH of 4.5–4.7 is reached 24 hours after manufacturing (Coulter *et al.*, 1938; Foster *et al.*, 1957). Metabolism of *P. roqueforti* and other microorganisms causes the pH of cheese to increase to 6.0–6.3 during ripening.

**c. Ripening.** Proteolysis and lipolysis of blue cheese results, to a great extent, from enzymes elaborated by *P. roqueforti* (Ernstrom and Wong, 1974). The water-soluble lipase system from this species hydrolyzes milk fat to yield fatty acids; caproic, caprylic, and capric acids and their salts impart a sharp, peppery flavor to blue cheese (Currie, 1914). Methyl ketones, formed by oxidation of fatty acids to β-keto acids and subsequent decarboxylation, are thought to be the principal components contributing to aroma of blue cheese (Girolami and Knight, 1955). A mixture of acetone, pentanone-2, heptanone-2, and nonanone-2 possessed an aroma similar to blue cheese (Patton, 1950). Heptanone-2 is the predominant ketone in this variety and Roquefort and Camembert cheeses (Anderson and Day, 1965; Schwartz and Parks, 1963; Schwartz *et al.*, 1963).

### 2. Cheese with Surface Mold

A wide variety of cheeses especially of French origin are made with surface mold growth. Typical varieties are Camembert, Coulommiers, and Brie.

**a. Camembert.** Initial stages of manufacture of this variety are technologically uncomplicated but strict sanitation is essential (Kosikowski, 1975;; Mocquot, 1955). Pasteurized, whole milk is inoculated with lactic bacteria (*S. cremoris* type) and *Penicillium camemberti* or *Penicillium caseicolum*. Curd is formed by milk-clotting enzymes and may be cut, when fairly firm, into large cubes. The temperature of curd and whey may be raised slightly to obtain desired moisture levels in cheese. Approximately 1.5–2 hours after the lactic culture is added, the curd and a portion of whey are dipped into draining forms. Special vats are used in some operations which allow dispensing of curd mass into draining forms. The diameter of forms is the same as that of the finished cheese so each form of curd yields one wheel of cheese.

After drainage in hoops, the wheels are salted by application of dry salt or immersion in a salt brine. A suspension of mold spores may be sprayed on the wheels prior to salting as an alternative or supplement to inoculation of milk. The cheese is placed on wire racks to allow maximum exposure of surface to facilitate mold growth. Ripening may be a sequence of storage at 14°C and relative humidity (RH) of 75% for 1–2 days to dry the surface, followed by storage at 14°C at an RH of 95% for 12 days. A cottony, white mat of mold develops on the cheese; the cheese is turned on the racks to obtain uniform growth. Initial storage at low RH is eliminated in some operations.

The mold-covered wheels of cheese are wrapped in foil and ripened at about 5°C. Ripening is rapid; the cheese is ready for consumption within 2–5 weeks depending upon consumer preference. Physical characteristics of cheese change dramatically during ripening. Initially, the cheese is brittle and has a white, grainy appearance which is transformed into a smooth consistency and translucent appearance. The change occurs from the surface inward, presumably from action of mold proteases.

**b. Coulommiers and Brie.**   Coulommiers is very similar to Camembert in methods of manufacturing and microbial population but is formed into larger wheels. Brie is also made in larger wheels and is a soft cheese but *B. linens* grows on the cheese surface with *P. camemberti*. This bacterial species, which also is important in ripening of Limburger cheese, produces distinctive flavor notes.

## V. PROCESS CHEESE

This type of cheese is made by melting and mixing desired blends of cheese, with the aid of heat, to give a smooth homogeneous product (Kosikowski, 1977). There are three general types of processed cheese defined under the Food, Drug and Cosmetic Act: pasteurized process cheese, pasteurized process cheese food, and pasteurized process cheese spread (U.S. Food and Drug Administration, 1974). These differ in moisture and fat contents and in number and kinds of food ingredients allowed.

## A. Manufacture

Process cheese can be made from most varieties of cheese, but Cheddar is used most commonly. A blend of different lots of cheese is selected to obtain desired composition and physical and flavor characteristics. The cheese is comminuted and heated in a steam-jacketed kettle or by infusion of culinary steam directly into the cheese mass. Two types of cookers have been used, namely, jacketed kettles with paddle agitators or horizontal, cylindrical cookers equipped with screw mixers. Processing is done on a batch basis through the cooking operation but a sufficient number of cookers are used to supply a continuous flow of cheese to subsequent packaging operations.

Emulsifying salts are mixed with cheese at cooking to increase smoothness and aid in dispersion of fat in the finished cheese. Citrates and phosphates of varying chain length are used depending upon the characteristics desired in process cheese. Softer cheese is produced when disodium phosphate is used rather than sodium citrate or certain polyphosphates.

## B. Packaging and Mechanization

Process cheese, cheese food, and firmer cheese spreads are commonly sold in 0.5–2.3 kg loaves, stacks of slices, or individually wrapped slices. Loaves or large stacks of slices are sold in institutional markets. Slices are formed by extruding molten cheese on a cooled belt and allowing the cheese to solidify before the slices are mechanically cut and stacked (Kosikowski, 1977). Softer process cheese spreads are packaged in jars, plastic tubes, and aerosol cans. The moisture level in such products must be sufficiently high for easy removal from the package.

## C. Cold-Pack Cheese

In addition to processed cheese, natural cheese is converted to cold-pack cheese and cold-pack cheese food by grinding or comminuting without the application of heat (U.S. Food and Drug Administration, 1974). Equipment with knives revolving at high speeds, adapted from the sausage industry, reduce the structure of natural cheese to a smooth spreadable paste. Additives such as dehydrated milk products, acidifying and sweetening agents, and gums can be added to modify the flavor and rheological characteristics of cold-pack cheese food.

# REFERENCES

Anderson, D. F., and Day, E. A. (1965). *J. Dairy Sci.* **48**, 248–249.
Bradley, R. L. (1969). "Milk," Encycl. Sci. Technol., pp. 499–510. McGraw-Hill, New York.
Breene, W. M., Price, W. V., and Ernstrom, C. A. (1964). *J. Dairy Sci.* **47**, 840–848.
Brown, L. W., and Price, W. V. (1934). *J. Dairy Sci.* **17**, 33–45.
Collins, E. B. (1952). *J. Dairy Sci.* **35**, 381–387.
Collins, E. B. (1961). *J. Dairy Sci.* **44**, 1989–1996.
Collins, E. B. (1977). *J. Dairy Sci.* **60**, 799–804.
Coulter, S. T., Combs, W. B., and George, S. T. (1938). *J. Dairy Sci.* **21**, 239–245.
Cousin, M. A., and Marth, E. H. (1977a). *Cult. Dairy Prod. J.* **12**, 15–18 and 30.
Cousin, M. A., and Marth, E. H. (1977b). *J. Dairy Sci.* **60**, 1048–1056.
Currie, J. N. (1914). *J. Agric. Res.* **2**, 429–434.
Czulak, J., Conochie, J., Sutherland, B. J., and Van Leeuwen, H. J. M. (1969). *J. Dairy Res.* **36**, 93–101.
Emmons, D. B., and Tuckey, S. L. (1967). "Cottage Cheese and Other Cultured Milk Products." Chas. Pfizer & Co., Inc., New York.
Ernstrom, C. A., and Kale, C. G. (1975). *J. Dairy Sci.* **58**, 1008–1014.
Ernstrom, C. A., and Wong, N. P. (1974). *In* "Fundamentals of Dairy Chemistry" (B. H. Webb and A. H. Johnson, eds.), 2nd ed., pp. 662–769. Avi Publ. Co., Westport, Connecticut.
Foster, E. M., Nelson, F. E., Speck, M. L., Doetsch, R. N., and Olson, J. C., Jr. (1957). "Dairy Microbiology." Prentice-Hall, Englewood Cliffs, New Jersey.
Fox, P. F., and Walley, B. F. (1971). *J. Dairy Res.* **38**, 165–170.
Frazier, W. C., Sanders, G. P., Boyer, A. J., and Long, H. F. (1934). *J. Bacteriol.* **27**, 539–549.
Frazier, W. C., Burkey, L. A., Boyer, A. J., Sanders, G. P., and Matheson, K. J. (1935). *J. Dairy Sci.* **18**, 373–387.
Gilliland, S. E. (1977). *J. Dairy Sci.* **60**, 805–809.
Gilliland, S. E., and Speck, M. L. (1974). *J. Milk Food Technol.* **37**, 107–111.
Girolami, R. L., and Knight, S. G. (1955). *Appl. Microbiol.* **3**, 264–267.
Green, M. L. (1977). *J. Dairy Res.* **44**, 159–188.
Harper, W. J., and Hall, C. W. (1976). "Dairy Technology and Engineering." Avi Publ. Co., Westport, Connecticut.
Hettinga, D. H., and Reinbold, G. W. (1972). *J. Milk Food Technol.* **35**, 358–372.
Holmes, D. G., Duersch, J. W., and Ernstrom, C. A. (1977). *J. Dairy Sci.* **60**, 862–869.
Jenness, R., and Patton, S. (1959). "Principles of Dairy Chemistry." Wiley, New York.
Kosikowski, F. V. (1975). *J. Dairy Sci.* **58**, 994–1000.
Kosikowski, F. V. (1977). "Cheese and Fermented Milk Foods," 2nd ed. Edwards, Ann Arbor, Michigan.
Kosikowski, F. V., and Mocquot, G. (1958). *FAO Agric. Stud.* **38**, 141–144.
Langhus, W. L., Price, W. V., Sommer, H. H., and Frazier, W. C. (1945). *J. Dairy Sci.* **28**, 827–838.
Langler, T., and Reinbold, G. W. (1974). *J. Milk Food Technol.* **37**, 26–41.
Langsrud, T., and Reinbold, G. W. (1973a). *J. Milk Food Technol.* **11**, 531–542.
Langsrud, T., and Reinbold, G. W. (1973b). *J. Milk Food Technol.* **36**, 593–609.

Lawrence, A. J. (1959). *Aust. J. Dairy Technol.* **14,** 169–172.

Lawrence, R. C., Thomas, T. D., and Terzaghi, B. E. (1976). *J. Dairy Res.* **43,** 141–193.

Lowrie, R. J. (1977). *J. Dairy Sci.* **60,** 810–814.

Lubert, D. J., and Frazier, W. C. (1955). *J. Dairy Sci.* **38,** 981–990.

Lundstedt, E., and Fogg, W. B. (1962). *J. Dairy Sci.* **45,** 149–150 (abstr.).

McGugan, W. A. (1975). *J. Agric. Food Chem.* **23,** 1047–1050.

Mather, D. W., and Babel, F. J. (1959). *J. Dairy Sci.* **42,** 1045–1056.

Micketts, R., and Olson, N. F. (1974). *J. Dairy Sci.* **57,** 273–279.

Mocquot, G. (1955). *J. Soc. Dairy Technol.* **8,** 17–24.

Mocquot, G. (1971). *Proc. Int. Symp. Conversion Manuf. Foodstuffs Microor., 1971.* pp. 191–197. SaiRon Publ. Co., Ltd., Tokyo, Japan.

Morris, H. A. (1969). Paper presented at *6th Annu. Marschall Invitational Ital. Cheese Semin.* April 30, 1969, Madison, Wisconsin.

Mulder, H. (1952). *Ned. Melk- Zuiveltijdschr.* **6,** 157–168.

National Research Council (1958). N.A.S.—N.R.C., *Publ.* **645,** 1–48.

Neelakantan, S., Shahani, K. M., and Arnold, R. G. (1971). *Food Prod. Dev.* **5,** 52–58.

Nelson, J. H., Jensen, R. G., and Pitas, R. E. (1977). *J. Dairy Sci.* **60,** 327–362.

Nelson, P. J., and Marshall, R. T. (1977). *Proc. 72nd Annu. Meet. Am. Dairy Sci. Assoc.* p. 35.

O'Keefe, R. B. *et al.* (1976). *J. Dairy Res.* **43,** 97–107.

O'Leary, J. O., Hicks, C. L., and Bucy, J. (1977). *Proc. 72nd Annu. Meet. Am. Dairy Sci. Assoc.* p. 55.

Olson, N. F. (1969). "Ripened Semisoft Cheeses." Chas. Pfizer & Co., Inc., New York.

Olson, N. F. (1971). *Int. Dairy Congr.* [Proc.], *Congr. Rep., 18th, 1970* Vol. 2E; pp. 334–347.

Olson, N. F. (1975). *J. Dairy Sci.* **58,** 1015–1021.

Olson, N. F. (1977). *Dairy Ind. Int.* **42,** 14–19.

Patel, M. C., Lund, D. B., and Olson, N. F. (1972). *J. Dairy Sci.* **55,** 913–918.

Patton, S. (1950). *J. Dairy Sci.* **33,** 680–684.

Peppler, H. J., Dooley, J. G., and Huang, H. T. (1976). *J. Dairy Sci.* **59,** 859–862.

Perry, K. D., and McGillivray, W. A. (1964). *J. Dairy Res.* **31,** 155–165.

Peters, I. I. (1964). *Dairy Sci. Abstr.* **26,** 457–461.

Price, W. V., and Buyens, H. J. (1967). *J. Dairy Sci.* **50,** 12–19.

Price, W. V., and Olson, N. F. (1965). *Wis., Agric. Exp. Stn., Spec. Bull.* **10,** 1–14.

Price, W. V., Calbert, H. E., and Olson, N. F. (1971). *Wis., Agric. Exp. Stn., Res. Bull.* **464.**

Quarne, E. L., Larson, W. A., and Olson, N. F. (1968). *J. Dairy Sci.* **51,** 527–530.

Ramet, J. P., Alais, C., and Weber, F. (1969). *Lait* **49,** 40–52.

Reinbold, G. W. (1963). "Italian Cheese Varieties." Chas. Pfizer & Co., Inc., New York.

Reinbold, G. W. (1972). "Swiss Cheese Varieties." Chas. Pfizer & Co., Inc., New York.

Reinbold, G. W. (1973). *Milk Ind.* **73,** 13–16.

Reiter, B. (1973). *J. Dairy Res.* **26,** 3–15.

Richardson, G. H., Gandhi, N. R., Divatia, M. A., and Ernstrom, C. A. (1971). *J. Dairy Sci.* **54,** 182–186.

Robertson, P. S. (1966). *J. Dairy Res.* **33,** 343–369.

Sandine, W. E. (1977). *J. Dairy Sci.* **60,** 822–828.

Schwartz, D. P., and Parks, O. W. (1963). *J. Dairy Sci.* **46,** 989–990.

Schwartz, D. P., Parks, O. W., and Boyd, E. N. (1963). *J. Dairy Sci.* **46,** 1422–1423.

United States Food and Drug Administration (1974). "Cheese, Processed Cheese, Cheese Foods and Related Foods," Federal Food, Drug and Cosmetic Act, Part 19, Title 21. Code of Fed. Regul., U.S. Dept. of Health, Education and Welfare, Washington, D.C.

Van Slyke, L. L., and Price, W. V. (1952). "Cheese." Orange Judd Publ. Co., Inc., New York.
Walter, H. E., and Hargrove, R. C. (1972). "Cheeses of the World." Dover, New York.
Wilson, H. L., and Reinbold, G. W. (1965). "American Cheese Varieties." Chas. Pfizer & Co., Inc., New York.
Wilster, G. H. (1969). "Practical Cheesemaking." 11th ed. Oregon State Univ. Bookstores, Inc., Corvallis.

# Chapter 3

# Distilled Beverages

W. F. MAISCH
M. SOBOLOV
A. J. PETRICOLA

## I. INTRODUCTION

"Bourbon Whisky" . . . is whisky produced at not exceeding 160° proof from a fermented mash of not less than 51 percent corn . . . and stored at not more than 125° proof in charred new oak containers. [Code of Federal Regulations, Title 27, Sub Part C 5.22(b) Class 2(1)(i).]

The above citation is representative of the large body of regulations which both protect and control the beverage alcohol producers of the world. It is important to understand this point when studying the technical data of this old and distinguished industry, for virtually all of the

**79**

MICROBIAL TECHNOLOGY, 2nd ed., VOL.II

distilled products consumed in the United States are manufactured according to legally defined production methods.

The history of the individual production region is also an important consideration. Scotch whisky is unique in taste and aroma because those distillers began with the malt, the peat, and the water that were uniquely available in Scotland. Law now protects that uniqueness, while technology has been assigned the role of controlling the quality, assuring the reproducibility, and collecting the taxes. These pursuits ultimately have produced a body of observations which has explained the processes and has recently permitted meaningful improvements.

One must be equally cautious when interpreting economic data of the distilling industry. New patterns are developing in consumer purchasing of beverage alcohol. This can produce real or imaginary surpluses or shortages among the various categories and skew raw material usage figures.

However, over the past 20 years, grain and grain products usage at all United States' distilled spirits plants has been roughly 40,000,000 bu (56 lb/bu) per year. Corn represents about 68% of the total; milo, 13%; and rye and malt approximately 8% each (Annual Statistical Review, 1973).

Worldwide, the total output of fermentation-based ethanol in 1976 is estimated at 85 million gal (Anonymous, 1977). The United States' total of public taxes and fees generated by beverage alcohol in 1976 was in excess of $10 billion (DISCUS Facts Book, 1976).

## II. PROCESSING

The production of beverage alcohol involves the interrelated operations of mashing, saccharifying, fermenting, and distilling. Each step contributes significantly to the success of the next.

### A. Mashing

The typical process begins with cleaned, sound grains (corn), which are milled to a predetermined particle size to increase the surface area for contact with water during cooking. Raw starch is quite resistant to the enzymatic action of most amylases. The mashing step causes hydration and swelling of the starch granules in the grain, resulting in a gel. Coarser grinds are more difficult to gelatinize, as are older, drier grains. The interrelatedness of the steps begins to be seen here. Too fine a grind can affect subsequent feed recovery, while too coarse a grind necessitates higher temperature in order to gelatinize the remaining 2–5% of the starch.

Liquefying amylases are necessary to reduce the gel strength in the cookers and prepare the starch for subsequent saccharification. For this, the enzyme (premalt) is added with the grain. After cooking, the mash is cooled and additional enzyme is added to the mash to achieve further liquefaction, ultimately cleaving 15–20% of the starch linkage in the case of malt enzymes. Approximately 95% of the starch linkages are $\alpha$1—4; the remaining 5% represent 1—6 linkage.

Mashing procedures are varied to meet conditions. For example, cooking times and temperatures can both be reduced if the saccharifying system is optimized. The source of the starch (type of grain) and the product type are also important variables.

The batch process for bourbon whiskey calls for separate cooking and conversion (saccharifying) of the corn and rye mash. For flavor considerations, this is generally low-temperature (atmospheric) cooking. Each batch of each grain receives its own malt enzyme slurry, sufficient for conversion. Nevertheless, the rye starch is not totally saccharified. Thus, flavor considerations can influence yields.

In the manufacturing of grain neutral spirits, one seeks to optimize yields and efficiency. Pressure cooking is usually chosen for this process because gelatinization is improved and more grain can be added to the mash. By coupling "continuous" cooking with high temperature, one can develop a shorter heating cycle than that of the batch process. A shorter conversion stage is also possible, for one needs only that drop in viscosity which facilitates the pumping of the mash. Saccharification must be continued in the fermentor in order to obtain optimum yields.

Scotch malt whisky is made using a wort rather than a whole-grain process. Here, the grains are cooked, converted, screened, and pressed to produce a liquid filtrate for fermentation.

Canadian whisky distillers employ generally the same practices as their counterparts in the United States.

## B. Saccharifying

### 1. Malt Enzymes

The experience of the distilling industry has clearly established the correlation between $\alpha$-amylase in malt and alcohol yield from grain. Thus, in addition to its initial thinning of the starch viscosity, it is the primary saccharifying enzyme in malt. $\beta$-Amylase is rapidly destroyed at temperatures used for most distillery saccharification. The major products of the malt amylases on starch are maltose, isomaltose, and isomaltotriose. Fermentation filtrates often contain appreciable residues of di-, tri-, and oligosaccharides, thereby suggesting there is still more to be learned.

Gibberellin treatment of malt has increased the $\alpha$-amylase content threefold over that in the original malt varieties. Activity in excess of 100 dextrinizing units (DU)/gm (dry basis) is now routine.

The bacterial load of malts is also very important to the distiller. Bacterial counts of $1 \times 10^6$/gm can be expected. Depending upon the heating/converting conditions selected, significant numbers of these organisms can survive and enter into the fermentation. Lactic acid organisms can compete there with the yeast, resulting in both flavor defects and yield reductions. This is especially serious in continuous conversion with shorter cooking times.

The pH of the mash is important at this stage. First, the optimum pH value for $\alpha$-amylase activity is approximately 5.0, with the minimum near 4.2. Lower values will stop $\alpha$-amylase conversion in the fermentor. This underlines the importance of controlling the lactic acid organism population.

### 2. Microbial Enzymes

Federal regulations permit the use of both malt and microbial amylase for the starch conversion. (Malt, however, must predominate for whiskey production.) Until the late 1950s there was little practical interest in using these enzymes.

Bacterial $\alpha$-amylase from *Bacillus* sp. is commercially available and has the advantage of more heat stability than malt $\alpha$-amylase. These enzymes permit the use of higher starch concentrations in the mash, also. However, there are disadvantages to their use for saccharifying from a distiller's standpoint. For example, they are more sensitive to the low pH conditions which prevail in mashes.

During the 1960s, distillers began in earnest to adapt "aerobic fermentation" techniques, and many now produce glucoamylase (amyloglucosidase) from a variety of fungal strains (*Aspergillus, Rhizopus*). The organisms are cultivated in submerged culture, much as in antibiotic production. Glucoamylase ferments containing 10–14 glucoamylase units (GU)/ml can be produced and stored under aseptic conditions on the distillery premises.

For spirits production, the addition of 3000 GU/bu of grain can reduce the malt usage to as little as 2% of the weight of the grain or one-half of 1% for gibberellin-treated malt (Van Lanen and Smith, 1968). A concomitant increase in yield is also experienced with this substitution. Enzymes of this group attack the starch by separating single glucose units and can break both $\alpha$1—4 and 1—6 links. Since the distillers yeasts utilize glucose more readily than maltose, fermentation can go to completion more rapidly.

A factor of interest is the stability of these enzymes under distillery fermentation conditions. In fermentors converted with glucoamylase, there is essentially no loss of enzyme activity through 96 hours, even under extreme pH conditions (less than 3.5). Malt enzymes become completely inactivated under ever less stringent conditions. Lower maltose, isomaltose, and oligosaccharide residues are the end result and likely account for the increased yield. Experience suggests that glucoamylase saccharifies the starch more rapidly and more completely than malt.

For flavor considerations and by law, distillers must use malt and rye (small grains) in the bourbon mash bills. Experimentation has demonstrated that glucoamylase is not detrimental to whiskey production, but today we know of no distillers who are producing bourbon in this way.

Though previously unreported, a new combination enzyme system is currently in use in the distilling industry. This improvement involved the cooperative efforts of Midwest Solvents Corporation and Hiram Walker & Sons, Inc.* In this process, a high $\alpha$-amylase and a high glucoamylase titer are achieved within the same *Aspergillus* fermentation broth. By measured feeding of gaseous ammonia into corn or milo mash, via the aeration system, one can increase the yield of glucoamylase to levels in excess of 20 GU/ml. The addition is balanced to maintain the pH of the fermentation above 4.5, and the yield increase is associated with the availability of nitrogen. Ammonium salts are also stimulative.

The pH control results in an accumulation of $\alpha$-amylase as well, and yields in excess of 100 DU/ml are easily maintained.

A finished ferment can be held under aseptic conditions in excess of a week at ambient temperatures and used for both thinning of starch slurry and saccharifying of mash, totally eliminating the need for malt in spirits production.

## C. Yeast

Most distillers add their yeast inoculum in the form of a grain mash. This mash is the result of yet another process. The operation begins in the laboratory where the yeast cultures are maintained and where they are generally checked daily for purity before transfer (Coulter, 1964). Trial fermentation may also be run to assure strain integrity. Malt syrup and wort agar are traditional yeast laboratory media.

Yeast propagation is an anaerobic operation in the distillery, and rye and barley malt make up the traditional medium. Other grains do not provide proper nitrogen for rapid yeast growth, which is the intent of this operation. There is a staged buildup in inoculum quantity throughout the

* Patent applied for.

process. A typical increase would be from 25 ml in the laboratory to 9000 gal in the plant in six steps. The time intervals are generally 12–24 hours for each step, depending upon incubation temperature.

With a population in excess of $1 \times 10^8$/ml of bacteria-free yeast, the mash is then ready to inoculate the distillery fermentation mash for ethanol production. A rate of 3% by volume of such a mash would be suitable for bourbon inoculation. One-half percent would be adequate for a glucoamylase converted corn mash for spirits production.

Distillers sour their yeast mash during its preparation to facilitate final sterilization and to aid thinning; rye mash can become quite viscous. The pH of this mash is lowered by culturing a lactic acid-producing organism in the mash for a few hours. These organisms are subsequently destroyed by the heat and acid during sterilization. The mash is then cooled and is ready for yeast propagation. Direct addition of acid can be substituted.

The introduction of glucoamylase production into the distillery has changed yeast production also. The glucoamylase ferment of *Aspergillus* can upgrade an inexpensive corn mash into an excellent yeast propagation medium (Van Lanen *et al.*, 1975). One increases the glucoamylase usage to 6000 GU/bu and adds the ferment to the corn mash along with the normal lactic culture. The mash (20° Balling) thus treated is then held at 50°C (122°F) for 4 hours to develop souring and to permit hydrolysis. During this time, most of the starch in the mash is converted to sugar by glucoamylase activity. Protease activity present in the same ferment modifies the corn protein, increasing the $\alpha$-amino nitrogen content. Combined, these actions free virtually all of the oils in the corn. At this point, the medium is sterilized and may be stored sterile as needed.

To prepare the yeast inoculum, this sterile mash is inoculated with yeast at 27°C (80°F) and is maintained at that temperature for a 4°–5° drop in Balling, usually 6–8 hours. At this point, the mash is cooled to 15°C (60°F) and held there for use. Yeast populations of $3–5 \times 10^8$/ml are obtained. This permits reduction of the volume of inoculum required for each fermentor. Also, this inoculum remains unusually stable for a yeast mash. Through several days of usage, these yeasts retain nearly 100% viability. The precise mechanism remains to be defined. Nevertheless, this represents a distinct advantage over traditional rye mash inoculum.

## D. Fermentation

The distillers mash must be cooled to fermentation temperature before the addition of yeast. Mash coolers, by the nature of their function, provide areas of optimum temperature for lactic acid bacteria development. Thus, the lighter the bacteria load entering the cooler, the cleaner the mash emerging. Cleaner malt-free mash offers a decided benefit for spirits production.

It is at this stage that spent stillage (alcohol-stripped, fermented mash) is added to the mash stream either at the cooler or at the fermentor. This dilutes the sugar mixture to the consistency the distiller has determined to be most suited for his particular process. Yeast strain, equipment, and final product all influence his decision. The pH of the grain mash is also lowered by this addition into the range for best yeast growth. Trace nutrients and buffering capacity from earlier fermentations thus benefit the current ones. However, all water added to the process must ultimately be removed again in the final drying of the stillage into distillers feed grains.

The initial temperature of the grain mash in the fermentor must also be set for the prevailing conditions. Highly converted corn mashes ferment rapidly with sudden $CO_2$ production, and they are strongly exothermic. Bourbons generally begin slowly. Unless the fermentor is open-topped and/or is provided with adequate cooling, a maximal temperature of some 32°C (90°F) will likely be exceeded, resulting in a loss of yield and/or abnormal flavor development.

Few whiskey distilleries are equipped to hold a constant optimum fermentation temperature. Some actively choose not to. The usual alternative is to set at a lower temperature (21°–24°C, 70°–75°F).

Very little *in vivo* work has been published by distillers on yeast enzyme patterns, etc. (Oura, 1977). Ethanol production can be detected almost from the onset, and a pH drop is characteristic of the fermentation. The number of yeast cells increases and ultimately reaches the same level regardless of inoculum size. There is a 30-billion cell increase associated with the production of each ounce of whiskey (De-Becze, 1964).

The sugar concentration of the mash is usually depleted within 48 hours, but to optimize spirit yields or to develop characteristic batch method whiskey flavors, most beverage alcohol undergoes a 3-day fermentation.

Distillery scheduling usually calls for some short-term (60–70 hours) and some long-term (approximately 90 hours) fermentation in order to fit distillation scheduling into a 5-day week. Fermentors scheduled for long term would be set at lower temperatures and held over the weekend, perhaps with cooling.

The mash in the fermentor is never bacteria-free, and rapid pH drops are possible with contaminated fermentors. A normal flora may contribute significantly to the final aroma profile of a distillate (Suomalainen and Lehtonen, 1976; Reazin et al., 1970).

The reaction products of yeast fermentations and the flavor development which accompanies these fermentations are beyond the scope of this chapter. However, many of these compounds have been researched (Suomalainen and Nurminen, 1975; Harrison and Graham, 1970). Troublesome off-odors are frequently associated with abnormal fermentations. Acrolein-producing bacteria can develop during fermentation if malt quality is not properly maintained (Serjak *et al.*, 1954). Diacetyl, a vicinyl diketone, and related compounds are a problem which distillers share with the brewing industry (Rice and Helbert, 1975; Branen and Keenan, 1971). Normal control procedures include rate and duration of fermentation and good sanitation. Improper yeast strain selection is also a contributing factor. Excessive aldehydes in the fermentation beer can be rectified by distillation techniques. Phenolic compounds are best controlled by adjusting the cooking temperatures of the grain.

## III. PLANT

Local grain surplus likely forced many of the early millers to attempt distilling. Present-day distillery operators are on a bit more solid footing (Prescott and Dunn, 1959). Technology is now available with which he can, and must, control his process. The water and the grains still are the essential raw material, energy and transportation still may determine profitability, but proper design will produce a quality product. The superior product requires skill and experience.

### A. Milling

Both product type and mode of operation are very important to design and ultimately determine the choice of process equipment. For example, hammer mills equipped with screens for particle sizing offer consistent operation over long periods with less operator attention, and they mesh nicely with continuous cooking equipment. A 12-mesh screen is standard for control, and approximately 5% of the grain particles should remain on the screen when milling corn for atmospheric cooking (100°C for approximately 40 minutes). Pressure cooking of corn (above 121°C for 5–15 minutes) permits a larger percentage, 12–25%, on the 12-mesh screen. Particle retention values for small grains average about one-half that of the corn standards.

Roller mills can cause striation of barley malt hulls, which tend to plug spiral plate mash coolers. However, they offer good control. Minimizing flour aids dried grain production later on. Roller mills find application primarily in batch cooking operations and where traditional mashing practices are used to preserve distillate flavors.

## B. Cooking

Batch cooking vessel design can be horizontal or vertical and includes mixers to suspend the grain particles. Typically, steam is injected into these vessels for heating, and cooling is provided via jacketing or flash evaporation under vacuum. Either barometric or surface condensers would be provided to condense the generated vapors. Operating capacities are approximately 10,000 gal (four hundred 56-lb bushels of grain per charge). The entire cycle may take from 2 to 3 hours.

Continuous cooking equipment draws a uniform, lower steam load than batch equipment. Space and time requirements are also less for the same throughput. The entire process of grain from the bin to mash into the fermentor can be 10 minutes when pressure cooking is used. Higher capacity distilleries, which mash over 5000 bu/day, are well-suited to this operation. In this equipment, a prewarmed grain slurry (15–18 gal/bu) enters the pipeline vessel and is instantly heated to over 149°C (300°F) in a steam jet. After 4–5 minutes of pipeline flow, the mash begins to cool. Careful design can develop considerable heat conservation at this point. A partial recycle of thin stillage (residues from the still bottoms which are screened or centrifuged to remove suspended particles) as liquid for slurrying the grain for the cooker will furnish prewarming and reduce the water load to the driers. Quality considerations or operating schedules may prevent these adaptations, however.

Recycling blowdown vapors from the continuous cooking holding tanks, as preheat for grain slurries, affords significant energy savings also.

## C. Saccharifying

After cooking, the gelatinized starch is enzymatically converted into sugars. Batch cooked mash is generally flashed to 63°C (145°F) at the end of the cycles while still in the cooker. Enzyme is added and the mixture is transferred into a receiving vessel (drop tub) for the conversion step. This stage lasts 30 minutes for malt enzyme.

Mash leaves a continuous cooker at a higher temperature and usually passes through a flashing vessel. An enzyme slurry stream is metered in at the cool end discharge, and conversion takes place as the mixture flows through a second pipeline to the final coolers.

Glucoamylase offers the distiller a tailored enzyme system for saccharifying grain mashes. The enzyme ferment is more tolerant of distillery variations in pH and heat, and it saccharifies more starch into a more utilizable form.

The production techniques are new to most distillers. The typical 5000-gal tub of mash would have set solids of about 18%, made up from approximately 10,000 lb of ground corn, 50 lb of malt, and 4000 gal of water or grain stillage. Such a mash would be cooked to 121°C (250°F) for 5 minutes, cooled to receive 100 lb of liquefying malt, held at 65.5°C (150°F) for 10 minutes, and finally sterilized at 121°C (250°F) for 60 minutes. The final pH runs to 4.5–5.0.

These culture vessels are generally pressurized with height two times the width and are equipped with cooling, air sparge, and a rotary blade agitator. After inoculation with an *Aspergillus* culture, the fermentor is incubated at 35°C (94°F) with aeration (0.25–1.0 v/v per minute) with sterile air. Within 4–8 days, the unitage of glucoamylase is sufficient for use and the vessel is cooled. Unless contamination occurs, the ferment is quite stable.

As in the case with yeast, commercial sources of the enzyme are also available. Standard enzyme usage rates are 3000 GU/56-lb bushel, regardless of the form. Addition may be made into the cooling mash or directly into the fermentor.

A new microbial technology for yeast propagation and the production of high $\alpha$-glucoamylase is outlined in Section II, B.

After conversion, mash must be further cooled to the fermentation temperature. Mash cooler designs are variable, and spiral plate exchangers, flash coolers, and double pipe exchangers are all employed. The mash has higher viscosity at this point and is vulnerable to bacteriological contamination, and both must be considered in design. Plate coolers require less heat transfer surface and less space than most other types.

### D. Yeasting

Microbial technology has progressed to the point that a distiller today need not design a yeast process at all. Active dry distillers yeasts are available commercially in several convenient package forms. These are stable and of respectable quality. However, considerations of distillate flavor and "real" costs lead most distillers to cultivate their own supply.

The pure culture yeast process begins in the laboratory, where the initial step is the selection of the type of yeast that is desired. Routine enrichment, isolation, and trial fermentation techniques produce the master culture(s) upon which the distiller depends. These cultures can be lyophilized or, more likely, maintained on wort agar medium slants and stored at reduced temperature until needed. Cultures are transferred at 6- to 12-month intervals. A standard regime follows below.

A fresh master culture is streak-plated on wort agar and individual isolated colonies are picked to produce the working slants.

Each day a working slant is inoculated into sterile, nondiastatic malt syrup (25 ml at 20° Balling), which is incubated for 24 hours at 24°–27°C (75°–80°F). The contents of this culture are, in turn, used to inoculate 900 ml of malt syrup, which is again incubated at 24°C (75°F). These cultures would typically contain 150–200 × $10^6$ viable cells per milliliter at 24 hours. In this time, the sugar content would be reduced to about 9%. These flasks are carefully screened for bacterial contamination, since once the culture leaves the laboratory and is cultivated in grain mash, detection is delayed and so is a very difficult process.

The plant medium traditionally is rye mash with not less than 10% barley malt. Distillery yeast propagation, in contrast to bakers' yeast production, does not work with sugar-to-nitrogen ratios; nor is it concerned with absolute yields. If the yield of yeast is below 75% of the weight of the raw material, the "loss" is recovered in subsequent alcohol production and/or feed recovery.

The requirements for available carbon, energy, and nitrogen sources are accounted for in the grain selection; phosphates and vitamins seldom become limiting, because cultivation until the Balling drops to one-half that in the original mash is the rule of thumb.

Plant propagation of yeast is also staged. The initial tank would contain 50 gal of mash; after 12–24 hours this would furnish inoculum for a 500-gal tank and, similarly, on to a 10,000-gal tank. Incubation temperatures are adjusted to yield an active inoculum of yeast at 1 × $10^8$ cells/ml in schedule with the distillery's fermentor sets. Inoculation rates vary as to product: 0.5% for spirit and 3.0% for whiskey would be average. A 20° Balling mash would be soured with lactic acid by permitting lactic acid-producing bacteria to incubate in the medium at about 49°C (120°F) until titratable acidity reached 1.2% (pH 3.7–4.0). Nine thousand gallons of such a mash can be sterilized in the propagation vessel by heating to 88°C (190°F) for 2 hours.

Obviously, for a pure culture system, all yeasting equipment, such as connecting lines and valves, should be of sanitary construction, avoiding pockets, dead-end lines, and pumps. Yeast culture delivery from vessel to vessel would be by gravity or air pressure (filtered), with transfer lines maintained under steam when not in use. Cleaning and sterilization programs vary, but 1 hour of hot water wash, followed by 2 hours at 100°C (212°F), is a common practice (Coulter, 1964).

### E. Fermenting

The typical set fermentor data for bourbon whiskey production averages: (1) temperature, 24°C (75°F); (2) Balling, 13–14; (3) pH, 4.5–4.9; and contents of approximately 25% back set stillage. Final data would show a (1) temperature of about 32°C (90°F); (2) Balling, 0.4–0.5; (3) pH, 4.0; (4) lactic acid, 0.3–0.5%; and (5) an alcohol proof over 15.5°.

Table I presents the data which are required to determine production efficiencies. The chemistry of the reactions dictates the maximum amount of product which can be formed (A and B). The input quantities forecast what should be formed (C). Recovered product permits the calculations (D).

**TABLE I.** Alcohol Yields

A. Stoichiometry
Starch + water → glucose
$C_6H_{10}O_5$ (162) + $H_2O$ (18) → $C_6H_{12}O_6$ (180)
Glucose → 2 alcohol + 2 carbon dioxide
$C_6H_{12}O_6$ (180) → 2 $C_2H_5OH$ (92) + 2 $CO_2$ (88)

B. Theoretical yields—dry basis (70% starch for calculation)
100 kg dry corn → 70 kg dry starch + 30 kg dry feed
70 kg starch → 77.7777 kg glucose (7.777 kg hydrolysis gain)
77.7777 kg glucose → 39.7531 kg alcohol + 38.0246 kg $CO_2$
*Summary:*
100 kg corn + 7.7777 kg hydrolysis gain → 107.7777 kg
39.7531 kg alcohol + 38.0246 kg $CO_2$ + 30.0 kg feed

C. Typical analysis of distillery grain receipts

| | | | Dry basis | |
| | % Moisture | Starch | Protein | Fat |
|---|---|---|---|---|
| Barley malt | 5.4– 6.0 | 57.0–61.5 | 12.2–13.8 | 1.6–2.1 |
| Rye | 11.6–12.9 | 64.2–67.7 | 13.3–14.6 | 1.3–2.0 |
| Corn | 13.5–16.1 | 71.2–73.0 | 9.4–10.3 | 3.9–4.9 |
| Milo | 13.1–14.6 | 72.4–75.9 | 9.7–11.9 | 2.2–3.4 |

D. Yield calculations
1. *Moisture*
Grain into fermentor × moisture in grain = moisture in mash

| Corn 99% | × | 14.08 | = | 13.94 |
| Malt 1% | × | 6.00 | = | 0.06 |
| | | | | 14.00% |

2. *Proof gallon (PG)*
Volume of mash × proof by sample distillation ÷ 100 = proof gallon
98,115 gal × 16.43 = 16,120

3. *Yield*
Proof gallons ÷ bushels in fermentor = PG/bu (56 lb)
16,120 ÷ 3060 = 5.27

4. *Dry basis*
Proof gallons ÷ dry matter
5.27 ÷ 0.86 (100 − 14) = 6.13 dry basis yield

5. *Alcohol/bushel (56 lb) (theoretical)*
56 lb Corn → 39.2 lb starch → 22.262 lb alcohol
22.262 ÷ 3.305 (conversion factor) = 6.76 PG

There is no typical vessel. Open-top, closed-top, wooden, and steel vessels all exist, often side by side. Successful operation may mean optimizing yields or repetitive production of critical flavor distillate. Most newer installations resemble Fig. 1; they are designed to facilitate temperature control and cleaning. Mechanical agitators, coupled with cooling coils or circulating mash coolers located alongside the tanks, are the two most common installations. Pressure washers with rotating heads and steaming aid in fermentor cleanup.

### F. Distilling

The pot still is considered the earliest form of distilling equipment. It is of the simplest design, consisting of a kettle to hold the mash, a steam coil, and a condenser. This was used for single-stage batch distillation. While there have been improvements, refluxing is still limited; rather large heads and tails cuts are generally removed to obtain the desired quality.

Today, most distillers employ continuous stills. In this design the fermented beer enters the upper end of the column and flows downward via an overflow weir and downpipes, criss-crossing a series of perforated trays. Steam is introduced from the bottom and rises with sufficient velocity to strip the beer of its volatile components. Vapors recondense higher up in the column, depending upon their volatilities. Thus, liquids can be drawn off or recycled at the various plates, as need be. If higher efficiency is desired, rectifying stills can be added which will concentrate volatile impurities within a narrow region of the still.

Heat exchangers are generally coupled with continuous stills to conserve energy by preheating the incoming beer with the outgoing stillage.

**FIGURE 1.** Fermentors used in distilled beverage production. (Photo courtesy of Hiram Walker & Sons, Ltd.)

## G. Drying

"Stillage" is the term applied to dealcoholized grain residues that remain after distillation. As this material emerges from the bottom of a still, it is typically sent by pipeline to the drying equipment, where the insoluble solids are separated from the soluble stream.

This liquid portion can be dried as distillers solubles, or, as is most common, all fractions are individually dewatered and then recombined and dried.

The initial step involves screening or centrifugation. The "throughs" can then be processed through a triple effect evaporator, where the moisture is reduced under vacuum from 95–97% down to 75%. A finishing pan completes the task and the resulting syrup, 35–40% solids, is then mixed back into the grain residues. Rotary drum or flash driers complete the drying. The resulting feed concentrate, distillers dried grains with solubles, is a valuable feedstuff, with annual production averaging 400,000 tons and a 1976 price of approximately $140/ton. Table II shows a typical analysis.

For total nutrient composition, the reader is referred to the Distillers Feed Research Council (Distillers Feed Research, 1970).

## IV. PRODUCT

The culmination of properly controlled process and well-designed plant is a quality product. The product type determines which process and which design are chosen.

*Vodka* is a neutral, unaged spirit distillate. The fermentation medium may contain virtually any starchy substance, but it is almost always corn or milo. Both the yeast and the fermentation conditions are chosen to maximize yield. Being a spirit, it is distilled at or above 190° proof, and the distillate may be treated with charcoal to render it more neutral in taste and odor.

**TABLE II.** Composition of Distillers Feed Dried Grains with Solubles (Corn)

| Composition | Percentage |
| --- | --- |
| Moisture | 9.0 |
| Protein | 27.0 |
| Fat | 8.0 |
| Fiber | 8.5 |
| Ash | 4.5 |

*Whiskies* are generally distilled at no more than 160° proof, and the distillates contain numerous flavor constituents (congeners). Depending upon which grain makes up the major portion of the mash bill, one produces rye, corn, or malt whiskey. At this time, no North American producer is utilizing microbial enzymes to produce whiskey, as far as is known. Thus, malt would make up a small but significant portion of all whiskey mash bills. *Bourbon whiskey* is further defined and must contain at least 51% corn, with rye and malt making up the remainder of the grain, generally 15–30% and 5–15%, respectively.

An aging process is necessary to fully develop the flavoring of whiskey. Only charred white oak, *Quercus alba,* will produce the desired mellowing effect. This is accomplished by prolonged storage in barrels or casks. The changes which occur during aging can be monitored chemically, and data show that the wood char removes some compounds and adds others. Still other flavor elements develop from the oxidation–reduction reactions which take place within the barrel as it breathes during aging. A significant loss of ethanol occurs during aging due to barrel soakage and some evaporation. Ultimately, barreling proof and solubility of the congeners together determine their final concentration. Scotch, Canadian, and American light whiskey may claim age for storage in reused cooperage, while bourbon whiskey must, by law, be aged in new barrels.

*Tequila* is a distinctive product of the Jalisco area of Mexico and is produced from agava cactus. *Rum* is the distillate from the fermented sugar cane products, distilled between 80° and 190° proof. *Brandy* is also distilled at less than 190° proof from fermented fruit or fruit juices.

Extensive chemical analyses have been made on aged as well as unaged distillates. Table III presents quantitative data for the different classes of compounds found in bourbon whiskey.

**TABLE III.**  An Example of Whiskey Analysis for 5-Year Bourbon

| | |
|---|---|
| Specific gravity 20°C/20°C | 0.93 |
| Proof | 100.00 |
| pH | 4.1 |
| Extract[a] | 136.0 |
| Tannins[a] | 42.0 |
| Furfural[a] | 1.3 |
| Aldehydes[a] | 5.2 |
| Esters[a] | 38.7 |
| Other congeners[a] | 218.0 |
| Total acids as acetic[a] | 54.0 |

[a] All values given are in grams per 100 liters.

## REFERENCES

Annual Statistical Review (1973). Distilled Spirits Inst., Washington, D.C.
Anonymous (1977). *Chem. Week.* **120,** 26–28.
Branen, A. L., and Keenan, T. W. (1971). *Appl. Microbiol.* **22,** 517–521.
Coulter, C. J. (1964). *Proc. Am. Soc. Brew. Chem.* pp. 64–69.
DeBecze, G. I. (1964). *Biotechnol. Bioeng.* **6,** 191–221.
DISCUS Facts Book (1976). Distilled Spirits Council U.S., Washington, D.C.
Distillers Feeds—A Progress Report (1970). Distillers Feed Res. Counc., Cincinnati, Ohio.
Harrison, J. S., and Graham, J. C. J. (1970). *In* "The Yeasts" (A. H. Rose and J. S. Harrison, eds.), Vol. 3, pp. 283–348. Academic Press, New York.
Oura, E. (1977). *Process Biochem.* **12,** 19.
Prescott, S. C., and Dunn, C. G. (1959). "Industrial Microbiology," 3rd ed. McGraw-Hill, New York.
Reazin, G., Scales, H., and Andreason, A. (1970). *J. Agric. Food Chem.* **18,** 585–589.
Rice, J. F., and Helbert, J. R. (1973). *Proc. Am Soc. Brew. Chem.* pp. 11–17.
Serjak, W. C., Day, W. H., Van Lanen, J. M., and Boruff, C. S. (1954). *Appl. Microbiol.* **2,** 14–20.
Suomalainen, H., and Lehtonen, M. (1976). *Kem.–Kemi* **3,** 69–77.
Suomalainen, H., and Nurminen, T. (1975). Some Aspects of the Structure and Function of the Yeast Plasma Membrane. Aikon Keskuslaboratorio Report 8101, Aiko, Helsinki, Finland.
Van Lanen, J. M., and Smith, M. B. (1968). U.S. Patent 3,418,211.
Van Lanen, J. M., Smith, M. B., and Maisch, W. F. (1975). U.S. Patent 3,868,307.

## GENERAL REFERENCES

Aiba, S., Humphrey, A. E., and Millis, N. F. (1965). "Biochemical Engineering," 1st ed. Academic Press, New York.
Gutcho, M. H. (1976). "Alcoholic Beverage Processes." Noyes Data Corp., Park Ridge, New Jersey.
Haas, G. J. (1976). *In* "Industrial Microbiology" (B. M. Miller and W. Litsky, eds.), pp. 165–191. McGraw-Hill, New York.
Miller, D. L. (1975). *Biotechnol. Bioeng.* Symp. **5,** 345–352.
Packowski, G. W. (1963). *Kirk-Othmer Encycl. Chem. Technol., 2nd ed.* **1,** 1.
Solomons, G. L. (1969). "Materials and Methods in Fermentation." Academic Press, New York.

# Chapter 4

# Mold-Modified Foods

HWA L. WANG
C. W. HESSELTINE

MICROBIAL TECHNOLOGY, 2nd ed., VOL.II

## I. INTRODUCTION

In earliest times, man was plagued with either feast or famine, so any means he could discover to conserve food when it was bountiful was a great step forward in his survival and his conquest of the earth. Since man was, of necessity, a wanderer and a hunter, he learned about the drying and smoking of meat. Certainly these methods not only conserved his supplies, but they also reduced weight, enabling him to carry more food with him. At the time of the discovery of North America by the Europeans, most of the Indians were in this stage of development. The discovery of these two methods—drying and smoking, like the invention of the wheel, was perhaps by serendipity. Early man might not know why foods spoiled, he only knew they did. Later we can speculate that he discovered the use of salt in combination with drying and smoking.

Man's next discovery for preserving food was the fermentation of foods; although he had no idea what happened when microbial growth occurred, he learned that meat and plant material could be kept for long periods of time. It was also essential that he knew how to use salt, a necessary agent in the successful fermentation of materials, because it inhibits the development of toxins from microorganisms. Undoubtedly, many an early ancestor died from botulism or was made ill by *Staphylococcus aureus*. When salt was added and natural fermentation occurred, he knew definite changes in odor, appearance, and taste occurred that helped the product remain wholesome. Probably the first fermentations were discovered accidentally when salt was incorporated with the food material, and the salt selected certain harmless microorganisms that fermented the product to give a nutritious and acceptable food. If we speculate along these lines, we might expect the first fermented food to have been fish. With the advent of certain religions in which meat was excluded from the diet, the use of salt and fermentation was adapted to certain plant products. For instance, Bush (1959) states that Buddhism was well established in China and Korea by the fourth century and was introduced into Japan between A.D. 500 and 600. It may very well be that the cultivation of soybeans and their use in food, including fermented foods, were then introduced in Japan. In fact, today, food fermentations in the Orient are spoken of as the "traditional fermentations," which include the very important soy sauce and miso (soybean paste). From these traditional food industries have developed the modern industrial enzyme industry and the flavoring agents such as monosodium glutamate and nucleotides. We would like to suggest that one reason Japan is now the leader in the industrial fermentation field is in large part due to the food fermentation base from which it launched its industrialization of microorganisms. This background in food fermentations supplied Japan with the knowhow and the technicians familiar with the handling of microorganisms.

Fish and soybean fermentations continue to be the most important fermented food industries in the Orient. Fish fermentations remain indigenous, and the products are strictly traditional and limited to local consumption. The soybean fermentations, on the other hand, are changing rapidly through modern microbiology and technology. Although fermentations carried out in the traditional way at home still exist, with the art passed on from one generation to the next, some of the most modern and sophisticated fermentations in the world are found in the Oriental fermented soybean foods industry, especially in Japan.

Some soybean fermentations are neither strictly fermentation of soybeans alone nor carried out by only one kind of microorganism; they may involve a substrate consisting of both cereals and soybeans and an inoculum consisting of bacteria, yeasts, and molds. In this chapter, we shall discuss an area of industrial microbiology that concerns the fermentation of raw agricultural products into foods mainly by the action of diverse species of fungi.

## II. SOY SAUCE

### A. General Description

Soy sauce is a dark-brown liquid with a salty taste and distinct pleasant aroma, which is made by fermenting soybeans, wheat, and salt with a mixture of mold, yeast, and bacteria. The fermentation is essentially a process of enzymatic hydrolysis of proteins, carbohydrates, and other constituents of soybeans and wheat to peptides, amino acids, sugars, alcohols, acids, and other low-molecular compounds by the enzymes of the microorganisms. In addition to the fermentation technique, a chemical method in which acid hydrolyzes the protein and the carbohydrates has been used in the United States. However, the flavor of the chemically hydrolyzed product is quite different from that of the product brought about by microbial enzymes.

The brewing of soy sauce originated in China many centuries ago and later was introduced to Japan and other Oriental countries. Consumed in almost every Oriental country, soy sauce is used as a seasoning agent in preparation of food and as a table condiment. The product is known as chiang-yu in China, shoyu in Japan, kecap in Indonesia, kanjang in Korea, toyo in the Philippines, and see-iew in Thailand. In the Western World, the product is often referred to as soy sauce.

Japan leads the soy sauce industry in the world. Japan not only has the largest fermentation plant but also employs the most advanced technology, developed through comprehensive research on all aspects of soy sauce fermentation. According to Ebine (1976), 14,000 metric tons of whole soybeans and 176,000 metric tons of defatted soybean products were used in Japan in 1974 to produce more than 1 million kl of soy sauce. Annual per capita consumption in Japan is estimated at about 10.2 liters. Although there are more than 4000 soy sauce producers in Japan, the largest four or five companies supply about 50% of the total produced. In 1972, Kikkoman, one of the largest producers in Japan, built a plant in Walworth, Wisconsin, and has since been in full operation to supply soy sauce for the American market.

## B. Method of Preparation

Many improvements have been made since the early development of the fermentation, but the basic method of manufacture is almost unchanged (Yokotsuka, 1960; Hesseltine and Wang, 1972; Yong and Wood, 1974) and is illustrated in Fig. 1.

### 1. Raw Materials and Their Treatment

**a. Soybeans.**   Traditionally, soy sauce is made from whole soybeans; however, in recent years, defatted soybean meals and flakes have taken their place. Today, more than 90% of soy sauce production in Japan is from defatted soybean products.

In preparation for fermentation, selected beans are washed, soaked overnight with several changes of water to prevent bacterial growth, drained, and steamed at 10 lb/in.$^2$ for several hours. Cooking conditions have been known to greatly affect enzymatic digestion of soybean protein which, in turn, affects nitrogen recovery. Tateno and Umeda (1955) and the Noda Soy Sauce Company, Ltd. (1955) claimed that soybean protein was best utilized by the enzymes when the beans were soaked in water for 10–12 hours at room temperature and then steamed at 10–13 lb for about 1 hour. So the current trend has been to cook the soaked beans at a higher temperature for a shorter time (Yokotsuka, 1971). For this purpose, a batch-type cooker was developed so that the temperature could be raised rapidly to 120°C and lowered quickly immediately after cooking.

When defatted soybean meals or flakes are used, they are first moistened by spraying with water amounting to about 130% of soybean weight and then are steamed at 13 lb/in.$^2$ for 45 minutes.

**b. Wheat.**   Whole wheat is generally preferred. It is first roasted and then coarsely crushed. The roasting process adds flavor and color to the

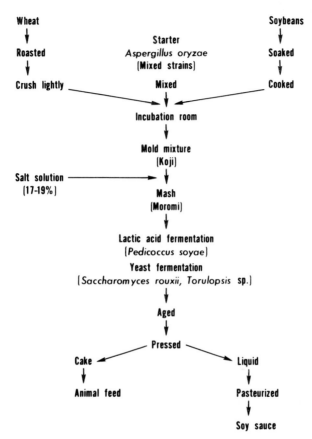

**FIGURE 1.** Flow sheet for the manufacture of soy sauce.

resulting soy sauce and in addition destroys surface microorganisms and facilitates enzymatic hydrolysis. According to Yokotsuka (1964), the addition of wheat to the fermentation mixture serves several purposes. First, the mold grows better and produces more enzymes on a mixture of wheat and soybeans than on wheat or soybeans alone. Second, the addition of roasted, crushed wheat to the cooked soybeans would minimize the growth of undesirable bacteria. Moisture of cooked soybeans is 60%, ideal for bacterial growth, whereas the moisture of the soybean–wheat mixture (1 : 1) is about 45%, adequate for mold growth but not for bacteria. Third, wheat serves as a precursor of sugars, alcohols, organic acids, and flavor compounds. Last, wheat is rich in glutamic acid.

**c. Salt.** Although commercial (not chemically pure) sodium chloride is generally used in making soy sauce, chemically pure sodium chloride has been used successfully in making soy sauce on a laboratory scale with added pure inoculum of yeasts and bacteria (Yong and Wood, 1974). It is possible that commercial salt may carry a larger inoculum of halophilic and halotolerant bacteria and yeasts. In addition to giving a salty taste, sodium chloride acts as a preservative and also has a selective action on the microorganisms that grow in the fermentation substrate. Salt, added in such quantities that dangerous spoilage micro-organisms or food poisoning bacteria are prevented from growing, per-mits the development of flavor- and aroma-forming yeasts and bacteria.

**d. Other Raw Materials.** Raw materials other than wheat and soy-beans have been tested in making soy sauce. Peanut press cake was used in place of the wheat and soybean mixture (Church, 1923). Baens-Arcega (1966) replaced wheat with copra meal, and Oda *et al.* (1949) substituted acorns for soybeans. Their results were unsuccessful com-mercially.

### 2. Soy Sauce Koji

*Koji* is a Japanese name given to a preparation consisting of mold growth on cooked cereals and/or soybeans. Koji serves as an enzyme source for converting complex plant constituents to simpler compounds. Usually this is the first fermentation stage in a process that also involves a second fermentation. Thus, koji has the same relationship to the Orien-tal food fermentations as malt has to the alcoholic fermentation of grain. Soy sauce koji is made from a mixture of roasted wheat and steamed soybeans, with a koji starter consisting of selected strains of *Aspergillus oryzae* or *A. soyae* (Fig. 1).

**a. Koji Starter.** Koji starter is known as tane koji in Japan. Different types of tane koji are available for commercial use in making soy sauce koji, miso koji, and others. Strains of *A. oryzae* and *A. soyae* have varying morphological and physiological characteristics. Each strain is selected for its special abilities by natural selection or by induced mutation to give a desirable starter for a particular fermentation. Although the strains of molds used for each type of tane koji may be different, the method for preparing tane koji is the same; it has been outlined as follows by Hesseltine and Wang (1972).

Polished rice is soaked in water overnight, drained, steamed for 1 hour, and mixed thoroughly with about 2% of wood ash as a source of trace elements. Addition of ash of certain trees is important because the RNA/DNA ratio of the spores varies; a high RNA/DNA ratio gives spores with greater viability and vigor. The rice is then inoculated with spores of a selected strain of *A. oryzae,* spread out in trays in a layer approximately 1.5-cm deep, and covered with damp cloths to favor the growth of mold mycelium. After incubation at 30°C for 5 days, the rice is well covered with mycelium and with green to yellowish green spores of *A. oryzae*. The spores are harvested, dried at 50°C, and stored at 15°C. A koji starter is usually composed of spores from several different strains of *A. oryzae* blended together in a definite proportion, so that various enzymes may be produced in the proper amounts during the preparation of koji. In Japan, each gram of commercial tane koji has about $25 \times 10^7$ viable spores.

The literature on preparation of mold spores for food fermentation has been reviewed by Hesseltine *et al.* (1976).

**b. Preparation of Koji.** Raw materials, soybeans, and wheat are treated as described above. The proportion of soybeans to wheat used in fermentation mixtures may vary from one manufacturer to another, but according to Yokotsuka (1960) the best soy sauce is generally made from a soybean-to-wheat ratio of 50 : 50 by weight, or 52 : 48 by volume based on that of raw materials. An appropriate amount of tane koji, which is about 0.1–0.2% of soybean–wheat mixture, is first mixed thoroughly with roasted wheat and then added to cooked soybeans. The inoculated mixture is distributed in shallow wooden boxes, covered with cloth, and then incubated at 30°C. After 24 hours of incubation, the mixture is covered with a thin white growth of mold. As mold growth continues, the temperature of the mixture could rise above room temperature to 40°C or higher. Therefore, the mixture should be turned or stirred periodically to maintain uniform temperature, moisture, and aeration. The mixture should also be free of lumps to minimize bacteria propagation. As the incubation time increases, the molds continue to grow and their growth turns yellow and dark green. Also, the moisture of the mixture gradually decreases. After about 72 hours of incubation, the soy sauce koji is matured. In recent years, automatic koji-making processes have been developed to replace the traditional way of making koji. The new equipment includes automatic inoculator, automatic mixer, large perforated shallow vats in closed chambers equipped with forced filtered-air devices, temperature controls, and mechanical devices for turning the substrates during incubation.

Soy sauce koji of superior quality has a dark-green color, pleasant aroma, sweet but bitter taste and, most important of all, high protease and amylase activity. The relationship between enzyme yield and culturing conditions varies with the organisms. *Aspergillus oryzae* (Maxwell, 1952) and *A. soyae* (Yamamoto, 1957) produce greater amounts of proteases at temperatures lower than their optimum growth temperature. Similar findings with other fungi were reported by Wang *et al.* (1974). The effect of temperature on enzyme production observed in these studies emphasizes the importance of a common practice in koji making, i.e., frequent turning of the growth mass or use of thin layers of solid substrates. Otherwise, the heating which results from active growth will increase incubation temperature and affect enzyme production. Harada (1951) found that it is necessary to keep the moisture content of koji at 27–37% in order to obtain high enzyme activity. Thus, moisture and temperature are the two most important factors in making koji of superior quality. Current industry practice is to increase moisture content of the soybean–wheat mixture and to lower incubation temperature.

### 3. Brine Fermentation

Soy sauce koji is transferred to a deep vessel to which an equal volume of salt solution (20° Be) is added to make a mash, or moromi as it is called in Japan. However, it was found (Yokotsuka, 1960) that moromi made with a ratio greater than 1:1 of salt solution to koji utilized total nitrogen of the raw material better. Also, the lower the concentration of the salt solution, the better the utilization of nitrogen. Therefore, the volume of salt solution has been increased to 1.1–1.2 times that of the koji and the concentration of salt solution has been reduced accordingly. The sodium chloride content of the mash is about 17–19%; below 16% putrefaction may occur.

Moromi is fermented in large concrete tanks or wooden vats for 8–12 months. Frequent agitation and aeration of moromi are necessary for normal fermentation to proceed and to prevent the growth of undesirable anaerobic microorganisms, to maintain uniform temperature, and to facilitate the removal of carbon dioxide. However, too much stirring hinders the fermentation. The change of temperature is said to be important for normal progress of fermentation. Therefore, shoyu fermentation in Japan usually starts in April and takes a year to complete. In general, low-temperature fermentation gives better results; because the rate of enzyme inactivation is slow, the enzymes remain active longer (Komatsu, 1968). Watanabe (1969) indicated that good-quality shoyu can be obtained by 6-month fermentation when the temperature of moromi is controlled as follows: start at 15°C for 1 month, followed by 28°C for 4 months, and finish at 15°C for 1 month.

Since koji is not prepared under aseptic conditions, one would expect the presence of yeasts and bacteria in the moromi. However, pure cultures of yeasts and bacteria are sometimes added to the mash to accelerate the fermentation and to improve the flavor of the final product. Strains of *Saccharomyces rouxii*, *Torulopsis* yeasts, and *Pediococcus soyae* were found to be important flavor producers. Representative strains maintained in the AR Culture Collection have been designated as *S. rouxii* NRRL Y-6681, *T. etchellsii* NRRL Y-7583, *T. versatilis* NRRL Y-7584, and *P. halophilus* NRRL B-4243 and NRRL B-4244. The initial pH of the mash (6.5–7.0) gradually decreases as the lactic acid fermentation advances, and at pH of around 5.5 yeast fermentation takes place.

### 4. Pressing

The matured moromi is separated by pressing into a liquid part, known as raw soy sauce, and a solid cake. When whole soybeans are used as raw material, soy sauce oil, consisting chiefly of ethyl esters of higher fatty acids, is produced during fermentation and appears at the upper layer of the raw soy sauce. This layer of oil must be removed and has no potential use. This is one of the reasons that defatted soybean products have replaced whole soybeans as raw material in soy sauce fermentation.

### 5. Pasteurization and Bottling

The raw soy sauce is pasteurized at 70°–80°C, filtered to remove precipitates, and bottled for market. In Japan either benzoic acid or butyl ester of *p*-hydroxybenzoic acid is added as a preservative.

### C. Composition

According to Watanabe (1969), about 5000 liters (specific gravity, 1.18) of soy sauce is obtained from 1 ton of defatted soybean meal, 1 ton of wheat, and 1 ton of salt. The composition of soy sauce varies with the raw materials used. Table I shows the average composition of soy sauce made from whole soybeans and from defatted soybean meal by 11 factories in Japan (Umeda, 1963; Umeda *et al.*, 1969). A good soy sauce has a salt content of about 18%. Its pH is between 4.6 and 4.8; below that the product is considered too acid, suggesting that acid has been produced by undesirable bacteria. It is also generally recognized in Japan that the quality and price of shoyu are determined by nitrogen yield, total soluble nitrogen, and the ratio of amino nitrogen to total soluble nitrogen. The nitrogen yield is the percentage of nitrogen of raw material converted to soluble nitrogen, which shows the efficiency of enzymatic conversion. The total soluble nitrogen is a measure of the concentration of nitrogenous material in the shoyu and indicates a standard of quality. A ratio of greater than 50% of amino nitrogen to total soluble nitrogen is also evidence of quality. These results can be affected by many factors such as raw materials, steaming conditions, tane koji, koji making, and brine fermentation. Technology to improve these values is constantly being sought. During the last 20–30 years, a shorter cooking time at higher temperatures has been adopted. Koji-making

**TABLE I.** Average Composition of Soy Sauce Made from Whole Soybeans and Defatted Soybean Meal[a]

|  | Raw material | |
| --- | --- | --- |
| Conditions | Whole | Defatted meal |
| Baume (°) | 22.7 | 23.4 |
| NaCl (%) | 18.5 | 18.0 |
| Total nitrogen (%) | 1.6 | 1.5 |
| Amino nitrogen (%) | 0.7 | 0.9 |
| Reduced sugar (%) | 1.9 | 4.4 |
| Alcohol (%) | 2.1 | 1.5 |
| Acidity I (ml)[b] | 10.1 | 14.0 |
| Acidity II (ml)[c] | 9.8 | 13.6 |
| pH | 4.8 | 4.6 |
| Glutamic acid (%) | 1.3 | 1.2 |
| Nitrogen yield (%) | 75.7 | 73.7 |

[a] From Umeda (1963); and Umeda et al. (1969).
[b] Milliliters of 0.1 $N$ NaOH required to neutralize 10 ml of soy sauce to pH 7.0.
[c] Milliliters of 0.1 $N$ NaOH required to bring the pH of 10 ml soy sauce from 7.0 to 8.3.

technology has been improved to increase enzymatic activity. New strains of koji molds, yeasts, and bacteria have been developed by induced mutations or by diploid formation. Temperatures for moromi fermentation have been controlled. As a result of these improvements, nitrogen recovery has increased from 60% in 1945 to about 90% in 1975 (Ebine, 1976), fermentation period has been reduced significantly, and flavor has been greatly improved. More recently, an enzymatic process for making soy sauce has been attempted. The raw materials were subjected to enzymatic treatment but did not change to koji (Nakadai et al., 1975).

The chemical changes in the production of soy sauce and in its flavor are complicated. Yokotsuka (1960) has written a complete review on these changes. He states that of the total nitrogen about 40–50% is amino acids, 40–50% peptides and peptones, 10–15% ammonia, and less than 1% protein. Seventeen common amino acids are present, with glutamic acid and its salts being the principal flavoring constituents. The organic bases, believed to be hydrolyzed products of nucleic acids, are adenine, hypoxanthine, xanthine, guanine, cytosine, and uracil. Sugars present are glucose, arabinose, xylose, maltose, and galactose, as well as two sugar alcohols, glycerol and mannitol. Organic acids reported in soy sauce are lactic, succinic, acetic, and pyroglutamic. The color of soy sauce is generally recognized to be the result of a nonenzymatic browning reaction. More than a hundred compounds have been reported as flavor components of soy sauce; however, the compounds that decisively characterize soy sauce remain unknown. The guaiacyl compounds seem to have an important effect upon the overall flavor of Japanese soy sauce.

## III. MISO

### A. General Description

Miso, the most important and popular fermented soybean food in Japan, is also made and consumed in other parts of the Orient. This product is generally known as bean paste. Like soy sauce, bean paste is made from cereal, soybeans, and salt by the action of molds, yeasts, and bacteria. With respect to microorganisms and fermentation principles, soy sauce and bean paste fermentations are also similar.

Bean paste has the consistency of peanut butter, some smooth and some chunky, and its color varies from light-yellow to reddish-brown. It has a distinctive pleasant aroma resembling that of soy sauce, and it is typically salty, although the degree of saltiness may vary; it sometimes even has a sweet taste. Like soy sauce, bean paste is used as a flavoring agent in cooking as well as a table condiment. These products blend well with varieties of foods, including seafoods, meats, and vegetables. They can be used in place of soy sauce and salt.

There are as many types of bean paste as there are different varieties of cheese in the United States. They are made by varying the cereal used, the ratios of beans to cereal, salt content, length of fermentation, and addition of other ingredients, such as hot pepper which is very popular in China and Korea. According to Ebine (1971), Japanese miso is categorized into three major types based on the raw materials used: rice miso made from rice, soybeans, and salt; barley miso from barley, soybeans, and salt; and soybean miso from soybeans and salt. These three types are further classified on the basis of salt content into sweet, medium salty, and salty groups; each of these groups is again divided by color into white, light yellow, and red depending upon the length of fermentation time. About 590,000 metric tons of commercial miso and 150,000 metric tons of homemade miso are produced in Japan annually. Of the total industrial production, approximately 80% is rice miso. The per capita yearly consumption in Japan, which probably is the highest in the world, is estimated at 6.7 kg.

### B. Method of Preparation

Although the manufacturing method may differ from country to country, or from variety to variety, the principle is believed to be the same. This account will be general in nature and follows the procedures for making

Japanese rice miso. The production process, as shown in Fig. 2, generally consists of cooking soybeans, preparing koji, mixing cooked soybeans with rice koji and salt, fermenting and ripening in a tank, blending, pasteurizing, and packing. Laboratory methods of making miso have been investigated in this country by Hesseltine and his co-workers (Shibasaki and Hesseltine, 1962; Hesseltine, 1965). The characteristics on three types of rice miso in relation to the fermentation conditions are presented in Table II (Ebine, 1967).

## 1. Treatment of Soybeans

Whole soybeans are generally used for making miso. Dehulled soybeans or full-fat soybean grits are sometimes used for making white or light-yellow rice miso. A patent by Smith et al. (1961) covers the use of grits. Unlike soy sauce fermentation, of which over 90% produced in Japan is made from defatted soybean products, defatted soybean products are not suitable for making miso of good quality.

In general, the miso industry prefers to use soybeans of a large size (more than 170 gm/1000 seeds) because the ratio of hull to cotyledon is lower in the large beans. The soybean should have high water-absorbing capacity and, when cooked under described conditions, the beans should be homogeneously soft with fine texture and bright color. The Japanese miso makers found the domestic soybeans to be the most suitable, followed by Chinese and then United States soybeans. The major difference between Japanese and United States soybeans is in the amounts of oil and carbohydrates (total sugar); the Japanese varieties have higher carbohydrate and lower oil content than United States varieties. Tests have shown no correlation between capacity to absorb water and protein or oil content, but there is a high correlation between carbohydrate content and water-absorbing capacity. Among the United States varieties tested, Kanrich, Mandarin, and Comet are the most promising ones (Ebine, 1967).

To prepare whole soybeans for fermentation (Fig. 2), they are washed, soaked in water for about 20 hours at 16°C, and drained. The soaked beans are then cooked in water (white miso) or steamed at a temperature of 115°C for about 20 minutes in a closed cooker and slightly mashed. The new batch-type cooker used in soy sauce fermentation, in which the temperature can be raised rapidly to 120°C and lowered quickly immediately after cooking, has also been introduced to the miso industry.

## 2. Preparation of Rice Miso Koji

Polished rice is soaked in water (15°C) overnight or until the moisture content is about 35%. Excess water is drained off, and the soaked rice is steamed at atmospheric pressure for 40 minutes. A continuous cooker

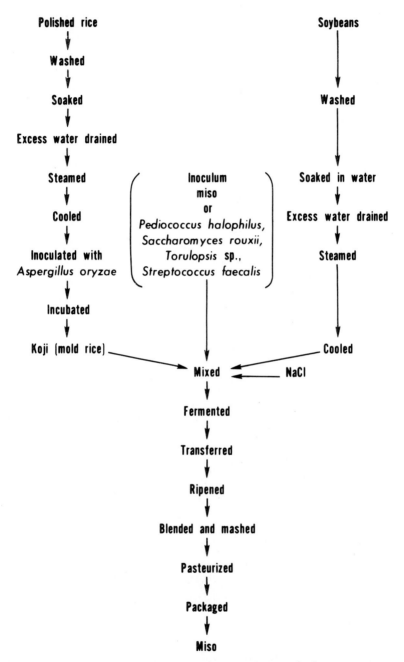

**FIGURE 2.** Flow sheet for the manufacture of miso.

**TABLE II.** Characteristics of Rice Miso in Relation to Fermentation Process[a]

| Properties | White miso | Light-yellow salty miso | Red salty miso |
|---|---|---|---|
| Fermentation process | | | |
| Soybean : rice : salt | 100 : 200 : 35 | 100 : 60 : 45 | 100 : 50 : 48 |
| Fermentation time | 2–4 days | 30 days | 60 days |
| Fermentation temperature | 50°C | 30°–35°C | 30°–35°C |
| Characteristics | | | |
| Color | Bright light-yellow | Light-yellow | Yellow–red |
| Taste | Very sweet | Salty | Salty |
| NaCl (%) | 5 | 12–13 | 12.5–13.5 |
| Moisture (%) | 43 | 48 | 50 |
| Protein (%) | 8 | 10 | 12 |
| Sugar (%) | 20 | 13 | 11 |
| Shelf-life | Short | Fairly long | Long |

[a] From Ebine (1967).

was developed and is now widely used for cooking rice and barley (Ebine, 1976). After cooling to 35°C, tane koji (koji starter), which is a blended mixture of several different strains of *A. oryzae*, prepared as described in Section II, is sprayed over the rice and mixed well. The inoculated rice is then incubated in a temperature- and humidity-controlled room. Koji fermentors of various types (Ebine, 1971) are now used by some miso producers instead of koji rooms; a rotary fermentor is the most popular. The temperature and humidity of the air in the fermentor can be regulated to promote growth and enzyme production of the mold. In about 40–48 hours at 30°–35°C, the rice is completely covered with white mycelium of the inoculated *A. oryzae* strains. Harvesting is done while the koji is white and before any sporulation has occurred. At this time, koji has a pleasant smell, lacks any musty or moldy odors, and is quite sweet in taste. The koji is removed from the fermentor and mixed well with salt to stop any further development of the mold.

### 3. Yeast and Bacterial Fermentation

Cooked and slightly mashed soybeans are mixed with the salted koji and inoculated with a starter containing pure cultures of yeasts and bacteria. Sufficient water is added so that the moisture content of the mixture is about 48%. The mixture, now known as green miso, is thoroughly blended and then tightly packed into a vat or tank for fermentation at 25°–30°C. During the fermentation period, the green miso is transferred from one vat to another at least twice to improve fermentation conditions.

In the past, miso from a previous batch was used as inoculum of bacteria and yeasts; however, the present pure culture starters speed up the fermentation from 6–12 months to 2–3 months and reduce the influence of weed yeasts and bacteria. Strains of *Saccharomyces rouxii, Torulopsis, Pediococcus halophilus,* and *Streptococcus faecalis* are the most important yeasts and bacteria in the miso fermentation.

### 4. Aging and Packaging

At the end of fermentation, the fermented mass is kept at room temperature for about 2 weeks to ripen. The aged product is then blended, mashed, pasteurized, and packaged. Generally, 3300 kg of light-yellow salty miso with a moisture content of 48% is made from 1000 kg of soybeans, 600 kg or rice, and 430 kg of salt.

Traditionally, miso is sold in wooden kegs of various sizes. Presently, miso is also sold in sealed polyethylene bags or tubes. For packaging in plastic bags, miso must first be pasteurized at 60°–70°C for 30 minutes to prevent swelling. Sorbic acid or its potassium salt is also added at a level of less than 1 gm/kg of miso (Watanabe, 1969).

Dehydrated miso powder has become increasingly popular. The dehydration, which is carried out by freeze-drying, does not affect the characteristic flavor of miso. Therefore, the dehydrated product has a great potential as an ingredient for instant mix products. The traditional miso can also be used as an ingredient for dips, salad dressings, and sauces in Western cookery.

### C. Composition

Since different types of miso are made by varying the ratios of the raw materials, the composition of miso varies with the type. Table III presents the standard chemical composition of several types of miso as reported by Ebine (1971). Miso has a fairly narrow moisture range of 44–50%. Protein content ranges from 8 to 19% and fat from 2 to 10%, reflecting the

**TABLE III.**  Composition of Various Types of Miso[a]

| Variety | Moisture (%) | Protein (%) | Reducing sugar (%) | Fat (%) | Sodium chloride (%) |
|---|---|---|---|---|---|
| White miso | 44 | 8 | 33 | 2 | 5 |
| Edo sweet miso | 46 | 10 | 20 | 4 | 6 |
| Salty light-yellow miso | 49 | 11 | 13 | 5 | 12 |
| Salty red miso | 50 | 12 | 14 | 6 | 13 |
| Salty barley miso | 48 | 12 | 11 | 5 | 12 |
| Soybean miso | 47 | 19 | 2 | 10 | 10 |

[a] From Ebine (1971).

increase of soybean used in the fermenting mixture. Except for the white and sweet types, miso contains more than 10% salt. Because of the high salt concentration, miso may be kept for considerable periods without refrigeration; but, for the same reason, the use of miso as a protein food is limited.

The Japanese soybean fermentation industry as well as government research institutes are actively engaged in research on all aspects of soy sauce and miso fermentation. There are literally hundreds of references on these subjects; therefore, no attempt is made to review the literature. Much of the literature being published has appeared in Japanese reports and journals, such as *Fermentation Technology, Agricultural and Biological Chemistry,* and annual reports from government research institutes.

## IV. HAMANATTO

Hamanatto is the Japanese name for a salty product made by fermenting whole soybeans with strains of *A. oryzae*. A similar product, however, is widely produced and consumed in other Oriental countries: tou-shih in China, tao-si in the Philippines, and tao-tjo in the East Indies. In the United States, the fermented beans are often referred to as black beans because of their color. The fermented beans have a salty taste and a pleasant flavor resembling soy sauce. They are used as a side dish to be consumed with bland foods such as rice gruel, or they can be cooked with meats, seafoods, and vegetables as a flavoring agent.

The methods of preparing soybeans for fermentation and the composition of the brine vary from country to country, but the essential features are similar. A typical process for making hamanatto in Japan based on information furnished by Dr. T. Kaneko (Hesseltine and Wang, 1972) of Nagoya University, Japan, is as follows. Soybeans are washed, soaked, steamed until soft, drained, cooled, and then mixed with parched wheat flour (soybeans : wheat flour, 2 : 1). The wheat flour-coated beans are inoculated with a short or medium stalked strain of *A. oryzae*. After incubation for 1–2 days, the molded beans (koji) are packed in a container with a desirable amount of salt, water, and other spices, e.g., 2.5 kg koji, 650 gm salt, and 3.6 liters water. The container is tightly covered and the beans are aged from 6 to 12 months. After that, the beans are dried. The composition of the brine solution may vary with each manufacturer; thus, the final product differs somewhat in taste and appearance. In general, Japanese hamanatto is rather soft, having a high moisture content. Chinese tou-shih has a much lower moisture content than hamanatto and therefore is not so soft. Tao-tjo tends to have a sweet taste because sugar is often added to the brine.

Investigations on the hamanatto fermentation are scanty; thus, methods of producing the product have not been modernized. Recently, Kon and Ito (1975) found that the main microorganisms considered responsible for the hamanatto fermentation are strains of *A. oryzae, Streptococcus,* and *Pediococcus.* The *A. oryzae* strain is dark olive-green and produces strong proteolytic, but not amylolytic, enzymes. Kiuchi *et al.* (1976) found that the gas chromatographic pattern of fatty acid composition of hamanatto was similar to that of soybeans; however, 78% of total lipids in hamanatto is free fatty acids.

## V. SUFU

### A. General Description

Sufu, a traditional Chinese food, is a soft cream cheese-type product made from cubes of soybean curd (tofu) by the action of a mold. In the Western world, sufu has been referred to either as Chinese cheese or as bean cake. Because of the numerous dialects used in China, the product is also known as fu-ju, tou-fu-ju, and others (Wang and Hesseltine, 1970a). Sufu has a salty taste and its own characteristic flavor and aroma that develops during the brining and aging process. It is consumed directly as a relish or is cooked with vegetables or meats. Either way, sufu adds zest to the bland taste of a rice–vegetable diet. Because sufu has the texture of cream cheese, it would be suitable to use in Western countries as a cracker spread or as an ingredient for dips and dressings.

### B. Method of Preparation

The process of making sufu had been considered as a natural phenomenon until 1929, when a microorganism believed to be responsible for sufu fermentation was isolated and described by Wai (1929). Almost 40 years later, Wai (1968) reinvestigated the sufu fermentation and developed a pure culture fermentation. The following information on sufu was obtained mainly from his study.

Three steps are involved in sufu fermentation: preparing soybean curd (tofu), molding, and brining and aging (Fig. 3).

#### 1. Preparing Soybean Curd

Soybeans are washed, soaked, and ground with water; a water-to-dry bean ratio of 10 : 1 is commonly used. After the ground mass is boiled or

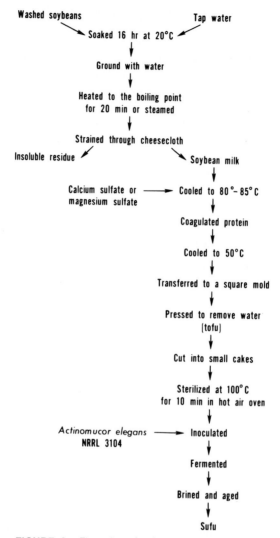

**FIGURE 3.** Flow sheet for the preparation of sufu.

steamed for about 20 minutes to facilitate filtering, to inactivate growth inhibitors, and to reduce the beany flavor, the hot mass is strained through a cloth bag of double-layered cheesecloth or a fine metal screen to separate the soluble fraction (soybean milk) from the insoluble residue. A coagulant, such as calcium sulfate, magnesium sulfate, or acetic acid, is then added to the milk to coagulate the proteins. The curd is transferred into a cloth-lined wooden box and pressed with weight on top to remove whey. A soft, but firm, cakelike curd (tofu) forms. Typically, tofu used for sufu fermentation has a water content of 83%; protein, 10%; and lipids, 4%.

### 2. Molding

The fungus responsible for sufu fermentation initially originated on rice straw. In the traditional method of preparing sufu, rice straw was used to cover the tofu cubes that were placed in a large bamboo tray. After 3–4 days at 10°–15°C, tofu cubes were overgrown by the mold mycelia.

To prevent the growth of contaminating bacteria, Wai (1968) suggested soaking the tofu cubes (2.5 × 3 × 3 cm) in a solution containing 6% NaCl and 2.5% citric acid for 1 hour, followed by hot air treatment at 100°C for 15 minutes. The treated tofu cubes are mounted on sticks, separated from one another, and placed in a tray with pinholes in the bottom and top to aid air circulation, because mycelia must develop on all sides of the cubes. After cooling, the cubes are inoculated over their surface by rubbing with pure culture of an appropriate fungus grown on filter paper impregnated with a culture solution, and then incubated at 20°C or lower for 3–7 days depending on the culture used. At this time, the cubes are covered with a luxurious growth of white mycelium and have no disagreeable odor. The molded cube has a water content of 74%; protein, 12.2%; and lipid, 4.3%.

The mold used in sufu fermentation has to have certain qualities. It must elaborate enzyme systems having high proteolytic and lipolytic activity to act upon the rich protein and lipid substrate. The organism probably utilizes the carbon in lipids as an energy source, because little carbohydrate is readily available in bean curd. The mold must have a white or light yellowish white mycelium to ensure an attractive final product. The texture of mycelial mat should be dense and thick to form a firm film over the surface of the fermented tofu cubes to prevent any distortion in their shape. Of course, it is also important that the mold growth does not develop any disagreeable odor, astringent taste, or mycotoxin. Wai (1968) confirmed that *Actinomucor elegans, Mucor hiemalis, M. silvaticus,* and *M. subtilissimus* possess all these characteristics and can be used to make sufu of good quality. Among these, *Actinomucor elegans* is the best species and the one always used commercially.

### 3. Brining and Aging

Conventionally, the molded tofu cubes are first salted by putting them into a tub in alternate layers with salt for a day. The salted cubes are rinsed with water and packed into glass or earthenware containers; these containers are then filled with a brine solution containing varying amounts of distilled liquor, salt, and flavoring agents, depending upon the type of sufu. After aging for 1–3 months or longer, the product is ready for market. The exact composition of the brine and the length of aging vary from one manufacturer to another, and from one type of sufu to another. According to Wai (1968), a basic and most common brine suitable for Chinese taste contains 12% NaCl and about 10% alcohol. Other additives, such as red rice, fermented rice mash, large amounts of wine, hot pepper, and rose essence are frequently incorporated into the brine either to give color or flavor. Therefore, the characteristically mild taste and aroma of sufu can be easily enhanced or modified by the ingredients of the brine solution.

## C. Changes Occurring during Fermentation

During the brining and aging process, the enzymes elaborated by the mold act upon their respective substrates and yield various hydrolytic products. After 30 days of aging at room temperature, total soluble nitrogen increased from 1.00 to 2.74% and total insoluble nitrogen decreased from 7.89 to 6.05%, whereas total nitrogen changed only slightly. Lipids in the molded cubes were also partially hydrolyzed through the aging period. Free fatty acids increased from 12.8 to 37.1% and total lipids remained unchanged. Wai (1968) also demonstrated that enzyme digestion occurred mostly during the first 10 days of aging.

Since sufu fermentation has a rather simple substrate—55% protein and 30% lipids on a dry basis—it is likely that the hydrolytic products of protein and lipids provide the principal constituents of the mild characteristic flavor of sufu. The added alcohol reacts with the fatty acids chemically or enzymatically to form esters that provide the pleasant odor of the product. Ethanol also prevents growth of some microorganisms.

The added salt imparts a salty taste to the product, as well as retarding mold growth and growth of contaminating microorganisms. Most important, the salt solution releases the mycelia-bound proteases. In sufu fermentation, the mold growth is limited to the surface of the cubes, and the mycelium does not penetrate into the tofu cubes. The enzymes produced by the mold, on the other hand, are not extracellular but are loosely bound to the mycelium, possibly by ionic linkage (Wang, 1967b), and can be easily eluted by NaCl or other ionic salt solutions. Therefore, the NaCl brine solution also serves to elute the enzyme from mycelia which, in turn, penetrate into the molded cubes and act upon the substrate protein. The minimal salt concentration to efficiently release the mycelia-bound proteases is about 3%. Thus, a less salty product than the traditional ones, which are too salty to suit Westerners' taste, may be obtained by reducing the salt content of brine solution from 12% to not less than 3%.

## VI. TEMPEH

### A. General Description

Tempeh, or tempe kedelee, originating in Indonesia, is made by fermenting dehulled and briefly cooked soybeans with a mold, *Rhizopus;* the mycelia bind the soybean cotyledons together in a firm cake. The raw tempeh has a clean, fresh, and yeasty odor. When sliced and deep-fat fried, it has a nutty flavor, pleasant aroma, and texture that are familiar and highly acceptable to almost all the people around the world. Tempeh does not possess the beany flavor some people find unpleasant in soybeans, and it is easy to cook and digest. Unlike most of the other fermented soybean foods, which are usually used as flavoring agents or relishes, tempeh is used as a main dish in Indonesia.

### B. Method of Preparation

#### 1. Traditional Method

Tempeh fermentation is characterized by its simplicity and rapidity. Making tempeh in Indonesia is a household art. The procedure may vary slightly from one household to another, but the principal steps are similar. Soybeans are soaked in tap water overnight until the hulls can be easily removed by rubbing between the hands. Some prefer to boil the soybeans for a few minutes to loosen the hulls and then to soak the beans overnight. After dehulling, the beans are boiled with excess water, drained, and surface-dried. Small pieces of tempeh from a previous fermentation, wrappings of the previous tempeh, or ragi tempeh (commercial starter) are mixed with the soybeans, which are then wrapped in wilted banana or other large leaves and incubated in a warm place for 24–48 hours until the beans are covered with white mycelium and bound together as a firm cake. The cake is either sliced into thin strips, dipped in a salt solution, and deep-fat fried in coconut oil or cut into pieces and used in soups much as we use chunks of meat.

#### 2. Pure Culture Method

In the late 1950s scientists at the New York Agricultural Experiment Station, Geneva, New York, and at the Northern Regional Research Center, Peoria, Illinois, began to study this centuries-old fermentation. As a result, a pure-culture fermentation method for laboratory or industrial scale was developed.

Earlier reports (Stahel, 1946; Van Veen and Schaefer, 1950; Steinkraus *et al.,* 1960) indicated that the mold responsible for tempeh fermentation was *Rhizopus oryzae.* We have received cultures isolated from different lots of tempeh in Indonesia, and we found that only *Rhizopus* could make tempeh in pure-culture fermentation. Of the 40 strains of *Rhizopus* received, 25 of them are *R. oligosporus* Saito; others are *R. stolonifer* (Ehren) Vuill, *R. arrhizus* Fischer, *R. oryzae* Went and Geerligs, *R. formosaensis* Nakazawa, and *R. achlamydosporus* Takeda. Apparently, *R. oligosporus* is the principal species used in Indonesia for tempeh fermentation; a strain identified as *R. oligosporus* Saito NRRL 2710 is one of the better producers of a good product (Hesseltine *et al.*, 1963b). This strain is characterized by sporangiospores showing no striations and very irregular in shape under any condition of growth. The sporangiophores are short and unbranched and arise opposite rhizoids that are very reduced in length and branching. All isolates show large members of chlamydospores (Hesseltine, 1965).

In making tempeh, Hesseltine *et al.* (1963b) used a spore suspension of *R. oligosporus* NRRL 2710 grown on potato–dextrose–agar slant at 28°C for 5–7 days. Steinkraus and his co-workers (1965) used a powdered lyophilized inoculum made by growing pure culture on sterilized, hydrated soybeans; Rusmin and Ko (1974) made an inoculum by growing pure culture on steamed rice, but the inoculum also contained a considerable amount of bacterial spores ($10^4$–$10^5$ per gm). Wang *et al.* (1975a) developed a tempeh inoculum having a high viable spore count that would maintain its viability for a long time with minimal attention. The spores of *R. oligosporus* Saito NRRL 2710 were made by fermenting rice at 40% moisture level for 4–5 days at 32°C. The fermenting mass was made into a slurry by blending with sterilized water and then was freeze-dried. On a dry basis, the viable spore count per gram of preparation was about $1 \times 10^9$ before freeze-drying and $1 \times 10^8$ after freeze-drying. When the freeze-dried preparation was kept in a closed plastic bag at 4°C for up to 6 months, the spore counts showed typical experimental variations and were comparable to their original counts. At room temperature, a significant decrease in viability was noted after 2 months (from $2 \times 10^7$ to $1 \times 10^6$); thereafter, no further decrease was observed. The bacterial count of the preparation was minimal; therefore, bacterial contamination was not found to be a problem either during the process of fermentation or in storage. Since the beginning of 1977, the Northern Regional Research Center has distributed nearly 30,000 samples of this inoculum and the starter is now available commercially.

The preparation of soybeans for pure-culture fermentation is similar to that for the traditional method: soaking, dehulling, cooking, and surface-drying. The soybeans are soaked in water overnight at room temperature, dehulled by hand or by a simple roller mill, and then boiled in water for 30 minutes (Hesseltine *et al.*, 1963b; Rusmin and Ko, 1974). In 1964, Martinelli and Hesseltine introduced full-fat soybean grits (soybean cotyledons that have been mechanically cracked into four to five pieces) for tempeh fermentation (Fig. 4). Since soybean grits absorb water easily, the soaking time can be reduced from more than 20 hours to 30 minutes. Furthermore, since the hulls are removed mechanically in producing grits, much labor can be saved. Steinkraus and his co-workers (1960) suggested soaking the beans in 0.85% lactic or acetic acid solution. The soaked beans are dehulled with an abrasive vegetable peeler and then boiled in the acid solution for 90 min. With this treatment, the cooked beans have an acidity around 4–5 which, according to Steinkraus et al. (1960), is necessary to prevent the development of the spoilage bacteria during the mold fermentation. However, other investigators have not encountered bacterial growth in their processes. Because R. oligosporus produces an antibacterial agent (Wang et al., 1969), and because this organism also has the unique characteristic of growing fast, there is probably little chance for bacteria to gain ground before the tempeh fermentation is complete.

The cooked beans are drained, spread to cool and to surface-dry, and thoroughly mixed with an inoculum of pure culture. A recommendation was made to use $10^6$ spores per 100 gm of cooked beans (Wang et al., 1975a). The inoculated beans are tightly packed into an appropriate container for incubation to obtain a final product in which a white mycelium has developed abundantly but black spores are minimal.

Like many other molds R. oligosporus requires air to grow, but it does

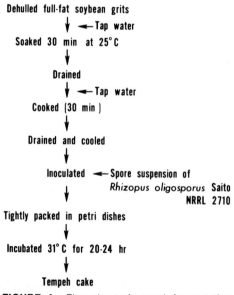

**FIGURE 4.**   Flow sheet of tempeh fermentation.

not require as much aeration as many other molds; in fact, too much aeration will cause spore formation and may also dry up the beans, resulting in poor growth. Therefore, it is important to properly pack the beans for fermentation.

Hesseltine *et al.* (1963b) have carried out pure-culture fermentation in petri dishes, a testing procedure which proved to be very satisfactory. Shallow aluminum foil or metal trays, with perforated or woven mesh bottoms and covers or perforated plastic film covers, or perforated plastic bags and tubings have also been successfully used in tempeh fermentation (Steinkraus *et al.*, 1960, 1965; Martinelli and Hesseltine, 1964). Thickness of the package and the size and the distance of the perforations are important. Containers of 2.5–3.0 cm in depth with holes having a diameter of about 0.6 mm and a distance apart of about 5 mm are most desirable.

Tempeh fermentation can be carried out at temperatures ranging from 25° to 37°C; however, the time required for fermentation decreases as temperature increases, 80 hours at 25°C and 15–18 hours at 35°–38°C (Martinelli and Hesseltine, 1964; Steinkraus *et al.*, 1965). At higher temperatures the soybeans tend to dry out; consequently, the mold growth is suppressed and there is a greater chance for bacteria to grow. We found (Wang *et al.*, 1975a) that at the inoculum level of $1 \times 10^6$ spores per 100 gm of cooked substrate tempeh fermentation can be best carried out at 32°C for 20–22 hours.

When fermentation is complete, the beans are covered with and bound together by white mycelia. Thus, raw tempeh looks like a firm white cake and has an attractive and slightly yeasty odor. Prolonged fermentation, on the other hand, often causes the product to become obnoxious due to the enzymatic breakdown of proteins. Although shelf-life of the product is also short, it can be prolonged by various methods, such as freezing, drying, and canning.

### C. Tempeh-like Products

In Indonesia, copra (pressed coconut cake) is sometimes used in tempeh fermentation; the product is then known as tempeh bongkrek. Tempeh can also be made (Gandjar and Slamet, 1972; Wang *et al.*, 1975a) from the water-insoluble fraction of soybeans which is the residue of making soybean milk and tofu, the two main food products derived from water extraction of soybeans (Wang, 1967a). However, the moisture content of this water-insoluble fraction must be reduced to less than 80% so that the texture appears crumbly before it is suitable for fermentation. We have also developed (Hesseltine *et al.*, 1967) new tempeh-like products by fermenting cereal grains such as wheat, oats, barley, rice, or mixtures of cereals and soybeans with *Rhizopus*. However, it is essential to slightly modify the surface of the grain or to crack the grain for good growth of the mold. Fermentation of starch tubes such as cassava with *Rhizopus* can also result in a food product with significant increases in protein content (Stanton and Wallbridge, 1969).

Another traditional Indonesian food closely related to tempeh is known as ontjom or oncom. Ontjom is made from peanut press cake by strains of *Neurospora* or *Rhizopus; Neurospora* fermentation results in a pink or orange cake, and *Rhizopus* in a cake of ash-grey color (Van Veen and Graham, 1968). The preparation and the use of ontjom are quite similar to that of tempeh.

## D. Biochemistry and Physiology of *Rhizopus oligosporus*

The utilization of various carbon and nitrogen compounds by *R. oligosporus* Saito NRRL 2710 was investigated by Sorenson and Hesseltine (1966). They found that the principal carbohydrates of soybeans, i.e., stachyose, raffinose, and sucrose, are not utilized as sole sources of carbon, whereas common sugars such as glucose, fructose, galactose, and maltose supported excellent growth, as does xylose. Various vegetable oils can be substituted for sugars as sources of carbon with excellent growth. Since the soybean sugars are not utilized by *R. oligosporus,* and since strong lipase activity has been reported for *Rhizopus* cultures used in tempeh fermentation (Wagenknecht *et al.*, 1961; Wang and Hesseltine, 1966), it is likely that lipid materials, and particularly fatty acids, are the primary sources of energy for the tempeh fermentation. Ammonium salts and amino acids such as proline, glycine, aspartic acid, and leucine are excellent sources of nitrogen. Other amino acids are less suitable, and tryptophan supports no growth at all. However, the fungus does not depend upon the presence of any specific amino acid in the medium for growth.

Hesseltine *et al.* (1963a) found a water-soluble, heat-stable mold inhibitor in soybeans which inhibits the growth of *R. oligosporus*. Later, Wang and Hesseltine (1965) found that the water-soluble and heat-stable fraction of soybeans also inhibited the formation of proteolytic enzymes by *R. oligosporus*. Therefore, soaking and cooking of soybeans in excess water, which is later discarded, are essential in making tempeh.

*Rhizopus oligosporus* is highly proteolytic, which is important in tempeh fermentation because of the high protein content of the substrate. The ability of *Rhizopus* to produce proteolytic enzymes varies greatly between different strains of the same species as well as between species (Wang and Hesseltine, 1965). The proteolytic enzyme systems have optimal pH at 3.0 and 5.5, with the pH 3.0 type predominating in submerged cultivation and pH 5.5 type predominating in tempeh fermentation. Both enzyme systems have maximum activities at 50°–55°C and are fairly stable at pH 3.0–6.0 but are rapidly denatured at pH below 2 or above 7. The proteolytic enzyme having optimum pH at 3.0 has been purified and separated into five active fractions; crystalline enzymes were obtained from the two major fractions (Wang and Hesseltine, 1970b).

In addition to high protease activity, the mold possesses strong lipase activity (Wagenknecht et al., 1961; Wang and Hesseltine, 1966) but low amylase activity and no detectable pectinase (Hesseltine et al., 1963b). Since starch is seldom found in mature soybeans, it is not particularly important that this species produces amylase during tempeh fermentation. Lipase is produced by the mold to hydrolyze soybean lipids.

Although antibiotics have frequently been isolated from fungal cultures, only a few have been reported from cultures of the class Phycomycetes. We found that strains of Rhizopus and Actinomucor do produce antibacterial agents (Wang et al., 1972a; Ellis et al., 1974). The compound produced by R. oligosporus (Wang et al., 1969) is especially active against some gram-positive bacteria, including both micro-aerophilic and anaerobic bacteria, e.g., Streptococcus cremoris, Bacillus subtilis, Staphylococcus aureus, Clostridium perfringens, and C. sporogenes. The compound contains polypeptides having high carbohydrate contents. Its activity is not affected by pepsin or R. oligosporus proteases, is slightly decreased by trypsin and peptidase, but is rapidly inactivated by pronase. Calloway et al. (1971) found that tempeh did not increase gas production over baseline values of healthy young men and caused a significant delay in the time of gas forming, which suggests temporary suppression of intestinal bacteria. The delay could well be due to the presence of the antibiotic substance produced by the mold, Rhizopus.

### E. Changes Occurring during Fermentation

The effects of R. oligosporus on soybeans have been studied by several investigators and reviewed by Hesseltine and Wang (1972) and Iljas et al. (1973). Steinkraus et al. (1960) found that the temperature of fermenting beans rises to above that of the incubators as fermentation progresses, but that it falls as the growth of mold subsides. The pH increases steadily, presumably because of the protein breakdown. After 69 hours of incubation, soluble solids rise from 13 to 28%; soluble nitrogen also increases from 0.5 to 2.0%, whereas total nitrogen remains fairly constant and reducing substances slightly decrease, probably due to utilization by the mold. Similar changes were observed when wheat was fermented by *R. oligosporus* (Wang and Hesseltine, 1966). A decrease in ether-extractable substances of soybeans after fermentation was reported by Murata *et al.* (1967) and Wang *et al.* (1968), indicating that the mold uses the soybean oil as its energy source (Sorensen and Hesseltine, 1966). Wagenknecht *et al.* (1961) reported that one-third of the total ether-extractable soybean lipid is hydrolyzed by the mold after 69 hours of incubation, and among all the fatty acids, 40% of the linolenic acid is utilized by the mold. The phytic acid content of soybeans is reduced by about one-third as a result of this fermentation (Sudarmadji and Markakis, 1977).

Although total nitrogen remains fairly constant during fermentation, the free amino acid in tempeh increases. The amino acid composition of soybeans, on the other hand, is not significantly changed by fermentation (Smith *et al.*, 1964; Stillings and Hackler, 1965). Perhaps the amount of mycelial protein present in tempeh is not high enough to alter greatly the amino acid composition of the soybeans, nor does the mold depend upon any specific amino acid for growth as suggested by Sorenson and Hesseltine (1966).

Niacin, riboflavin, pantothenic acid, and vitamin $B_6$ contents of soybeans increase after fermentation, whereas thiamin has no significant change (Roelofsen and Talens, 1964; Murata *et al.*, 1967). Wang and Hesseltine (1966) also noticed in fermenting wheat with *R. oligosporus* that the amount of niacin and riboflavin of the wheat tempeh greatly exceeds that of unfermented wheat, while thiamin appears to be less. Apparently, *R. oligosporus* has a great synthetic capacity for niacin, riboflavin, pantothenic acid, and vitamin $B_6$, but not for thiamin.

Lipids in tempeh were found to be more resistant to autoxidation than those in control soybeans. The peroxide value of tempeh was 1.1, whereas that of control soybeans was 18.3–201.9 (Iljas, 1969). Ikehata *et al.* (1968) found that the peroxide value of lyophilized tempeh stored at 37°C for 5 months increased from 6 to 12 compared with the increase of from 6 to 426 in unfermented soybeans. The antioxidant activity of tempeh was further substantiated by Packett *et al.* (1971), who reported that corn oil containing 50% tempeh showed higher antioxidant potential than oils containing 25% tempeh, 0.01% $\alpha$-tocopherol, or 0.03% $\alpha$-tocopherol. In 1964, György *et al.* isolated a new isoflavone from tempeh designated as "Factor 2," which was then identified as 6,7,4'-trihydroxyisoflavone. 6,7,4'-Trihydroxyisoflavone was later chemically synthesized and proved to be a potent antioxidant for vitamin A and for linoleate in aqueous solution at pH 7.4 (Ikehata *et al.*, 1968). However, when the isoflavone was mixed with soybean powder or soybean oil, it did not prevent their autoxidation. These authors speculated that the insolubility of the isoflavone in the oil and the difficulty of dispersing into soybean powder may be some of the reasons for its failure to prevent autoxidation.

Although *R. oligosporus* does not synthesize substances that inhibit the activity of trypsin, the trypsin-inhibitory activity of the soybeans was found to increase as the tempeh fermentation progressed (Wang *et al.*, 1972b). The compounds responsible for this increased trypsin-inhibitory activity have been identified as free fatty acids, primarily oleic, linoleic, and linolenic acids (Wang *et al.*, 1975b). The free fatty acids are liberated from oil in the soybeans by fungal lipase, and they differ from other reported soybean trypsin inhibitors that are protein in nature. Free fatty acids have been previously reported to inhibit various enzymes, such as glycolytic, glyconeogenic, lipogenic, and also proteolytic enzymes. Their effect appears to be a nonspecific type of inhibition.

In the production of tempeh, soybeans are only partially cooked and they remain nearly as firm as the soaked beans. After fermentation the beans are soft and similar in texture to completely cooked soybeans. An earlier cytological study (Steinkraus *et al.*, 1960) showed only slight penetration of the mycelia into the underlying tissue of the bean, suggesting that the digestion was mainly enzymatic. However, a recent study (Jurus and Sundberg, 1976) revealed hyphae infiltration to a depth of 742 $\mu$m or about 25% of the average width of a soybean cotyledon. These authors speculated that the extreme depth of mycelial infiltration partially explains the rapid physical and chemical changes occurring during tempeh fermentation. The hyphae may mechanically push the bean cells apart prior to, or in conjunction with, enzymatic digestion; thus, the beans become soft. Likewise, the penetration of enzymatic activity could also be enhanced, since the distance over which diffusion of enzyme must occur is greatly reduced.

### F. Nutritional Value

Indonesians consider tempeh to be a nourishing and easily digestible food, and it is used as a main dish. Van Veen and Schaefer (1950) observed beneficial effects of tempeh on patients with dysentery in the prison camps of World War II, and they suggested that tempeh was much easier to digest than soybeans. However, animal feeding experiments have not substantiated this conclusion (Hackler *et al.*, 1964; Smith *et al.*, 1964; Murata *et al.*, 1967), even though more than half of the soybean protein, fat, and N-free extract could be solubilized by 72-hour fermentation (Van Buren *et al.*, 1972). The increase in vitamin content, as previously mentioned, must have a great nutritional significance to the people of those countries whose food industries do not enrich their foods with vitamins as do those in the United States.

The protein quality of tempeh as measured with the protein efficiency ratio (PER) is not significantly different from that of unfermented beans (Hackler et al., 1964; Smith et al., 1964; Wang et al., 1968; Murata et al., 1971). However, the quality of tempeh protein can be improved by making tempeh from mixtures of cereals and soybeans. For example, the PER value of wheat–soybean (1 : 1) tempeh was comparable to that of casein (Wang et al., 1968). Not only does a mixture of wheat and soybean (1 : 1) have a balanced pattern of amino acid, but the authors also found that the PER of wheat was greatly increased by Rhizopus fermentation.

The superior nutritive value of tempeh over unfermented soybeans has been noted by György (1961) on animals fed low-protein diets. His results resemble those obtained with animals fed antibiotics added to their protein source. Therefore, the finding of antibacterial agents produced by R. oligosporus may offer a better understanding of the value of tempeh in the diet of Indonesians, and, perhaps, of fermented foods in the diets of all Orientals.

The cooking procedures also affect the nutritional value of tempeh, and the PER value of tempeh significantly declined after more than 3 minutes of frying in oil; on the other hand, steaming for up to 2 hours had no effect (Hackler et al., 1964).

Because of the high acceptability of taste and texture, lack of beany flavor, nutritional advantages, and its simple, low-cost processing techniques, tempeh appears to be a good candidate for any country searching for a low-cost, high-protein food. With the recently increasing interest of vegetarians in foods of vegetable origin, tempeh consumption has been on an upsurge in this country. In addition to tempeh making as a home project, several commercial tempeh producers have been established. Tempeh may soon be a regular item in the United States market.

## VII. ANG-KAK

### A. General Description

Unlike other mold-modified products, which are usually used as a flavoring agent or as a protein source, ang-kak or red rice is a color agent. It was originated in China and is a product made by fermenting rice with strains of Monascus purpureus. Because of its red color, ang-kak is used for coloring various foods and as a color additive for manufacturing fermented foods such as sufu, red wine, fish sauce, and fish paste in the Orient.

## B. Cultures

Only those strains (Church, 1920) that produce a dark-red growth throughout the rice kernels, at low enough moisture levels to allow the individual grains to remain separate from one another, are suitable for the fermentation. *Monascus purpureus* NRRL 2897 maintained in the AR Culture Collection was isolated from an ang-kak sample bought in the Philippines market and has demonstrated the ability to carry out the fermentation successfully. Palo *et al.* (1960) studied various conditions of the fermentation and found that the optimum temperature for pigment formation is about 27°C. Growth will occur at as low as 20°C and as high as 37°C, but at these extremes poor pigmentation results. The mold will produce the pigment over a wide range of pH values (3–7.5).

Lin (1973) isolated a strain of *Monascus* sp. F-2 from kaoliang koji and found that it produced large amounts of pigment by submerged culture with rice as a sole carbon source. *N*-Methyl-*N'*-nitro-*N*-nitrosoguanidine treatment and successive isolation (Lin and Suen, 1973) greatly improved the yield of the pigment production by *Monascus* sp. F-2. Two hyperpigment-producing strains, R-1 and R-2, were isolated which produced five times more red pigment than the original parent strain F-2 did.

## C. Method of Preparation

All varieties of rice are suitable except the glutinous ones, which are unsatisfactory because the rice becomes gluey and the grains stick together. The procedure for preparing ang-kak on a laboratory scale as developed by Palo *et al.* (1960) is to first wash the rice, soak it in water for 24 hours, and then drain it thoroughly. The rice is placed in a beaker or other suitable container that is large enough to have considerable air space above the rice. It is then covered with a milk filter disk, autoclaved for 30 minutes at 121°C, cooled, and inoculated with a sterile water suspension of ascospores removed from 25-day-old cultures of *M. purpureus* grown on Sabouraud's agar. After the suspension is thoroughly mixed with the rice, it should be incubated at 25°–32°C. At the time of inoculation the rice will seem rather dry; in fact, the rice should never be wet or mushy. This is the most important state for the good production of ang-kak. In 3 days, the rice should begin to redden. At this time, the material should be stirred and shaken to redistribute the rice in the center and bottom of the fermentor and to retain an even moisture at the surface. We have found that an occasional vigorous shaking of the fermentation jar is quite adequate. Addition of some sterile water may be required from time to time to replace lost moisture. Care must be taken only to moisten and not to soak the rice. In 3 weeks, the rice should be deep purplish red, with each rice kernel unattached to its neighbors. The material is then dried in an oven at 40°C. When the kernels are broken, the pigment should be completely through the rice. Each grain should crumble easily between one's fingers.

## D. Pigments

The ang-kak studied by Nishikawa (1932) consisted of two pigments, principally monascorubrin, a red coloring material with a formula of $C_{22}H_{24}O_5$, and a small amount of monascoflavin, a yellow pigment ($C_{17}H_{22}O_4$). The pigments tend to accumulate in the microorganisms because of their poor solubility. Recently Yamaguchi et al. (1973) reported that the pigment can be solubilized by reacting it with a water-soluble protein, a water-soluble peptide, an amino acid, or a mixture of these. The reaction is carried out at a pH from 5 to 8.5 and requires from several minutes to several hours. As the reaction progresses, the color of the pigment changes from a brownish red to a deep scarlet and the pigment becomes water soluble. The solubility of the pigment in water varies with the type of protein, peptide, or amino acid used. The same authors also indicated that the water-soluble pigment can be produced directly by the mold grown in a medium containing soluble protein, peptide, or amino acid, at a pH of 7–9 under aerobic conditions, and at a temperature of about 27°C for 48–72 hours. The mechanism of this reaction is not clear; Yamaguchi et al. (1973) speculated that the water-soluble pigment is a complex resulting from binding between the pigment and the chosen amino group. The water-soluble pigment is suitable for coloring foodstuffs, particularly sour-type drinks, candy, milk products, and meat products.

## VIII. ABSENCE OF MYCOTOXINS IN FERMENTED FOODS

The mycotoxin in which most interest has developed is the carcinogen, aflatoxin, produced by *Aspergillus flavus* and *A. parasiticus*. These species are in the same group as *A. oryzae*, a species widely used in Oriental food fermentations. Therefore, the Japanese scientists have looked extensively at the aflatoxin problem in fermented products.

Yokotsuka et al. (1967a,b) investigated 73 industrial strains of *Asper-*

*gillus,* including *A. oryzae, A. sojae, A. niger,* and *A. usami,* for production of fluorescent compounds after cultivation on a Czapek–Dox medium containing zinc. No aflatoxin was found, but about 30% of these strains produced compounds that were very similar to those of aflatoxin in fluorescence spectra and $R_f$ values in thin-layer chromatograms. However, the fluorescent compounds proved to be nontoxic (Manabe *et al.,* 1968, 1972; Manabe and Matsuura, 1972a,b). Some strains produced aspergillic acid in modified Mayer's medium at 30°C by surface culture after more than 10 days, and on solid soy sauce koji fermentation after 4–10 days (Yokotsuka, 1971). However, the koji is usually harvested within 2–3 days in manufacturing soy sauce and miso. The same author also reported that a small amount of $\beta$-nitropropionic acid was produced by one of the tested strains in the beginning of the culturing, but it rapidly disappeared. Kojic, oxalic, or formic acids were formed by some molds during koji making, but only in trace amounts. Considering the low toxicity of these compounds, they create no problems. Similar results were also obtained by other investigators (Murakami *et al.,* 1967, 1968a,b; Matsuura *et al.,* 1970).

Earlier, our laboratory (Hesseltine *et al.,* 1966) examined samples of soy sauce, miso, and tou-shih and of soybean, rice, and wheat tempehs, but we found no aflatoxin. Rice koji, miso, and soy sauce samples collected from various producers in Japan were examined and found negative for aflatoxin (Matsuura, 1970).

Maing *et al.* (1973) prepared soy sauce with a strain of *A. oryzae* NRRL 1988 and with an aflatoxin-producing strain of *A. parasiticus* NRRL 2999. No aflatoxin was produced in the soy sauce prepared with *A. oryzae,* but large amounts of aflatoxin were formed by *A. parasiticus* when it was grown with *A. oryzae.* After brining for 6 weeks, large amounts of aflatoxin remained in the fermentation in which *A. parasiticus* was grown. Obviously, if an aflatoxin-producing strain of *A. flavus-* or *A. parasiticus*-contaminated culture was used to make koji for soy sauce, the soy sauce would contain toxin.

### IX. CONCLUSIONS

Among many mold-modified foods consumed around the world, we have been concerned only with those that have been scientifically studied. Information on the nature of those products, the methods of manufacture, the correctly identified cultures, and the properties of the microorganisms that make them unique in their particular fermentation have been discussed. Although future studies are needed to develop uniform high-quality products and well-defined economical processes to manufacture them, pure culture methods for preparing those foods have been developed which would facilitate further studies.

Obviously, the preparation of foods by fermentation has certain advantages that can be summarized as follows: Fermentation (1) produces desirable enzymes, (2) destroys or masks undesirable flavor and odors, (3) adds flavor and odor, (4) preserves, (5) synthesizes desirable constituents such as vitamins and antibiotics, (6) increases digestibility, (7) changes physical states, (8) produces color, and (9) reduces cooking time.

If properly fermented, these foods are not hazardous to health. The microorganisms responsible for these processes are not toxin producers.

## REFERENCES

Baens-Arcega, L. (1966). Philippines Patent 2,553.
Bush, L. (1959). "Land of the Dragonfly," pp. 28–51. Robert Hale, Ltd., London.
Calloway, D. H., Hickey, C. A., and Murphy, E. L. (1971). *J. Food Sci.* **36,** 251–255.
Church, M. B. (1920). *J. Ind. Eng. Chem.* **12,** 45–46.
Church, M. B. (1923). *U.S., Dep. Agric., Bull.* **1152,** 1–26.
Ebine, H. (1967). U.S. Dep. Agric. Final Tech. Rep., Public Law 480. Project UR-A11-(40)-2. Natl. Agric. Libr., Beltsville, Maryland.
Ebine, H. (1971). *In* "Conversion and Manufacture of Foodstuffs by Microorganisms—A Symposium," pp. 127–132. Saikon Publishing Co., Ltd., Kyoto, Japan.
Ebine, H. (1976). *In* "Expanding the Use of Soybeans for Asia and Oceania—A Conference," (R. M. Goodman, ed.), INTSOY Ser. No. 10, pp. 126–129. University of Illinois, Urbana.
Ellis, J. J., Wang, H. L., and Hesseltine, C. W. (1974). *Mycologia* **66,** 593–599.
Gandjar, I., and Slamet, D. S. (1972). *Penelitian Gizi dan Makanan Jilia* **2,** 70–79.
György, P. (1961). *N. A. S.–N. R. C., Publ.* **843,** 281–289.
György, P., Murata, K., and Ikehata, H. (1964). *Nature (London)* **203,** 870–872.
Hackler, L. R., Steinkraus, K. H., Van Buren, J. P., and Hand, D. B. (1964). *J. Nutr.* **82,** 452–456.
Harada, Y. (1951). *Rep. Tatsuno Inst. Soy Sauce* **2,** 51–55.
Hesseltine, C. W. (1965). *Mycologia* **57,** 149–197.
Hesseltine, C. W., and Wang, H. L. (1972). *In* "Soybeans: Chemistry and Technology" (A. K. Smith and S. J. Circle, eds.), Vol. 1, pp. 389–419. Avi Publ. Co., Westport, Connecticut.
Hesseltine, C. W., Decamargo, R., and Rackis, J. J. (1963a). *Nature (London)* **200,** 1226–1227.
Hesseltine, C. W., Smith, M., Bradle, B., and Djien, K. S. (1963b). *Dev. Ind. Microbiol.* **4,** 275–287.
Hesseltine, C. W., Shotwell, O. L., Ellis, J. J., and Stubblefield, R. D. (1966). *Bacteriol. Rev.* **30,** 795–805.
Hesseltine, C. W., Smith, M., and Wang, H. L. (1967). *Dev. Ind. Microbiol.* **8,** 179–186.
Hesseltine, C. W., Swain, E. W., and Wang, H. L. (1976). *Dev. Ind. Microbiol.* **17,** 101–115.
Ikehata, H., Wakaizumi, M., and Murata, K. (1968). *Agric. Biol. Chem.* **32,** 740–746.
Iljas, N. (1969). M.S. thesis, Ohio State University, Columbus.
Iljas, N., Peng, A. C., and Gould, W. A. (1973). *Ohio Agric. Res. Dev. Cen., Hortic. Ser.* **394.**

Jurus, A. M., and Sundberg, W. J. (1976). *Appl. Environ. Microbiol.* **32,** 284–287.
Kiuchi, K., Ohta, T., Itoh, H., Takabayashi, T., and Ebine, H. (1976). *J. Agric. Food Chem.* **24,** 404–407.
Komatsu, Y. (1968). *Seas. Sci.* **15,** No. 2, 10–20 (in Japanese).
Kon, M., and Ito, H. (1975). *Rep. Natl. Food Res. Inst. (Tokyo)* **30,** 232–237.
Lin, C. F. (1973). *J. Ferment. Technol.* **51,** 407–414.
Lin, C. F., and Suen, S. J.-T. (1973). *J. Ferment. Technol.* **51,** 757–759.
Maing, I. V., Ayers, A. C., and Koehler, P. F. (1973). *Appl. Microbiol.* **25,** 1015–1017.
Manabe, M., and Matsuura, J. (1972a). *J. Food Sci. Technol. (Tokyo)* **19,** 268–274.
Manabe, M., and Matsuura, J. (1972b). *J. Food Sci. Technol. (Tokyo)* **19,** 275–279.
Manabe, M., Matsuura, S., and Nakano, M. (1968). *J. Food Sci, Technol. (Tokyō)* **15,** 341–346.
Manabe, M., Ohnuma, S., and Matsuura, J. (1972). *J. Food Sci. Technol. (Tokyo)* **19,** 76–80.
Martinelli, A. F., and Hesseltine, C. W. (1964). *Food Technol.* **18,** 167–171.
Matsuura, S. (1970). *JARQ* **5,** 46–51.
Matsuura, S., Manabe, M., and Sato, T. (1970). *Proc. U.S.-Jpn. Conf. Toxic Micro-Org.: 1968,* pp. 48–55.
Maxwell, M. E. (1952). *Aust. J. Sci. Res., Ser. B* **5,** 43–55.
Murakami, H., Takase, S., and Ishii, I. (1967). *J. Gen. Appl. Microbiol.* **13,** 323–334.
Murakami, H., Takase, S., and Kuwabara, K. (1968a). *J. Gen. Appl. Microbiol.* **14,** 97–110.
Murakami, H., Sagawa, H., and Takase, S. (1968b). *J. Gen. Appl. Microbiol.* **14,** 251–262.
Murata, K., Ikehata, H., and Miyamoto, T. (1967). *J. Food Sci.* **32,** 580–588.
Murata, K., Ikehata, H., Edani, Y., and Yojanagi, K. (1971). *Agric. Biol. Chem.* **35,** 233–241.
Nakadai, T., Nasuno, S., and Iguchi, N. (1975). U.S. Patent 3,914,436.
Nishikawa, H. (1932). *J. Agric. Chem. Soc. Jpn.* **8,** 1007–1015.
Noda Soy Sauce Company, Ltd. (1955). "The Treatment of Soy Sauce Fermentation." Noda-shi, Chiba-ken, Japan.
Oda, M., Ikeda, K., and Tanimoto, M. (1949). *J. Ferment. Technol.* **27,** 16–23.
Packett, L. V., Chen, L. H., and Liu, J. T. (1971). *J. Food Sci.* **36,** 798–799.
Palo, M. A., Vidal-Adeva, L., and Maceda, L. M. (1960). *Philipp. J. Sci.* **89,** 1–22.
Roelofsen, P. A., and Talens, A. (1964). *J. Food Sci.* **29,** 224–226.
Rusmin, S., and Ko, S. D. (1974). *Appl. Microbiol.* **28,** 347–350.
Shibasaki, K., and Hesseltine, C. W. (1962). *Econ. Bot.* **16,** 180–195.
Smith, A. K., Hesseltine, C. W., and Shibasaki, K. (1961). U.S. Patent 2,967,108.
Smith, A. K., Rackis, J. J., Hesseltine, C. W., Smith, M., Robbins, D. J., and Booth, A. N. (1964). *Cereal Chem.* **41,** 173–181.
Sorenson, W. G., and Hesseltine, C. W. (1966). *Mycologia,* **58,** 681–689.
Stahel, G. (1946). *J. N. Y. Bot. Gard.* **47,** 261–267.
Stanton, W. R., and Wallbridge, A. (1969). *Process Biochem.* **4,** No. 4, 45–51.
Steinkraus, K. H., Hwa, Y. B., Van Buren, J. P., Provvidenti, M. I., and Hand, D. B. (1960). *Food Res.* **25,** 777–788.
Steinkraus, K. H., Van Buren, J. P., Hackler, L. R., and Hand. D. B. (1965). *Food Technol.* **19,** 63–98.
Stillings, B. R., and Hackler, L. R. (1965). *J. Food Sci.* **30,** 1043–1049.
Sudarmadji, S., and Markakis, P. (1977). *J. Sci. Food Agric.* **28,** 381–383.
Tateno, M., and Umeda, I. (1955). Japanese Patent 204858.
Umeda, I. (1963). U.S. Dep. Agric. Final Tech. Rep., Public Law 480. Project UR-A11-(40)-(C). Natl. Agric. Libr., Beltsville, Maryland.
Umeda, I., Nakamura, K., Yamato, M., and Nakamura, Y. (1969). U.S. Dep. Agric. Final Tech. Rep., Public Law 480 Project UR-A11-(40)-21. Natl. Agric. Libr., Beltsville, Maryland.

Van Buren, J. P., Hackler, L. R., and Steinkraus, K. H. (1972). *Cereal Chem.* **49,** 208–211.

Van Veen, A. G., and Graham, D. C. W. (1968). *Cereal Sci. Today* **13,** 96–99.

Van Veen, A. G., and Schaefer, G. (1950). *Trop. Geogr. Med.* **2,** 270–281.

Wagenknecht, A. C., Mattick, L. R., Lewin, L. M., Hand, D. B., and Steinkraus, K. H. (1961). *J. Food Sci.* **26,** 373–376.

Wai, N. (1929). *Science* **70,** 307–308.

Wai, N. (1968). U.S. Dep. Agric. Final Tech. Rep., Public Law 480. Project UR-A6-(40)-1. Natl. Agric. Libr., Beltsville, Maryland.

Wang, H. L. (1967a). *Food Technol.* **21,** 115–116.

Wang, H. L. (1967b). *J. Bacteriol.* **93,** 1794–1799.

Wang, H. L., and Hesseltine, C. W. (1965). *Can. J. Microbiol.* **11,** 727–732.

Wang, H. L., and Hesseltine, C. W. (1966). *Cereal Chem.* **43,** 563–570.

Wang, H. L., and Hesseltine, C. W. (1970a). *J. Agric. Food Chem.* **18,** 572–575.

Wang, H. L., and Hesseltine, C. W. (1970b). *Arch. Biochem. Biophys.* **140,** 459–463.

Wang, H. L., Ruttle, D. I., and Hesseltine, C. W. (1968). *J. Nutr.* **96,** 109–114.

Wang, H. L., Ruttle, D. I., and Hesseltine, C. W. (1969). *Proc. Soc. Exp. Biol. Med.* **131,** 579–583.

Wang, H. L., Ellis, J. J., and Hesseltine, C. W. (1972a). *Mycologia* **64,** 218–221.

Wang, H. L., Vespa, J. B., and Hesseltine, C. W. (1972b). *J. Nutr.* **102,** 1495–1499.

Wang, H. L., Vespa, J. B., and Hesseltine, C. W. (1974). *Appl. Microbiol.* **27,** 906–911.

Wang, H. L., Swain, E. W., and Hesseltine, C. W. (1975a). *J. Food Sci.* **40,** 168–170.

Wang, H. L., Swain, E. W., Wallen, L. L., and Hesseltine, C. W. (1975b). *J. Nutr.* **105,** 1351–1355.

Watanabe, T. (1969). "UNIDO Expert Group Meeting on Soybean Processing and Use." USDA, Peoria, Illinois.

Yamaguchi, Y., Ito, H., Watanabe, S., Yoshida, T., and Komatsu, A. (1973). U.S. Patent 3,765,905.

Yamamoto, K. (1957). *Bull. Agric. Chem. Soc. Jpn.* **21,** 319–324.

Yokotsuka, T. (1960). *Adv. Food Res.* **10,** 75–134.

Yokotsuka, T. (1964). "Oilseed Protein Foods—a Symposium," pp. 31–48. Mt. Fuji, Japan.

Yokotsuka, T. (1971). "Conversion and Manufacture of Foodstuffs by Microorganisms—a Symposium," pp. 117–125. Saikon Publishing Co., Ltd., Kyoto, Japan.

Yokotsuka, T., Sasaki, M., Kikuchi, T., Asao, Y., and Nobuhara, A. (1967a). *In* "Biochemistry of Some Foodborne Microbial Toxins" (R. I. Mateles and G. N. Wogan, eds.), pp. 131–152. MIT Press, Cambridge, Massachusetts.

Yokotsuka, T., Sasaki, M., Kikuchi, T., Asao, Y., and Nobuhara, A. (1967b). *J. Agric. Chem. Soc. Jpn.* **41,** 32–38.

Yong, F. M., and Wood, B. J. B. (1974). *Adv. Appl. Microbiol.* **17,** 157–194.

# Chapter 5

# Wine

HERMAN J. PHAFF
MAYNARD A. AMERINE

MICROBIAL TECHNOLOGY, 2nd ed., VOL. II

## I. INTRODUCTION

### A. Definition

Wine is, by definition, "produced by normal alcoholic fermentation of the juice of sound ripe grapes (including restored or unrestored pure condensed grape must), with or without added fortifying grape spirits or alcohol, but without other addition or without the addition or abstraction except as may occur in cellar treatment: Provided, that the product may be ameliorated before, during or after fermentation." (U.S. Internal Revenue Service, 1974). The regulations also provide for maximum limits for volatile acidity, amount of sulfur dioxide, and ethanol limits for various types of wines. Amelioration (addition of sugar and/or water) is permitted in most states, but not in California.

Fruit wines are also provided for in the regulations. In this country, few have been produced. Fruit and citrus wine must also be "produced by the normal alcoholic fermentation of the juice of sound, ripe fruit (including restored or unrestored pure condensed fruit must), with or without the addition, after fermentation, of pure condensed fruit must, and with or without added fruit brandy or alcohol, but without other addition or abstraction except as may occur in cellar treatment: Provided, that . . . the product may be ameliorated." (U.S. Internal Revenue Service, 1974).

Federal and state regulations (see, for example, California, 1970) in this country also control many winery operations, wine labeling, tax levels, etc. In other wine-producing countries, regulations of varying degrees of complexity apply (Amerine *et al.*, 1973).

### B. Size of the Industry

While the wine industry includes fruits other than grapes, the volume of nongrape wine is very small—probably less than 1%. World wine production (from grapes) has been increasing since World War II. This is due to greater yields per acre as well as to large new acreages (as in Australia, California, and the Soviet Union). Currently about 9 billion gal (or 335 million hectoliters) of wine are produced annually. Nearly 80% of this is produced in Europe, with 14% from North and South America, over 4% from Africa, and only 1% each from Asia and Oceania. To supply the raw material, more than 22 million acres of grapes (10 million ha) are planted. Some of this acreage, perhaps 15%, is used for table grapes and raisin production.

The economic value of the world wine industry has not been calculated. It would include grape producers, wineries, auxiliary industries (producing bottles, corks, etc.), transportation to national and international markets, wholesalers, retailers, restaurant wine services, and local, state, and national taxes and law enforcement. The economic value of the small California wine industry (FOB the winery) has been estimated at nearly a billion dollars annually. The retail value is, of course, much greater.

## C. Historical Background

Fermentation predates recorded history. Honey and grains were probably fermented as household beverages before grapes. The early wine industry of the Fertile Crescent, about 3500 B.C., spread west (around the Mediterranean) and north (to Hungary, Germany, and France), and in the post-Columbus period to the Americas, South Africa, and Oceania. The Romans advanced the art of wine making, but it was an industry of large risks due to spoilage until the mid-nineteenth century. The research of Louis Pasteur revolutionized the wine industry. From a high-risk labor-intensive industry, it has been converted into a low-risk industry with all the accoutrements of a large-scale modern food production. While more and more technology has been introduced into the industry, in variety selection, location of vineyards, crop level, process control, aging, and blending, considerable art remains in quality evaluation.

The technology of the industry has been summarized in books by Amerine *et al.* (1973), Amerine and Joslyn (1970), Joslyn and Amerine (1964), Ribéreau-Gayon and Peynaud (1972–1976), Troost (1971), and others. A useful list of bibliographies on wines and related subjects has been published by Amerine and Singleton (1971).

## II. MICROBIOLOGICAL ASPECTS

## A. Nature of the Microbial Process

Wine fermentation is basically the transformation of the various sugars of grapes by yeast under anaerobic conditions into ethanol, carbon dioxide, and small amounts of by-products. D-Glucose and D-fructose, the two principal sugars of grape juice, yield essentially equimolar proportions of ethanol and carbon dioxide. Desirable strains for this process are representatives of the species *Saccharomyces cerevisiae* Hansen or several closely related species, such as *Saccharomyces oviformis* Osterwalder. The nomenclature of yeast species mentioned in this chapter is based on the usage in "The Yeasts—A Taxonomic Study" (Lodder, 1970). Yeast species and strains vary considerably in their alcohol tolerance, rate of fermentation, completeness of sugar to ethanol conversion, and nature and proportion of desirable by-products.

Natural fermentations, in which a pure-culture inoculum of a wine yeast is not used and where the operator relies on the natural yeast flora of the crushed grapes and fermentation equipment, are now less commonly used. In such fermentations a succession of yeast and bacterial populations is found in the fermenting grape must. Strains of *Kloeckera apiculata* (and its sporogenous counterparts *Hanseniaspora guilliermondii* and *H. uvarum*) and *Candida pulcherrima* (sporogenous form *Metschnikowia pulcherrima*) are common during the early stages of fermentation. Because of their low alcohol tolerance (5–7%) they are overgrown later on by more alcohol-tolerant species such as *Saccharomyces rosei* and *S. cerevisiae*. Numerous species of yeasts have been isolated from grapes in various parts of the world (Amerine and Joslyn, 1970). As pointed out by these authors, under uncontrolled conditions a high-quality wine is produced only occasionally, i.e., when these different microorganisms occur in proper proportion and sequence and undesirable spoilage organisms are absent or present in very low numbers. Because of the difficulty of controlling the natural microbial population and the danger of a high relative proportion of undesirable organisms, most wines, especially in new viticultural areas, are fermented with the aid of essentially pure culture inocula using strains of wine yeast of proven quality. Such strains are usually used in conjunction with proper concentrations of sulfur dioxide (see Section II, E) to which the wine yeast is resistant and wild yeasts and bacteria are sensitive. Under these conditions the fermentation starts rapidly and goes to completion in a relatively short time. The yeast starter may be taken from a slant culture in a suitable container and the growth is introduced into a gallon of pasteurized must. After active fermentation ensues the fermenting liquid is transferred to progressively larger containers and finally into large tanks of must. In recent years some yeast companies have introduced compressed yeast or active dried yeast of suitable wine yeast strains. The latter product, with about 7.5% moisture content, is stable and can be stored for many weeks. After rehydration at the proper temperature (usually 43°C) it can be added in high concentration to the must to be fermented; fermentation then starts rapidly. Since a heavy inoculum is used, yeast growth at the expense of sugar from the grape is reduced.

## B. Fermentation and Its By-Products

Alcoholic fermentation takes place in the yeast cell by a series of reactions, usually referred to as the Embden–Meyerhof–Parnas pathway (Sols et al., 1971). The overall reaction $C_6H_{12}O_6 \rightarrow 2\ C_2H_5OH + 2\ CO_2$ theoretically yields 51.1% ethanol by weight and 48.9% $CO_2$. In practice the ethanol yield is somewhat lower, approximately 48%. The formation of several by-products and conversion of about 1% of the sugar into yeast cell mass account for the lower yield. In addition some alcohol is lost by entrainment in the carbon dioxide and by evaporation.

Compounds normally recognized as by-products of alcoholic fermentation of grape must are the following (percentages are based on fermentable sugar transformed): glycerol (2.5–3.0%), acetic acid (0.05–0.65%), acetaldehyde (0.01–0.04%), 2,3-butanediol (0.06–0.10%), succinic acid (0.02–0.05%), and higher alcohols (highly variable but ranging from approximately 0.01 to 0.04%). The level of all of these by-products is strongly influenced by yeast strain and environmental conditions, especially temperature.

Glycerol, lactic acid, acetaldehyde, 2,3-butanediol, and succinic and acetic acids are constant by-products of alcoholic fermentation. Glycerol accumulates mostly during the initial stages of fermentation and is formed by the reduction of dihydroxyacetone phosphate to glycerol phosphate followed by enzymatic hydrolysis to free glycerol and phosphate. The enzymes involved are an NADH-dependent dehydrogenase and $\alpha$-glycerophosphatase. Lactic acid is thought to arise by reduction of pyruvate during the course of fermentation. Acetaldehyde represents primarily a small fraction of that intermediate not reduced by NADH and alcohol dehydrogenase to ethanol. Some acetaldehyde is assumed to undergo condensation with another molecule to form acetoin ($CH_3 \cdot CO \cdot CHOH \cdot CH_3$) and the latter is reduced to 2,3-butanediol.

Succinic acid is thought to be formed by the action of phosphoenolpyruvate carboxykinase, which enzyme with the aid of high-energy phosphate catalyzes the condensation of carbon dioxide and pyruvate to form oxaloacetic acid, which is then converted to succinic acid via malic acid. Under anaerobic conditions acetic acid is formed only during the initial stages of fermentation by dismutation of acetaldehyde. During this reaction one molecule of acetaldehyde is oxidized to acetate at the expense of a second molecule that is reduced to ethanol.

Higher alcohols, formerly believed to be formed solely in wine by deamination, decarboxylation and reduction of preformed amino acids, are now known to arise also from keto-acid precursors of amino acids. A small portion of these keto-acids undergoes decarboxylation followed by reduction to the corresponding alcohol. The alcohols most prevalent in wine are 1-propanol, 2-methyl-1-propanol (isobutanol), 3-methyl-1-butanol (isoamyl alcohol), and 2-methyl-1-butanol (optically active amyl alcohol).

## C. Other Compounds Formed by Yeast Fermentation

Hydrogen sulfide, methyl mercaptan, and ethyl mercaptan are undesirable odoriferous compounds found in some wines. Their levels vary greatly with the use of different strains of wine yeast. Hydrogen sulfide is formed primarily by reduction during fermentation of elemental sulfur sprayed on grapes as a fungicide. Mercaptoethanol is believed to be formed by a reaction between hydrogen sulfide and acetaldehyde. These sulfur compounds impart an objectionable flavor to wines in very low concentration (1–5 ppm). Low levels of hydrogen sulfide can often be removed by aeration of the wine. Higher concentrations can be reduced by addition of sufficient sulfur dioxide to provide free sulfur dioxide which is capable of oxidizing hydrogen sulfide to sulfur.

### D. Effect of Temperature on Fermentation

Alcohol yield and the rate of fermentation, as well as the concentrations and proportions of the fermentation by-products, are affected by temperature. Temperature is also among the most important factors in flavor formation. Most enologists agree that more bouquet is formed in a wine by a long, slow fermentation at a low temperature, than by a short rapid fermentation at a temperature nearer to the maximum temperature for fermentation. Fermentation at relatively low temperatures produces more aromatic compounds than at higher temperatures. There is no single optimum temperature for this process since it varies with the strain of yeast used. It has been suggested (Amerine and Joslyn, 1970) that a desirable temperature schedule for white wines is a starting temperature between 7° and 16°C and a finishing temperature of about 19°–20°C.

Alcoholic fermentation generates not only 2 moles of ATP per mole of glucose (used for cell mass biosynthesis) but also a large amount of heat. The generation of 2 moles of ATP represents only 14 kcal out of a total of 56 kcal free energy loss per mole of glucose fermented. Thus it has been estimated that 75% of the energy of anaerobic metabolism is lost in the form of heat (Sols et al., 1971). Large fermentation tanks, in particular, need cooling, since without it the temperature of the fermenting must rises to a point where the fermentation sticks (i.e., ceases because of damage to the yeast cells). Control of fermentation temperature can be achieved by cooling coils in the fermentation vessel through which cold water or a refrigerant is circulated. Other procedures use a refrigerant in double-walled tanks, or the fermenting must may be pumped through a heat exchanger.

### E. Use of Sulfur Dioxide

This chemical has been used in wine making since antiquity to disinfect containers, control contaminating microorganisms, and protect wine against excessive oxidation during storage and aging. Its use is worldwide and if properly employed is beneficial to the quality of even the finest wines, especially white table wines. In wine making sterilization by sulfiting is done prior to the addition of a pure wine yeast starter.

Sulfur dioxide, besides binding to microorganisms and causing their death or inhibiting their growth, also combines readily with carbonyl compounds (e.g., acetaldehyde), unsaturated aliphatic compounds, and proteins of the must. In the combined form it is much less toxic to microorganisms and less effective as an antioxidant. Analytical procedures (Amerine and Ough, 1974) are available which distinguish between free and bound sulfur dioxide in a wine.

In large wineries sulfur dioxide is added to must or wine from pressure cylinders where the gas is stored in liquefied form. For smaller quantities of wine potassium bisulfite is often used. The concentration of free $SO_2$ in finished wines is generally 10–40 mg/liter or less. The effectiveness of sulfur dioxide increases with lowering of the pH, and wines with high total acidity are therefore given the lowest level of sulfur dioxide. Freshly pressed musts are supplied with approximately 100–200 mg/liter of sulfur dioxide. The concentrations normally used in must have little if any effect on the activity of the cultured strains of wine yeast. This resistance has been acquired by adaptation of these strains to certain levels of free sulfur dioxide. The inhibitory action of sulfur dioxide on spoilage bacteria and wild yeasts is attributed to its inhibition of metabolic enzymes containing S—H groups and to its inhibition of the action of the important coenzyme nicotinic acid adenine dinucleotide (NAD).

## F. Yeast Nutrition

Yeast multiplication during complete fermentation of grape must depends to some extent on the amount of inoculum used. For growth under anaerobic conditions, yeast requires a fermentable source of carbon and energy, a source of nitrogen, minerals (both macronutrients and trace elements), and several accessory factors, such as vitamins, traces of fatty acids, and sterols (Phaff et al., 1978). Nutrients such as potassium salts, phosphates, and other minerals, as well as accessory factors, are rarely limiting in grape must, but nitrogenous compounds occasionally are. Small amounts of ammonium salts [e.g., $(NH_4)_2HPO_4$] are therefore added sometimes to stimulate sluggish yeast growth. Ammonium ions are rapidly assimilated by yeasts and in addition are known to stimulate fermentation.

Wine yeasts normally ferment hexose sugars, sucrose, and maltose although at different rates. It is also known that most strains ferment glucose faster than fructose but some do the reverse and prefer fructose. This appears to be related to the rate of uptake of the different sugars or the affinity they have for the common carrier responsible for their facilitated diffusion into the cell. Wine yeasts as well as other yeasts cannot ferment pentose sugars.

## G. Yeast Autolysis

Autolysis is defined as self-digestion of yeast cells. It starts usually with the breakdown of the cell envelope resulting in cell death and this is followed by hydrolysis of the cytoplasmic proteins and nucleic acids by proteases A, B, and C and nucleases, respectively, contained in the vacuoles of the cell. The released amino acids and nucleotides impart an undesirable yeasty flavor to wine. Autolytic tendencies are promoted by high fermentation temperatures and autolysis also varies greatly with the strain of yeast used. An added problem of autolysis is that the released amino acids, nucleotides, and vitamins from yeast can serve as nutrients for fastidious spoilage bacteria, in particular, lactic acid bacteria. For this reason wine should be separated from the sedimented yeast as early as possible upon completion of fermentation. A possible exception to early racking may be in cases of high-acid wines, where autolysis of yeasts is desired in order to release amino acids to stimulate the growth of the bacteria involved in the malolactic fermentation (see Section IV, C).

## H. Deacidification of Wines

Organic acids affect the sensory properties of wines, particularly tartness. Tartaric and L-malic acids are the major acids in grapes, the former being quantitatively the most important. However, grapes grown in cool regions, such as the eastern United States, sometimes contain high levels of L-malic acid, leading to an excessive titratable acidity in the wine produced.

Two methods are known to reduce by fermentation excessive tartness due to L-malic acid. The first involves the conversion of malic acid (a dicarboxylic acid) to lactic acid (a monocarboxylic acid) and carbon dioxide during the so-called bacterial malolactic fermentation. The second method depends on the ability of certain yeasts to convert L-malate to ethanol and carbon dioxide in a mixed fermentation together with a strain of wine yeast.

The malolactic fermentation has been recognized since 1890 and is used in many countries (Kunkee, 1967). It is not recommended in regions where wines are low in total acidity, but this is not always avoidable, since the fermentation often occurs spontaneously by bacteria present in the winery or on grapes. The taxonomy of the lactic acid bacteria associated with the malolactic fermentation is quite confusing and probably more than a single species has the capacity of decomposing malic acid. Species have been placed in *Lactobacillus, Pediococcus, Leuconostoc,* and various other genera. Some of the complex growth requirements for the malolactic bacteria are supplied by yeast cells carrying out the alcoholic fermentation. Some of these are probably the result of yeast autolysis (see Section II, G). The malolactic fermentation most commonly occurs during the middle to late stages of alcoholic fermentation. The conversion of malic to lactic acid takes place in two steps as shown in Eqs. (1) and (2).

$$\text{L-Malic acid} + \text{NAD} \underset{\substack{\text{malic} \\ \text{enzyme}}}{\overset{\text{Mn}^{2+}}{\rightleftharpoons}} \text{pyruvic acid} + CO_2 + \text{NADH} + H^+ \tag{1}$$

$$\text{Pyruvic acid} + \text{NADH} + H^+ \underset{\substack{\text{lactic} \\ \text{dehydrogenase}}}{\rightleftharpoons} \text{L-lactic acid} + \text{NAD} \tag{2}$$

Others have presented evidence that malolactic fermentation is a simple decarboxylation process without intermediates (Peynaud *et al.*, 1968). In either of the two mechanisms there is no production of energy (ATP) which can support the growth of the malolactic bacteria. It is therefore the view of most investigators that these bacteria obtain their energy for growth from the metabolism of small amounts of residual sugars or citric acid present in the fermenting must or wine (Amerine and Kunkee, 1968).

Yeast fermentation of malic acid constitutes another procedure for lowering the total acidity of wine. Several species of the genus *Schizosaccharomyces* (the fission yeasts), e.g., *Sch. pombe* and *Sch. malidevorans,* have the interesting potential to convert L-malic acid under fermentative conditions to 1 mol of ethanol and 2 moles of carbon dioxide. It appears that some strains of *Saccharomyces* can also carry out this reaction but with low efficiency. Little is known about the biochemistry of this conversion. This principle has been applied in recent years to fermentation of wines of excessive titratable acidity with mixed cultures of a normal wine yeast and a strain of *Schizosaccharomyces.* According to Gallander (1977) the best procedure was to inoculate the must initially with *Sch. pombe* (which can also carry out alcoholic sugar fermentation) and to reinoculate the fermenting must with a strain of *Saccharomyces cerevisiae.* According to several authorities, however, the effect of such mixed fermentations on the quality of the wine needs further investigation.

Chemical reduction of acidity using calcium carbonate is also used. This treatment primarily reduces the level of tartaric acid by precipitation as the calcium salt.

## I. The Fermentation of Flor Sherries

When newly fermented wine containing about 15% ethanol is exposed to the air, a thick surface film of yeast may develop with the capacity of slowly oxidizing a small portion of ethanol to acetaldehyde. This compound and its reaction product with ethanol, acetal, are the most important flavor compounds in Spanish flor sherry and similar products made in other countries. The levels of free aldehyde may rise to 200 or 300 mg/liter and total aldehyde (including bound) may reach 400 mg/liter. The procedure of making flor sherries consists of two fermentation steps, (1) fermentation of the must from desirable grape varieties (e.g., Palomino grapes) to an ethanol level of close to 15% and (2) development of a film yeast to develop the required aldehyde level.

If the sugar content of the grapes is insufficient to reach 15% ethanol, fortification with distilled wine spirits is necessary to avoid the growth of spoilage bacteria (e.g., *Acetobacter*) or wild yeasts which would raise the acetic acid level of the wine. The two basic types of Spanish sherry are the *fino* type, whose flavor is developed largely by a secondary film (flor) fermentation, and the *oloroso* type, which is not subjected to this secondary fermentation. The secondary flor fermentation used for the *fino*-type of Jerez sherries of Spain requires from one to several years for completion. The secondary fermentation is done in partially filled (about three-quarters full) oak barrels where the wine is covered by a thick layer (1- to 1.5-cm thick) of film yeast. The barrels are arranged so that progressive fractional blending during maturation can be done (the so-called solera system) with as little disturbance of the film yeast layer as possible. The wine undergoes a continuous aging and improvement in quality during its prolonged storage in the oak casks. The details of the solera process are described by Joslyn and Amerine (1964).

In recent years a number of procedures have been worked out to reduce the long oxidative secondary fermentation. Considerable success has been achieved by using a submerged fermentation process in which a strain of flor yeast is mixed with wine of 15% alcohol in pressure tanks where air or oxygen is introduced in finely dispersed form to promote the required aldehyde formation. This process can be made semicontinuous by removing 10% of the volume each day and replacing it by fresh wine. The submerged process has the advantage of saving a considerable amount of time, less alcohol is lost by evaporation, and the long-term investment in the product is reduced. However, an acceptable, good-quality sherry still requires aging in full oak casks although not nearly as long as in the *fino* solera process. For further details of the process see Ough and Amerine (1960).

The taxonomy of the yeasts found in the flor is quite confusing. Numerous investigators in various countries have attached names to these film-forming yeasts. These include *Saccharomyces beticus* (various races), *S. cheresiensis, S. montuliensis,* and even strains of the better known species *S. oviformis, S. fermentati, S. delbrueckii,* and even *S. cerevisiae.* It seems evident that flor yeasts are made up of a number of different species of yeast either in pure culture or as a balanced population of different strains. For further information on this interesting microbiological process and commercial practice the reader is referred to Joslyn and Amerine (1964).

## J. Wine Spoilage Organisms

Bacteria as well as yeast may be responsible for the spoilage of wines. Important factors influencing microbial spoilage are the composition of the wine (sugar, alcohol, and sulfur dioxide content primarily), storage conditions (e.g., temperature and amount of air space in the container), and, perhaps most important, the extent of initial contamination during storage in vats or bottles.

Spoilage by yeasts is often more of an aesthetic problem than a marked deterioration of the flavor and aroma of a wine. Occasionally a yeast may be desirable or undesirable depending on the conditions. For example, *Saccharomyces oviformis,* which is very effective in producing a very dry wine, becomes an undesirable contaminant when it is found in wines where a residual sugar is desired. *Saccharomyces bayanus* also has been found in bottled dry table wines with 0.5–1.0% residual sugar where it causes turbidity or a sediment. In young table wines *Pichia membranaefaciens* (an aerobic yeast) is sometimes found when the bottle contains sufficient oxygen to permit limited development of this species (or its nonsporogenous counterpart *Candida valida*). The most serious spoilage yeasts are species of *Brettanomyces* (sporogenous forms are placed in the genus *Dekkera*). These are slow-growing yeasts which can develop in both white and red table wines causing turbidity and slight off-flavors. If oxygen is present they can oxidize ethanol to acetic acid. They are more prevalent than had been suspected for many years since plating of a turbid wine may not produce visible colonies for 10–12 days on agar media. A suitable differential medium is malt agar with 100 mg/liter of cycloheximide (Actidione). Wine yeasts are very sensitive to this antibiotic while *Brettanomyces* species are resistant. The most common species of that genus in wines is *B. intermedius.*

Bacterial spoilage of wine (Vaughn, 1955) is less of a problem today than it was in the past. This is due to observance of good sanitary practices in most wineries and the proper use of sulfur dioxide.

Aerobic spoilage may be caused by acetic acid bacteria which are common on grapes. They oxidize ethanol to acetic acid and some ethyl acetate may be formed as well. Of greater importance is anaerobic spoilage by various lactic acid bacteria. These organisms are capable of producing metabolic products other than lactic acid which can profoundly affect the quality of a wine. For further details see the excellent review on this subject by Vaughn (1955).

## III. KEY DEVELOPMENTS

From the microbiological point of view, the most significant recent developments have been large-scale use of dried pressed yeasts, industrialization of the flor sherry process, and greater knowledge of and control of the malolactic fermentation. Most recently great interest has been shown in yeasts producing lower sulfide and greater flavor. Should the wine industry have to cut back on use of sulfur dioxide, the need for a new antimicrobial agent would become acute. Since sulfur dioxide has a variety of useful properties other than its antiseptic and preservative value (antioxidative and enzyme inhibition; clarification; bleaching and discoloration; promotion of solubilization of the coloring matter of grape skins; increasing the fixed acidity, the extract, and the alkalinity of the ash because of its solvent effect on potassium acid tartrate), a single agent may not be available and agents for specific needs may be needed. In any case grapes will have to have less microbial infection than at present and fermentation and aging must be conducted under increasingly stringent sanitary conditions.

Since the advent of dried pressed yeast, few wineries have used yeast culture tanks. Those that employ such vessels use stainless steel equipment with built-in sterilizing and cooling facilities. Most fermentors built in recent years have been of plastic, stainless steel, or epoxy-lined iron. Except for small fermentors, they have built-in or auxiliary cooling equipment. Many are operated manually, but a few have automatic controllers or even attached computers. Many wineries now routinely centrifuge musts for white wine production (Fig. 1). Many wines are now germ-proof filtered into the bottles (Fig. 2). Pasteurization of high-quality table wines is not an acceptable practice in the wine industry because of its unfavorable effect on wine quality. To prevent microbial spoilage of any origin all winery equipment, including cooperage, must be kept scrupulously clean. The wine must be inspected periodically for contamination by direct microscopic inspection and especially by filtration of measured volumes of wine through membrane filters followed by plating of the filters on a suitable agar medium.

Semidry table wines (with 1–3% reducing sugar) have become in-

**FIGURE 1.** Large centrifuge (rear left) in a California winery, used for clarifying musts and wines. (Courtesy Paul Masson Vineyards, Saratoga, California.)

**FIGURE 2.** Large tubular membrane (Millipore) filter in California winery used for germ-proof filtration of wine. (Courtesy Guild Wineries and Distilleries, San Francisco, California.)

creasingly popular. Such wines are particularly subject to recurrence of yeast fermentation.

The wine industry is making increasing use of filtration through large membrane filters to remove microbial contamination before bottling (Fig. 2). In addition, a number of chemical preservatives (not including sulfur dioxide) have been advocated in recent years for the preservation of wine. The latter include diethyl pyrocarbonate (DEPC) and sorbic acid. The former compound, claimed to decompose into ethanol and carbon dioxide when added to wine, is a potent inhibitor of yeast growth. Although still used in some countries, it has been prohibited in the United States since 1972 because of some evidence for the formation of harmful complexes between DEPC and organic components of wine.

Sorbic acid is an effective inhibitor of yeast growth but it does not affect bacterial development. Since the ban of DEPC, sorbic acid or its potassium salt has become the only practicable fungistatic agent legally authorized for wine use in the United States. The usual concentration of sorbate in wine is 150–200 ppm. A disadvantage of sorbate is the possibility of its conversion by the action of lactic acid bacteria to an ethyl ether (2-ethoxyhexa-3,5-diene) which has an intense odor of crushed geranium leaves (Crowell and Guymon, 1975). Such an off-odor associated with bacterial spoilage of sorbate-containing wines is easily detected by sensory analysis. For further details on chemical preservatives, their use, and appropriate literature references see Amerine and Kunkee (1968).

## IV. PROCESS TODAY

### A. Yeast Selection

Pure yeast cultures are available from private and public laboratories in wine-producing countries. Some wineries maintain stock cultures and propagate them as pure cultures. Others purchase dried yeast strains from commercial producers. A variety of yeast strains of S. cerevisiae is being used: Burgundy, Montrachet, Chablis, Champagne, Steinberg, etc. Mixed cultures of different species have been tried but are not widely used.

### B. Process Flow

Almost all wine is produced by a batch process. The main reason for use of a batch rather than a continuous process is the short period of availability of the raw product (usually only 6–10 weeks). The expensive equipment and controls needed for a continuous process would thus be unused for most of the year—a heavy overhead. Furthermore, the raw material is of variable suitability for producing different types of wines. However, there are several cases where continuous systems have been used. Where the wine is refermented for a new product, continuous systems may be considered. Sparkling wine production in a series of tanks is used in the Soviet Union, but the process has not found favor in the West. Submerged culture of flor yeasts in a continuous fashion has been successful in California, the Soviet Union, and elsewhere. For ordinary quality wines, continuous fermentors have been used in Argentina and Italy. Distilling material has also been successfully produced in continuous systems. Producers have avoided such systems for quality wines because of the problems of controlling the types of grapes entering the system and because (unless strict controls are maintained) of the possibility of spreading bacterial and yeast contamination.

Another possibility for continuous fermentation would be the conversion of grape juice to grape concentrate by vacuum concentration and using diluted concentrate for wine production throughout the year. This might be particularly useful for wine production in areas where grapes are not grown. It would be most appropriate for white wines and sherry stock.

The process of white wine production involves crushing the grapes in crusher–stemmers and separation of the seeds and skins from the juice. The most common method of separation is pressing, usually in screw-presses, although a number of hydraulic and other types are employed. Where grapes are cheap and there is a large demand for wines for distilling, only the juice which drains off from the crushed grapes—the free-run—is normally used for wine. The rest is fermented and distilled for wine spirits.

Red wine production differs in that the grapes must be fermented on the skins for 3–10 days or longer. This is necessary for the alcohol formed to extract the red anthocyanin pigments from the skins. Unfortunately, the skins tend to float on the surface of the juice and reduce contact of the skins with the fermenting juice. The traditional method of securing contact has been to push the floating skins down into the fermenting liquid. In large containers this is very difficult and at present the liquid is pumped from the bottom of the container over the skins. When grapes contain mold, a number of polyphenoloxidase enzymes may be present. These cause an oxidation of the red anthocyanin pigments to compounds with a brown color. To prevent this there is an increasing practice of heating red grapes to inactivate the enzymes before fermentation. Heating also helps to release the color from the skins.

Pink (or rosé) wines are produced in three ways: from pink-colored grapes, by separating the skins of dark grapes from the juice after 12–36 hours of fermentation on the skins, or by blending white and red wines. Some are finished without residual sugar; others may be sweet.

In the production of white or of red dessert wines, the process is the same except that wine spirits are added after pressing, during fermentation. The time of addition of the spirits is determined by the level of residual sugar which is desired in the finished wine. For "dry" sherry production, this may be less than 1%. For muscatels, it may be 12–14% sugar. The level of fortification also varies: for dry sherry intended for flor-yeast fermentation, 15% ethanol is the maximum. For sherries that are to be baked (i.e., heated at 55°C for 90 days), 17–18% is the usual level. For muscatels and ports, 18–19%, or more, is common.

Following the primary fermentation, table wines which are not protected from air may develop four types of surface growth (i.e., a film): of *Acetobacter,* of *Pichia* sp. (usually *P. membranaefaciens*), *Candida vini,* or of *Saccharomyces* sp.

*Acetobacter* is sensitive to alcohol and develops slowly on wines with more than 12% ethanol. These bacteria produce acetic acid and other acids, ethyl acetate, and minor products. The various strains of *Pichia* and *Candida* are likewise sensitive to ethanol. They do not carry out alcoholic fermentation, but they will grow (flower is the colloquial term) aerobically on the surface of wines of 10–13% ethanol. They produce acetaldehyde and traces of organic acids.

As a beneficial process we are only interested in the aerobic film stage of *Saccharomyces* sp. These yeasts can grow on the surface of wines of up to 15 and 16% ethanol. This is of great industrial importance because, by keeping the ethanol content at 15%, growth of *Acetobacter* sp., *Pichia* sp., and other undesirable film-forming yeasts is inhibited. As discussed in Section II in more detail, the primary oxidative metabolic product of *Saccharomyces* species is acetaldehyde.

The domestication of the film stage of *Saccharomyces* as the producer of a special type of wine dates from the nineteenth century in southern Spain. Doubtless it existed earlier and elsewhere (Armenia, for example), but it was first used for producing large amounts of dry fino-type sherries in Spain.

Besides the specialty product of flor sherry (by the film or submerged culture process), sparkling wines require special processing. The excess carbon dioxide needed in sparkling wines can come from direct carbonation or by a secondary fermentation of added sugar. Very few carbonated wines are produced. The fermentation-produced sparkling wines are made by two basic procedures: fermentation in large closed tanks (500–25,000 gallons; 19–950 hl) or fermentation in bottles of less than 2 liters capacity. Fermentation in tanks is by far the cheaper process since following fermentation the wine is simply filtered into special bottles, labeled, and sold.

The bottle fermentation process may take two forms. Following fermentation the wine can be transferred from the bottle into a tank and then filtered into another bottle. As with the tank process, this filtration is necessary to remove the yeast which was added to secure the secondary fermentation. The producer is able to label such wine as "bottle-fermented." In the more traditional bottle process, the wine is left in the bottle for 1–3 years. The yeast cells autolyze during this aging period and this is believed responsible for the development of the additional bouquet in the wine. To remove the yeast cells it is necessary to get the yeast deposit onto the cork (done by turning the bottles on end and by a complicated process taking 1–6 weeks of getting the yeast deposit to slide down through the neck of the bottle onto the cork). The neck of the bottle, including the deposit, is then frozen and the cork removed. The ice plug is then disgorged. The small amount of wine lost by disgorging is then replaced and a new cork inserted. About 1–2 atmospheres of pressure are lost. Nevertheless, commercial sparkling wines usually retain 2–3 atmospheres of pressure. In this case the producer can label the wine "fermented in this bottle." For further details see Amerine and Joslyn (1970).

## C. Processing

Following fermentation the treatment of table and dessert wines may take a variety of forms. The new wines are usually separated from the yeast sediment (racked) soon after fermentation. In the case of red wines where a malolactic fermentation is desired, the initial racking may be delayed. This is believed due to the desire to secure release of amino acids by autolysis of the yeast. The additional amino acids favor the growth of the bacteria that are responsible for the malolactic fermentation.

Large-scale rack-and-cloth filters are used for the initial clarification. Recalcitrant wines (i.e., wines that do not clarify well) may be centrifuged (Fig. 1). Following the rough filtration, the wine is "fined." This involves addition of some agent which physically and by adsorption helps clarify the wine of suspended material. A slurry of clay material, bentonite, is now the main agent used. Other fining agents that have been used in special cases are PVPP (polyvinylpolypyrrolidone), gelatin, casein, isinglass, and others.

The wines are again filtered (Fig. 3) following fining. They are aged in wood, plastic, or stainless steel or epoxy-lined metal tanks. These vary in

**FIGURE 3.** Large plate filter for filtration of finished wines in a California winery. (Courtesy E. & J. Gallo Winery, Modesto, California.)

size from about 50 gal (2 hl) to 250,000 gal (9500 hl), or more. The time of aging may vary from a few weeks to two or more years, depending on the type of wine.

Most new wines are supersaturated with potassium acid tartrate (cream of tartar). With time and especially when the wine is stored at low temperatures, the excess potassium acid tartrate crystallizes and settles. However, since many wines are sold when very young, there may not be sufficient time for their precipitation. In this case, prior to bottling, the wines may be cooled to a temperature just above freezing for one to several weeks to hasten the precipitation of the excess potassium acid tartrate.

### V. PACKAGING AND DISTRIBUTION

In many parts of the world, a large amount of wine is distributed directly in bulk (e.g., wooden barrels) to the consumer. In Austria, this may be as much as 50%. Shipment to bottlers may be in wooden barrels or even in glass- or plastic-lined tank cars, trucks, or, for long distances, in tankers.

However, increasingly, wines are sold in glass bottles of about 750-ml capacity. Mechanically operated bottling machines are commonly employed. The bottles may be closed with metal screw caps or with corks. The labels and capsules (to cover the cork) are usually applied at the same time as the wine is bottled. In fact, in the larger wineries, the wines may be bottled, labeled, packaged, and placed on shipping pallets in a continuous mechanically operated line. Shipment of the cases is by truck or rail. During shipment, the cased wines should not be exposed to extremes of cold or warm temperature.

## VI. EVALUATION OF WINES

Examination of wines in the laboratory is essential to assure a sound product of high quality. The procedures include tasting and other sensory observations, microbiological examination, and determination of such constituents as ethanol, total acidity, volatile acidity, sugar, extract, tannin, sulfur dioxide, and depth of color. Such analyses also assist the wine maker in correcting certain deficiencies by blending of different lots of wine. The details of the sensory evaluation and various analytical procedures are described by Amerine and Ough (1974), Amerine and Roessler (1976), and Amerine et al. (1973). Below we present briefly the highlights of the various procedures.

### A. Sensory Examination

Sensory examination is an important aspect of cellar operation. A good winery operation requires at least one individual who is responsible for collecting and evaluating the records of the sensory evaluation of its wines. Persons on the evaluation panel should have a keen palate and discriminating judgment, should know wine types, and should be able with reasonable accuracy to classify the cellar's wines. Wines are judged by the following criteria: (1) Appearance (e.g., clarity and freedom from sediment) and color (appropriateness for a particular type of wine). (2) Odor, aroma, and bouquet. Particular attention is paid to the bouquet (which may reveal the extent of aging) and to the presence of odors of sulfur dioxide, hydrogen sulfide, acetic acid, diacetyl, and other volatiles resulting from bacterial action. (3) Taste. Here the main variables are acidity (sourness), sugar (sweetness), bitterness, and astringency (caused by the tannin content of wine). (4) Flavor. Flavor expresses the overall impression of the wine and is based on a combination of taste and bouquet plus aroma. It includes the aftertaste, the sensory notes that linger on the palate after the wine is tasted. Sensory evaluation is of great importance in blending and in deciding when a wine is ready for bottling or shipment.

### B. Microbiological Examination

This includes both microscopical inspection and plating. Its main purpose is to detect excessive numbers of spoilage bacteria (mainly acetic and lactic acid bacteria) and wild yeasts. Their presence can be detected amid a large preponderance of wine yeasts by supplementing suitable agar media with 100 mg/liter of cycloheximide. Bacteria and many wild yeasts (e.g., *Kloeckera* and *Brettanomyces*) are resistant to this antibiotic, whereas wine yeasts are totally inhibited. Low degrees of contamination in wines can be detected by filtering a liter or so of wine through a Millipore or similar filter and placing the filter on a suitable agar medium.

## C. Chemical and Physical Analyses

Detailed procedures have been published (Amerine and Ough, 1974) to determine various important constituents of both must and wines. Space does not permit describing the various procedures in detail. We will merely present a list of the most important determinations normally performed in wineries having good laboratory facilities.

### 1. Soluble Solids

Most of the soluble solids in grape must are sugars. A determination of this property indicates ripeness of the grapes and the maximum alcohol level that can be expected after fermentation. The reducing sugar content of wines is a measure of their dryness or sweetness.

### 2. Acidity and Individual Acids

Methods differentiate between volatile and nonvolatile acids; together they represent the total (titratable) acidity. Estimation of changes in malic and lactic acids provides information on the course and extent of the malolactic fermentation. The pH is also determined.

### 3. Alcohols.

It is obvious that the determination of ethanol (Fig. 4) is important in connection with required maximum and minimum levels in table wines, dessert wines, and appetizer wines. Less routine is the determination of glycerol, 2,3-butanediol, methanol, and higher alcohols.

### 4. Carbonyl Compounds

The principal carbonyl compound is acetaldehyde, which is particularly important in flor sherries. Minor compounds include acetoin, diacetyl, and hydroxymethylfurfural.

### 5. Phenolic Compounds

Among the compounds in this group that are important in winery operations, the following may be mentioned: grape pigments (including anthocyanins), flavonoids and nonflavonoid phenols, hydrolyzable tannin phenols, catechins, and others.

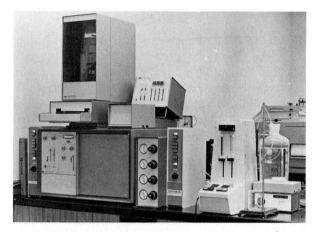

**FIGURE 4.**  An automated gas chromatography system constitutes one method for the analysis of ethanol. Samples of wine are precisely diluted with a diluent containing an internal standard. The diluted samples are loaded in the tray of the automatic sample injector. The peak areas for ethanol and the internal standard are electronically integrated. Corrections are made for response factors and the ethanol content is calculated and printed out for the sample by the integrator. The integrator is not shown in the photograph. (Courtesy E. & J. Gallo Winery, Modesto, California.)

### 6. Nitrogen Compounds

These include total nitrogen, ammonium ion, amino acids, $\alpha$- amino nitrogen, protein, and nitrate. The presence of excessive levels of amino acids, protein, and nucleotides indicates that yeast autolysis may have occurred during late stages of fermentation.

### 7. Mineral Constituents and Gases

Analyses may be needed for cations (potassium, sodium, calcium, magnesium, iron, and copper), anions (bromide, chlorine, phosphate, and sulfate), and dissolved gases (oxygen, carbon dioxide, hydrogen sulfide).

### 8. Color Determination

Color is determined most accurately by using tristimulus values. However, for white wines a simple measurement of light absorbance in a spectrophotometer is adequate to follow development of a brown color due to oxidation and storage at high temperature.

### 9. Chemical Additives

Such analyses include the determination of free and total sulfur dioxide. If other chemical preservatives are used the chemist may have to test at least for benzoic and sorbic acids.

## VII. USES

Probably no other food product has as many cultural patterns of use as wine. These arise from historical precedents and the nature of wine itself. The Mediterranean concept of wines is for their consumption as table beverages, that is, as accompaniments of food. These are wines of moderate ethanol and with little sugar. Since many white table wines are tart (have an appreciable acid taste), they are generally considered to be a good accompaniment of fish and moderately flavored meats (such as chicken). Red wines, with their stronger flavor, are considered best with the stronger flavored meats and cheeses. However, personal predilections vary, and who is to say that a dry white table wine is better with shrimp, chicken, chops, or roasts?

Later sweet wines of low or high alcohol content were produced. The latter are made either by harvesting of specially ripened grapes or by addition of distilled spirits to wines during fermentation. These were more properly considered as dessert wines—that is, wines to be consumed with desserts or after meals. Traditionally wines have been used in the kitchen, as for *coq au vin* or *boeuf bourguignonne*. Small amounts of sherry are added to clear soups. Sparkling wines or dessert wines are sometimes added to fruits.

## REFERENCES

Amerine, M. A., and Joslyn, M. A. (1970). "Table Wines: The Technology of Their Production," 2nd ed. Univ. of California Press, Berkeley and Los Angeles.
Amerine, M. A., and Kunkee, R. E. (1968). *Annu. Rev. Microbiol.* **22,** 323–358.
Amerine, M. A., and Ough, C. S. (1974). "Wine and Must Analysis." Wiley, New York.
Amerine, M. A., and Roessler, E. B. (1976). "Wines: Their Sensory Evaluation." Freeman, San Francisco, California.
Amerine, M. A., and Singleton, V. L. (1971). "A List of Bibliographies and a Selected List of Publications That Contain Bibliographies on Grapes, Wines, and Related Subjects," pp. 1–39. University of California, Agricultural Publications, Berkeley.
Amerine, M. A., Berg, H. W., and Cruess, W. V. (1973). "The Technology of Wine Making," 3rd ed. Avi Publ. Co., Westport, Connecticut.
California (1970). "Regulations Establishing Standards of Identity, Quality, Purity, Sanitation, Labeling, and Advertising of Wine." California Administrative Code, Title 17, Chapter 5, Article 14, Sections 17,000–17,105. California State Printing Office, Sacramento.
Crowell, E. A., and Guymon, J. F. (1975). *Am. J. Enol. Vitic.* **26,** 97–102.
Gallander, J. F. (1977). *Am. J. Enol. Vitic.* **28,** 65–68.
Joslyn, M. A., and Amerine, M. A. (1964). "Dessert, Appetizer and Related Flavored Wines—The Technology of Their Production." University of California, Division of Agricultural Sciences, Berkeley.
Kunkee, R. E. (1967). *Adv. Appl. Microbiol.* **9,** 235–279.
Lodder, J., ed. (1970). "The Yeasts—A Taxonomic Study." North-Holland Publ., Amsterdam.

Ough, C. S., and Amerine, M. A. (1960). *Food Technol.* **14,** 155–159.

Peynaud, E., Lafon-Lafourcade, S., and Guimberteau, G. (1968). *Mitt. hoeheren Bundeslehr- Versuchsanst. Wein- Obstbau, Klosterneuburg* **18,** 343–348.

Phaff, H. J., Miller, M. W., and Mrak, E. M. (1978). "The Life of Yeasts," 2nd ed. Harvard Univ. Press, Cambridge, Massachusetts.

Ribéreau-Gayon, J., and Peynaud, E. (1972–1976). "Sciences et techniques du vin," Vols. 1, 2, and 3. Dunod, Paris.

Sols, A., Gancedo, C., and DelaFuente, G. (1971). *In* "The Yeasts: Physiology and Biochemistry of Yeasts" (A. H. Rose and J. S. Harrison, eds.), Vol. 2, pp. 271–307. Academic Press, New York.

Troost, G. (1971). "Die Technologie des Weines," 3rd ed. Ulmer, Stuttgart.

United States Internal Revenue Service (1974). "Wine," Part 240, Title 26, Code, Federal Regulations, U.S. Govt. Printing Office, Washington, D.C. (ATF Publ. 5120.2, 12/74).

Vaughn, R. H. (1955). *Adv. Food Res.* **6,** 67–108.

# Chapter 6

# Vinegar

## G. B. NICKOL

## I. HISTORY AND DEVELOPMENT

## A. Derivation of Vinegar

It is very likely that the first vinegar used by ancient peoples was simply spoiled wine. In fact, the word *vinegar* comes from the French words *vin* ("wine") and *aigre* ("sour"). Wine making dates back at least 10,000 years, and it can be assumed that vinegar making goes back just as far. Evidence exists indicating that most ancient peoples who used

**155**

MICROBIAL TECHNOLOGY, 2nd ed., VOL.II

vinegar derived it from wines made from the numerous fermentable materials available in the areas where they lived (Hildebrandt et al., 1967; Allgeier et al., 1974a, b). Both the Old and New Testaments contain references to vinegar. The best known of these tells of vinegar offered as a drink to Jesus on the cross.

Early records indicate that vinegar had many uses other than as a food or food preservative. Medicinal uses were numerous and varied from the use of vinegar as an antibiotic in wet compresses to its use in treatment of plague. In early times most households made their own vinegar, but eventually its production was taken over by wineries and breweries. Vinegar produced by breweries was known as alegar.

## B. Advances in Vinegar Making

Initially vinegar was commercially produced by the *let-alone process,* which was simply a matter of holding wine in open, partially filled containers (Allgeier et al., 1960). An advance was made when it was discovered that some finished vinegar added to partially filled casks of wine substantially speeded up the wine's transformation into vinegar. This procedure was known as *field process,* since the barrels were held in open fields. The *Orleans process,* developed in 1670, was a mechanical adaptation of the field process that made it a continuous one. The cask was provided with a spigot for withdrawing finished vinegar, and a tube was inserted through the bung to the bottom of the container permitting the addition of wine. When the contents of the cask had changed to vinegar, only part of it was drawn off, leaving some vinegar behind to mix with the new wine added. This *slow process,* as it is also called, is still used in some parts of the world.

The idea of making vinegar by trickling the liquid through packing materials such as branches and stems dates to the early eighteenth century and is credited to both H. Boerhaave and Kastner. It was improved by J. S. Schüzenbach in 1823 to make it the *quick process.* His improvement was nothing more than the provision of ventilation holes near the bottom of the generator. Sufficient warmth is maintained in the generator to cause a continuous draft up through the packing, thus supplying the oxygen required for the chemical reactions that produce vinegar (see Section II). Schüzenbach's version of the quick process predominated in commercial vinegar production for about 100 years.

The next major improvements of the quick process were made in 1929. At that time, forced aeration and temperature controls were introduced, and the trickling generator widely used today resulted. The most recent process to be devised and developed for commercial use was first publicized in 1949. This *submerged culture process,* as it is called, involves the aeration of the liquid itself, and its success hinges greatly on the efficiency of that aeration. The improved quick process and the submerged culture process are both used extensively today and will be described in detail in Section IV.

## II. MECHANISM OF ACETIC ACID FERMENTATION

Vinegar is a solution of which acetic acid is the essential ingredient, the others being water and various impurities. By FDA standards, vinegar must contain at least 4% acetic acid (see Section V). In all the processes described in Section I, the acetic acid is produced from ethanol by microorganisms. The conversion of ethanol to acetic acid is relatively simple compared to other biological processes. It is a two-step reaction. Ethanol is converted to acetaldehyde and concurrently hydrated, as shown in Eqs. (1) and (2) below. The resulting hydrated acetaldehyde is then dehydrogenated to acetic acid, as shown in Eq. (3). The conversion of ethanol to acetic acid is exothermic. One gram of ethanol theoretically produces 1.304 gm of acetic acid and 0.391 gm of water.

$$CH_3CH_2OH + [O] \longrightarrow CH_3CHO + H_2O$$

Ethanol        Oxygen                                    Acetaldehyde      Water

$$CH_3CHO + H_2O \longrightarrow \begin{array}{c} H\ \ H \\ | \ \ \ | \\ H-C-C-OH \\ | \ \ \ | \\ H\ \ OH \end{array}$$

Hydrated acetaldehyde

$$\begin{array}{c} H\ \ H \\ | \ \ \ | \\ H-C-C-OH \\ | \ \ \ | \\ H\ \ OH \end{array} + [O] \xrightarrow[\substack{\text{dehydro-}\\\text{genase}}]{\text{aldehyde}} CH_3COOH + H_2O$$

Acetic acid

## III. ACETIC ACID ORGANISMS

### A. Classification of *Acetobacter*

For thousands of years, people who produced vinegar were making use of acetic acid-producing organisms, although the microbiological nature of the vinegar process was not suspected until F. T. Kützing theorized in 1837 that microorganisms converted ethanol to acetic acid. His theory was confirmed by Pasteur in 1868.

Although the acetic acid-producing organisms, or *Acetobacter*, were among the first bacteria to be studied, vinegar production has essentially been accomplished with chance contaminants. The reason for this has been the relative difficulty of isolating pure cultures from active vinegar processes. Also, the nature of the industry up until very recently has been conducive to operating profitably in a nontechnical manner.

Over 100 species, subspecies, and varieties of the genus *Acetobacter* have been classified. Many of these species probably would be acceptable in a vinegar generator with the exception of *A. xylinum,* a heavy slime former, and any species that oxidizes acetic acid to carbon dioxide (overoxidation). The production of vinegar by pure-culture techniques is, strictly speaking, not done. However, in the submerged culture process some operations have been started with pure cultures, although aseptic conditions have not been maintained to prevent contamination. One of the first discussions of pure-culture vinegar production was published by Shimwell (1954). The most extensive treatise on the biochemical activities and taxonomy of acetic acid bacteria is that of Asai (1968).

## B. Laboratory Techniques for Culturing *Acetobacter*

### 1. Slant Cultures

Techniques used quite successfully by workers at U.S. Industrial Chemicals Co. for a strictly practical approach to culturing the organisms for acetic acid production have been published (Nickol *et al.*, 1964; Hildebrandt *et al.*, 1968). A condensed version of one of the relatively simple techniques follows below.

Agar slants are prepared from tomato juice agar in screw-capped tubes. These are autoclaved and then cooled in a slanted position, after which a portion of a sterile special broth is added to cover approximately the lower one-third of the slant. This broth is a solution of yeast nitrogen base, dextrose, ethanol, acetic acid, and peptone. Inoculations are made from stock cultures or vinegar using a wire loop and streaking the culture through the liquid up over the agar slant. The slants are then incubated at 30°C until good growth is attained, after which they are stored at 5°–10°C. Cultures must be transferred monthly to retain viability.

### 2. Culture Purification

To purify a culture, tomato juice agar petri dishes are numbered from 1 to 6. One inoculating loop of culture is placed on plate 1. Cultures are spread over plate 1 with a sterile bent glass rod using a circular motion about five times so as to cover the entire plate. Without flaming the bent rod, plate 2 is smeared uniformly with the culture in a similar manner. Still without flaming the rod, plates 3 to 6 are smeared. The effect is to diminish the inoculum with each succeeding plate. Plates are inverted, placed in polyethylene bags, closed with a rubber band, and incubated at about 28°C for approximately 48 hours or until colonies appear. If growth is confluent and single colonies cannot be isolated, the procedure is repeated using fresh plates with fewer bacteria in the initial inoculum.

A typical well-isolated colony is fished to a slant, as above, and incubated 48 hours or until satisfactory growth appears. Once a culture is purified and shown to be efficient in vinegar production, it need not be repurified unless it is inadvertently contaminated. Care is exercised in purifying cultures so as to avoid genetic variation. This may be somewhat overcome by picking several typical colonies and recombining them in a purification process to get rid of an obvious contaminant.

All cultures, purified or nonpurified, are gram stained and checked under the microscope. Most good vinegar cultures are gram-negative short rods.

### 3. Maintenance of Cultures

Although vinegar bacteria survive under natural conditions in orchards and vineyards, under pure-culture conditions they tend to die rather quickly. Even under refrigeration at 5°–10°C, cultures normally should be transferred each month. The development of the lyophilization process enables cultures to be maintained for extended periods without transfer (Hildebrandt et al., 1968). Such cultures remain unaltered both qualitatively and quantitatively for all practical purposes. Contamination is precluded, once the ampules have been sealed. Lyophilized cultures can be shipped without concern as to viability. Therefore, subculturing is not necessary. Further, there is no need for special storage upon receipt.

Two to five years is a reasonable time to expect lyophilized cultures to survive, although one culture of *Acetobacter oxydans* was viable after 20 years. Care, however, must be exercised in ascertaining that the lyophilization process is carried out properly.

### 4. Shake-Flask Cultures

The acid-producing characteristics of cultures can be evaluated in shake flasks. A convenient medium is sweet cider (no preservative added) with sufficient alcohol added to give an initial alcohol concentration of about 4% by volume. Additional alcohol must be added as acetification proceeds to prevent complete depletion of alcohol. If alcohol is unavailable for the organisms, the culture will die. However, the 4% alcohol concentration should not be exceeded. A good acid producer will reach a 10% acidity within 3 days. *Acetobacter* cultures that attain a 10% acidity in shake flasks will produce 12% or higher acidities in trickling or submerged culture generators. Laboratory-scale generators can be readily started with active shake-flask cultures.

## IV. COMMERCIAL VINEGAR PRODUCTION

### A. Trickling Generators

#### 1. Description of Operation of Frings-Type Generator

The most widely used generator is of the Frings type (Fig. 1). It is a tank constructed of cypress or redwood and is packed with about 2000 ft³ of curled beechwood shavings. A false bottom supports the shavings above the lower one-fifth of the tank, which is a storage reservoir with about a 3500-gal capacity. A pump circulates an ethanol–water–acetic acid mixture from this storage space up through a cooler to a distributing sparger arm in the top of the tank. The flow of the liquid through the sparger propels it and keeps it turning continuously, resulting in uniform distribution of the in-process vinegar over the packing. The liquid trick-

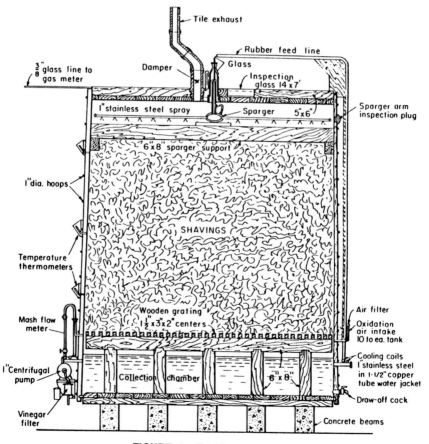

**FIGURE 1.** Trickling generator.

les down through the packing and returns to the storage reservoir. The temperature of the generator is regulated by flow rate and cooled to be between 29°C at the top of the generator to a maximum of around 35°C at the bottom of the packing. A blower is used to force air into ports in the false bottom and up through the packing. The air rate is valve-controlled and monitored by means of a rotameter. The top of the generator is covered but vented to the outside air. Excessive air rates waste alcohol and vinegar through the exhaust. An air-flow rate of about 80–90 ft³/ hour/100 ft³ of packing is considered adequate. The oxygen content of the exhaust can be measured with an oxygen meter. If this is done, the air rate should be regulated to maintain an oxygen concentration in the exhaust slightly above 12%.

The packing used in most generators consists of shavings made of air-dried beechwood planed to form coils about $1\frac{1}{4}$ inches in diameter and about 2 inches long. Although these coils tend to soften somewhat with age, the degree of settling is relatively minor compared to that obtained with cheaper packing material, such as corn cobs or other forms of wood shavings. Extractives from the fresh beechwood also are far more acceptable than those from other woods. While some supplemental packing may be required to compensate for settling, the life of a properly operated packed generator is well over 20 years.

Ethanol concentration is another critical factor in the generator operation. The amount of alcohol charged must be sufficient to yield the desired acetic acid concentration plus an amount that allows a residual of over 0.2% unconverted alcohol. If all alcohol is depleted from the charge, *Acetobacter* die and the generator is inactivated. The amount of finished vinegar withdrawn from the reservoir should be limited to an amount that, when replaced with the ethanol-contained charge (mash), will have a combined alcohol concentration of less than 5% by volume. This is especially important when the finished acidity is 12% or higher.

### 2. Efficient Operation of Generators

A common way of expressing acid production rates for this type generator is in terms of grain gallons per 100 ft³ of packing per day. The term grain gallon used in the industry in the United States is defined as 1 gal of 100 grain (10% acetic acid) vinegar. Percentage acidity is expressed as grams of acetic acid per 100 ml of vinegar. Production rates vary greatly in commercial generators; however, when producing 120 grain vinegar, a rate of 30 grain gal/100 ft³ of packing per day would be very good. These production rates may be calculated as follows:

$$P = \frac{24GV}{PH}$$

where $G$ = grain of draw-off
     $V$ = draw-off volume in gallons
     $P$ = packing volume in 100 ft³
     $H$ = cycle time—feed to draw-off in hours
    24 = factor converting hours to days

Calculations are applicable only when all conditions such as volumes and draw-off acidities are the same for extended periods. In addition to good operating practices, the final acidity or grain strength of the vinegar greatly affects the attainable acid production rate. With low acidities, high rates are readily produced. When acidities exceed 12%, however, rates tend to drop drastically. Generators are at times maintained on standby by operating at 13–14% acidities.

In trickling generators, wine and cider vinegar can be produced without adding supplementary nutrients, since sufficient nutrients already exist naturally in wine and cider. However, when a pure ethanol–water–acetic acid mixture is used, supplementary nutrients are necessary. A number of good prepared nutrients are commercially available. These very often contain yeast as a source of vitamins and other growth factors (Hildebrandt *et al.*, 1961). The formula for a simple, easily prepared, and completely soluble nutrient developed by USI researchers is as listed in the following tabulation.

| Ingredient | Weight/1000 gallon charge |
|---|---|
| Corn sugar | 1 lb–13 oz |
| Ammonium phosphate, dibasic | 13 oz |
| Magnesium sulfate | 3¼ oz |
| Potassium citrate | 3¼ oz |
| Calcium pantothenate (*d* form) | 1 gm |

The efficiency of the conversion of ethanol to acetic acid is another important factor for consideration in the operation of the trickling generator. From the reactions shown in Section II, it can be seen that 1 mole of ethanol is oxidized to 1 mole of acetic acid and 1 mole of water. From this it can be calculated that 1 gal of 12% ethanol by volume theoretically yields 1.035 gal of 11.98% acetic acid, which is equivalent to 1 gal of 12.4% acetic acid.

In actual practice some residual alcohol must always be retained. Thus, in making efficiency calculations this must be considered by either deducting it from the ethanol in the charge or adding its acid equivalent to the final product. A good efficiency in commercial trickling generators is considered to be in the range of 88–90% of the theoretical. Losses by evaporation and some conversion of alcohol to compounds other than acetic acid cannot be avoided.

### 3. Common Disorders and Problems

The vinegar eel (*Anguillula aceti*), a nematode, is a frequent source of trouble during vinegar manufacture, occurring more commonly in trickling generators than submerged culture operations. Eels are undesirable, for in large numbers, putrefaction of the dead organisms can cause off-odors. They feed on the zoogleal mat and organic matter in the generator, gaining access via dirt and insects. At one time they were so common in generators that they were thought to be essential for good acid production. Good filtration and pasteurization eliminates them from bottled vinegar. Only sterilization of the generators with live steam will eliminate them from infected tanks.

The troublesome vinegar mites may also appear in large numbers in trickling generators (Schierback, 1951). Species commonly found are *Tyroglyphas longior* and *T. siro*. These, too, can only be eliminated by steaming.

Infection with *Acetobacter xylinum* can result in coating the packing in generators with slime which can eventually lead to complete blockage. This is much more likely to occur when the generators are operated at acidities under 10%.

Excessive amounts of phosphates added with the nutrients can result in turbidity of the finished vinegar if hard water is used for dilution.

### B. Submerged Culture Generators

### 1. Description of Frings Acetator

The most widely used submerged culture vinegar generator is the Frings Acetator (Ebner, 1976) produced by the Heinrich Frings Company of Bonn, Germany (see Fig. 2). It is a skillfully engineered stainless steel fermentor that is equipped with a high-speed bottom-entering agitator, temperature control, and a foam breaker. The unique feature of this equipment is its highly efficient method of supplying air. This is accomplished by means of a high-velocity self-aspirating rotor on the bottom of the tank that pulls air in through a pipe that is connected to a rotameter and control valve on the outside. The rotor is hollow and star-shaped, with slots cut on the trailing side of the star points, and on turning distributes finely divided bubbles of the aspirated air. Stator vanes

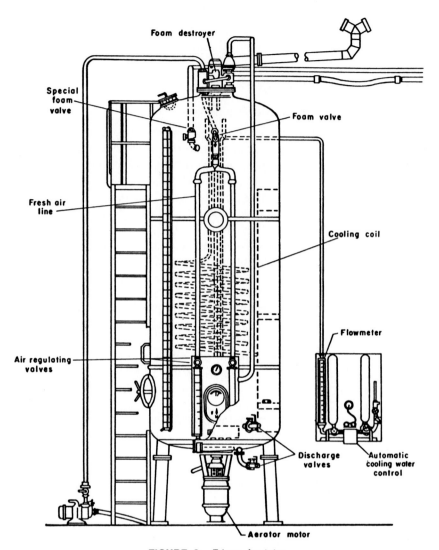

**FIGURE 2.**  Frings Acetator.

critically placed around the circumference of the rotor direct the bubbles so that even distribution of air is obtained over the entire bottom. The air bubbles rise, and the agitated liquid in the tank is uniformly supplied with air. This action is highly essential, for if oxygen starvation occurs at any time, the *Acetobacter* organisms are adversely affected immediately.

The temperature of this type of generator is controlled with cooling water circulated through internal coils. With the rapid production rates obtained in these generators, an adequate means of dissipation of the heat of oxidation is necessary. Temperature of operation is generally about 30°C, although the optimum temperature appears to vary inversely with the total concentration of alcohol and acid in the mash. Some strains of *Acetobacter* are more tolerant than others of higher operating temperatures, but most strains will die above 38°C.

Foam formation is a common problem in submerged fermentation generators, and foam-breaking mechanisms are absolutely essential. In the Acetator a centrifugal foam breaker is used that is automated by an electrode that senses both type and volume of foam.

The Acetator is operated batchwise and can be regulated by means of an Alkograph, an instrument that constantly monitors the alcohol concentration of the mash and is used to signal the completion of the batch. The automated system can be set to be triggered when the residual alcohol drops to 0.2% by volume. About 35% of the finished product is then pumped out. Fresh mash replaces this vinegar, and the cycle starts again. Cycle times for 12% vinegar are usually about 35 hours. The rate of production can be as much as ten times as great as that obtained in trickling generators of equivalent size.

Vinegar produced by the submerged process is extremely turbid, since it contains the bacteria that produced it. For filtration, large-capacity filters, filter aids, fining agents, and settling tanks are required. In contrast, when vinegar is produced in trickling generators, the vinegar organisms are retained on the shavings, and relatively small filter capacities are required.

### 2. Operation of the Acetator

To start a submerged culture generator, "warm" vinegar from a similar operation can be used as seed. This is freshly prepared vinegar that will contain some live organisms, which are activated under ideal conditions of aeration, alcohol concentration, and nutrients. This startup could take as much as 7–21 days. The Frings Company manufactures small laboratory (1.7 liters) and pilot-plant (5 liters) fermentors, which can be used for the cultivation of bacteria starting from lyophilized cultures of proven acid producers. The pilot units are constructed to be battery operated when normal current supply is interrupted. This assures the availability of a supply of organisms if the commercial units are shut down because of electrical failures.

The formula for a soluble nutrient that was developed in the USI vinegar laboratory for submerged culture follows in the tabulation below.

| Ingredient | Submerged culture weight/1000 gallon charge |
|---|---|
| Corn sugar | 9 lb |
| Ammonium phosphate, dibasic | 4 lb |
| Magnesium sulfate | 1 lb |
| Potassium citrate | 1 lb |
| Calcium pantothenate (d form) | 0.16 oz (5 gm) |

In this type of operation, where the bacteria are drawn off with the finished vinegar, a considerably greater concentration of food is required. In the formula above, the same ingredients are used as were given in Section III for trickling generators, but five times the amounts are required.

### 3. Comparison of Trickling and Submerged Operations

The desirable features and advantages of the Acetator over the trickling generator are numerous: (1) about one-sixth the amount of space is required to produce an equal quantity of vinegar; (2) invested capital as compared with productive capacity is much lower than for packed generator operations; (3) the efficiency is greater, and yields are 5–8% higher; (4) it is easy to switch from one type of vinegar to another and to start up to full production; and (5) generators can be highly automated.

## C. Other Vinegar Generators

### 1. Yeomans Cavitator

The Yeomans Cavitator (Beaman, 1967) is a submerged culture-type system somewhat similar to the Acetator but differing in the manner in which air is supplied. In this generator, which is no longer being manufactured, the rotor–agitator is top driven and withdraws liquid and air from a centrally located draft tube. The continuous liquid overflow into the draft tube is mixed with the finely divided air in an upward current and given uniform distribution of air through the tank. Some cavitators are still in use in the United States and Japan.

### 2. Bourgeois Process

The Bourgeois process is a submerged culture system used in Spain and Italy in which compressed air is distributed to the tanks through a rotor. It is a two-generator system wherein one is used to supply inoculum for the other. Batches are finished when all alcohol has been converted to acetic acid. Then the entire contents of the tank are withdrawn. Since vinegar bouquet and flavor are attributed to esters, some alcohol is subsequently added to the finished product to enhance its quality through esterification. The Bourgeois process is satisfactory only where low acidities and slow production rates can be tolerated.

### 3. Fardon Process

Fardon process generators are also of the submerged culture type. Liquid is pumped from the tank through a venturi nozzle for aeration and returned. Like the Bourgeois process, it is slow and inefficient and used mainly in Africa for making malt vinegar.

### 4. The Tower Fermentor

The Tower Fermentor is a relatively recent development in aeration systems patented in England (Greenshields and Smith, 1974) and used in the United Kingdom. It is a continuous trickling-type generator about 2 ft in diameter and 20-ft tall. Plastic perforated plates covering the cross-section are used for the distribution of air. It has been reported that this is satisfactory for producing all types of vinegar.

## V. FINISHED VINEGARS

### A. Federal Regulations

There are numerous types of vinegar on the market, including many with added spices and flavors. Basically, the major types are those listed in Food and Drug Administration Compliance Policy Guide 7109.04, Chapter 9—Condiments.

Policy:
FDA considers the following to be satisfactory guidelines for the labeling of vinegars:
Natural vinegars as they come from the generators normally contain in excess of 4 grams of acetic acid per 100 ml. When vinegar is diluted with water, the label must bear a statement such as "diluted with water to ____ percent acid strength," with the blank filled with the actual percent of acetic acid—in no case should it be less than 4 percent. Each of the varieties of vinegar listed below should contain 4 grams of acetic acid per 100 ml (20°C). Maximum residual alcohol concentration permitted is 0.5% by volume of the total product as sold.

Vinegars:
1. VINEGAR, CIDER VINEGAR, APPLE VINEGAR. The product made by the alcoholic and subsequent acetous fermentations of the juice of apples.
2. WINE VINEGAR, GRAPE VINEGAR. The product made by the alcoholic and subsequent acetous fermentations of the juice of grapes.
3. MALT VINEGAR. The product made by the alcoholic and subsequent acetous

fermentations, without distillation, of an infusion of barley malt or cereals whose starch has been converted by malt.

4. SUGAR VINEGAR. The product made by the alcoholic and subsequent acetous fermentations of sugar sirup, molasses or refiners sirup.

5. GLUCOSE VINEGAR. The product made by the alcoholic and subsequent acetous fermentations of a solution of glucose. It is dextrorotatory.

6. SPIRIT VINEGAR, DISTILLED VINEGAR, GRAIN VINEGAR. The product made by the acetous fermentation of dilute distilled alcohol.

7. VINEGAR, made from a MIXTURE OF SPIRIT VINEGAR AND CIDER VINEGAR. The product made by preparing a mixture of distilled vinegar and cider vinegar. This product should be labeled as a blend of the products with the product names in order of predominance. This labeling is applicable to a similiar product made by acetous fermentation of a mixture of alcohol and cider stock.

8. VINEGAR MADE FROM DRIED APPLES, APPLE CORES OR APPLE PEELS. Vinegar made from dried apples, apple cores or apple peels should be labeled as "vinegar made from _____," where the blank is filled in with the name of the apple product(s) used as the source of fermentable material.

## B. Popular Types of Vinegar

### 1. White Distilled Vinegar

For white distilled vinegar it is the ethanol used as the raw material that has been distilled. In the United States, ethanol purchased for vinegar making is denatured (U.S. Industrial Chemicals Company, 1969). The specially denatured alcohol (SDA 29) used by most manufacturers contains 1 gal of ethyl acetate in 100 gal of 190° proof ethanol. Other SDA formulas permitted but seldom used for vinegar production are SDA 18 and SDA 35-A. In 1975, 19.6 million gal of SD alcohol were sold for vinegar processing. The presence of the ethyl acetate denaturant in SDA 29 has no deleterious effect on acetic acid production. In fact, it is considered as an essential flavoring in vinegar. It is always present whether added or not, its concentration varying directly with residual alcohol content.

The alcohol can be either synthetic ethanol, derived from natural gas, or fermentation alcohol, usually from corn or molasses. Price normally determines the type of ethanol that is used, since both synthetic and fermentation alcohol can be of excellent quality. Vinegar produced from the two types can be differentiated by their $^{14}C$ contents using a liquid scintillation counter (Hildebrandt et al., 1969). Relatively high concentrations of $^{14}C$ are found in vinegar made from alcohol originating from recently photosynthesized carbohydrates, while low concentrations are found in vinegar from synthetic alcohol from fossilized materials. This method cannot distinguish between vinegar produced from fermentation ethanol and acetic acid (synthetic) produced by the destructive distillation of wood.

The flavor of distilled vinegar produced from high-quality ethanol, whether from the trickling or the submerged culture processes, is generally unaromatic in comparison with fruit or grain vinegars. Unaromatic vinegar is desirable in products that contain vinegar solely for its acid content. Its flavor and aroma can be even more subdued by holding the residual alcohol at a bare minimum so as to limit the ester content of the vinegar. On the other hand, if a highly aromatic distilled vinegar is desired, it can be obtained by adding to the 190° proof ethanol about 0.25–0.5% grain fusel oil. The higher alcohols that make up the fusel oil are partially esterified and add a fruity aroma to the vinegar.

Few studies have been made on the composition of distilled vinegar. Kahn et al. (1972), using gas chromatography–mass spectrometry, identified 27 compounds in samples of submerged and trickling vinegars. Other than the acetic acid these were mainly esters, alcohols, halogenated compounds, hydrocarbons, carbonyls, ethers, and acetals, with no significant differences attributed to the production method.

### 2. Cider Vinegar

Cider vinegar is made from the fermented juice of apples or apple-juice concentrate. Apple juice containing 11% soluble solids yields vinegar of approximately 55 grain. Most cider vinegar is bottled for retail trade and must be sparkling clear. Thus, special care is required in its filtration. Even after careful filtration, cider vinegar has a tendency to form a certain amount of colloidal turbidity, which in time will form a sediment. This is sometimes overcome by storage of the vinegar for a considerable period of time prior to filtration and bottling. Tannins from the apple that are present in cider vinegar make it extremely sensitive to blackening from iron contamination. As little as 1 ppm iron can result in a pronounced color change (Conner and Allgeier, 1976).

As would be expected, gas–liquid chromatography (GLC) analysis of cider vinegars shows the presence of many compounds not found in distilled vinegar. Unfortunately, none of these is in sufficient and uniform concentration to make possible the detection of adulteration of cider vinegar with the cheaper distilled vinegar. Complaints that some adulterated products are on the market are occasionally directed at vinegar producers.

### 3. Wine Vinegar

Lower quality grape wines are commonly used for acetous fermentation to wine vinegar. Both white and red wines are used. The comments made above regarding cider vinegar are also applicable to wine vinegar. Treatments that are specifically permitted for wine vinegar and not for others are the addition of sulfur dioxide (140 ppm) and the addition of wine (to give a maximum of 3.0% residual alcohol) to the finished 40-grain vinegar. The sulfur dioxide is added for preservation of color and for stability, and the wine is added for bouquet.

### 4. Malt Vinegar

Malt vinegar is made by the alcoholic and acetous fermentation of a malt mash or mash containing corn and barley in addition to the malt. Enzymes in the malt convert the grain starches to sugar, and the sugar is fermented by yeast to ethanol. This semiclarified fermented beer is then subjected to the usual acetous fermentation.

## VI. PROCESSING OF VINEGAR

### A. Filtration and Clarification

Distilled vinegar from trickling generators is relatively free of insoluble material, so filters of very small capacity suffice. However, other vinegars from the trickling process and all vinegars from the submerged process require the use of filter aids and fining agents to accomplish the desired clarity. Large-capacity filters are required when these additives are used.

### B. Bottling

In order to prevent bacterial growth subsequent to bottling, vinegar must be pasteurized. The filled bottles are tightly sealed, heated to 60°–65°C, and held at that temperature for 30 minutes prior to cooling. A somewhat less desirable procedure is sometimes used that consists of heating the vinegar to 65°–70°C and filling the bottles with this hot vinegar and sealing immediately.

### C. Concentration of Vinegar

Vinegar can be concentrated by a freezing process. The vinegar is frozen to a slush, which is centrifuged to separate the concentrated vinegar from the low-acid-content liquid. The water from this melted ice contains about 1–2% acid and can be used for the dilution of 190° ethanol in making up mash for the generators. This process was developed and commercialized by Girdler Process Equipment Division, Chemtron Corporation, Louisville, Kentucky. Vinegars of 200-grain strength are readily obtainable from 120-grain raw vinegar. Although 300-grain vinegar can be made, the production cost is relatively high.

There are a number of advantages to 200-grain vinegar. Transportation costs are substantially reduced by the reduction in volume; also, it is of special value to pickle processors, since it enables them to build up the acid strengths and reuse brine solutions that otherwise would have to be discarded because of dilution by the juice of the pickles.

## VII. ANNUAL PRODUCTION AND USES

The total sales of all types of vinegar in 1976 in the United States amounted to about 148 million gal. Broken down as to types, the values are given in the tabulation below.

| Type | Millions of gal (approximate) |
| --- | --- |
| Distilled | 110 (100 grain) |
| Cider | 26 (50 grain) |
| Wine | 6 (50 grain) |
| Others | 6 (50 grain) |

Approximately 68% of all vinegar produced is used in products commercially prepared (in order of the volume sold to each) by the dressing and sauce industry, the pickle industry, tomato products manufacturers, and the mustard industry.

## VIII. APPENDIX

The Vinegar Institute* is a trade organization of vinegar manufacturers. It is made up of approximately 25 active members, who are the principal manufacturers and bottlers of vinegar. The organization has worked closely with the FDA to establish high standards and exacting specifications for all types of vinegar. Environmental Protection Agency standards and labeling regulations applicable to vinegar manufacture are closely followed and explained to members.

The principal vinegar producers in the United States are: California-Omega Foods, Emeryville, California; Heinz U.S.A., Pittsburgh, Pennsylvania; Hunt Foods, Inc., Fullerton, California; Indian Summer, Inc., Evansville, Indiana; National Vinegar Company, St. Louis, Missouri; A. M. Richter Sons Company, Manitowoc, Wisconsin; Speas Company, Kansas City, Missouri; and Standard Brands, Inc., Baltimore, Maryland.

* 64 Perimeter Center East, Atlanta, Georgia 30346.

## REFERENCES

Allgeier, R. J., and Hildebrandt, F. M. (1960). *Adv. Appl. Microbiol.* **11,** 163–181.

Allgeier, R. J., Nickol, G. B., and Conner, H. A. (1974a). *Food Prod. Dev.* **8**(5), 69–71.

Allgeier, R. N., Nickol, G. B., and Conner, H. A. (1974b). *Food Prod. Dev.* **8**(6), 50–53 and 56.

Asai, T. (1968). "Acetic Acid Bacteria." Univ. Park Press, Baltimore, Maryland.

Beaman, R. G. (1967). *In* "Microbial Technology" (H. J. Peppler, ed.), pp. 344–376. Van Nostrand-Reinhold, Princeton, New Jersey.

Chemetron Corporation (undated). "Crystallization and Concentration. Subject: Vinegar," Tech. Bull. Votator Div., Chemetron Corp., Louisville, Kentucky.

Conner, H. A., and Allgeier, R. J. (1976). *Adv. Appl. Microbiol.* **20,** 82–133.

Ebner, H. (1976). *Ullmans Encyklo. Tech. Chemi.* **11,** 41–55.

Greenshields, R. N., and Smith, E. L. (1974). *Process Biochem.* **9,** 11–13, 15, 17, and 28.

Hildebrandt, F. M., Nickol, G. B., Dukowicz, M., and Conner, H. A. (1961). "Vinegar Newsletter," No. 33. U.S. Ind. Chem. Co., New York.

Hildebrandt, F. M., Nickol, G. B., and Conner, H. A. (1967). "Vinegar Newsletter," No. 50. U.S. Ind. Chem. Co., New York.

Hildebrandt, F. M., Weber, G. R., and Nickol, G. B. (1968). "Vinegar Newsletter," No. 54. U.S. Ind. Chem. Co., New York.

Hildebrandt, F. M., Nickol, G. B., and Kahn, J. H. (1969). "Vinegar Newsletter," No. 56. U.S. Ind. Chem. Co., New York.

Kahn, J. H., Nickol, G. B., and Conner, H. A. (1972). *Agric. Food Chem.* **20,** 214–218.

Nickol, G. B., Conner, H. A., Dukowicz, M., and Hildebrandt, F. M. (1964). "Vinegar Newsletter," No. 43. U.S. Ind. Chem. Co., New York.

Schierback, J. (1951). "The Manufacture of Vinegar," Tech. Inf. Serv., Rep. No. 17. Natl. Res. Counc., Ottawa, Canada.

Shimwell, J. L. (1954). *J. Inst. Brew.* **60,** 136–141.

U.S. Industrial Chemicals Company (1969). "Ethyl Alcohol Handbook." U.S. Ind. Chem. Co., New York.

Chapter 7

# Ketogenic Fermentation Processes

## D. PERLMAN

## I. INTRODUCTION

The ketogenic fermentation processes to be reviewed in this chapter are those in which polyhydric alcohols are converted into ketoses. The most important processes are the oxidation of D-sorbitol into L-sorbose, and of glycerol into dihydroxyacetone, both operations being carried out on an industrial scale using strains of *Acetobacter suboxydans*.

The type reaction for this bacterial oxidative ketogenic process is:

$$-\overset{|}{\underset{|}{C}}-OH \longrightarrow -\overset{|}{\underset{|}{C}}=O$$

The product of the reaction is governed by the choice of substrate and by the specificity of the enzyme system of *Acetobacter suboxydans*. The major requirements are high oxygen levels in the solution and an adequate nutrient supply.

## II. THE SORBOSE FERMENTATION

L-Sorbose is produced from the polyhydric alcohol D-sorbitol by the action of several species of bacteria of the genus *Acetobacter*.

**173**

MICROBIAL TECHNOLOGY, 2nd ed., VOL. II
Copyright © 1979 by Academic Press, Inc.
All rights of reproduction in any form reserved. ISBN 0-12-551502-2

*Acetobacter suboxydans* appears to be the species most widely used industrially, although many other species, including *A. xylinum, A. gluconicum, A. xylinoides, A. pasteurianum, A. kutzingianum, A. orleanse,* and *A. hoshigaki,* also produce sorbose readily.

$$
\begin{array}{ccc}
\text{CHO} & \text{CH}_2\text{OH} & \text{CH}_2\text{OH} \\
\text{H}-\text{C}-\text{OH} & \text{H}-\text{C}-\text{OH} & \text{C}=\text{O} \\
\text{HO}-\text{C}-\text{H} & \text{HO}-\text{C}-\text{H} & \text{HO}-\text{C}-\text{H} \\
\text{H}-\text{C}-\text{OH} \longrightarrow & \text{H}-\text{C}-\text{OH} \longrightarrow & \text{H}-\text{C}-\text{OH} \\
\text{H}-\text{C}-\text{OH} & \text{H}-\text{C}-\text{OH} & \text{HO}-\text{C}-\text{H} \\
\text{CH}_2\text{OH} & \text{CH}_2\text{OH} & \text{CH}_2\text{OH} \\
\text{D-Glucose} & \text{D-Sorbitol} & \text{L-Sorbose}
\end{array}
$$

The starting material in the sorbose fermentation is the sorbitol obtained when the aldehydic terminal group of the D-glucose molecule is reduced to a primary alcohol by either electrolytic or catalytic methods. When a nickel catalyst is used for the hydrogenation, care must be used to remove it from the purified product since *A. suboxydans* is very sensitive to nickel ion.

A procedure for the production of L-sorbose from D-sorbitol by *A. suboxydans* described by Wells *et al.* (1937, 1939) uses sorbitol solutions of 20–30% concentration, supplemented with 0.5% yeast extract (or the cheaper corn steep liquor) and sufficient $CaCO_3$ to neutralize the acidity of the medium (mainly from the corn steep liquor and a small amount of gluconic acid formed by oxidation of the traces of glucose remaining in the sorbitol syrup). Inoculum was a cell suspension of *A. suboxydans* grown in a medium containing sorbitol, glucose, yeast extract, and $CaCO_3$.

The fermentation conditions found optimal for this oxidative process included vigorous aeration, agitation, operation of the vessel at 30 psig air pressure, and incubation at 30°C. Wells *et al.* (1937, 1939) used small amounts of octadecanol to control foaming. They noted almost quantitative conversions of 20, 25, and 30% solutions of sorbitol in fermentation periods of 24, 30.5, and 45 hours, respectively.

At the completion of the incubation period, the sorbose-containing solution was mixed with decolorizing charcoal and filter aid and the suspension was filtered through a press. This was followed by concentration of the filtrate *in vacuo* to a syrup, which crystallized on cooling to 15°C. The crystals were separated by centrifugation, washed with ice water, and dried. A second crop of sorbose was obtained from the mother liquor and washings. The total recovered yield was about 70% based on the L-sorbose in the fermented medium.

More recently, attention has been focused on shortening the incubation time and increasing the process efficiency. Elsworth *et al.* (1959) have pointed out the potential of a continuous fermentation process as being more economical than a batch operation, and this has been explored by Muller (1966). Bull and Young (1977) reported enhanced efficiency in continuous fermentations on a laboratory scale with microbial cell recycle, obtaining both higher yields and reduced incubation intervals. Aeration using oxygen-enriched air was required to obtain maximal efficiency of the fermentation process.

Nearly all of the 70,000,000 lb of L-sorbose produced annually by this process is used as an intermediate in the chemical synthesis of L-ascorbic acid. The major manufacturer is Hoffmann-LaRoche Inc. (with operations in the United States and West Germany), and significant quantities are produced by E. Merck (West Germany), Pfizer, Inc. (United States), and Takeda Chemical Industries Ltd. (Japan).

## III. THE DIHYDROXYACETONE FERMENTATION

Dihydroxyacetone is produced by the microbial oxidation of glycerol with selected strains of *Acetobacter suboxydans*.

$$H-\underset{\underset{CH_2OH}{|}}{\overset{\overset{CH_2OH}{|}}{C}}-OH \longrightarrow \underset{\underset{CH_2OH}{|}}{\overset{\overset{CH_2OH}{|}}{C}}=O$$

This conversion was noted as early as the end of the 19th Century (Bertrand, 1898) and some realization of its use as a pharmaceutical intermediate followed (Cathcart and Markowitz, 1927; Lambie, 1927; Underkofler and Fulmer, 1937). The most recent interest (Andreadis and Miklean, 1960; Anonymous, 1960) is derived from its utility as a suntanning agent.

Initial studies (Visser t'Hooft, 1925) showed that aeration of the fermentation medium for 6 days increased the rate of conversion of glycerol to dihydroxyacetone as well as the efficiency of conversion—90% with aeration and 57% under static conditions. Virtanen and Barlund (1926) also noted highest conversions when the cultures were aerated and the pH was maintained at about pH 5.

Although a series of studies showed that increasing the initial concentration of glycerol from about 6 to 12% resulted in decreased efficiency of conversion (Underkofler and Fulmer, 1937; Low and Krema, 1928), this may now be explained on the basis of limited oxygen levels in the medium and also limited nutritional requirements (Flickinger and Perlman, 1977). Underkofler and Fulmer (1937) have also noted that conversions of 91% of the glycerol (6% solution) could be obtained in 3–7 days using submerged aeration when the basal medium contained 0.5% yeast extract and 0.25% $KH_2PO_4$. On the other hand, Bernhauer and Schon (1928) noted that a 30-day incubation period was needed to convert a 12% glycerol solution with only a 43% conversion efficiency.

The current process requirements for commercial production of dihydroxyacetone include a conversion of a 10% glycerol solution efficiently, e.g., 95% or better, to dihydroxyacetone in a short fermentation cycle. Green (1960; Green et al., 1961) has described a process in which the mash contains 0.5% $KH_2PO_4$, 0.5% brewers' yeast, 2% $CaCO_3$, and 10% glycerol. If the initial pH was set at 5.5, about 60% of the fermented glycerol was converted in the 40-hour incubation period to dihydroxyacetone. He also commented that pilot-plant trials provided assurance that dihydroxyacetone yields could reach 93% at a plant level.

Flickinger and Perlman (1977) reported that *Gluconobacter melanogenus* IFO 3293 converted glycerol to dihydroxyacetone in a 25-liter batch fermentor when grown in a yeast extract–phosphate medium. The efficiency of conversion was increased over threefold when oxygen tension was controlled by increasing the partial pressure of oxygen in the aeration. Under optimal conditions, 0.05 atm dissolved oxygen, quantitative conversion of 10% glycerol solutions were obtained.

Current commercial manufacture of dihydroxyacetone involves a recovery system where carbon treatment is used to decolorize the fermented medium, the cells are removed by filtration, and the clarified liquor is concentrated *in vacuo*. Suitable solvents, e.g., acetone, isopropyl alcohol, are added to induce crystallization of the desired product and care is taken to prevent polymerization. The white crystalline dihydroxyacetone produced commercially has an ash content of not over 0.2% and a purity of over 98%.

Companies currently in commercial operations include G. B. Fermentation Industries, Inc. (United States), E. Merck (West Germany), S. B. Penick Company (United States), Roussel-UCLAF (France), Schering Corporation (United States), Tanabe Seiyaku Company, Ltd. (Japan), and Wallerstein Laboratories (United States).

## REFERENCES

Andreadis, J. T., and Miklean, S. (1960). U.S. Patent 2,949,403.
Anonymous (1960). *Chem. Eng. News.* Feb. 22; Aug. 15.
Bernhauer, K., and Schon, K. (1928). *Hoppe-Sayler's Z. Physiol. Chem.* **177,** 107.
Bertrand, G. (1898). *C. R. Hebd. Seances Acad. Sci.* **126,** 842.

Bull, D. N., and Young, M. D. (1977). *Abstr. Pap. Meet., 174th, Am. Chem. Soc.* p. 01.
Cathcart, E. P., and Markowitz, J. (1927). *Biochem. J.* **21,** 1419.
Elsworth, R., Telling, F. C., and East, D. N. (1959). *J. Appl. Bacteriol.* **22,** 138.
Flickinger, M. C., and Perlman, D. (1977). *Appl. Environ. Microbiol.* **33,** 706.
Green, S. R. (1960). U.S. Patent 2,948,658.
Green, S. R., Whalen, E. A., and Molokie, E. (1961). *J. Biochem. Microbiol. Technol. Eng.* **3,** 351.
Lambie, C. G. (1927). *J. Soc. Chem. Ind.* **46,** 300.
Low, A., and Krema, A. (1928). *Klin. Wochenschr.* **7,** 2432.
Muller, J. (1966). *Zentralbl. Bakteriol., Parasitenkd., Infektionskr. Hyg., Abt. 2,* **120,** 349.
Underkofler, L. A., and Fulmer, E. I. (1937). *J. Am. Chem. Soc.* **59,** 301.
Virtanen, A. I., and Barlund, B. (1926). *Biochem. Z.* **169,** 169.
Visser t'Hooft, F. (1925). Dissertation. Technical University, Delft, Holland.
Wells, P. A., Stubbs, J. J., Lockwood, L. B., and Roe, E. T. (1937). *Ind. Eng. Chem.* **29,** 138.
Wells, P. A., Lockwood, L. B., Stubbs, J. J., Roe, E. T., Porges, N., and Gastrock, E. A. (1939). *Ind. Eng. Chem.* **31,** 1518.

Chapter 8

# Mushroom Fermentation

RANDOLPH T. HATCH
STANLEY M. FINGER

## I. INTRODUCTION

The worldwide demand for edible mushrooms has increased significantly in the last decade to the point that there are now in excess of 1.5 billion lb produced annually. The demand for fresh mushrooms in the United States has been particularly strong. The per capita consumption has increased from 1.6 to 2.2 lb over the last decade. This is reflected in the nationwide production of mushrooms which has grown from 108 million to 399 million lb since 1959 as shown in Table I. The demand for imported mushrooms (primarily canned) has grown even faster to the point that imports now account for 24% of the United States market. This has led to a recent shift in the nature of the industry from that of the small

**179**

MICROBIAL TECHNOLOGY, 2nd ed., VOL.II
Copyright © 1979 by Academic Press, Inc.
All rights of reproduction in any form reserved. ISBN 0-12-551502-2

**TABLE I.**  United States Mushroom Production and Imports[a]

| | | Canned imports | |
|---|---|---|---|
| Crop year | U.S. Production | Production | Percentage of U.S. Market |
| 1959–1960 | 108 | 3 | 2.7 |
| 1960–1961 | 115 | 5 | 4.2 |
| 1961–1962 | 127 | 15 | 10.6 |
| 1962–1963 | 132 | 16 | 10.8 |
| 1963–1964 | 131 | 21 | 13.8 |
| 1964–1965 | 140 | 16 | 10.3 |
| 1965–1966 | 156 | 21 | 11.9 |
| 1966–1967 | 165 | 26 | 13.6 |
| 1967–1968 | 181 | 34 | 15.8 |
| 1968–1969 | 189 | 31 | 14.1 |
| 1969–1970 | 194 | 42 | 17.8 |
| 1970–1971 | 207 | 43 | 17.2 |
| 1971–1972 | 231 | 62 | 21.2 |
| 1972–1973 | 254 | 74 | 22.6 |
| 1973–1974 | 279.5 | 70 | 20.0 |
| 1974–1975 | 299.1 | 77 | 20.5 |
| 1975–1976 | 309.8 | 88 | 22.1 |
| 1976–1977 | 347.1 | 107 | 23.6 |
| 1977–1978 | 399 | — | — |

[a] Million pounds, fresh weight.

independent grower toward an industry dominated by the large producer. Although there are more than 500 independent mushroom growers, the Butler County Mushroom Farm, Ralston Purina, Castle & Cook, and Campbell Soup account for more than 50% of the total production. The geographic distribution of the producers as shown in Fig. 1 is nationwide, with the center of the largest growing area at Kennett Square, Pennsylvania.

A number of different mushrooms are cultivated worldwide. The most common of western countries are *Agaricus bisporus* and *Agaricus bitorques*. The variety Golden White of *A. bisporus* is grown by the Butler County Mushroom Farm due to the ease of production and quality of mushroom produced in its underground caves. The optimum fruiting temperature lies in the range of 14°–18°C. The *Agaricus bitorques,* which fruits at a higher temperature, is grown in Holland. The *A. bitorques* has the commercial advantages of freedom from viruses and ease of hybridization. Because of its higher fruiting temperature, *A. bitorques* is ideally suited to above-ground production in climatically controlled rooms.

The mushroom traditional to Japan is *Lentinus edodes* (Hayes, 1974), otherwise known as Shii-take. This mushroom is unique in the method of

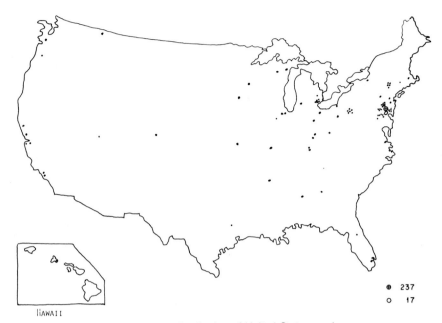

HAWAII

**FIGURE 1.** Geographical distribution of United States mushroom growers.

its culture. Hardwood logs are inoculated through plugs drilled into the logs. After 8 months the logs are moved to a site which is optimum for fruiting body formation. The mushroom caps can then be harvested for up to 6 years. This is a low-cost process which takes advantage of natural environmental conditions. The fruiting bodies form in the spring and fall when the ambient temperatures are in the range of 12°–20°C.

Other traditional mushrooms of the orient are the oyster (*Pleurotus oestreatus*) and padi-straw (*Volvaria volvaceae*) mushrooms (Hayes, 1974). The oyster mushroom can be grown on a variety of composted plant residues and industrial wastes. Its production differs from that of *Agaricus* in that fruiting body formation requires low light intensity and no casing layer. The name is derived from the oyster shape of the fruiting bodies. The padi-straw species is a tropical mushroom which is well suited to the high temperatures of the tropical climates. The traditional method of production utilizes rice straw as the substrate with minimal composting before inoculation with *V. volvaceae* mycelium. The modern methods of cultivation used in *A. bisporus* production are now being adopted for *V. volvaceae* due to its ability to achieve much higher yields.

A great deal of controversy exists concerning the way mushrooms are, and should be, grown. This is due to the differing requirements of the wide variety of mushrooms produced around the world and the ease in producing mushrooms (but at low yields). Fruiting body formation is often possible for temperature ranges up to 10°C and in a moist atmosphere. The mushroom mycelium usually exhibits some enzyme activity against cellulose and lignin and can assimilate a wide variety of crude carbon sources. Once established, the mushroom mycelium produces antibacterial and antimycotic substances which protect it from microbial attack. Although mushroom growing may appear to be foolproof, the production of high yields of consistent quality is quite challenging.

During the 6-week production period an experienced grower will harvest ½–1 lb of mushrooms per pound dry weight in the bed or 2.5–4.5 lb/ft² of bed area. Large producers, such as the Butler County Mushroom Farm, have developed highly successful processes through the use of a professional approach to the art of mushroom production. In the following sections the variables which influence mushroom production will be discussed in detail and illustrated with the process at the Butler County Mushroom Farm as shown in Fig. 2.

## II. MUSHROOM FERMENTATION

It is possible to grow mushroom mycelium on a wide variety of substrates in submerged culture (Litchfield, 1967) and on semisolid mate-

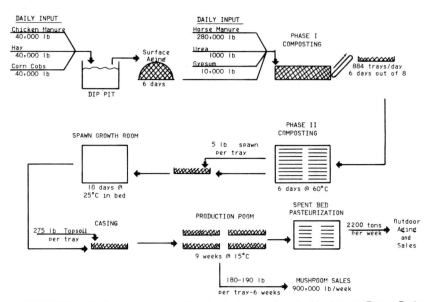

**FIGURE 2.** Mushroom process flowsheet, Butler County Mushroom Farm, Butler County, Pennsylvania.

rials. Since the mycelial growth is quite slow, the most practical and most common method of growth is on a composted bed of semisolid materials. Although the nutritional requirements are not completely understood, it has been established that the mushroom can metabolize a wide variety of carbon sources, including cellulose and lignin (Rancheva, 1971).

The production of mushrooms consists of essentially three steps: (1) Substrate preparation, (2) inoculation, and (3) induction of fruiting body formation. Each of these steps will be examined in detail.

## A. Substrate Preparation

Mushroom growing on a commercial scale always starts with the production of a suitable compost material. Proper composting is essential for successful mushroom growing since the resulting yield of mushrooms is strongly affected by the availability of nutrients and the presence of toxic substances, competing weeds, and disease microorganisms. In a general sense the composting process can be said to convert the starting materials to a form suitable for mushroom growth and unsuitable for the growth of contaminants. The substrate materials usually consist of manure, corn cobs, straw, and other agricultural residues, although municipal garbage has also been used (Franz, 1972). The one notable exception is the use of cow manure which has been found to result in poor mushroom yields. (This is probably due to the efficiency of the cow's digestive system.) The composting is accomplished in two steps: (1) Phase I, usually conducted outdoors for the initial breakdown of the raw materials; and (2) phase II, conducted indoors at controlled elevated temperatures for the conversion of free ammonia to protein and the growth of thermophilic microorganisms (Regan and Jeris, 1970; Shilesky and Maniotis, 1969; Siu, 1951; Tribe, 1962) (possibly with an amino acid profile closer to the optimum for mushroom growth).

During the initial stages of a compost pile, decomposition is quite rapid, causing temperatures to rise to the thermophilic region and stay there for an extended period. During this time materials that are easily converted are preferentially reacted. Easily transformed materials include such carbohydrates as sugars and starches (Keller, 1961; Gray, 1967; Braun, 1962), hemicelluloses, polysaccharides, and protein (Fuller and Bosma, 1965). As much as 80–95% of the easily decomposed portion of the substrate in the pile is converted (Keller, 1961). As the easily decomposed material is being used up, the microorganisms begin to attack the more resistant parts of the substrate, such as cellulose and lignins (Gotaas, 1956; Keller, 1961; Gray, 1967; Fuller and Bosma, 1965; Braun, 1962; Fergus, 1969).

## B. Carbon-to-Nitrogen Ratio (C/N)

Carbon and nitrogen are both needed for cellular growth. Living organisms utilize about 30 parts of carbon for every part of nitrogen (Mercer et al., 1962; Waksman, 1938). An excess of one or the other of these elements during aerobic microbial growth causes undesirable results. If carbon is in excess, several generations of organisms are needed to consume the additional carbon (Gotaas, 1956; Fuller and Bosma, 1965), thereby extending the time necessary for decomposition. If nitrogen is in excess, all the carbon is consumed and the extra nitrogen is released as ammonia (Gotaas, 1956; Fuller and Bosma, 1965). Several groups have studied the question of what is the optimum value of C/N for the most rapid and effective composting. It is not surprising that most of the research shows an optimum of about C/N = 30, considering this is approximately the stoichiometric proportion at which these elements are utilized by living organisms (Mercer et al., 1962; Waksman, 1938). It is generally agreed that composting time "increases considerably with increases in C/N ratios above 40." (Fuller and Bosma, 1965). On the low side, a C/N of 10 results in satisfactory composting, while values of 14–20 are common (Gotaas, 1956). It should be pointed out that the carbon and nitrogen may exist in unbiodegradable states and the unavailable portion does not belong in a calculation of C/N. For example, carbon in the form of fuel remains (coal, coke) and nitrogen in plastics, rubber, etc., are very resistant to aerobic decomposition (Gotaas, 1956; Keller, 1961).

## C. Water

Microorganisms require sufficient water in order for metabolism to proceed. However, if the moisture content of a compost pile is too high, the water will displace the air in the voids, causing anaerobic conditions (Gotaas, 1956). Thus, there is an optimal range of water content for most effective composting. This range has been reported to be 40–60% (Gotaas, 1956), 60% (Shulze, 1961), 52–58% (Anon., 1965), 55–69% (Wiley and Peace, 1955), and 30–75% (California, 1953). In general, it is desirable to have as much water as possible adsorbed into the straw and other materials being composted, but free water in the voids of the pile should be kept to a minimum.

Anyone who has seen an active compost pile cannot help but notice the prodigious amounts of water vapor generated. Water loss measurements have shown that the water content in a compost pile decreases quite significantly during composting unless additional water is added (Mercer et al., 1962; Kochitzky et al., 1969; Rose et al., 1965).

## D. Acidity

As with many of the other physiological variables, the pH of the compost changes during the composting process. Initially the pile becomes more acidic as long as the substrate is not highly buffered (Keller, 1961). This is due to the fact that the decomposition products of the easily attacked components of the substrate are simple acids (Gray, 1967). As the composting progresses, the pile becomes alkaline and ammonia is produced. If the initial pH of the substrate is low, however, as in the case with fruit wastes (pH of 4.5–5.0), there is an initial lag in the development of microbial flora and hence a lag in the rise of the temperature to the thermophilic region (Mercer et al., 1962; Rose et al., 1965).

## E. Phase I Composting

The first phase is usually carried out in windrows, which are simply large piles of the substrate material. A windrow is made up each day of the following mixture:

| Component | Quantity (lb) |
| --- | --- |
| Race horse manure | 280,000 |
| Corn cobs | 40,000 |
| Hay | 40,000 |
| Chicken manure | 40,000 |
| Urea | 1,000 |
| Gypsum | 10,000 |

The hay, corn cobs, and chicken manure are first dipped in water and allowed to age 6 days before being mixed with the other ingredients. The compost is mixed and formed into rectangular piles 6.5 ft high × 7.5 ft wide × 200 ft long, as shown in Fig. 3. The compost piles are turned four times over a period of 8–9 days, 50 hours between turnings. The turning or mixing of the compost can be done with agricultural equipment or, in large-scale composting, with equipment especially designed to mix and reform the pile in one continuous operation is often used. After forming the pile, the temperature increases rapidly to the thermophilic region (>45°C) and will reach temperatures of about 60°–70°C (Gotaas, 1956). During this time the pile shrinks and the color and physical appearance of the substrate changes. The thermophilic stage is also marked by a rise in the carbon dioxide concentration and a drop in the oxygen concentration often down to 1% or less (Keller, 1961).

**FIGURE 3.** Compost windrows. (Courtesy of Butler County Mushroom Farm, Butler County, Pennsylvania.)

### F. Temperature

Aerobic microbial growth during composting generates a good deal of heat. Since composting materials are relatively good insulators, a temperature distribution occurs within the pile with the highest temperatures being in the interior. Temperature distributions in actual compost piles have been measured as illustrated in Fig. 4 (Lambert, 1934). It is interesting to note that measured temperature distributions are quite irregular. This is due to the nonhomogeneity of the pile from one point to another in terms of composition, density, water content, etc. Recently, a mathematical model of the composting process was developed (Finger, 1975; Finger *et al.*, 1976) for the prediction of conditions and reaction rates

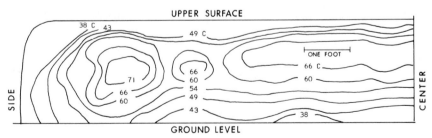

**FIGURE 4.** Measured temperature distribution of a compost pile.

over the cross-section of the pile. It was assumed in the development of the model that the rate of substrate decomposition is limited by the transfer rates of heat out of and oxygen into the pile. Extensive data taken from compost piles were then used to establish the rate constants in the model. Computer predictions of the temperature distributions within the pile at various times after turning are shown in Fig. 5a–d.

The temperatures reached in the pile depend on the ambient temperature, water content, degree of aeration, substrate materials, and pile size. The maintenance of aerobic conditions is very important in achieving thermophilic conditions as anaerobic decomposition processes generate much less heat than aerobic processes. The common way to maintain aerobic conditions is by periodic turning of the pile. After the turning, the temperature in the pile drops somewhat but returns to its original high value within a few hours (Gotaas, 1956).

The effect that pile size has on temperature is quite significant; a moderate increase in pile size will increase the temperatures measured within the pile. In one experiment the height of a compost pile was changed from 5 ft to 2 ft and back again. When the pile was 5-ft high the maximum interior temperature was 70°C (ambient temperature was 15°–20°C). When the height was reduced to 2 ft the maximum temperature dropped to 65°C within 3 hours. When the pile was reconstructed to its original 5-ft height, the temperature returned to 70°C (Gotaas, 1956). Thus, the interior temperatures in the pile can be controlled somewhat by varying pile size and, indeed, this technique is used commercially to obtain suitable interior temperatures. The effect of size on compost temperatures is directly related to the fact that compost substrates are good thermal insulators. Thus, the larger the pile, the more insulation there is to keep in the heat generated during the decomposition reactions. Convective heat loss at the surface should also be minimized by using a tightly packed outer shell. This leads to more uniformly high temperatures over the cross-section.

## G. Oxygen

Aerobic microbial growth requires oxygen to be present for the enzymatic reactions to proceed (Aiba et al., 1965). This oxygen comes from the surrounding atmosphere and diffuses into the compost pile. Although it might seem that the pile could be made sufficiently porous, analyses of the air in compost piles "showed that even with high porosity the gas exchange was retarded" (Tietjen, 1964); that is, concentration gradients were observed. Since the biochemical reactions occur in an aqueous environment, the oxygen must diffuse into the aqueous layer around the substrate and perhaps even diffuse into the pores of the substrate before

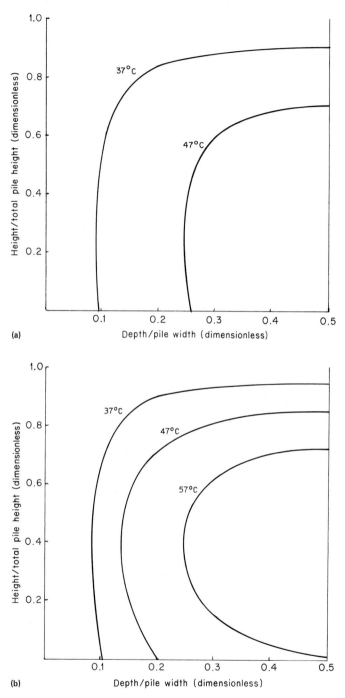

**FIGURE 5.** Predicted temperature distribution (a) 12 hours after turning; (b) 24 hours after turning.

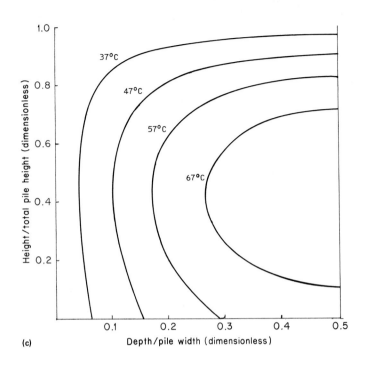

(c)

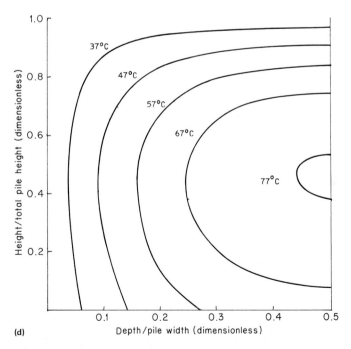

(d)

**FIGURE 5.** (c) 38 hours after turning; (d) 50 hours after turning.

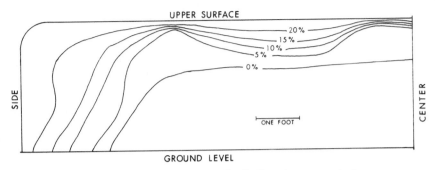

**FIGURE 6.** Measured oxygen distribution of a compost pile.

it reacts. Thus, even if the pile is loosely packed, diffusion of oxygen is still an important factor. These resistances to diffusion cause an oxygen distribution across the pile with the lowest oxygen concentrations in the interior. An experimentally measured distribution is shown in Fig. 6 (Lambert and Davis, 1934). Computer predictions of the progression of oxygen concentration distributions at various times after turning based on the model of phase I composting discussed above are presented in Fig. 7a–d (Finger, 1975; Finger et al., 1976).

The need for aeration of the pile has been shown both experimentally and by practical experience. Laboratory experiments using a pulverized substrate showed that the oxygen concentration fell to less than 1% unless the substrate was rotated (Gray, 1967). Practical experience has shown that frequent turning of the pile is necessary to maintain aerobic conditions.

The oxygen requirements of an aerobically decomposing substrate increase markedly with temperature. The oxygen consumption was found to be five to ten times greater at thermophilic temperatures as compared to completely decomposed compost (Schulze, 1957).

### H. Phase II Composting

At the end of the phase I process, the compost is loaded into ricks or trays for phase II composting. The main objectives of phase II composting are to (1) eliminate free ammonia from the phase I material, and (2) pasteurize the compost to eliminate the microbial and insect pests.

The free ammonia, which is toxic to mushrooms, is either converted to microbial protein by thermophilic microorganisms (Tschierpe, 1973) or evaporated and thereby removed from the compost. The portion of the ammonia converted to protein is, of course, available for utilization by the mushroom mycelium. Pasteurization is required to kill pests and

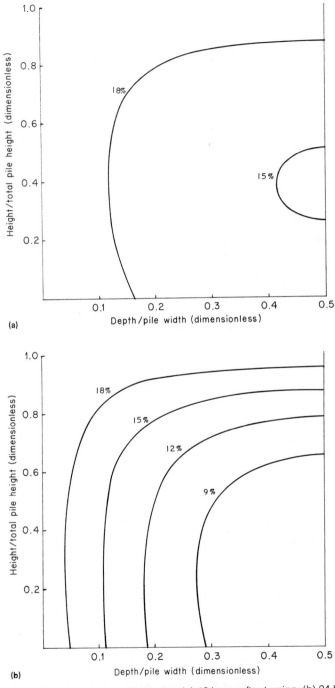

**FIGURE 7.** Predicted oxygen distribution (a) 12 hours after turning; (b) 24 hours after turning.

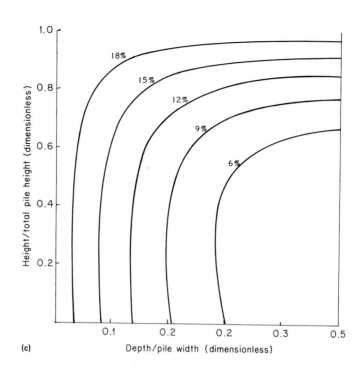

(c)

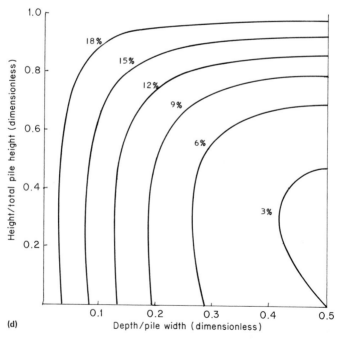

(d)

**FIGURE 7.** (c) 38 hours after turning; (d) 50 hours after turning.

weeds which could compete with mushroom growth or injure the mushroom.

The phase II process is much more controlled than the first phase of composting. Phase II is generally performed in an enclosed chamber or room. Environmental conditions, such as temperature, humidity, and gas content of the air, are controlled. Room temperatures up to 71°–77°C have been reported (Franz, 1972). The humidity is high (close to 100% relative humidity) and is controlled by blowing air through the room. This also controls the oxygen and carbon dioxide levels and removes any vaporized ammonia. Ricks or trays are used for the phase II process so that relatively uniform compost conditions can be achieved. The temperatures reached by the compost during phase II are sufficient to promote the growth of thermophilic microorganisms. It is important, therefore, that sufficient carbohydrate remain after phase I composting to promote strong thermophilic growth during phase II.

The compost is loaded into deep wooden trays (10 in × 10 ft × 4.5 ft). The first day live steam is used to bring the room temperature up to 60°C. Thermophilic growth is then usually sufficient to maintain the tray temperature for the next 6 days. The room temperature is then lowered and after 48 hours the temperature in the bed (tray of compost) is 21°–29°C. Close control of the phase II process is very important in maximizing mushroom yield. Once the ammonia has been eliminated and pasteurization has taken place, the temperature should be dropped to the spawning temperature. Prolonged phase II composting will adversely affect nutrients in the compost, resulting in lower mushroom yields. The nitrogen content of the bed prior to spawning is approximately 2–2.5% of dry weight.

## III. SPAWN PRODUCTION

Once the phase II composting is complete, the compost bed is ready for inoculation. The inoculum is prepared from preserved mushroom spores selected from a variety according to the following criteria: (1) Type of mushroom growth, (2) mushroom yield, (3) growth rate, (4) mushroom size, and (5) mushroom sensitivity to environmental conditions.

Mushroom spores are collected on a card placed under the mushroom cap. Once collected, the spore viability can be preserved indefinitely under refrigeration. [*Agaricus bisporus* spores have been reported by J. Sinden (Personal communication, 1977) to remain viable after storage at

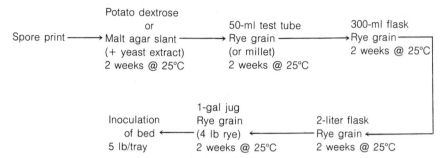

**FIGURE 8.**   Spawn propagation from mushroom spores.

room temperature since 1937.] The mushroom spores are then propagated as illustrated in Fig. 8.

The rye grain medium is quite simple and composed of the following: Rye, 49%; water, 50%; and calcium carbonate, 1% (precipitated chalk). The calcium carbonate is necessary to remove organic acids such as oxalic acid in order to prevent autolysis. The rye acts as a support for mushroom growth as much as a nutrient source since very little rye is actually consumed by the mushroom mycelium.

The spawn is used to inoculate the phase II compost bed once the bed temperature has dropped to approximately 27°C. The spawn is added at the ratio of 5 lb spawn to each tray (1500 lb total weight per tray, 360 lb dry weight). The trays are then placed in an incubation room with a controlled air temperature of 17°C and a bed temperature of 21°–29°C.

After inoculation of the bed, the mycelial growth goes through a lag phase of several days. By the end of the first week, however, there is sufficient growth to produce large amounts of carbon dioxide and heat. In a typical growth room 36,000 ft³/m of saturated 13°C air are required to maintain the room air temperature at 17°C (100% humidity) for 450 trays. This level of air flow is required for the last 11 days of the 18-day spawn growth phase.

## IV. MUSHROOM FORMATION

Once the spawn is fully developed in the tray (usually after 18 days), the tray is ready to support the healthy production of mushrooms. The tray is removed from the spawning room and cased. This involves the addition of a 2-inch (~6 lb/ft²) layer of top soil to the top of the bed. The casing is moisturized to about 50% saturation to allow the even spreading over the surface of the tray. (At the Butler County Mushroom Farm the casing is made from the spent mushroom trays which are aged outdoors for 2 years.)

After casing, the trays are moved into production rooms where the air temperature is maintained at 14°–17°C. The first step in the formation of a mushroom cap is mycelial cell differentiation to the growth of fruiting body cells. The fruiting body starts as a pinhead-sized mass of mushroom tissue at or near the surface of the bed which is the start of the mushroom.

Although the various factors which affect the development and growth of fruiting bodies are only partially understood, the effect of carbon dioxide has been found to be quite important as shown in Table II. Usually the carbon dioxide level in the production room is raised to 0.5% in order to suppress fruiting for the first week until the mycelium completely penetrates the casing.

Other factors which influence the fruiting include the air temperature and the nature of the casing. At a temperature above 22°C, *Agaricus bisporus* stops fruiting. Some varieties of *Agaricus bitorques* will, however, form fruiting bodies at higher temperatures. The function of the casing in the formation of fruiting bodies has been determined to be a combination of effects (Hayes, 1973), including (1) starvation of mycelium; (2) removal of mycelial metabolites; (3) diffusion barrier for mycelial metabolites; (4) carbon dioxide gradient from bed to casing; and (5) microbial activity which produces substances required for fruiting, such as reduction of $Fe^{3+}$ to $Fe^{2+}$ by *Pseudomonas putida,* and/or removal of mycelial by-products. Although sterile casing does not promote fruiting, the use of activated carbon has been found to work as well or better than topsoil.

After 3 weeks in the production room, the mushrooms are ready to be picked from the trays as shown in Fig. 9. Mushrooms are all hand picked and trimmed. It is necessary to remove the butt of the stem (~20% of the total mushroom) due to the casing particles which are intertwined with the mushroom tissue. Approximately 180–190 lb of salable mushrooms are harvested per tray over a 6-week period. During this time the surface of the tray is moistened with a fine water spray several times a week to

**TABLE II.**  Effect of Carbon Dioxide on Fruiting Body Formation

| Percentage $CO_2$ content of air | Effect |
| --- | --- |
| 0.5 | Fruiting inhibited |
| 0.2 | Fruiting disturbed |
| 0.1–0.15 | Optimum for good fruiting |
| 0.1 | Fruiting is slower |

**FIGURE 9.** Mushroom picking in production room. (Courtesy of Butler County Mushroom Farm, Butler County, Pennsylvania.)

maintain the moisture content of the casing. Approximately half of the water content of the mushroom comes out of the casing and the remainder from the original bed.

At the end of the 9-week period in the production room, the trays are removed and steamed to kill mushroom viruses, any insects, nematodes, and weed molds which may have entered the bed. Although the spent bed has a nitrogen content of 2.5% (dry weight) the residual nutrient content is not generally exploited. The primary uses are (1) top dressing for nurseries, (2) topsoil, and (3) casing soil for mushroom production. Approximately one-half of the spent bed material is aged outdoors for 2 years before recycle as casing soil. The remainder is sold immediately as top dressing or aged and sold as topsoil.

## V. PROCESSING PROBLEMS

A wide variety of difficulties can affect the production of mushrooms. The process must be carried out in a consistent manner to assure reasonable yields and high quality. The most difficult variable to control in this regard is the compost. The raw materials vary in their nutrient content, water-holding ability, and microbial activity. The phase I composting step is undertaken year-round outside and is affected to some extent by the weather. The raw materials are also stored unprotected outside. The specific problems with phase I composting tend to be (1) too much water, (2) too large an anaerobic region, (3) too loose a pile, (4) too large a pile, (5) too low or too high a nitrogen content, and (6) overcomposting. The first four problems affect the uniformity of the compost. The nitrogen content of the starting materials is usually adjusted to 1.5% (by dry weight). This will rise to 2.0–2.4% nitrogen by the end of the composting step due to the weight loss from carbon dioxide evolution and drainage of the pile. If the starting materials are overdosed with nitrogen, the free ammonia content will be too high (>0.05%) or the level of free amines will be too high. If the phase I composting is carried out too long, the remaining nutrient level will be too low for sufficient thermophilic microbial growth during phase II composting.

After spawning, a number of contaminants can infect the mycelial bed and the mushrooms. The most common problems are caused by (1) viruses, (2) bacterial diseases, (3) fungal diseases, (4) weed molds, and (5) insects (small flies, mites, etc.) and (6) nematodes. There are three different viruses which have been found to infect the mushroom and cause the following effects: (1) complete inhibition of mycelial growth; (2) formation of long, spindly mushrooms with small caps; (3) reduction of shelf-life; (4) reduction in mushroom yield; and (5) death of mushroom cap followed by softness and rotting. The viral particles tend to concentrate in the mushroom cap and enter the mushroom spores. The spores released by the infected cap will then act as the vector for transmission of the viruses.

The bacterial diseases are most commonly caused by *Pseudomonas* species. Two of the more common diseases are known as brown blotch and mummy disease. Brown blotch is a mahogany colored surface infection caused by *Pseudomonas tolaasi*. The mummy disease (also caused by a *Pseudomonas* species) can stop all cap formation and result in dried-out mushrooms and curved stems. The mummy disease is not transmitted through spores but has been shown to be carried by water drops from ceilings of spawning and pasteurizing rooms.

## VI. SUMMARY

Mushrooms have become an important, high-protein food in the United States; and annual production exceeded 349 million pounds in 1977–1978. The annual growth in sales has averaged 8.5% over the last 5 years based upon the pounds produced domestically and 9.2% when imports are included. The price has also risen significantly with a doubling of the fresh sales price since 1968. This sales growth has resulted in an increased competition for the raw materials. The annual consumption of horse manure now exceeds 350,000 tons and is rapidly approaching the supply limit. This is reflected in a doubling of the horse manure price over the last 2 years. The mushroom industry is countering the raw material problems by the use of a more scientific approach to the entire process in order to increase mushroom yields. Over the next decade, the composting step will undergo significant change. Out of necessity, less horse manure will be used and new raw materials will be substituted.

## ACKNOWLEDGMENT

The authors express their appreciation to John Yoder, president of the Butler County Mushroom Farm, and Dr. James Sinden for their significant contributions to the background information of this process.

## REFERENCES

Aiba, S., Humphrey, A. E., and Millis, N. F. (1965). "Biochemical Engineering," 1st ed. Academic Press, New York.
Anonymous, W. (1965). "Preliminary Report on a Study of the Composting of Garbage and Other Solid Wastes," Civ. Sanit. Eng. Rep. Michigan State University, Ann Arbor.
Braun, R. (1962). Compost Sci. **3,** 36.
Bulboaca, M. (1970). Lucr. Stiint., Inst. Agron. **13,** 53.
California (1953). "Composting for Disposal of Organic Refuse," Sanit. Eng. Res. Proj., Tech. Bull. No. 9. University of California, Berkeley.
Fergus, C. L. (1969). Mycologia **61,** 120–129.
Finger, S. M. (1975). Ph.D. thesis, University of Maryland, College Park.
Finger, S. M., Regan, T. M., and Hatch, R. T. (1976). Biotechnol. Bioeng. **18,** 1193.
Franz, M. (1972). Compost Sci. **13,** 6.
Fuller, W. H., and Bosma, S. (1965). Compost Sci. **6,** 26.
Gotaas, H. B. (1956). "Composting." World Health Organ., Geneva.
Gray, K. R. (1967). Compost Sci. **7,** 29.
Hayes, W. A. (1973). Mushroom News.
Hayes, W. A. (1974). Process Biochem. **9,** 21–28.
Keller, P. (1961). Compost Sci. **2,** 20.
Kochitzky, O. W., Seaman, W. K., and Wiley, J. S. (1969). Compost Sci. **9,** 5.
Lambert, E. B. (1962). Mushroom Dig. **47.**
Lambert, E. B., and Davis, A. C. (1934). J. Agric. Res. **48,** 587.
Litchfield, J. H. (1967). In "Microbial Technology" (H. J. Peppler, ed.), p. 107. Van Nostrand-Reinhold, Princeton, New Jersey.
Mercer, W. A., Chapman, J. E., Rose, W. W., Katsuyama, A., and Dwinnel, F., Jr. (1962). Compost Sci. **3,** 9.
Rancheva, T. (1971). Gradinar. Lozar. Nauka **8,** 135–140.

Regan, R. W., and Jeris, J. S. (1970). *Compost Sci.* **11,** 17.

Rose, W. W., Chapman, J. E., Roseid, S., Katsuyama, A., Porter, V., and Mercer, W. A. (1965). *Compost Sci.* **6,** 13.

Schulze, K. L. (1957). "Aerobic Decomposition of Organic Waste Material," Interim Reports No. 1, No. 2, and No. 3. College of Engineering, Michigan State University, East Lansing.

Schulze, K. L. (1961). *Compost Sci.* **2,** 32.

Shilesky, D. M., and Maniotis, J. (1969). *Compost Sci.* **9,** 20.

Siu, R. (1951). "Microbial Decomposition of Cellulose." Van Nostrand-Reinhold, Princeton, New Jersey.

Tietjen, C. (1964). *Compost Sci.* **5,** 8.

Tribe, H. T. (1962). *Compost Sci.* **2,** 38.

Tschierpe, H. J. (1973). *Mushroom J.* No. 1, p. 30.

Waksman, S. A. (1938). "Humus," 2nd ed., Williams & Wilkenson Co., Baltimore, Maryland.

Wiley, J. S., and Peace, G. W. (1955). *Proc. Am. Soc. Civ. Eng.* **81,** Paper No. 846.

Chapter 9

# Inocula for Blue-Veined Cheeses and Blue Cheese Flavor

GERARD J. MOSKOWITZ

## I. INTRODUCTION

Blue-veined cheeses, such as Roquefort, Gorgonzola, and Stilton, have been known for centuries. Their discoveries have been attributed to fortunate accidents where shepherds, having left sheep or goats' milk curd in a moist and cool cave, returned sometime later to find that a greenish blue mold had developed throughout the cheese. The mold was responsible for imparting the desirable flavor of the cheese. References to blue-veined cheeses were reported as early as A.D. 79 while

**201**

MICROBIAL TECHNOLOGY, 2nd ed., VOL.II

records of the Monastery of Conques mention Roquefort as early as the year 1070. Stilton cheese was of more recent vintage, the first records dating back to 1720 in England (Davis, 1965).

In the United States, blue cheese production has been primarily concentrated in Minnesota, Iowa, Illinois, and Wisconsin. In 1976, the quantity of blue variety cheese produced in the United States was 33,875,000 lb. This showed an increase of 19% over the 1975 figure of 28,506,000 lb and was comparable with an overall cheese production increase of 19%. The quantity of blue cheese produced in the United States has remained at about 1% of the total United States cheese production for a number of years (Dairy Products Annual Summary, United States Department of Agriculture, 1976).

The characteristic flavor of blue cheese has been attributed to the metabolic products of blue molds such as *Penicillium roqueforti*. Recent research on strain selection has resulted in the isolation of several *P. roqueforti* strains that have desirable properties for cheese manufacture. These include the ability of the organism to produce an adequate quantity of low molecular weight ketones and alcohols, which are responsible for the characteristic flavor. This, however, must be balanced with an appropriate amount of proteolytic activity, which has been shown to be important for the texture and background flavor of the cheese.

## II. PRODUCTION OF BLUE CHEESE

### A. Manufacture

In the United States, blue-veined cheeses are produced primarily from cow's milk, although goat milk is occasionally used. Milk is generally pasteurized and homogenized. Pasteurization destroys most of the microorganisms native to the milk, thereby allowing the cheese maker to more closely control the microbial flora of his cheese. Homogenization uniformly distributes the milk fat and makes the fat more susceptible to enzymatic hydrolysis. Milk fat is the substrate that provides the fatty acids from which the characteristic blue cheese flavor develops. This is described in greater detail in Section IV.

Lactic acid bacteria are first added to the milk. At the cheese maker's option, mold spores may also be added. When acid development has reached the desired stage, rennet is added to form the curd. After 60–90 minutes, the coagulum is cut, the whey drained, and the curd transferred to perforated hoops. Some cheese makers split the mold spore inoculum, adding one portion to the milk and the remainder to the drained curd with 1–2% salt.

The hoops are drained at room temperature, with frequent turning, usually overnight. The wheels of cheese are rubbed with salt daily, up to 5 days, or bathed in saturated brine for 24–48 hours. At 1 or 2 days of age the loaves are pierced, top and bottom, with a gang of stainless steel needles. The penetration of air is essential for mold growth and flavor development. Blue-veined cheeses are cured for at least 60–90 days at about 8°C in 95% relative humidity. For additional details see this volume, Chapter 2, and Davis (1976).

## B. Standards of Identity

The 1977 Code of Federal Regulations, Section 133.106, defines blue cheese as a cheese produced from cows' milk by methods similar to those described above and that contains a bluish green mold throughout (Fig. 1). Moisture should not exceed 46% and the cheese solids must contain not less than 50% milk fat. The mold prescribed for use is *Penicillium roqueforti*. In addition, a harmless preparation of enzymes may be added to help develop the flavor during curing (less than 0.1% of the weight of the milk used). The standards also allow for the use of $CaCl_2$, at not more than 0.02%, to aid curd formation and for the use of benzoyl peroxide to bleach the milk.

## III. PRODUCTION OF *Penicillium roqueforti* SPORES FOR BLUE-VEINED CHEESES

### A. Strain Selection

In 1972, Dairyland Food Laboratories purchased the Midwest Blue Mold Company of Stillwater, Minnesota. Midwest produced a selected spore inoculum for blue cheese. Through years of experimentation, two strains were chosen because they possessed superior cheese-making properties. The Minnesota strain was first isolated at the University of Minnesota and the Iowa strain was first isolated at Iowa State University,

**FIGURE 1.** Blue cheese.

Ames. Both were sold by Midwest Blue Mold and constitute part of the current product line.

## B. Spore Inoculum Production

### 1. Procedures

The techniques used at Dairyland Food Laboratories to produce blue mold start with the propagation of pure cultures on agar slants. Spores are washed from the slants and used to inoculate the substrate. These substrates include sterilized bread (Hussong and Hammer, 1935) or sterilized bran. Growth occurs under controlled conditions of time, temperature, and humidity. After the mold has covered the substrate and the spore count is optimal (usually 16–22 days), the material is harvested, dried, and milled to produce a fine powder. The product is then blended and standardized to produce a uniform product and packaged at a concentration of not less than $5.0 \times 10^8$ spores/gm.

Bread was the original substrate used by Hussong and Hammer (1935) for mold growth. Work by Nelson and Zanzig at Dairyland Food Laboratories indicated that bran would also adequately serve as a substrate for mold growth and that the mold produced on bran had similar properties to bread-grown mold (J. H. Nelson, unpublished observations).

### 2. Production Facility Design

Conditions for blue mold production must be carefully controlled to produce a uniform and consistent product. Special precautions must also be taken to prevent contamination of both the production facility and the external environment. A specially designed facility has been built to accomplish this (Zanzig, 1975). The 2750-ft² area contains both laboratories and a sophisticated production area. The inoculation room contains a vertical laminar air-flow system that prevents outside air penetration by positive pressure. Specially designed perforated tables allow the air to flow uninterrupted and without turbulence. One room is set aside exclusively for packaging the product. The laboratory is equipped with a vertical air flow system and filters to prevent contamination. All air is incinerated prior to exhausting to the outside. In addition, this area has its own air conditioning and heating systems which isolate it almost completely from the general plant.

## C. Liquid Suspension of Spores

Concentrates of *Penicillium* spores in liquid suspension have been developed primarily for Gorgonzola cheese by Centro Sperimentale del

Latte, Milan, Italy. Four strains of *Penicillium,* differing in ripening characteristics, are available in bottles containing 25 ml, sufficient for the inoculation of 1000 gal of milk. Shelf-life of these preparations is reported to be 2 months when stored at 2°–5°C.

## IV. BLUE CHEESE FLAVOR

### A. Chemistry

The flavor of blue cheese is a chemically complex system. A great deal of effort has gone into the elucidation of the flavor components of blue cheese.

Hammer and Bryant (1937) identified the methyl ketone 2-heptanone in blue cheese. A number of workers, including Patton (1950a,b), Morgan and Anderson (1956), Coffman *et al.* (1960), and Scarpellino and Kosikowski (1961), have described the major role methyl ketones play in the flavor of blue cheese. The characteristic flavor of blue cheese is primarily due to the presence of various methyl ketones and secondary alcohols containing from 5–11 carbon atoms (Nelson and Richardson, 1967; Arnold *et al.*, 1975). The most abundant methyl ketones identified are 2-heptanone, followed by 2-nonanone, 2-pentanone, and 2-undecanone. Higher molecular weight ketones may also be present (Schwartz and Parks, 1963). Morris *et al.* (1963) and Anderson and Day (1965) identified the free fatty acids, acetic through stearic, along with several unsaturated $C_{18}$ fatty acids in blue cheese.

Jackson and Hussong (1958) have identified the secondary alcohols 2-pentanol, 2-heptanol, and 2-nonanol in blue cheese. These are probably produced by reduction of the above-mentioned methyl ketones. The products of proteolysis are important in the background flavor. However, the role of these peptides, amino acids, and other metabolites in flavor development has not been well characterized.

Table I (Nelson and Richardson, 1967) lists a number of compounds identified in blue cheese. It is quite likely that all of these are responsible for the characteristic flavor although, as indicated previously, the methyl ketones are probably the major contributors.

### B. Mold Metabolism

*Penicillium roqueforti* produces both a protease and a lipase. The protease is, in large part, responsible for the smooth, full flavor and texture. The amino acids produced by proteolysis act as buffers for the cheese at pH 6.5 and are undoubtedly used by the mold for growth.

**TABLE I.**  Summary of Compounds Positively or Tentatively Identified in Blue Cheese[a,b]

| | |
|---|---|
| Acetic acid* | Ethyl formate |
| Butanoic acid* | Ethyl acetate |
| Hexanoic acid* | Ethyl butanoate* |
| Octanoic acid* | Ethyl hexanoate |
| | Ethyl octanoate |
| Acetone* | Ethyl decanoate |
| 2-Pentanone* | Ethyl 2-methylnonanoate |
| 2-Hexanone | |
| 2-Heptanone* | Isopropyl hexanoate |
| 2-Octanone | Isopropyl decanoate |
| 2-Nonanone* | Pentyl hexanoate |
| 2-Decanone | 3-Methylbutyl butanoate |
| 2-Undecanone* | |
| 2-Tridecanone | Acetaldehyde |
| | 2-Methyl propanal |
| 2-Propanol | 3-Methylbutanal |
| 2-Pentanol* | Furfural |
| 2-Heptanol* | |
| 2-Octanol | Hydrogen sulfide |
| 2-Nonanol* | Methyl mercaptan |
| | Diacetyl |
| Methanol | Diethyl ether |
| Ethanol | Benzene |
| 2-Methylbutanol | Toluene |
| 3-Methylbutanol | Dimethylcyclohexane |
| 1-Pentanol | Cresyl methyl ether |
| 2-Phenylethanol* | δ-Octalactone |
| | δ-Decalactone |
| Methyl acetate | |
| Methyl butanoate | |
| Methyl hexanoate* | |
| Methyl octanoate* | |
| Methyl decanoate | |
| Methyl dodecanoate | |

[a] From Nelson and Richardson (1967).
[b] Compounds marked with an asterisk indicate blends of compounds which contribute to the flavor of blue cheese (Anderson, 1966).

Excessive proteolysis results in a soft, bitter cheese while insufficient proteolysis produces tough, crumbly cheese (Kinsella and Hwang, 1976).

The lipase hydrolyzes the triglycerides of milk fat to produce the free fatty acid precursors of the low molecular weight methyl ketones and secondary alcohols. Both the milk lipase, if it has not been inactivated by pasteurization, and the *P. roqueforti* lipase are involved. *Penicillium roqueforti* lipase has been isolated and studied.

The properties of the enzyme vary depending on the substrate used, the conditions of assay, and the source of the enzyme (Morris and Jezeski, 1953). Thus, Shipe (1951) found that the enzyme had a temperature optimum of 30°–35°C and a pH optimum of 5.0. Morris and Jezeski (1953) found that the enzyme had a pH optimum on butterfat of 6.5–6.8 and a temperature optimum of 30°–32°C. Further variations were found depending on whether the enzyme was isolated from the medium (soluble) or was released by disrupting the mycelia. This observation, that soluble lipase seemed to differ from mycelial lipase, indicated the possible existence of more than one lipase. Imamura and Kataoka (1967) have confirmed this through the isolation of both an intracellular and an extracellular lipase from *P. roqueforti*.

The soluble enzyme has a greater specificity for low molecular weight fatty acids than high molecular weight fatty acids (Shipe, 1951; Morris and Jezeski, 1953; Arnold *et al.*, 1975). These fatty acids serve both as flavor components and as precursors for methyl ketones, which in turn can be reduced to secondary alcohols as shown in Fig. 2 (Kinsella and Hwang, 1976).

The fatty acid is first activated with coenzyme A to form an acyl-CoA which is then oxidized via the fatty acid β-oxidative pathway to β-ketoacyl-CoA. The β-ketoacyl-CoA is deacylated and decarboxylated to form a methyl ketone with one less carbon atom than the fatty acid precursor.

## C. Production of Blue Cheese Flavor by Submerged Fermentation

### 1. Theoretical Aspects

Because of the differences in the physical form of the substrate and the need to develop the flavor in a matter of hours, it was not possible to

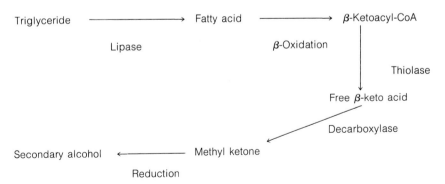

**FIGURE 2.**   Production of methyl ketones and secondary alcohols from free fatty acids by *P. roqueforti*.

directly translate the conditions found in blue cheese to a submerged fermentation.

The flavor in blue cheese is due, in part, to the metabolism of the mycelia or vegetative cells. The use of mycelia in submerged fermentations would result in a heavy fungal growth that would be difficult to remove from the product. Heavy mycelial growth is inhibited and spore development enhanced in both the spore inoculum and submerged fermentation by the addition of salt (Watts and Nelson, 1963). Knight (1963) found that in the presence of free fatty acids, *P. roqueforti* spores were more capable of converting fatty acids to methyl ketones than were vegetative cells. However, spores themselves did not produce lipase enzyme and therefore could not be efficiently used with nonlipolyzed milk fat.

Dairyland Food Laboratories, under license from the Wisconsin Alumni Research Foundation, has developed a commercial process for blue cheese flavor production based on Knight's discovery (Watts and Nelson, 1963). The process closely parallels conditions found in the manufacture of blue cheese, namely the use of similar dairy substrates, salt levels, aeration conditions, *Penicillium* mold, and lactic acid.

Free fatty acids are provided either by activation of the milk lipase, or preferably by pretreatment of the milk with pregastric esterase (Italase), to produce low molecular weight free fatty acids (Watts and Nelson, 1963). The use of the pregastric esterase more closely parallels the mold lipase activity pattern as both are more specific for low molecular weight fatty acids than for higher molecular weight fatty acids. Milk lipase is not specific for low molecular weight fatty acids (Nelson, 1972). Watts and Nelson (1963) further refined this procedure by using pregastric esterase-treated butterfat (LBO-50, Farnham, 1950, 1957) in programmed additions to provide concentrated substrate. High levels of free fatty acids inhibit spore metabolism. This programmed addition of lipolyzed butterfat more closely simulated the natural process that occurs during cheese ripening, namely the continuous release of free fatty acids (Arnold *et al.*, 1975; Watts and Nelson, 1963).

Watts and Nelson (1963) also developed a medium for a spore inoculum that produced little mycelial development. Using the white mutant of *P. roqueforti* (Knight *et al.*, 1950), which eliminated the development of blue color in the fermentor, they described the following method.

### 2. Production of the Spore Inoculum

Cultures maintained in V-8* juice containing agar slants are propagated further in V-8 juice to increase the inoculum concentration. The

---

*V-8 is the Trademark of the Campbell Soup Company, Camden, New Jersey.

suspension is transferred to a salt-containing milk substrate, its pH is adjusted to 4.5–5.5 with lactic acid, and it is incubated on a rotating shaker for up to 4 days to further develop the culture and condition it to a dairy substrate. After 4 days, additional salt is added and the incubation is continued for another 24 hours.

### 3. Fermentation of the Blue Cheese Flavor

The fermentation medium consists of nonfat dry milk or other milk products, such as skim milk, whey, or whole milk, salt, and lipolyzed milk fat (LBO-50). The medium is blended to produce a homogeneous mass and then sterilized. After cooling to 24°–26°C, the fermentor is inoculated with the spore preparation and fermented at 78°F for up to 66 hours. Programmed additions of LBO-50 are made periodically. The product has 7–12 times the ketone content of blue cheese and four times the flavor level (Arnold et al., 1975). The product is subjected to 130°C for 4 seconds to kill the mold and inactivate the enzymes. The final product contains 4% fat, 5% protein, 8% carbohydrates, 5% salt, and 78% water.

Dwivedi and Kinsella (1974a,b) described a procedure in which P. roqueforti mycelia, acting on a lipolyzed milk fat substrate, produced methyl ketones in a continuous fermentation. They further found that the inhibition due to high concentrations of free fatty acids could eventually be overcome by the mold. In addition, the concentration of free fatty acids present affected the relative ratio of methyl ketones produced.

Jolly and Kosikowski (1975) recently described a whey-based submerged fermentation supplemented with coconut oil. A microbial lipase plus P. roqueforti spores were used.

### 4. Chemical Composition of Dairyland Food Laboratories' Blue Cheese Flavor

Dairyland Food Laboratories' blue cheese flavor contains 2-pentanone, 2-heptanone, 2-octanone, 2-nonanone, and 2-undecanone, much the same as that found in blue cheese itself.

### 5. Fortified Blue Cheese Flavor

A fortified blue cheese flavor is prepared by the addition of compounds normally found in blue cheese. This product can have between 6 and 20 times the flavor level of cheese itself but must be considered an artificial flavor. This addition of flavor compounds to a fermented base provides a superior product as compared to simple chemical compounding due to the presence of the suitable background components.

Blue cheese flavor can be used in salad dressing, dips, appetizers and blue cheese powder. The flavor can be used to increase the potency of blue cheese solids and to reduce cheese solids' content in formulations.


## REFERENCES

Anderson, D. F. (1966). Ph.D. thesis, Oregon State University, Corvallis.
Anderson, D. F., and Day, E. A. (1965). *J. Dairy Sci.* **48,** 248–249.
Arnold, R. G., Shahani, K. M., and Dwivedi, B. K. (1975). *J. Dairy Sci.* **58,** 1127–1143.
Coffman, J. R., Smith, D. E., and Andrews, J. S. (1960). *J. Food Res.* **25,** 663–669.
Davis, J. G. (1965). "Cheese," Vol. I, pp. 3–15. Am. Elsevier, New York.
Davis, J. G. (1976). "Cheese," Vol. III, pp. 621–622. Am. Elsevier, New York.
Dwivedi, B. K., and Kinsella, J. E. (1974a). *J. Food Sci.* **39,** 83.
Dwivedi, B. K., and Kinsella, J. E. (1974b). *J. Food Sci.* **39,** 620.
Farnham, M. G. (1950). U.S. Patent 2,531,329.
Farnham, M. G. (1957). U.S. Patent 2,794,743.
Hammer, B. W., and Bryant, H. W. (1937). *Iowa State Coll. J. Sci.* **11,** 281.
Hussong, R. V., and Hammer, B. W. (1935). *J. Dairy Sci.* **18,** 599.
Imamura, T., and Kataoka, K. (1967). *Dairy Sci. Abstr.* **29,** 215.
Jackson, H. W., and Hussong, R. V. (1958). *J. Dairy Sci.* **41,** 920.
Jolly, R., and Kosikowski, F. V. (1975). *J. Food Sci.* **40,** 285–287.
Kinsella, J. E., and Hwang, D. (1976). *Biotechnol. Bioeng.* **18,** 927–938.
Knight, S. G. (1963). U.S. Patent 3,100,153.
Knight, S. G., Mohr, W. H., and Frazier, W. C. (1950). *J. Dairy Sci.* **33,** 929.
Morgan, M. E., and Anderson, E. O. (1956). *J. Dairy Sci.* **39,** 253.
Morris, H. A., and Jezeski, J. J. (1953). *J. Dairy Sci.* **36,** 1285–1298.
Morris, H. A., Jezeski, J. J., Combs, W. B., and Kuramoto, S. (1963). *J. Dairy Sci.* **46,** 1.
Nelson, J. H. (1972). *J. Am. Oil Chem. Soc.* **49,** 559–562.
Nelson, J. H., and Richardson, G. H. (1967). *In* "Microbial Technology" (H. J. Peppler, ed.),
    pp. 82–106. Van Nostrand-Reinhold, Princeton, New Jersey.
Patton, S. (1950a). *J. Dairy Sci.* **33,** 400.
Patton, S. (1950b). *J. Dairy Sci.* **33,** 680.
Scarpellino, R., and Kosikowski, F. V. (1961). *J. Dairy Sci.* **44,** 10.
Schwartz, D. P., and Parks, O. W. (1963). *J. Dairy Sci.* **46,** 989.
Shipe, W. F. (1951). *Arch. Biochem.* **30,** 165–179.
United States Department of Agriculture (1976). *Dairy Prod. Annu. Summ.* p. 21.
Watts, J. C., and Nelson, J. H. (1963). U.S. Patent 3,072,488.
Zanzig, C. E. (1975). *Food Process* **36**(12), 98–99.

Chapter 10

# Microorganisms for Waste Treatment

## LARRY L. GASNER

## I. INTRODUCTION

The increased sophistication of our society in the past several decades has led to dramatic increases in the volume and complexity of treatment of waste effluent streams and spills. This is only now beginning to be matched by increased sophistication of waste stream and spill biodegradation microbiology. Inoculum cultures for this use are marketed by about 20 companies with a 1978 total market value of about $2–4 million. One estimate of the total potential domestic market is $200 million (Anonymous, 1976, 1977a,b).

Although some role of microorganisms was recognized since the

**211**

MICROBIAL TECHNOLOGY, 2nd ed., VOL. II

activated sludge process was first developed (Andern and Lockett, 1914), the process is still not well enough understood to be able to define the respective roles of the various components of the microbial population (Jones, 1976). Indeed, a form of "spontaneous generation" is frequently relied upon to develop the appropriate complex population mix. The development of the art of inoculation of sewage systems is a recent phenomenon that is rapidly increasing in activity and dollar volume. This experience is also leading to increased efficacy of some of the microbial cultures sold. In like manner, it was the huge oil spill of the *Torrey Canyon* near the coast of England in 1967 that led to extensive efforts in the United States to develop enriched "seed" cultures for hydrocarbons (Anonymous, 1970), although septic tank additives and drain cleaner mixtures containing microorganisms had been marketed considerably earlier.

## A. Scope

This chapter is limited to the technology of isolation, production, and application of those microbial cultures which are commercially available or imminent as inocula for the treatment of wastes. This does not include description of processes which do not involve the marketing of microbial cultures. For example, *Candida utilis* production from sulfite waste liquors, ethanol production from whey wastes or waste cellulose, or single-cell protein production from hydrocarbon, food, or other wastes are not included. Also not included are novel waste-processing schemes or fermentor designs that have been proposed for waste treatment.

## B. Products

### 1. Characteristics

Products are available as freeze-dried or air-dried solids or as stabilized microbial suspension in liquid form. Solids are available in various size shipments and containers with the most common container size being 25 lb. Costs vary between about $7.00 and $22.00 per lb, FOB, which frequently includes technical service. Stabilized liquid formulations typically contain of the order of $5 \times 10^{10}$ microbes per liter and sell in the above price range per gallon in plastic gallon containers. Products can be stored at room temperature for years and retain useful potency. Dry products are activated by mixing at least 2 gal of water with each dry pound. Both liquid and activated dry products can be sprayed onto the surface or otherwise well mixed into the system. Drip feeders are available for continuous feeding of microbial suspensions.

## 2. Types of Cultures

Formulations including live microorganisms are available with specific activity for one specific component such as phenol, aromatics or glycol, or one specific enzyme activity such as amylase activity, protease activity, lipase activity, or cellulase activity. On special order specific biodegradation activities can be obtained against DDT, polychorinated biphenyls, and other more exotic compounds. The above are generally nonclassified isolates; however, pure identified microbial cultures are also available, for example, *Bacillus subtilis* KJ, which produces high levels of amylolytic and proteolytic enzymes; *Bacillus thuringiensis*, which produces high levels of lipolytic enzymes and low levels of amylolytic enzymes; and *Nocardia corallina*, which degrades phenol and phenolic compounds. Other unidentified products have broader ranges of activity, such as the degradation of cutting oils and crude oil, and the treatment of ironworks wastewater.

Most preparations consist of mixtures of microbes. These mixtures are of three types as follows: (1) natural mixtures, for example rumen bacteria with high cellulase activity, (2) selected mixtures which have been propagated together in a single fermentor, or (3) mixtures which have been propagated individually and combined in the final formulation. As many as 20 microorganism types can be found in a general purpose-type preparation. Also available are products of unspecified contents and unspecified preparation methods.

Aerobic, anaerobic, or facultative organisms are available as are low (4.5) or high (9.5) pH-tolerant preparations.

## C. Application Guidelines

Actual applications must be guided by several considerations. Mixing with the waste stream is necessary. A wetting agent in a few cases is incorporated in the product. For application to oil spills a natural dispersant, such as a lignosulfonate, may be supplied as a component of the product or should be applied separately. For oil spills on land it may be necessary to build a dike and flood the area. Topical preparations can be applied directly to surfaces, such as in dumpsters for odor control.

Necessary nutrients are required for any microbial growth. In certain cases supplementary nitrogen and phosphorus may be provided in the formulation or may be added separately.

The idea of providing an inoculation requires adding an initial high dose followed by a continuous maintenance low dose or several maintenance periodic doses. Actual recommended dose size varies widely. For activated sludge systems cost estimates for a 1 million gal/day plant range between \$20 and \$100 per million gallons. Because periodic fluctuations upset small waste treatment facilities more than larger waste treatment plants, the recommended maintenance dosage level may be severalfold larger for the smaller vessel.

A major problem area is that most waste material is not well defined, nor is it easily classified. The mechanical, nutritional, or chemical environment may not be identical from site to site. Also, interactions between species within the inoculum and those native to the site are influenced by subtle, little understood differences in the environment. Therefore, it cannot be reliably predicted that any given culture mixture will reproducibly perform well under superficially identical conditions. Before any product should be used in significant amounts, therefore, it is necessary to have verification tests performed and to view as speculation any advance claims as to efficacy on an untested waste. Lack of confidence on the part of the purchaser due to prior overzealous claims on the part of some suppliers is the biggest weakness in this entire industry. This lack of confidence is a combination of disappointment with the microbial enzyme products which had promised similar performances and over-promotion by those firms which are not soundly based in the area.

There should be caution in dealing with firms that have limited documented similar case histories and cannot refer prospective customers to several other satisfied clients with recent success in similiar systems. Gimmicks in the promotion of products should also lead to caution. Advertising the storage of stock cultures in a bank vault, emphasizing the remoteness of the earth from which certain cultures were isolated, or enriching cultures the normal way but adding a trace of mutagenic agent so the resulting cultures are called "mutants" are gimmicks to imply unproved uniqueness and undemonstrated value.

## II. PRODUCTION

### A. Origin of Cultures

Sources of soil and water samples vary according to the type of waste degradation expected. For example, soil samples were taken which were contaminated by a natural oil seep in Cape Simpson, Alaska (Cundell and Traxler, 1974). Fifteen hydrocarbon-degrading bacteria were isolated from natural saline waters of low temperature (0°–25°C). The facultative psychrophilic bacteria represented a wide range of genera and grew on aliphatic, aromatic, and naphthenic materials. In fact, the coastal waters of the world contain a large fraction of microorganisms, up to 50%, that are capable of oxidizing hydrocarbons (Anonymous, 1973). A good medium for marine hydrocarbon microbial isolation and growth has been demonstrated to be enriched seawater medium with a small volume of seawater added (Katos et al., 1971).

Soil organisms have also been found to readily degrade oily wastes under prevailing conditions. A rate of 1 lb/month-ft$^3$ of soil was reported for soil at a Houston refinery when excess nitrogen and phosphorous were added (Franche and Clark, 1974). It appears that microorganisms are widely distributed in oil field soils and anywhere that is particularly subject to frequent petroleum pollution. Once a population becomes established in an oil-enriched soil, it will remain, either dormant or utilizing other substrates, and it will readily respond to the reapplication of oil.

Because organisms generally degrade hydrocarbons according to type, paraffinic first, naphthenic second, and aromatic third, it is generally believed that mixed cultures of organisms of various selectivities, such as are found in nature, are the best to use for rapid and complete degradation. On the other hand, the possibility of genetically introducing multiple selectivities into a single strain, producing a "supermicrobe," has resulted in some exciting new prospects, which will be discussed in Section IV. Mixed cultures of hydrocarbon- and petroleum-oxidizing microbes are used in petroleum spills on water and on land, treating industrial and refinery wastes, and cleaning ship tanks.

Additional isolates with lipolytic activity and proteolytic activity can be obtained from grease traps of meat packing plants, sewage treatment plants, septic tanks, and food-processing waste treatment, in addition to common sources of samples. Cultures obtained are aerobic, facultative aerobic, or anaerobic. Selection of facultative aerobic microorganisms has advantages in that they can be effective both in anaerobic sludges, grease trap lagoons, lift stations, and tanks as well as maintain activity when the lagoon or grease trap becomes aerobic. Facultative aerobes can be selected to not produce malodorous $H_2S$ when anaerobic. These sources of microbes combined with the hydrocarbon-utilizing microbes are used in food-processing waste streams, grease traps, waste lines, municipal sewer lines, and places where deposits of food fats and oils occur. They are also effective with newly refined petroleum oils.

Microbial isolates with activity against specific chemical or industrial wastes come from soil, water, and sediment samples where repeated or continuous exposure to similar chemistries is suspected. Waste treatment systems where a particular chemical, such as a phenol, a cyclic or aromatic compound, or mixture are being successfully treated is a particularly good source of such isolates. Cultures can be isolated that will degrade phenol, pesticides, and other chemicals, including complex industrial wastes characteristic of an industry. For example, pulp and paper mill effluent treatment can be enhanced by microbial isolates that attack the various forms of lignin (Stern and Gasner, 1973, 1974).

## B. Isolation and Enrichment

In general, microbial cultures found in these products are isolates from natural soils and waters where normal enrichment techniques have been used. A normal enrichment method acclimates the organisms to a synthetic or actual waste environment, selecting those to grow which can best accommodate to the chemistry and the nutrients and which reproduce most rapidly. Temperature, ionic composition, pH, and dissolved $O_2$ are among the variables. Selection is generally for those cultures where the desired degradative enzyme system is constitutive and not subjected to substrate inhibition, repression, or toxicity. Isolation of pure cultures is usually on selective media streak plates. Stock cultures are maintained on slants or as lyophilized powder according to standard techniques.

Specific procedures and tests include acclimatization of cultures and treatability tests of specific wastes. This may use the enrichment cultures mentioned earlier, municipal sewage inocula, or natural soil samples, water samples, or other natural sources of microorganisms. These same techniques can be used to obtain microbial cultures for any specific waste.

The first step in a "treatability" test is the development of a "synthetic sewage" medium. A typical formula consists of a stock salt solution containing (per liter) 18.6 gm $MgCl_2 \cdot 6H_2O$, 36.4 gm $CaCl_2 \cdot 2H_2O$ and 0.2 gm $FeCl_3 \cdot 6H_2O$; and a nutrient stock solution containing (per liter) 20 gm glucose, 20 gm nutrient broth, 30 gm $KH_2PO_4$, and 10 gm sodium benzoate. One milliliter of the stock salt solution and 20 ml of the nutrient stock solution are added to each liter final volume. The pH is adjusted to within 6.5–7.5 with 1.0 $N$ $H_2SO_4$ or 1.0 $N$ NaOH. Distilled water is used throughout.

Toxicity is commonly assessed in a series of parallel shake flasks with the suspected toxic component or the whole raw chemical waste added in geometrically increasing concentrations to the above "synthetic sewage" medium. Inoculation of each flask, including the control flasks, is uniform. Microbial growth is measured daily, often by turbidity. The concentration at which the toxic component begins to inhibit growth is noted and is later used as a warning of an upper concentration limit. A certain amount of experience is helpful in interpreting these experiments and repeat experiments are frequently carried out to verify the preliminary results. In some cases the initial shock to the inoculum may dominate the response or a long acclimatization lag may delay the growth response.

Treatability tests are commonly carried out in a laboratory-sized continuous flow treatment apparatus. Batch test information is much less useful. A 3- to 5-liter chamber with a sloping bottom is used, fitted with an air-stone sparger. About three-fourths of the chamber is aerated. This is separated from the overflow sludge settling zone by an underflow vertical partition.

Synthetic sewage plus a nontoxic level of wastes are added and inoculated with up to 2000 mg/liter suspended solids inoculum. An initial influent flow rate of about 0.3–0.5 v/v vessel is used. This flow rate is increased stepwise by two- to fourfold during the test. The toxic component or raw waste stream content is increased slightly each day. The pH and salt content are allowed to change in the direction of the natural tendency. The variables are gradually changed individually in order to approach a natural optimum. Additives, such as nutrients, dispersants, or flocculants, may be used. Influent and effluent samples are taken daily for tests. Tests run include suspended solids (dry weight), volatile solids (dry weight minus ash), microscopic examination (cell growth observation), BOD (biochemical oxygen demand), COD (chemical oxygen demand), TOC (total organic carbon), and dissolved oxygen. Suspended solids are particularly susceptible to toxicity. Total organic carbon reductions are also generally accompanied by BOD and COD reductions. Therefore, all of the above tests may not be necessary.

Treatability tests such as these take at least 1 month to complete. Because of the variability of many industrial wastes and the cyclic chemical nature of the influent, it may be necessary to operate the laboratory unit for up to 2 years. This type of test is also used by industrial wastewater design engineers to simulate an actual wastewater treatment plant and aid in the design of new facilities.

## C. Cultivation and Processing

It should be noted that many suppliers do not produce their own cultures. Rather, the fermentation is done by contract fermentation companies. In a few cases, commercially available bulk-produced pure-culture dried biomass is simply purchased, as is, or is marketed directly by the producer as a sideline business.

For those suppliers who carry out their own fermentations, procedures vary, but the basic slant, loop, inoculum preparation, inoculation sequence is generally unremarkable. Fermentors up to 2000 gal are used. Aseptic conditions are maintained even where mixed cultures are simultaneously propagated, largely as a guard against *Salmonella, Staphylococcus,* and *Streptococcus* contamination. Some care is taken to select fermentation conditions, temperature, and substrates that will

produce microbes derepressed, induced, and conditioned to their ultimate environment. Separation methods are by centrifuge or filter. Air-
drying is less expensive than freeze-drying; it is used for those cultures
whose viability is not severely affected. Spore formers, such as *Bacillus*
species, can be air-dried, whereas non-spore formers usually are
freeze-dried.

Details of fermentation and preservation and application of a live
mixed culture liquid product are provided in a patent (Horsfall and
Gilbert, 1976). Seven bacteria with complementary activities were isolated from soil and water samples. These have been identified as
*Aerobacter aerogenes, Bacillus subtilis, Nitrobacter winogradskii, Cel-
lulomonas biazotea, Rhodopseudomonas polustris, Pseudomonas stut-
zeri,* and *Pseudomonas* species. Of these only the *Bacillus* is a spore
former. The *Pseudomonas* species contribute to denitrification. Individual inocula are prepared and inoculated into a sterile 400-gal steam-
jacketed fermentor. The medium contains 0.1% $K_2HPO_4$, 0.1% $NH_4$ Cl,
0.05% $MgSO_4 \cdot 7H_2O$, 0.2% $CH_3COONa \cdot 2H_2O$, and 0.1% yeast. This is
a minimal medium with acetate as the carbon source designed specifically to derepress the organisms to a maximum number of substrates.
Aerobic aseptic growth continues for about 24 hours at 20°–25°C.

Following this, the addition of 0.1% sodium sulfide, $Na_2S \cdot 9H_2O$,
causes the cessation of aerobic metabolism and growth. The broth is
then transferred to translucent bottles and anaerobic growth under warm
light is continued for 24–72 hours. Protosynthetic growth of the *Rhodopseudomonas* during this time is indicated by the development of its
characteristic red color. The product contains $5 \times 10^7$ viable microorganisms per milliliter, which has been found to be maintained for 3 years
when stored in plastic gallon containers at about room temperature. The
viability maintained is such that upon dilution in a nutrient medium
growth begins immediately without a lag phase. In one variation it is
formulated with a trace of polymeric flocculant.

After final preparation of the cultures they may be blended with additives before packaging. Wetting agents, emulsifiers, and dispersants
can be added to aid in dispersing the microorganisms throughout a
hydrocarbon phase and to aid in dispersing the hydrocarbon into droplets such that oxygen transfer is facilitated. The addition of flocculants is
occasionally practiced. Most frequently nitrogen and phosphorus are
added to supply nutrients.

The final stage in production is to take a sample for microbiological
testing. The U.S. Department of Agriculture requires a certification that
any product used in or about a food-processing plant to be free of
*Salmonella.* Most products carry this certification.

## III. APPLICATIONS

### A. Municipal Sewage Treatment

Modern sewer systems deal with large amounts of food wastes, laundry wastes, and petroleum wastes. The greasy materials build up a coating on the inner surface of the sewer lines and become embedded with nonsoluble materials, reducing the design capacity of the line. Particularly at locations where the flow velocity is low this leads to foul odors, overflow, and backup. Mechanical removal is difficult and hazardous.

Microbial cultures can be drip fed into manholes at the extremities of the sewer system. Populations of microbes increase rapidly, consuming the grease. When the grease has been significantly removed, the population decreases. A maintenance level of microorganism feed is maintained. A 1974 "rough and conservative" estimate for the Washington Suburban Sanitary Commission for 3000 miles of pipe with a 150 million gal/day flow was that mechanical cleaning would cost approximately $4 million, whereas the use of bacteria would cost $1.3 million (Hoover, 1974). Extrapolation of these estimates from a service area of 3 million population to the entire nation gives an idea of the total magnitude of this market.

These cultures have been used in a number of cities with good results. Arvada, Colorado which uses a digester for production of methane gas, reported that the production of gas tripled within 60 days after cleanup. The Perry, Florida system which consists of primary treatment, normally removed 28–30% of the BOD (biochemical oxygen demand) and suspended solids. This increased to 50–70% within 1 year of the use of bacterial inoculation. In Erie, Pennsylvania the energy requirements for operation of the pumping stations was reduced by a factor of 3 and all offensive odors were eliminated. Stockton, California also reported a tripling of methane production from 6.5 to 18.8 ft$^3$/lb volatile solids while the BOD reduction of the primary stage of the treatment plant increased from 18.9 to 26.4%.

Numerous reports of improvements in operation of a normal activated sludge plant can be provided by the various suppliers of inoculation cultures. These claim digestion of grease and oil, elimination of odors, improved BOD removal rates, improved settling, greater treatment capacity in the lagoon, reduction of sludge volume, increased removal of complex substrates, energy savings through reduced aeration, improved predictability of results, reduced corrosion, and improved safety. Some references are Baig and Grenning (1976), Bower (1972), Chambers (1977), and Kirkup and Nelson (1977).

## B. Edible Oils and Greases

Food processors, meat processors, restaurants, army bases, schools, homes, and other places where food is processed may have grease buildups in their drain lines, downpipes, grease traps, or septic tanks leading to backups and odors. This problem can also be treated with microbial cultures. In addition refined oils often are not consumed by those microorganisms that attack crude oil and tars. Therefore, special cultures have been developed which degrade these materials.

The major problem to be confronted with respect to small installations is that the water does not stay in the piping system long enough to establish a biodegradation flora. In addition, these pipes may be frequently flushed with hot water containing toxic bleaches and detergents which flush out the cultures which have been established. Therefore, it may be necessary to add an inoculum at the low-flow time of day, i.e., at closing time. It also may require adding nitrogen (5–20 ppm) and phosphorus (1–4 ppm) to the water to supply nutrients. Regular retreatment (three to seven times/week) is usually specified. The microorganisms in the liquid products may revive faster than dried microorganisms, which is an advantage.

## C. Heavy and Crude Oils

Oily wastewaters aboard ships cause problems because they must be contained and treated before release. The Queen Mary contained 800,000 gal of oily wastewater (20% was lubricating oil and waste fuel) when it was docked at Long Beach in 1967 (Queen Mary Report, 1971). One hundred fifty pounds of dried microbial cultures were mixed with water (1 gal/lb) and poured into the bilge compartments without agitation. Within 6 weeks the water was pure enough to discharge into the harbor. Since then, CHT (collection, holding, and transfer) tank degreasers have been routinely used by the U.S. Navy (McPhee and Geyer, 1977). For 10,000-gal tanks, 3.5 lb of dried microbial culture are added each day for 5 days followed by a 1.0 lb/day maintenance dose.

Refinery wastes can be handled in lagoons or flow-through systems. The process requires inoculation with microorganisms followed by emulsification of the oil to increase interfacial area. Mixing is usually required. Full decomposition can be achieved in several months (Caswell, 1971). Oil-contaminated soil can be treated by damming and flooding; inoculation and continuous mixing speed the process. Inoculation by spraying directly on the soil surface can be used to seed soils that receive occasional spills with hydrocarbon-utilizing microbial populations. Fertilization with nitrogen and phosphorus is desirable.

## D. Chemical and Industrial

The wastes of the chemical industry usually can be handled by dilution and pH adjustment along with the inoculation of appropriate acclimatized cultures in an activated sludge system. Recalcitrant wastes, such as pesticides and halogenated aromatics, often are concentrated enough to support microbial growth and to be biodegradeable, whereas they are not in the dilute form found in nature. In certain cases the inoculum supplier will provide some microbial isolation and enrichment services.

Other industrial wastes consist of mixtures of materials that are characteristic to that particular industry. Cultures are available for many industries although the recommended method of application varies widely. Certain fungi, for example, were found to be efficient and practical converters of organic materials in commercial corn and soy food-processing waste streams (Church and Nash, 1970).

## IV. FUTURE PROSPECTS

As the technology grows in sophistication and integrity, the market penetration will also undoubtedly increase. The application of selected microorganisms to just a few large municipal systems will have a large impact on the total market size.

The possibility of genetic engineering of cultures to achieve specific activities and multiple activities is an exciting concept which could lead to improved culture applicability to new systems. In March, 1978 the Washington D.C. Court of Customs and Patent Appeals held that a corporation may patent a new form of life that has been bred in its laboratories (Anonymous, 1978). This case represents a landmark in that it now is possible to patent a microbial species if it can be shown that it has been genetically manipulated to give a novel organism which could not be found in nature. The case involves a General Electric Co. patent of a *Pseudomonas putida*-type bacteria into which was incorporated multiple compatible plasmids from other strains, giving it unique multisubstrate capability (Chakradarty, 1974). The specific plasmids were for octane, xylene, metaxylene, camphor and salicylate degradation. The plasmids were specifically bound to the genome by ultraviolet (uv) radiation (see this volume, Chapter 19).

It can be speculated that many systems now requiring multiple organisms for attack or those having complex interrelationships can be manipulated genetically to give more efficient, patentable microorganisms. The patentability also would lend exclusivity to the culture marketer and may make this a more attractive market in which to compete.

## REFERENCES

Andern, E., and Lockett, W. T. (1914). *J. Soc. Chem. Ind.* **33,** 1122–24.
Anonymous (1970). *Chem. & Eng. News.* Sept. 7, pp. 48–49.
Anonymous (1973). *Oceanology (Engl. Transl.)* **13,** 725.
Anonymous (1976). *Bus. Week,* July 5, p. 28D.
Anonymous (1977a). *Chem. Week* **121,** 47.
Anonymous (1977b). *Food Eng.* **49,** 138.
Anonymous (1978). Newsweek, March 13, p. 71.
Baig, N., and Grenning, E. M. (1976). *In* "Biological Control of Water Pollution" (J. Tourbier and R. W. Piersow, eds.), pp. 245–252. Univ. of Pennsylvania Press, Philadelphia.
Bower, G. C. (1972). *Proc. Ann. Meeting Chesapeake Water Poll. Control Assoc.*
Caswell, C. A. (1971). *Purdue Ind. Wastewater Conf., 6th, 1971* pp. 1020–1022.
Chakradarty, A. (1974). U.S. Patent 3,813,316.
Chambers, J. (1977). In *5th Annu. Ind. Pollut. Conf., Water Wastewater Equip. Manuf. Assoc.*
Church, B. D., and Nash, H. A. (1970). "Use of Fungi Imperfecti in Waste Control," Final Rep., Grant No. 12060 EHT. Federal Water Quality Administration, U.S. Dept. of the Interior; US Printing Office, Washington, D.C.
Cundell, A. M., and Traxler, R. W. (1974). *Dev. Ind. Microbiol.* **15,** 250–255.
Franche, H., and Clark, F. (1974). *Nucl. Sci. Abstr.* **30/31,** 6590.
Hoover, J. R. (1974). Montgomery County Planning Board, News release dated July 12, 1974; Silver Spring, Maryland.
Horsfall, F., and Gilbert, B. (1976). U.S. Patent 3,963,576.
Jones, G. L. (1976). *Process Biochem.* **11,** 3–5 and 24.
Katos, H., Oppenheima, C., and Miget, R. (1971). *Biol. Conserv.* **14,** 12.
Kirkup, R. A., and Nelson, L. R. (1977). *Public Works* **10,** 74–75.
McPhee, A. D., and Geyer, A. T. (1977). "Aeration, Bottom Turbulence and Bacteriological Studies of Naval Ship Sewage Collection, Holding and Transfer Tanks," Rep. 77–0043. Commander, Nav. Ship Eng. Cent., Washington, D.C.
"Queen Mary Report" (1971). Bower Industries, Orange, California.
Stern, A. M., and Gasner, L. L. (1973). U.S. Patent 3,737,374.
Stern, A. M., and Gasner, L. L. (1974). *Biotechnol. Bioeng.* **16,** 789–805.

Chapter 11

# Elementary Principles of Microbial Reaction Engineering

GEORGE T. TSAO

## I. INTRODUCTION

Efficiency and reproducibility are major concerns of reaction engineering. Applications of chemical engineering principle and practice in microbiological processes constitute the main activities of fermentation engineers. In the development of an industrial process, good engineering could mean the following tasks undertaken in a logical manner:

1. Identification of main products and substrates and, if possible, also main intermediates. In the case of a fermentation process, such information is usually provided by microbiologists and biochemists working with laboratory microbial cultures.

**223**

MICROBIAL TECHNOLOGY, 2nd ed., VOL. II

2. Stoichiometry of the process. For a fermentation, an exact material balance of the process is not always possible. Nevertheless, information should be gathered in as much detail as possible regarding the ratios among products, substrates (e.g., carbon source, nitrogen source, oxygen supply), and known intermediates and regarding the variations of these ratios responding to environmental changes. In some fermentation processes such as those for producing single cell protein (SCP) from hydrocarbon, energy balance is also of great importance.

3. Kinetics and process rate. Often, problems of (1) and (2) cannot be fully answered without the consideration of a time scale. In batch processes, the accumulated changes in products, intermediates, and substrates are of great concern. However, the time span and the manner by which such changes take place, and thus the kinetics and rate of the involved processes, are also necessary information. In the case of continuous fermentation which is still gradually gaining in its popularity, the design and analysis of the reactors are usually based upon the rate of change of these quantities and dilution rate.

4. Reactor design. Information on (1), (2), and (3) is a prerequisite for the ultimate objective of microbial reaction engineering which deals with the design and analysis of fermentors. Even though stirred tanks still represent the most popular geometry of fermentors, an increasing number of other physically shaped vessels, such as tower fermentors, do appear in the industry. Even with the same stirred tank, considerations regarding a proper choice from different modes of operation, including batchwise, fed-batchwise, continuous, and others, are also a part of the reaction engineering activity.

## II. IDEALIZED REACTOR DESIGN I: PERFECTLY MIXED BATCH FERMENTOR

An industrial fermentation carries out microbiological changes in a finite physical container. The design of a fermentor to carry out a fermentation in an efficient and predetermined manner is the most important objective and function of fermentation engineering. In chemical engineering curricula the teaching of reactor design is often started by analyzing three idealized situations: perfectly mixed batch reactor, perfectly mixed continuous flow reactor, and plug flow reactor. By examining how reactors are analyzed, one can identify the types of information that are needed in order to have good engineering of the fermentation system.

A material balance of a selected chemical species (e.g., a product, a substrate, or an intermediate) is usually the starting point of reactor analysis.

Accumulation = (input + production) − (output + consumption) (1)

In the case of a perfectly mixed batch fermentor, where input = 0 and output = 0, Eq. (1) can be rewritten as

$$V \, \Delta C = VR_p \, \Delta t - VR_c \, \Delta t \tag{2}$$

where $V$ = volume of the fermentation broth; $R_p$ and $R_c$ = rates of production and consumption of the chemical species under considera- tion, respectively; $\Delta C$ = the change of concentration; and $\Delta t$ = time span over which the change takes place. If the chemical species under con- sideration is the carbon source, say, glucose, and $R_p = 0$, Eq. (2) becomes, with $G$ to be glucose concentration and $R_G$ to be the rate of consumption of glucose,

$$\Delta G = -R_G \, \Delta t \tag{3}$$

or in a differentially small time span, $dt$,

$$dG = -R_G \, dt \tag{4}$$

In theory, if Eq. (4) can be integrated from $t = 0$ to any $t$, we will have the knowledge of how glucose concentration varies.

$$\int_0^t dt = t = \int_{G_0}^{G_t} \frac{dG}{-R_G} \tag{5}$$

In order to integrate Eq. (5) we need to know the functional form of $R_G$, i.e., how $R_G$ varies with other parameters of the fermentation system. Knowledge of $R_G$ in a batch fermentation is, however, often incomplete, and thus integration of Eq. (5) is not always an easy matter. Similarly, if the chemical "species" is the total microbial biomass in the fermentor, Eq. (2) will yield, with $X$ to be biomass per unit volume of fermentation broth,

$$dX = R_x \, dt \tag{6}$$

where $R_x$ = rate of production of biomass. In writing Eq. (6) microbial cell lysis and other processes that destroy the biomass are assumed to be absent; i.e., the rate of consumption of biomass is zero. Again, integration of Eq. (6) from 0 to $t$ will, in theory, predict the biomass production during the time span.

$$\int_0^t dt = t = \int_{X_0}^{X_t} \frac{dX}{R_x} \tag{7}$$

Equations (5) and (7) thus identify knowledge of $R_G$ and $R_x$ as important prerequisites to good fermentor design.

In practice, how $G$ and $X$ vary with $t$ are determined in experimental batches in laboratory or pilot-plant fermentors and expressed as "growth curves." From the slopes of growth curves, $R_G$ and $R_x$ can be extracted. When the results do not yield some simple mathematical expressions for $R_G$ and $R_x$, as often is the case, one then uses graphical or numerical techniques to integrate Eqs. (5) and (7) to predict $G$ and $X$ at various $t$. How accurately one can use $R_x$ and $R_G$ from laboratory or pilot-plant fermentors to predict "growth curves" (i.e. $G$ and $X$ at various $t$) in a different fermentor is the problem of "scaleup" (or "scaledown"). Since $R_G$ and $R_x$ depend upon many parameters of the fermentation system in addition to $G$ and $X$, our incomplete knowledge of these parameters is the source of difficulty in translating laboratory culture results to industrial operations. It does not appear that we can overcome this difficulty, in the near future, before we develop the necessary capability of defining and monitoring many more parameters of a fermentation system than we can presently.

## III. MICROBIAL CELL GROWTH

The Monod equation, even though it resembles those of Michaelis–Menten in enzyme kinetics and Langmuir in surface adsorption, is basically an empirical equation which correlates surprisingly well with many sets of published data of microbial cell growth. The Monod equation states, when the growth is limited by one single substrate [Eq. (8)],

$$\mu = \frac{\mu_{max}S}{K_s + S} \tag{8}$$

where $\mu = R_x/X$ = specific rate of biomass production, $\mu_{max}$ = maximum value of $\mu$ when $S$ is very large, $S$ = substrate concentration, and $K_s$ = a constant. Similar to the case of Michaelis–Menten equation, by plotting $1/\mu$ against $1/S$, the values of $\mu_{max}$ and $K_s$ can be determined from experimental data. When $S$ is much larger than $K_s$, $\mu$ approaches $\mu_{max}$ and becomes independent of substrate concentration; i.e., when

$$S >> K_s \quad \text{and} \quad \mu \to \mu_{max} \tag{9}$$

$$R_x = dX/dt = \mu_{max}X$$

Equation (9) leads to

$$\ln X = (\mu_{max})t + \ln X_0 \tag{10}$$

where $X_0 = X$ at $t = 0$. Equation (10) predicts exponential growth in microbial batch cultures when substrate is in good supply.

Equation (8) has been expanded to Eq. (11) to account for microbial cell growth that shows multiple exponential growth phases.

$$\mu = \mu_{max}\left(\frac{k_1 S_1}{K_1 + S_1} + \frac{k_2 S_2}{K_2 + S_2} + \cdots + \frac{k_i S_i}{K_i + S_i}\right)^{\frac{1}{\sum\limits_{j=1}^{i} k_j}} \tag{11}$$

When substrates $S_1$ through $S_i$ are all in excess; i.e., $S_1 \gg K_1$, $S_2 \gg K_2$, . . . , $S_i \gg K_i$, $\mu = \mu_{max}$ and the growth will show one exponential phase. When all but one substrate are in excess and when the one (for instance, glucose) is exhausted (i.e., $S_1 = 0$)

$$\mu^1 = \left(\sum_{j=2}^{i} k_j \ \sum_{j=1}^{i} k_j\right) \mu_{max} \tag{12}$$

cell growth will show a second exponential phase with a different specific growth rate, $\mu^1$.

In some practical cases, such as methanol fermentation for SCP production, the specific growth rate is subjected to substrate inhibition. A commonly written equation for this situation is

$$\mu = \mu_{max}\frac{S}{(K + S + S^2/K_i)} \tag{13}$$

where $K_i$ is known as the inhibition constant.

When microbial growth is inhibited by a metabolic product, the following Eq. has been written:

$$\mu = \mu_{max}\frac{S}{K + S}\frac{K_p}{K_p + p} \tag{14}$$

where $p$ = product concentration. Another equation for product inhibition is given below:

$$\mu = [S/(K + S)] (\mu_{max} - K_a A) \tag{15}$$

where $A$ = product concentration and $K_a$ = the toxicity index of the product.

All equations listed above are empirical in nature. They are useful in quantifying the effects of substrate and product concentration on cell growth rate. From these empirical equations, expressions of $R_x$ can be derived and substituted in, for instance, Eq. (7) in calculating biomass production.

Modeling of cell growth is an area of active research from which one gains additional insight into the complicated cellular metabolic activities which are the basis of all industrial fermentation processes. All equations given here treat biomass as an undifferentiated whole. An additional number of growth models do exist in the literature in which biomass is differentiated according to biological activities. In some recent cases, enzyme repression and induction are also considered in modeling cell growth. As one can predict, such sophisticated models will become increasingly useful in future engineering design of fermentation systems.

## IV. YIELD, METABOLIC QUOTIENT, AND PRODUCTION RATE

Growth yield is defined by the quotient

$$\Delta X / \Delta S = Y_{X/S} \qquad (16)$$

where $\Delta X$ is the increase in biomass and $\Delta S$ is the consumption of substrate. As $\Delta S$ approaches zero,

$$Y_{X/S} = \frac{dx}{ds} = \frac{dx/dt}{ds/dt} = \frac{R_x}{R_s} \qquad (17)$$

where $R_s$ is the rate of substrate consumption.

The yield concept can be easily broadened to specific products other than the total biomass. If $P$ is a specific product,

$$Y_{P/S} = \frac{dP}{dS} = \frac{R_p}{R_s} \qquad (18)$$

where $R_p$ is the rate of $P$ formation. The $Y$ factors are useful parameters in quantifying growth and product formation. It would be a mistake to read much more meaning into a $Y$ factor than what is defined above. For instance, even if we define a $Y_{A/G}$ to be the "yield" of an antibiotic compound when glucose is consumed by the microbes, this definition does not necessarily imply a direct metabolic linkage between glucose and the compound. However, the manner by which $Y_{A/G}$ responds to environmental changes, including the change in glucose concentration in a fermentor, may help reveal some metabolic linkage between $A$ and $G$.

The specific consumption rate of a substrate is also known as the metabolic quotient of the substrate:

$$q = (1/X)\,(ds/dt) \qquad (19)$$

Well-known examples are $q_{O_2}$ and $q_G$ for oxygen and glucose consumption, respectively. Metabolic quotient, like the yield factor, is again a useful tool in quantifying fermentation phenomena.

Depending upon the type of fermentation, the rate of product formation may relate to biomass in more than one way. The most commonly quoted relationship is the following:

$$dP/dt = k_1 (dx/dt) + k_2 x \qquad (20)$$

If $k_1$ is a constant and $k_2$ is equal to zero, the product formation is said to be directly linked to cell growth. Again, however, a word of caution is in order. Until justified by other evidence, one should not read too much physiological meaning into the relationship of product formation, Y factor, and $q$ factor. Even though they can be helpful in quantifying fermentation phenomena, they are basically empirical parameters.

## V. IDEALIZED REACTOR DESIGN II: PERFECTLY MIXED CONTINUOUS FLOW REACTORS

In addition to the perfectly mixed batch reactor discussed previously, two idealized continuous flow reactors will be analyzed. This section deals with the perfectly mixed continuous flow reactor.

The phrase "perfectly mixed" requires that concentration and temperature are uniform throughout the liquid fermentation broth in the reactor. As in the case of the idealized batch fermentor, material balances can be conveniently done over the whole broth in a perfectly mixed continuous flow reactor.

Applying Eq. (1) and assuming steady state,

$$\text{Input + production = output + consumption} \qquad (21)$$

For biomass $X$ (see Fig. 1),

$$FX_0 + VR_x = FX \qquad (22)$$

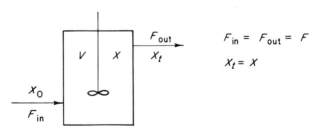

**FIGURE 1.** Perfectly mixed continuous flow reactor.

with the consumption term set to be zero, and $F$ being the liquid volumetric flow rate in and out of the fermentor, and $X_0$ the microbial cell concentration in the feed. The condition of "perfectly mixed" requires that the cell concentration in the outlet, $X_f$, is identical to the cell concentration $X$ in and throughout the fermentation broth.

Equation (22) can be changed to

$$D(X - X_0) = R_x = X\mu \tag{23}$$

where $D = F/V$ = dilution rate. With $\tau = 1/D$ = average residence time in the reactor,

$$1/D = \tau = (X - X_0)/R_x \tag{24}$$

If $X_0 = 0$, i.e., no cells in the feed, Eq. (23) reduces to

$$D = \mu \tag{25}$$

Equation (25) states that a steady state, perfectly mixed continuous flow fermentation can be maintained as long as the specific growth rate is equal to the dilution rate. According to Eq. (25), however, the dilution rate alone does not determine the cell concentration $X$ in a perfectly mixed continuous flow reactor.

When Eq. (21) is applied to a specific substrate $S$ which is consumed but not produced in the fermentor,

$$FS_0 = FS + VR_s \tag{26}$$

or

$$D(S_0 - S) = R_s \tag{27}$$

or

$$\frac{1}{D} = \tau = \frac{S_0 - S}{R_s} \tag{28}$$

Combining Eqs. (17) and (27),

$$D(S_0 - S) = \frac{R_x}{Y_{X/S}} = \frac{\mu X}{Y_{X/S}} \tag{29}$$

When Eq. (21) is applied to a specific product $p$ which is produced but not consumed in the fermentor,

$$FP_0 + VR_p = FP \tag{30}$$

or

$$D(P - P_0) = R_p \tag{31}$$

or

$$1/D = \tau = (P - P_0)/R_p \tag{32}$$

Combining Eqs. (18) and (31),

$$D(P - P_0) = Y_{P/S}R_S \tag{33}$$

Combining with Eq. (17),

$$D(P - P_0) = Y_{P/S}\frac{RX}{Y_{X/S}} = Y_{P/X}R_X \tag{34}$$

where

$$\frac{Y_{P/S}}{Y_{X/S}} = Y_{P/S}Y_{S/X} = Y_{P/X} \tag{35}$$

Equation (34) can be rewritten as

$$D(P - P_0) = Y_{P/X}\mu X \tag{36}$$

Equations (25), (29), and (36) describe the steady state concentrations of biomass, substrate, and product in a perfectly mixed continuous flow reactor. For microbial cell growth limited by one substrate obeying the Monod equation, a unique set of steady state conditions is determined by the dilution rate and the substrate concentration in the feed. By substituting Eq. (8) into Eq. (25),

$$D = (\mu_{max}S)/(K_S + S) \tag{37}$$

or

$$S = S \text{ at steady state} = \frac{DK_S}{\mu_{max} - D} \tag{38}$$

Combining Eqs. (25) and (29),

$$X = Y_{X/S}(S_0 - S) \tag{39}$$

Combining Eqs. (25), (36), and (39) and letting $P_0 = 0$,

$$P = Y_{P/X}X = Y_{P/X}Y_{X/S}(S_0 - S)$$

or

$$P = Y_{P/S}(S_0 - S) \tag{40}$$

When the dilution rate $D$ is increased, $S$ will increase and approach $S_0$ ($S \leq S_0$) according to Eq. (38). When $S$ approaches $S_0$, according to Eq. (39) $X$ will approach zero, which is the condition of "washout."

The biomass productivity of the reactor can be measured by $FX/V = DX$, i.e., the amount of biomass produced per unit time per volume of fermentation broth. According to Eq. (23), with $X_0 = 0$,

$$DX = \mu X - (\mu_{max}SX)/K_S + S) \qquad (41)$$

Substituting Eq. (39) into Eq. (41), one will have

$$DX = \frac{\mu_{max}S}{K_S + S} Y_{X/S}(S_0 - S) \qquad (42)$$

To optimize biomass productivity, Eq. (42) is differentiated with respect to $S$ and the first derivative set to zero:

$$\frac{d(DX)}{dS} = 0 = (\mu_{max}Y_{X/S}) \left[ \frac{S_0 - S}{K_S + S} - \frac{S}{K_S + S} - \frac{S(S_0 - S)}{(K_S + S)^2} \right]$$

The condition of maximum productivity is, therefore,

$$S_{max} = \sqrt{(K_S^2 + S_0 K_s)} - K_s \qquad (43)$$

Substituting Eq. (43) into Eq. (38), $D_{max} = D$, at which maximum $DX$ occurs $= \mu_{max} [1 - \sqrt{(K_s/K_s + S_0)}]$. $D = \mu_{max}$ is the condition of washout. There seems to be a misconception often stated in textbooks; i.e., the maximum biomass productivity of a perfectly mixed continuous flow reactor always exists "near" washout. According to Eq. (44), this is not necessarily true because how close the maximum productivity condition is to that of washout depends upon the ratio of $K_s/S_0$ of which, at least, the value of $S_0$ can be arbitrarily set.

The biomass yield of the fermentor can be measured by $FX/FS_0 = X/S_0$, i.e., the amount of biomass produced by unit of substrate fed into the reactor. According to Eq. (39), the maximum $X$ and thus the maximum $X/S_0$ occurs at the trivial condition of $S = 0$, the condition of batch growth (i.e., $D = 0$).

$$\max \frac{X_f}{S_0} = \lim_{S \to 0} Y_{X/S}(S_0 - S) \frac{1}{S_0} = Y_{X/S} \qquad (44)$$

If the microbial cell is subjected to substrate inhibition (for instance, growth on methanol) according to Eq. (13), then at steady state,

$$D = \mu = \mu_{max} \frac{S}{K + S + S^2/K_i} \qquad (45)$$

$$S = S \text{ (at steady state)} = \frac{K_i}{2} \frac{\mu_{max} - D}{D}$$

$$\pm \frac{1}{2} \left[ k_i^2 \left( \frac{\mu_{max} - D}{D} \right)^2 - 4KK_i \right]^{1/2} \qquad (46)$$

Therefore, there are two possible steady state values of $S$ at the same dilution rate.

A further analysis of these two potential steady states will reveal another interesting problem of reactor design. This deals with the problem of stability, which can best be explained with the diagram in Fig. 2. The curve in the diagram is a graphical representation of Eq. (45). At a particular specific growth rate $\mu_1$ ($= D$), the two possible steady states are represented by $P_1$ and $P'_1$ with $S_1$ and $S'_1$, respectively. First, we will consider steady state $P_1$. Suppose a small disturbance in the reactor conditions takes place resulting in a new $\mu_2$ which is slightly smaller than $\mu_1$ which is equal to D that remains unchanged. The rate of substrate consumption is smaller at $\mu_2$ than that at $\mu_1$. Consequently, substrate concentration will increase in the fermentor. Because of substrate inhibition, at a higher $S$ value a smaller $\mu$ will result which will further decrease the substrate consumption rate and lead to an even higher $S$ value. Therefore, $P_1$ represents an unstable steady state at which a small decrease in $\mu$ will enlarge itself and lead to cessation of growth completely at $S = S_2$.

Suppose a small disturbance of $P_1$ results in a new $\mu_3$ which is slightly larger than $\mu_1$ which equals $D$ that remains unchanged (see Fig. 2). The rate of substrate consumption is larger at $\mu_3$ than that at $\mu_1$. Consequently, substrate concentration will decrease, leading to an even faster specific growth rate, which will result in an even smaller $S$. Therefore, with a small increase in $\mu$ the reactor condition deviates away from that of $P_1$ and, in fact, will eventually change to $P'_1$ which represents a stable steady state.

Point $P'_1$, being stable, can be analyzed in an exercise identical to that described above for analyzing $P_1$. In the case of $P'_1$, a small disturbance will decay itself and the reactor will return automatically to $P'_1$.

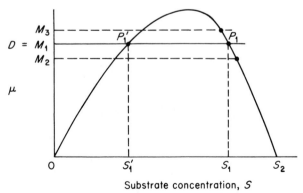

**FIGURE 2.** Diagram for stability analysis.

When a number of perfectly mixed continuous flow reactors are operated in series, the overall performance can be analyzed by referring to the diagram in Fig. 3. At steady state.

$$F = F_1 = F_2 = \cdots = F_n \tag{47}$$

Let

$$V_1 = V_2 = \cdots = V_n = V/n \tag{48}$$

Making a biomass balance of the $n$th reactor,

$$FX_{n-1} + V_n R_{x,n} = FX_n \tag{49}$$

or

$$D_n(S_n - X_{n-1}) = R_{x,n} = X\mu \tag{50}$$

or

$$\frac{X_n - X_{n-1}}{R_{x,n}} = \frac{1}{D_n} = \tau_n \tag{51}$$

where $R_{x,n}$ is rate of biomass production which is a function of the conditions in the $n$th reactor. Equation (51) can be written for each of $n$ reactors. Summing all the $n$ equations will yield

$$\sum_{i=1}^{n} \frac{X_i - X_{i-1}}{R_{x,i}} = \sum_{i=1}^{n} \tau_i = \tau = \frac{1}{D} \tag{52}$$

where $D = F/V = F/V_i n$. Similarly, one can write:

$$\sum_{i=1}^{n} \frac{S_i - S_{i-1}}{-R_{s,i}} = \tau \tag{53}$$

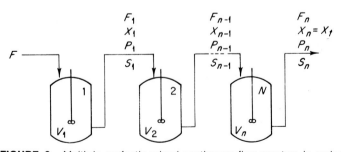

**FIGURE 3.** Multiple perfectly mixed continuous flow reactors in series.

and

$$\sum_{i=1}^{n} \frac{P_i - P_{i-1}}{R_{p,i}} = \tau \tag{54}$$

## VI. IDEALIZED REACTOR DESIGN III: PLUG FLOW REACTOR

Idealization of a plug flow reactor is made in that no backmixing takes place. Consequently, every infinitesimal quantity of feed that enters into the reactor will remain a separate entity, reside in the reactor for a fixed length of time, and then exit from the reactor. Numerous such infinitesimal entities move through the reactor, with no interchange of material or energy among different entities, each of which behaves like a separate perfectly mixed batch reactor. In the real physical world, idealized plug flow conditions are approached by those in packed bed reactors and reactors in the shape of long tubes. Therefore, plug flow reactors are also called idealized tubular reactors. A plug flow reactor may be visualized as a column of soldiers marching with a uniform pace down an avenue through a city block; each soldier remains a separate entity and spends an equal length of time passing through the particular city block.

A plug flow reactor is diagrammed in Fig. 4, with a length $L$ from the entrance to the exit. The volumetric liquid feed rate is designated by $F$ and the total liquid volume of the reactor by $V$. Assuming that the tubular reactor has a uniform liquid cross-section area of $A$, the liquid feed that has moved from the entrance and reached the differential reactor volume marked $dV$ in Fig. 4 will have resided in the reactor for a time of $AL/F$; and by the time it exits, its total residence time in the reactor will be $\tau = V/F = AL/F$. Therefore, at any instance, the liquid content in the reactor is of nonuniform residential age, with the oldest about to exit, the youngest just at the entrance, and the inbetween residential ages increasing from

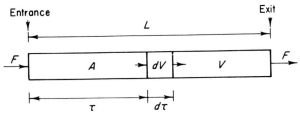

**FIGURE 4.** Plug flow reactor with constant liquid volumetric flow rate $F$ and cross-sectional area $A$.

the entrance end to the exit end. Liquid feed that has resided for different lengths of time in the reactor naturally has chemical and/or biological reactions carried out to different extents. It should be obvious that the concentrations of substrates, intermediates, and products in a plug flow reactor are thus nonuniform and vary from the entrance to the exit of the reactor. When material balance equations are written in analyzing a plug flow reactor, they are usually first written not for the entire reactor but only for a differential section (say $dV$ in Fig. 4) of the reactor. Applying the steady state Eq. (21) to $dV$ of Fig. 4, one can write the following biomass balance equation:

$$FX + (dV)R_x = F(X + dX) \tag{55}$$

In writing Eq. (55), $X$ is the biomass concentration in the inlet to $dV$ and $(x + dx)$ is the biomass concentration in the outlet from $dV$. Equation (55) simplifies to

$$dV/F = dX/R_x \tag{56}$$

Equation (56) is then integrated to describe the variation of biomass concentration through the whole plug flow reactor:

$$\int_0^v \frac{DV}{F} = \frac{V}{F} = \tau = \int_{x_0}^{x_f} \frac{dx}{R_x} \tag{57}$$

where $X_0$ = original $x$ at inlet to the whole reactor and $x_f$ = final $x$ at exit of the reactor. Similarly, for a substrate, from Eq. (21),

$$FS = (dV)R_s + F(S + dS) \tag{58}$$

or

$$dV/F = dS/(-R_s) \tag{59}$$

or

$$\int_0^v \frac{dV}{F} = \frac{V}{F} = \tau = \int_{S_0}^{S_f} \frac{dS}{-R_s} \tag{60}$$

The close respective resemblance between Eqs. (57) and (60) written for plug flow reactor and Eqs. (7) and (5) written for perfectly mixed batch reactor should come· as no surprise. A plug flow reactor has been analyzed as the totality of numerous infinitesimal batch reactors with a total liquid volume $V$, each of which is allowed a time $\tau$, the residence time, in the plug flow reactor, to carry out reactions from $S_0$ to $S_f$ and from $X_0$ to $X_f$, while a perfectly mixed batch reactor of a liquid volume $V$ carries out reactions over a time t to change from $S_0$ (or $G_0$) to $S_f$ (or $G_f$) and from $X_0$ to $X_f$. All changes in $S$, $X$, and $P$ that take place in a perfectly mixed batch reactor at different times $t$ can be found in a plug flow reactor at different $L$ from the entrance, and thus at different residence times in the reactor. In theory, therefore, conditions in a perfectly mixed batch reactor along a time axis can be exactly reproduced in an idealized plug flow reactor along an axis representing the distance from reactor entrance. As discussed previously in Section V (Eq. 44), the maximum biomass yield, $max(X_f/S_0)$, is achieved at zero dilution rate to a perfectly mixed continuous flow reactor, i.e., when it is operated as a batch reactor. The same maximum biomass yield can be achieved, therefore, in a plug flow reactor, which a perfectly mixed continuous flow reactor is, in theory, unable to achieve (unless $D = 0$).

Even though a plug flow reactor can in theory outperform a perfectly mixed continuous flow reactor as we have just described, in reality, a truly plug flow fermentor does not exist. The idealization of "plug flow" requires absolutely no backmixing, while in a real fermentation, requirements of aeration and/or agitation destroy this condition. In order to approach the performance of a plug flow reactor, multiple perfectly mixed continuous flow reactors in series provide a practical compromise. The performance of multiple-stage, perfectly mixed, continuous flow reactors of a total liquid volume $V$ will exactly equal to that of a plug flow reactor having a liquid volume $V$ when the number of stages approaches infinity. This behavior can be seen from Eq. (52) by letting $\Delta X = X_i - X_{i-1}$, and $n \to \infty$, and comparing with Eq. (57).

$$\tau = \lim_{n \to \infty} \sum_{i=1}^{n} \frac{X_i - X_{i-1}}{R_{x,i}} = \lim_{n \to \infty} \sum_{i=1}^{n} \frac{\Delta x}{R_{x,i}} = \int_{x_0}^{x_f} \frac{dx}{R_x} \tag{61}$$

In reality, of course, only a finite number of stages of perfectly mixed continuous flow reactor can be used, which should have a performance in $X_f/S_0$ intermediate to those of a plug flow reactor and a one-stage perfectly mixed continuous flow reactor.

When a different performance index, the specific volumetric biomass output (biomass productivity $DX_f$) is used in comparing the two idealized flow reactors, the relative superiority of the two can be different. This can be seen by the following analysis.

Growth is a special case of autocatalytic reactions. The growth rate $R_x$ behaves in general according to the curve in Fig. 5 with a minimum at an intermediate extent of substrate conversion. $R_x$ equals $X\mu$ and is a function of both $X$ and $S$. When the extent of conversion is low, $X$ is low but $S$ is high, and when the extent of conversion is high, $X$ is high but $S$ is low. Therefore, generally speaking, at an intermediate extent of conver-

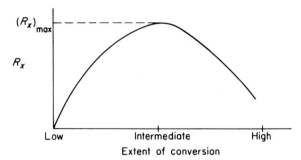

**FIGURE 5.**   Growth rate at different conversion.

sion, a maximum $R_x$ exists. The extent of conversion can be measured and expressed in many ways. One convenient way is to express conversion in the concentration of substrate left. This could be the value of $S$ in a batch reactor after a reaction time $t$ or in a plug flow reactor after a residence time $\tau$. This could also be the value of $S$ in the exit stream of a perfectly mixed continuous flow reactor or in the exit stream of several stages of such reactors in series.

Taking the curve in Fig. 5, and expressing the extent of conversion in $S$, one can plot $1/R_x$ versus $S$ in Fig. 6. From Eqs. (17) and (60), assuming $Y_{X,S}$ to be a constant and $X_f \gg X_0$,

$$\frac{V}{F} = \int_{S_0}^{S_f} \frac{dS}{-R_s} = Y_{X/S} \int_{S_0}^{S_f} \frac{dS}{-R_x} \tag{62}$$

and

$$Y_{X/S}(S_f - S_0) = X_f - X_0 = X_f \tag{63}$$

The biomass productivity of a plug flow reactor is, therefore,

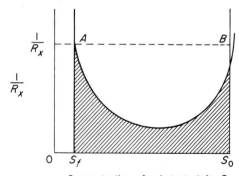

Concentration of substrate left, $S$

**FIGURE 6.**   Inverse growth rate at different conversion.

$$DX_f = \frac{FX_f}{V} = X_f / \left[ Y_{X/S} \int_{S_0}^{S_f} dS/(-R_x) \right] \qquad (64)$$

The shaded area in Fig. 6, according to the rules of graphical integration, is directly proportional to the expression in the denominator on the right-hand side of Eq. (64) and is inversely proportional to biomass productivity $DX_f$.

Similarly, the biomass productivity of a perfectly mixed continuous flow reactor can be shown as follows, by combining Eqs. (17) and (27), to be inversely proportional to the area of the rectangle $ABS_0S_f$ [area $ABS_0S_f = (S_0 - S_f)X \, 1/R_x$]:

$$DX_f = \frac{FX_f}{V} = \frac{X_f}{Y_{X/S}[(S_f - S_0)/(- R_x)]} \qquad (65)$$

knowing that for a perfectly mixed continuous flow reactor

$$X_f = X \text{ and } S_f = S$$

In the case shown in Fig. 6, the area of rectangle $ABS_0S_f$ is obviously much larger than the shaded area. This means a plug flow reactor has a better performance in biomass productivity, $DX_f$, than a perfectly mixed continuous flow reactor operating under the same $S_0$, $S_f$, and $Y_{X,X}$. However, being autocatalytic and having a maximum $R_x$ at an intermediate extent of conversion, a perfectly mixed continuous flow reactor can give a better biomass productivity than a plug flow reactor. This is shown by a comparison between the shaded area and the area of rectangle $A°B°S_0S_b$ in Fig. 7, if the desired fermentation is to run from $S_0$ and stop at $S_b$.

In practical cases, we usually would like to finish a fermentation at a high extent of conversion, i.e., a low concentration of substrate left, $S_f$ ($S_f$

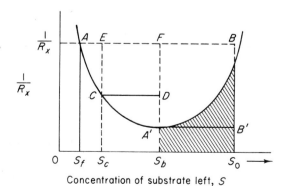

**FIGURE 7.** Comparison of reactor performance.

$< S_b$). The situation as shown in Fig. 7 suggests the use of a perfectly mixed continuous flow reactor to bring $S_0$ to $S_b$ and finish the conversion from $S_b$ to $S_f$ in a plug flow reactor with its biomass productivity inversely proportional to the area bounded by $AA^\circ S_b S_f$.

As we described previously, it is very difficult to realize truly plug flow conditions in a real-world fermentor which is often aerated and agitated. A compromise solution could be the use of a two- or multistage perfectly mixed continuous flow reactors in series to bring conversion from $S_b$ to $S_f$. In Fig. 7, an overall three-stage perfectly mixed continuous flow reactor series is shown with $S_b$, $S_c$, and $S_f$ being the concentrations of substrate left after the first, second, and third stage, respectively. The total area of three rectangles $A'B'S_0S_b$, $CDS_bS_c$, and $AES_cS_f$ is inversely proportional to the overall biomass productivity $DX_f$ of the reactors in series.

## VII. FED-BATCH REACTOR

In situations shown in Figs. 6 and 7, a multiple-stage, perfectly mixed, continuous flow reactor has been offered as a compromise solution for obtaining high biomass productivity. Another solution could be the use of a "fed-batch reactor." A fermentor operated in a fed-batch mode is first charged partially full with a broth volume of $V_0$. After a controlled length of time liquid feed is added continuously at a controlled rate until the fermentor is full at $V_f$. The fermentor is then left to run as a batch until $S_f$, after which the fermentor is discharged to $V_0$ and the cycle is repeated. The fermentor is cyclic in fed and batch phases with a short discharge phase in between cycles. A fully detailed analysis of a fed-batch fermentor is very intriguing and is beyond the scope of this chapter. By controlling the liquid feed, the fed phase may be made to behave like a perfectly mixed continuous flow reactor with a nearly constant $S_b$. For instance, one can add the liquid feed at an exponential rate to sustain biomass production in this phase at an exponential rate. During the batch phase, conversion is to be carried out from $S_b$ to $S_f$.

## VIII. AERATION

Finally, a few words need to be said about aeration which, due to dispersion of air or oxygen bubbles, creates a heterogeneous mixture in a fermentor. This author has been careful, in this chapter, in distinguishing the liquid volume from the total fermentor volume. The latter includes those of the head space and gas bubbles. All material balances that we have described apply to the fermentation liquid volume which is assumed to be homogeneous and ignoring the fact that microbial cells actually form yet a third material phase. Aeration is, in a way, a special problem. No batch fermentation is truly "batch" because oxygen is always continuously added and carbon dioxide often continuously removed. Similar complications also exist in liquid hydrocarbon fermentation, in which an additional liquid phase exists. The analysis of this chapter applies to the volume of aqueous broth. In analyzing the overall fermentation system involving several material phases, one can write material balance equations for each of all the phases. These equations are coupled by mass transfer in between phases. These more complicated problems of modeling and reactor design and analysis are not included here.

## GENERAL REFERENCES

Levenspiel, O. (1972). "Chemical Reaction Engineering," 2nd ed. Wiley, New York.

Lim, H. C., Chen, B. J., and Creagan, C. C. (1977). *Biotechnol. Bioeng.* **19,** 425.

Perlmutter, D. D. (1972). "Stability of Chemical Reactors," Prentice-Hall, Englewood Cliffs, New Jersey.

Pirt, S. J. (1975). "Principles of Microbe and Cell Cultivation." Halsted Press, New York.

Smith, J. M. (1970). "Chemical Engineering Kinetics," 2nd ed. McGraw-Hill, New York.

# Chapter 12

# Microbial Culture Selection

RICHARD P. ELANDER
L. T. CHANG

**243**

MICROBIAL TECHNOLOGY, 2nd ed., VOL.II
Copyright © 1979 by Academic Press, Inc.
All rights of reproduction in any form reserved. ISBN 0-12-551502-2

## I. INTRODUCTION

Microbial culture selection is involved in the initial isolation of microorganisms from nature and later in the purification of strains of interest
due to the synthesis of desirable metabolic products. In most large-scale
screening programs, rare, unusual cultures are selected by taxonomists
since the probability of new interesting activities is maximized in sensitive specific assays using rare, fastidious microorganisms (Conover,
1971). The purification of unusual organisms is difficult and involves
considerable technique in culture selection. Often interesting cultures
must be constantly repurified, because of slowly-growing commensal
organisms found associated with the organism of interest.

Pure cultures of microorganisms are inherently variable in their growth
characteristics and metabolic activities. Therefore, the initial activity of
the microbiologist is to minimize the genetic variability of the microorganisms by selecting out stable and genetically uniform isolates which
produce a minimum number of unwanted metabolites and copious
amounts of the desired component. The second important activity is to
develop high-producing strains. At first, a simple genetic approach is
used such as the selection of strains from untreated spore populations,
or the exposure of spores or cells to mutagens followed by the random
selection of isolates via monospore isolation and their testing in limited
fermentation test systems. In order to find the few strains capable of
higher productivity, a large number of cultures must be evaluated. This
approach probably has resulted in the discard of interesting strains, but
the methodology is simple and requires relatively little genetic experience.

Although natural selection and mutation selection are generally successful in improving titers of described metabolites, culture stability,
growth and sporulation characteristics, etc., often represent additional
critical problems to be resolved. These characteristics are also inherent
genetic problems and must be resolved in a culture improvement program. Loss in titer (termed "culture rundown" or "strain degeneration")
often occurs with improved strains and such variants are frequently
encountered during growth following vegetative transfer, storage, or
long-term preservation. Another associated problem is the biosynthesis
of undesirable metabolites in a multicomponent complex of related
metabolites. The elimination or decrease in titers of unwanted compo-

nents can be resolved by culture selection or by changes in the chemical and physical environment of the fermentation.

A high-producing stable strain is essential both to fundamental studies carried out to establish pathways of biosynthesis and to fermentation optimization through control of the chemical and physical environments. The use of selection to develop a stable productive culture and the acquisition of knowledge of the strain–environment relationships are important to allow the fermentation specialist to control and direct the microbe to more efficient productivity of the desired metabolite.

The purpose of this chapter is to review genetic and selection methodology which is currently utilized in the selection, purification, and stabilization of high-productivity mutants as exemplified by currently important fermentation microorganisms. We also review the present status of genetic recombination with regard to the generation of improved fermentation microorganisms. We also highlight the recent progress attained in cloning foreign DNA through plasmid, phage, or transformation systems, to generate new "hybrid" organisms. This new technology of genetic engineering further expands man's control of microorganisms and allows further exploitation of the microbial world to better serve man's needs.

## II. MUTAGENESIS AND IMPROVED PRODUCT YIELD

Mutagenesis followed by the subsequent selection and purification of superior strains represents the most important initial activity in improving the yield of a fermentation product. Mutation programs are vital to fermentation companies in the higher productivities are essential in reducing or maintaining product cost in periods of inflation and increasing labor and raw material costs. Therefore, mutation and genetic programs are being expanded in both industrial and applied microbiological laboratories. It is clear that mutation and selection to enhanced strain productivity represents the major factor involved in the manyfold improvement of fermentation products.

### A. Natural Selection

As soon as a metabolite shows clinical potential, programs are initiated in fermentation development areas for strain improvement and the subsequent long-term preservation of improved variants. It is well established that the most effective method initially for improving the yield of a fermentation product is the application of mutation followed by selection.

General reviews of this subject are available and recent reports by Fantini (1976) and Elander and Espenshade (1976) may be useful.

Most microorganisms exhibit a great capacity for natural variation. Therefore, the initial approach in selection work is to survey this background variability for improved cultures. Colonies are generally examined in laboratory fermentations which originate from untreated spore suspensions (also termed monospore isolates). Cultures isolated from nature are generally heterokaryotic and the selection of single spores with uninucleate conidia generally leads to a variety of colony types (termed "population pattern"), many of which may be superior in their capacity to synthesize a fermentation product. A classic paper dealing with penicillin and initial strains of *Penicillium chrysogenum* is by Backus and Stauffer (1955).

## B. Mutation Selection

The exposure of spores to mutagenic agents and the subsequent selection of colony populations is known as "mutation selection;" it has two aspects when applied to industrial microorganisms. One aspect, *major mutation,* involves the selection of mutants which manifest a profound change in a biochemical character of practical interest. These strains are usually isolated from spores surviving prolonged exposures to mutagens and generally exhibit a distinct change in some morphological or biochemical character. In contrast, *minor mutations* in strains represent only subtle changes in a particular character and are superficially indistinguishable from the parent phenotype. Usually the biochemical changes are so slight that only the trained expert can distinguish these variants. Minor variants are commonly exploited in programs devoted exclusively to yield improvement.

### 1. The Role of Major Mutation in Culture Selection

Examples of major mutation in strain development are numerous. Alikhanian (1962) cites a role for major mutation in the streptomycin-producing organism *Streptomyces griseus.* The initial strain synthesized large amounts of a substance with low activity in addition to small quantities of streptomycin. The low-activity substance is mannosidostreptomycin (streptomycin B), which competes with streptomycin for biosynthetic intermediates and also interferes with the isolation of the antibiotic. A variant was isolated which produced negligible amounts of the undesired substance, thus allowing greater synthesis and recovery of the desired moiety. The development of nonpigmented strains of *P. chrysogenum* is another example of major mutation in an industrially important fungus. Strains blocked in the synthesis of the yellow pigment chrysogenin were isolated which allowed for a more efficient recovery of penicillin (Backus and Stauffer, 1955).

A careful study of variants exhibiting impaired antibiotic productivity may elucidate biosynthetic pathways and contribute to the identification of precursors. Studies on cobalamin by Barchielli *et al.* (1960), and on tetracycline by Miller *et al.* (1965), and by Hošťálek and Vaněk (1973), reveal interesting data on precursor molecules involved in reaction steps prior to terminal ring closure.

The use of major mutation has acquired particular significance and, in certain instances, has led to new and more efficacious products. The tetracycline-producing organisms appear to be particularly amenable to this approach. McCormick *et al.* (1957) described an interesting modification of tetracycline synthesized by a mutant strain of *Streptomyces aureofaciens*. The antibiotic was shown to be changed at the C-5a position and was almost devoid of antibiotic activity. However, they also reported that *S. aureofaciens* (strain S-604) synthesized 6-demethyltetracycline, a new antibiotic material not elaborated by the Duggar strain A-377. The new molecule has several advantages over the methylated form and today is one of the leading commercial forms of tetracycline.

The employment of mutants for the synthesis of modified antibiotic molecules appears to be a fertile area for major mutation in strain development. Mutants have been mentioned already that may accumulate presursor molecules which aid in the elucidation of pathways. A minor modification of an antibiotic molecule could lead to new biological and therapeutic properties for a known antibiotic. Some years ago, Kelner (1949) published a paper which has great interest in this connection. He examined a series of streptomycete cultures which failed to inhibit certain test bacteria. The negative strains were then exposed to heavy doses of ultraviolet (uv) and x-ray radiation. After examining several thousand irradiated strains, Kelner found mutant lines which exhibited antimicrobial activity. In fact, certain weak antibiotic producers then showed a change in spectrum. The difference in antimicrobial spectrum indicated that in certain cases this change might have resulted in the qualitative modification of an antibiotic. Structural modifications with a resulting change in spectrum have been demonstrated for oxytetracycline and chlortetracycline.

### 2. The Role of Minor Mutation in Culture Selection

Minor mutation usually plays the dominant role in yield improvement programs. These mutations generally affect only quantitatively the amount of product synthesized. Such mutations are subtle, and variants exhibiting such features are generally similar in characteristics, other than yield, to the parent form. Quantitative definition of what constitutes a significant yield increase is somewhat relative and depends on the reliability and accuracy of the assay procedure and the productivity level of the metabolite under investigation. Usually, a 10–20% increase is implied. The variants are usually selected from spore suspensions exposed to small or moderate doses of a mutagen. If one repeatedly selects "minor" (positive gain) variants and uses each succeeding improved strain for further mutation and selection work, after many stages of "stepwise selection," a large yield increase may be obtained (see Section VII, A).

Minor variants exhibiting a slight change in a quantitative feature may vary also in other features due to pleiotropic effects, i.e., pigmentation, conidiation, and radial growth rate. However, since the variation in yield is slight, great dependence must be placed on efficient and accurate selection techniques. The population to be tested must be large and the assay must be highly accurate and specific for the desired product. Problems of this sort are statistical and are discussed in reviews by Davies (1964), Brown and Elander (1966), and Fantini (1975).

Examples of gradual stepwise improvement in antibiotic production are numerous and are reviewed in detail for penicillin and cephalosporin later in this chapter (see Section VII, B). Vezina et al. (1976) have recently summarized the development of a series of improved antimycin A-producing strains at the Ayerst Research Laboratories. Their study emphasized the power of selection procedures. First, much attention was focused on developing a rapid, accurate, and convenient assay system. Such assay procedures are now usually chemical in nature and are often automated spectrophotometric or high-pressure liquid chromatographic systems capable of handling large numbers of samples per day with computerized data handling systems. Second, at each round of mutagenesis (usually a cyclical use of different mutagens having differing modes of action), several improved strains were compared not only for productivity but also for early dense conidiation, absence of pigment, and growth rate; only one strain, which represented a distinct advantage over the parent, was chosen for further selection. Third, since the fermentation environment plays an important role in the expression of genetic determinants, the fermentation medium and conditions were periodically revised to best suit the organism for further enhanced titer improvement. Fourth, strains showing productivity in laboratory flasks were later evaluated in small pilot-scale aerated–agitated fermentors. Throughout their antimycin improvement program, a rigid maintenance, propagation, and preservation scheme was adhered to. The results of this industrial strain improvement program resulted in a major yield improvement from 0.75 to 9.75 gm/liter, an overall 125-fold augmentation. This program is summarized in Fig. 1.

Many refinements in the techniques of mutation and selection have been introduced during the past two decades. Today, there are scores of mutagens available, and for many of them a mode of action is known. Excellent reviews on the chemical basis of mutation are available (Orgel, 1965; Drake and Baltz, 1976). Unfortunately, the phenomenon of mutation is random and not directable, so that mutation as applied to strain

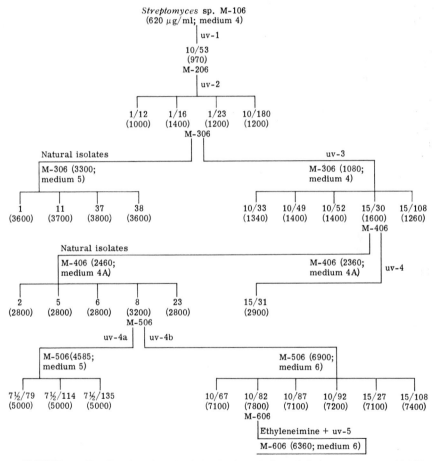

**FIGURE 1.** Family of antimycin A-producing strains of *Streptomyces* sp. M-106. Yields in micrograms per milliliter are given in parentheses. (Adapted from Vezina *et al.*, 1976.)

development is largely empirical. Little is known of the kind of mutants desired. The limited knowledge of antibiotic biosynthesis is shown by the fact that a biosynthetic pathway is not fully elucidated for a single antibiotic.

In contrast, considerable refinement in mutation and selection techniques has been demonstrated over the past two decades. Refinements in evaluation have been numerous, and these probably represent the key to a successful strain development program. The importance of evaluation is illustrated by the problem of selecting relatively few enhanced variants from a population of hundreds of thousands of individuals showing control yields or less. Many of these techniques are peculiar to organisms of commercial interest, and their publication is restricted.

Examples of sophistication in techniques are numerous and are discussed in detail in Section III of this chapter. One such example is that of James et al. (1956) on an interesting method of selection with the citric acid-producing fungus Aspergillus niger. The preliminary test involved the propagation of mold colonies on filter paper impregnated with a suitable production medium containing an appropriate indicator dye. After a prescribed incubation period, the acidity was measured as citric acid. The best strains were then evaluated in shake-flask fermentations. Many of the strains which were superior in the preliminary test synthesized greater quantities of citric acid in submerged fermentation. Alikhanian (1962) studied colony morphology with reference to subsequent productivity of penicillin, and he recommended the rejection of poorly growing, poorly conidiating strains of P. chrysogenum. He also suggested screening variants of Penicillium in the absence of phenylacetic acid on the basis that the enzymes generating the penicillin nucleus (6-aminopenicillanic acid) might be rate limiting under these conditions with respect to total penicillin G yield. Screening of new strains on an inferior medium was suggested, because a strain superior on such media might exhibit marked superiority under more optimal fermentation conditions. Ostroukhov and Kuznetsov (1963) described a unique plate method for the detection of superior penicillin producers on the basis of increased oxidation–reduction potential. They also mentioned that strains may exhibit more negative potential under conditions of reduced aeration.

The use of actinophage for the selection of phage-resistant strains first arose with the violent outbreak of phage infection in the streptomycin industry (Carvajal, 1952). Carvajal observed that an occasional resistant colony which survived phagolysis produced more streptomycin than the sensitive production strain. A pronounced mutagenic effect of actinophage was observed by Alikhanian (1962), who claimed that 99% of the *Actinomyces olivaceus* colonies cultivated from a phagolyzate were similar morphologically to uv mutants and certain of the mutant types were not found in usual uv-survivor populations.

Sophistication at the selection stage is essential for effective strain improvement; the following generalities can be made:

1. Strains selected as obvious variants following exposure to mutagens usually are inferior in their capacity to elaborate metabolites. Those strains with enhanced capacity for accumulation of products are extremely few in number, and selection and evaluation play extremely important roles in their detection.

2. Mutagen dose is important in strain selection methodology. The rate of mutation is a function of dose. Strains experiencing major mutation are best isolated from populations surviving prolonged doses of mutagens. Variants to be employed for increased productivity are generally isolated from populations surviving intermediate dose levels.

3. Strains with enhanced capacity for synthesis generally exhibit wildtype morphololgy and growth habit. Strains with altered morphology, etc., may be inherently better producers but may require considerable fermentation development. Since positive variants are extremely few in number, it is better to screen a large number of variants under a few fermentation conditions rather than to screen a small number of variants under a wide variety of environmental conditions. This is especially true when one seeks only quantitative change in the variants.

4. The philosophy of stepwise selection accepts small increments in increased productivity. The range of increased productivity at each step is generally 10–15% after the initial strains are obtained by natural selection. As productivity increases, the probability of finding superior strains decreases; hence accurate and more sophisticated evaluation procedures play increasingly important roles. The development program is only as effective as the mutation, selection, and evaluation procedures coupled to it.

5. Variant strains often require special propagation and preservation procedures. Actual production gains depend largely on stability and reliability of performance. Maintenance through continued selection and purification plays an important role in production laboratories.

6. Although a strain may meet the numerous necessary criteria of superiority in the laboratory, there is no guarantee that enhanced productivity will occur in production fermentors. Aeration–agitation patterns, nutrient availability, etc., are often unbalanced for variant cultures. Scaleup experimentation in pilot-plant studies is often necessary before any enhanced strain potential may be realized in actual production.

## III. ISOLATION OF MUTANT CLASSES AND THEIR USE IN MICROBIAL PROCESSES

Although the application of a suitable mutagenic treatment procedure is essential to effective mutation methodology, the development of rapid, selective isolation procedures and the tests employed to identify the desired class of mutants is of far greater importance. Procedures used for the isolation of mutant classes can be considered as variations on the proverbial problem of "discovering the needle in the haystack." The simplest procedure, and probably the most tedious, is to examine a large number of clones in liquid fermentations in the hopes of finding one mutant which shows some improvement in a particular biochemical character over the parent strain. This approach has often been utilized with industrial microorganisms. However, it has become increasingly important to consider new methods which allow for the selective isolation of particular classes of mutants, including overproducers of desired products. The following discussion will be concerned with the isolation of stable mutants exhibiting robust growth which can be used on an industrial scale.

The chance of discovering the desired mutant may be improved by increasing the frequency of mutation by a cyclical application of mutagens, either singly or in combination. It is also possible to increase the number of potentially interesting mutants by utilizing selective enrichment procedures which allow the mutants to grow more rapidly than the parent culture. These procedures include alternate subculture in selective and nonselective media either in stationary or in continuous culture. Selection procedures can employ color reactions to identify mutant colonies and are very useful. The use of replica-plating procedures allows for the efficient transfer of hundreds of colonies in a single transfer on a variety of media which may be selective in themselves or reveal the presence of higher concentrations of desirable products.

Positive selection procedures utilizing nutrient media which restrict the growth of undesirable clones, but which allow for the expression of particular mutant classes, are highly desirable. Selection for growth on novel substrates often restricts the growth of the parent and allows the mutants to grow, albeit at reduced rates. Often, leaky auxotrophs (bradytrophs) may be more sensitive to inhibition by metabolic analogs and this represents a way to discover derepressed mutants for the production of biosynthetic intermediates. The use of synchronized cells grown in chemostat culture is an excellent procedure for obtaining mutants resistant to high concentrations of toxic inhibitors, metabolic analogs, antibiotics, etc. The theoretical basis for mutant isolation is that mutant cells capable of producing larger quantities of enzymes or possessing greater permeability to substrates will have a distinct growth advantage and eventually outgrow and replace the less efficient parental strain.

We now review the procedures for the enrichment and isolation of distinct classes of industrially important mutants and describe how these mutant classes have been utilized in microbial fermentation processes. For background reading, we recommend important reviews by Hopwood (1970), Abe (1972), Clarke (1976), and Hopwood and Merrick (1977).

## A. Auxotrophic Mutants

Auxotrophic mutants are recognized by their growth on a complex multicomponented, "complete" medium and their lack of growth on a chemically defined "minimal" medium. The mutants are then classified into categories by propagation on minimal medium supplemented with amino acids (amino acid-requiring), vitamins (vitamin-requiring), or nucleic acid components (purine- or pyrimidine-requiring). Holliday (1956) developed a single-step characterization procedure which includes the replication of a masterplate of mutant strains to a number of petri dishes, each containing a pool of substances, arranged in such a manner that each substance is present in two different pools arranged in such a way that only 12 plates are needed to test 36 different compounds.

Auxotrophic mutation generally leads to lower productivity in antibiotic-producing microorganisms. This is especially true when one selects mutants which are deficient in growth factors which are also precursor molecules for antibiotic synthesis. Antibiotic productivity is usually low for auxotrophic mutants, even when tested in nutritionally complex media containing copious amounts of the required growth factor. Auxotrophy often affects antibiotic synthesis in the negative direction, even when the required growth factor is not a precursor of the secondary metabolite. Often such auxotrophs grow poorly and, therefore, exhibit decreased antibiotic formation. Other auxotrophic strains appear to have multiple genetic lesions affecting growth and antibiotic synthesis. Despite all these problems, high-producing auxotrophic strains have been obtained for producers of tetracycline, erythromycin, tylosin, actinomycin, and streptomycin. Multiple-mutant cultures can be readily selected from populations following long exposure to potent mutagenic agents, i.e., diazomethane and nitrosoguanidine.

Conidial color variants and biochemical mutants are routinely encountered in survivor populations of *P. chrysogenum*. In one detailed study, albino and yellow conidial mutants were reported to produce 30% less penicillin in flask fermentations. Auxotrophic mutants exhibited reduced vegetative development, with yields varying from 60 to 90% of the productivity of the prototrophic parental culture (Elander, 1967). A large number of morphological and biochemical variants of the cephalosporin C-producing fungus, *Cephalosporium acremonium* (also classified as *Acremonium chrysogenum),* showed a decrease in relative potency with increasing mutagenesis (Elander, 1975).

Auxotrophic mutants of the glutamic acid bacterium, *Corynebacterium glutamicum,* have been exploited in Japan for the commercial production of a number of amino acids listed in Table I.

The lysine-producing strain is a homoserine (or threonine plus methionine) auxotroph and is blocked in that part of the branched pathway leading to methionine, threonine, and isoleucine. The total world market for L-lysine is in excess of 15,000 tons/year, the amino acid being used primarily as a feed supplement. Fermentation potencies of 44 gm/liter of L-lysine·HCl utilizing cane molasses have been reported accompanied by yields of 30–40% in relation to the initial sugar concentration (Nakayama, 1976). The blocking of homoserine synthesis via mutation in the homoserine dehydrogenase gene results in the release of concerted feedback inhibition by threonine plus lysine on aspartokinase as long as threonine is restricted in the medium. In this case, aspartic semialdehyde is converted to lysine and the amino acid is over-produced.

*Phe, tyr,* and *phe tyr* mutants of *C. glutamicum* have been utilized for the commercial production of L-tyrosine, L-phenylalanine, and L-tryptophan, respectively. Figure 2 shows the remarkable progress in productivity of mutants producing high yields of these amino acids when subsequent analog-resistance mutations are imposed upon the auxotrophs.

## B. Analog-Resistant Mutants

The first step in isolating mutants resistant to structural analogs of metabolites is to determine, by plating on a graded series of inhibitor levels, the minimal concentration that prevents growth of the wildtype strain. Resistant mutants are then selected by plating out large numbers of mutagenized cells on plates containing a series of increasing concentrations of the inhibitor. Gradient-plating techniques represent useful alternative procedures for obtaining resistant mutants (Hopwood, 1970).

Mutants resistant to amino acid analogs have been shown to owe their resistance to mutation (a) in a regulatory or operator gene, which has resulted in derepression of biosynthetic enzymes, thereby overproducing the antagonized amino acid and thus overcoming competitive inhibition by the analog; (b) in the structural gene of a permease for the antagonized amino acid, which has resulted in reduced capability to incorporate the analog; (c) in the structural gene of an amino acyl-tRNA

**TABLE I.** Amino Acid Accumulation by Auxotrophic Strains of *Corynebacterium glutamicum*[a]

| Auxotrophic markers | Amino acid accumulated |
|---|---|
| *arg/cit* | Ornithine |
| *hom; thr; leu; ile; met + thr; ile + leu* | Lysine |
| *ile; leu; thr; arg* | Valine |
| *thr* | Homoserine |
| *tyr* | Phenylalanine |
| *lys* | Diaminopimelic acid |
| *leu* | Isoleucine |
| *phe* | Tyrosine |
| *arg* | Citrulline |

[a] From Abe (1972).

synthetase, which has resulted in discrimination between the antagonized amino acid and the analog during protein synthesis; and (d) in the structural gene of a biosynthetic enzyme subject to end-product inhibition, which has resulted in loss of feedback control.

Selection of mutants resistant to analogs of primary metabolites often overproduce end product since the mutants are no longer subject to feedback repression. Such a concept has been applied by Elander *et al.* (1971) to the pyrrolnitrin fermentation, an antifungal phenylpyrrole derived from tryptophan in strains of *Pseudomonas*. D-Tryptophan is a direct precursor of this antibiotic but was impractical to use in large-scale fermentation because of its high cost. An effort was made to select fluoro- and methyltryptophan-resistant mutants no longer subject to feedback inhibition by tryptophan. Resistant mutants were ultimately selected which produced nearly threefold more antibiotic than the sensitive parent culture and, in addition, D-tryptophan was no longer necessary for pyrrolnitrin formation (Fig. 3).

Nüesch *et al.* (1976) utilized analogs of methionine in studies on the relationship between methionine and cephalosporin C synthesis in strains of *C. acremonium*. They isolated a series of mutants both sensitive and resistant to selenomethionine and found reduction in methionine uptake which correlated with reduction in cephalosporin C synthesis and excretion of 2-hydroxy-4-methiolbutyric acid, a deamination product of methionine.

The use of auxotrophic mutants which require one or more amino acids in the synthesis of a desired amino acid and the additional induction of resistance to an analog of the desired metabolite often results in increased overproduction levels of amino acids. As shown in Fig. 2 for the production of L-tyrosine by mutants of *C. glutamicum*, first-stage screening was initiated by isolating phenylalanine-requiring mutants (Phe⁻).

The *Phe⁻* mutant was then further mutagenized and selected for resistance to a series of analogs, including 3-aminotyrosine (3AT), *p*-aminophenylalanine (PAP), *p*-fluorophenylalanine (PFP), and tyrosinehydroxamate (TyrHx). A mutant strain was eventually isolated (Pr-20) which produced 17.6 gm/liter of L-tyrosine, a sixfold increase when compared to the KY10233 (*Phe⁻*) strain (Arima, 1977). The stepwise increases in productivity of L-phenylalanine and L-tryptophan are also summarized in Fig. 2.

## C. Reversion Mutants

### 1. Reversion of Nonproducing Strains

To increase the efficiency of selection methods utilizing tetracycline-producing strains of *Streptomyces rimosus* and *Streptomyces viridifaciens,* procedures were developed based on the selection of high-producing variants from various types of mutant populations. The greatest number of these variants were obtained from revertants of non-producing mutants, i.e., mutants in which a previous mutagenic treatment had induced lesions in loci controlling tetracycline biosynthesis. Reversion mutants which cannot produce tetracycline can be detected easily by plating the nonproducing culture after mutagen treatment, followed by an overlay assay using a sensitive assay organism. Revertants will produce clear zones, whereas nonproducers will show no zones of inhibition. Utilizing this technique, Dulaney and Dulaney (1967) obtained many revertants of a nonproducer parent which produced more than twice as much chlortetracycline as the original grandparent culture. One revertant strain showed more than a sixfold increase over its grand-parent strain.

### 2. Reversion of Auxotrophic Mutants

Overproduction of precursor molecules is another potential means to increase antibiotic formation. This overproduction of the primary metabolite precursor can be achieved by relieving feedback regulation of its biosynthetic pathway. Modification of the structure of a feedback-sensitive enzyme through auxotrophic mutation followed by enzyme replacement via a reversion mutation is a common method for relieving such regulation. Dulaney and Dulaney (1967) used this technique to increase chlortetracycline production in a low-producing strain of *S. viridifaciens.* Reverse mutation of a homocysteine auxotroph resulted in 88% of the revertants producing threefold more tetracycline than the original prototrophic culture (Fig. 4).

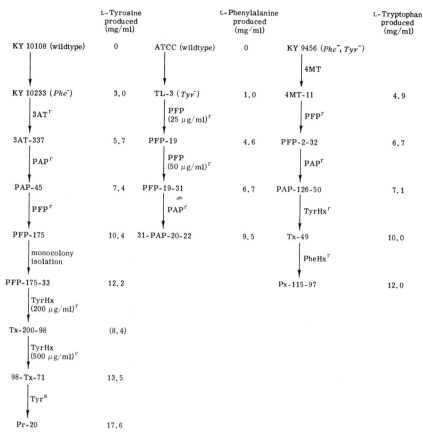

**FIGURE 2.** Overproduction of L-tyrosine, L-phenylalanine, and L-tryptophan by mutants of *Corynebacterium glutamicum*. 3AT = 3-aminotyrosine, PAP = *p*-aminophenylalanine, PFP = *p*-fluorophenylalanine, PyrHx = tyrosinehydroxamate, 4MT = 4-methyltryptophan, PheHx = phenylalaninehydroxamate. (After Arima, 1977.)

## D. Phage-Resistant Mutants

The use of phage as a mutagenic agent has often been claimed, although evidence for discrete mutagenesis based on a well-designed model has not been reported. However, microbial populations surviving phagolysis show extreme variability in both morphology and productivity. Viruses are known to induce chromosomal aberrations and breakage in mammalian cells and to induce point mutations in *Drosophila*. Mutagenic effects of actinophage have been demonstrated in production strains of *S. aureofaciens*. Phage-resistant variants showed increased

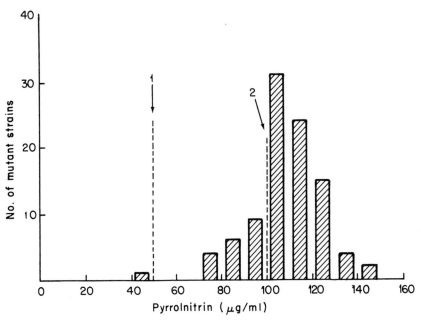

**FIGURE 3.** Pyrrolnitrin production by strains of *Pseudomonas fluorescens* resistant to 5-fluorotryptophan (5FT, 1 mg/ml). (1) Yield of drug-sensitive grandparent A10338.5 (2) Yield of first step 5-FT resistant parent A10338.7. (From Elander *et al.*, 1971.)

variability in tetracycline production, as compared to an untreated control strain, of 60–200% depending on the phage type and on the activity of the strain from which the phage had been isolated (Alikhanian and Iljina, 1958). Perlman and Ogawa (1976) reported that streptomycin-negative strains of *S. griseus* resistant to phage isolated from producing strains synthesized streptomycin, which can be interpreted as an example of transduction of antibiotic-producing capacity. Increase in antibiotic titer could well be a pleiotropic effect of resistance mutations selected by phage action. The selection of phage-resistant populations in screening for gain variants has proved to be of utility in a number of actinomycete-based antibiotic programs.

It should be reiterated that phage-resistant strains are inevitably subject to the possibility of infection by new contaminating phages and by host range mutants. Even though the use of resistant strains may contribute to stabilization of the fermentation process, it is important to eradicate all sources of contaminating phages from the fermentation area. At present, the most effective method aside from rigorous tank sterility procedures is to keep in the fermentation culture repository a variety of resistant strains with differing sensitivities to each of the phage types

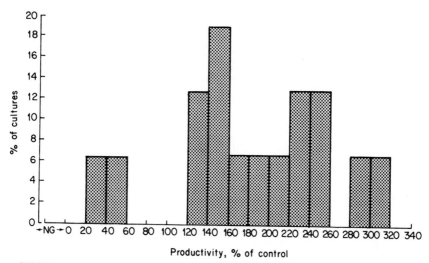

**FIGURE 4.** Spread in tetracycline productivity of prototrophs recovered from a homocysteine-requiring mutant of *Streptomyces viridifaciens*. (From Dulaney and Dulaney, 1967.)

detected in the fermentor. The strains can then be utilized alternately during periods of repeated contamination.

## E. Mutation to Nonproduction of Undesirable Metabolites and the Use of Blocked Mutants

Biosynthesis of antibiotic mixtures containing a variety of related or unrelated antibiotics is commonly encountered with producer strains. The problem is to produce the desired antibiotic component exclusively or to select variants which selectively produce more of the desired component and less of the others. Mutation and medium manipulation with the nebramycin-producing organism *Streptomyces tenebrarius*, a culture producing a number of nebramycin components, yielded a variety of strains, some of which elaborated a single component in fermentations (Stark *et al.*, 1971). A methionine auxotroph of *Streptomyces erytheus*, the producing culture of erythromycin, when fed varying concentrations of the required amino acid, synthesized increased levels of erythromycin C relative to the erythromycin A component (Ostrowska-Krysiak, 1971). *Streptomyces noursei* produces cycloheximide and the unrelated metabolite, nystatin. Mutants were selected which no longer synthesized cycloheximide and which concomitantly produced markedly increased levels of nystatin (Spizek *et al.*, 1965). Strains of *C. acremonium* produce a variety of different antibiotic moieties. Intensive development programs culminated in strains producing dramatically increased levels of cephalosporin C with proportionately lower levels of the related but unwanted component, penicillin N (Elander et al., 1976). In a series of mutants which were blocked in the synthesis of cephalosporin C, strains were encountered which synthesized enhanced levels of deacetoxycephalosporin C as well as deacetylcephalosporin C (Queener et al., 1974). Strains producing enhanced levels of deacetoxycephalosporin C at the expense of cephalosporin C could have utility in producing an intermediate important for the manufacture of orally active cephalosporin antibiotics.

## F. Metabolite-Resistant Mutants

Often, metabolites are toxic to growing cultures of the producer organism. Fortunately, this toxicity is not an important factor in large-scale fermentation production since the secondary metabolite is usually produced in idiophase after the growth phase of fermentation is essentially completed. Antibiotics which inhibit the growth of the producer organism include tetracycline, novobiocin, actinomycin, streptomycin, and nystatin. Differences in biosynthetic activity may be affected by genetically determined resistance to the producing culture's own antibiotic. Therefore, increasing resistance either by adaptation to gradually increasing concentrations of antibiotics or by subjecting mutagen-treated populations to increasing gradients of antibiotic is a useful procedure for improving the biosynthetic capacity of production strains. Katagiri (1954) obtained a fourfold increase in chlortetracycline (CTC) productivity in S. aureofaciens by repeated transfers in media containing 200–400 $\mu$g CTC/ml. The sensitivity of the surviving strains to increasing concentrations of chlortetracycline was in direct proportion to their capacity to synthesize antibiotic. Whereas low concentrations of tetracycline led to increased frequency of low-producing variants, exposure to high concentrations resulted in an increased number of active mutants. These results can be explained by the lethal effect of the antibiotic which selectively deprived the natural population of low-producing sensitive clones. This method is most effective during the early phase of strain improvement or in cases when a population of a high-producing strain must be repurified.

## G. Mutations Affecting Permeability

The production of large amounts of glutamic acid in bacteria can be attained by selecting mutants having an altered cell membrane with increased outward permeability of the amino acid. The increased permeability can be attained by biotin deficiency or addition of penicillin or fatty acid derivatives to wildtype strains of *C. glutamicum,* or by oleic acid limitation of oleic acid-requiring auxotrophs or by glycerol limitation of glycerol-requiring mutants (Nakayama, 1976). A change in membrane chemistry correlates with increased permeability of cells and with excretion of high levels of glutamic acid (Arima, 1977).

Permeability may also be important in the production of L-lysine. Mutants of *Escherichia coli* resistant to S-($\beta$-aminoethyl)cysteine excrete lysine and are defective in the lysine transport system (Halsall, 1975). The overproduction of lysine was not due to mutations affecting either of two regulated enzymes, aspartokinase and dihydrodipicolinic acid synthetase, but rather to active transport systems which enable an efflux of lysine against a fivefold higher concentration of extracellular lysine compared to the intracellular concentration (Ibuki and Takinami, 1975).

## H. Miscellaneous Classes of Mutants

The metabolic poison monofluoracetate (MFA) is known to inhibit aconitate hydratase, an enzyme which converts citrate to isocitrate in strains of *Candida lipolytica.* Mutants sensitive to MFA produced less aconitate hydratase and, when used to produce citric acid on *n*-paraffin, synthesized citric acid and isocitric acid at a ratio of 97 : 3 compared to the usual ratio of 60 : 40 (Akiyama *et al.*, 1972).

Selenomethionine-sensitive strains of *C. acremonium* have increased capacity to synthesize cephalosporin C due to enhanced methionine uptake (Nüesch *et al.*, 1976). Okanishi and Gregory (1970) using mutant strains of *Candida tropicalis* obtained cells whose protein content had an increased methionine content by the ingenious selection of small colonies on a sulfur-deficient medium. They postulated that methionine-rich mutants would have a greater sulfate requirement than parental cells and would, therefore, form smaller colonies on a medium in which sulfate was limiting.

The production of glucose isomerase (GI) in strains of *Streptomyces,* used commercially for the conversion of glucose to fructose, was repressed by glucose and required the presence of D-xylose as an inducer. Mutants resistant to D-lyxose, an analog of glucose, were found to produce GI constitutively without xylose but were still repressed by glucose. Additional selection for resistance to a glucose analog, 3-*o*-methylglucose, resulted in the isolation of a unique mutant which produced GI in the presence of glucose, thereby being insensitive to catabolite repression (Sanchez and Quinto, 1975).

Manipulation of the parasexual cycle in filamentous fungi can be used to select for high-yielding diploid or polyploid strains. Heterozygous diploid strains of *A. niger* have been reported by Ikeda *et al*. (1957) to be superior producers of citric acid and kojic acid, and Ilczuk (1971) reported that other higher ploidy strains of *A. oryzae* and *A. sojae* were superior amylase and protease producers. Pathak and Elander (1971) reported that diploid strains of *P. chrysogenum* were heterotic with respect to *β*-galactosidase, glucose oxidase, and alkaline protease.

## I. Blocked Mutants and Product Cosynthesis

Cosynthesis depends upon the use of two blocked mutants which, when propagated together, are able to synthesize the end product. Intermediates which accumulate in the blocked mutants diffuse out of the cell and are used by mutants blocked earlier in the pathway. The phenomenon has been reported for *S. aureofaciens* (Delic *et al*., 1969) and *S. rimosus* (Alikhanian *et al*., 1961), which synthesize tetracyclines. Kahler and Noack (1974) have recently reported cosynthesis of turimycin in strains of *S. hygroscopicus*. Delic *et al*. (1969) demonstrated cosynthesis between *S. rimosus* and *S. aureofaciens* and arranged the mutants into complementation classes. McCormick (1967) and Mitscher (1968) utilized cosynthesis in the elucidation of the biosynthetic pathway of tetracyclines.

Cosynthesis has the potential of detecting new antibiotics, as demonstrated in strains producing tetracyclines. Growing a mutant of *S. aureofaciens* blocked in the synthesis of demethyltetracycline together with a mutant of *S. aureofaciens* blocked in the synthesis of chlortetracycline resulted in the synthesis of an antibiotic not produced by *S. aureofaciens:* tetracycline (Mitscher, 1968). Čižmek (1976) cultivated two blocked mutants of *S. psammoticus* together and detected a compound active against a tetracycline-resistant strain of *Bacillus subtilis*. The uncharacterized compound was reported to be an unknown product of tetracycline biosynthesis. Cosynthesis between colonies of two mutants separated by a distance of 5 mm on agar eliminates genetic exchange as a mechanism for the phenomenon. A unique technique of sandwiching filter membranes between two layers of agar, each inoculated with one of two blocked mutants, indicates the diffusability of the compound.

## J. Localized Mutagenesis and Comutation

Localized mutagenesis in small, selected regions of the chromosome offers a promising new approach for the industrial microbial geneticist. Mutation programs can be directed for maximizing mutations in any marked area on the chromosome, especially areas known to affect the formation of end products. *N*-Methyl-*N'*-nitro-*N*-nitrosoguanidine (NG) induces multiple mutations on the bacterial chromosome which tend to occur in clusters near the replication points. Localized mutagenesis, achieved by treating synchronized bacterial cultures with NG, was accomplished in *E. coli* (Hohfeld and Vielmutter, 1973) and extended to the genus *Streptomyces* by Randazzo *et al.* (1973), Godfrey (1974), and Matselyukh and Mukvic (1973).

The comutation procedure, discovered by Guerola *et al.* (1971) in *E. coli,* depends upon NG-induced mutations occurring in clusters regardless of culture synchronization. A specific mutation is selected, such as reversion of an auxotroph, and the revertants are scored for a second mutation. A high frequency of nonselected mutants is usually found. The nonselected mutations are referred to as *comutations* and regularly occur in genes closely linked to the selected mutation.

The comutation technique has been effectively applied to the isolation of temperature-sensitive mutants by Oeschger and Berlyn (1974). Randazzo *et al.* (1973) used comutation to select a large number of nutritional mutants within the histidine operon of *Streptomyces coelicolor*. In a study of four mutations in *S. coelicolor* (*cys* A1, *met* A2, *his* A1, and *arg* A1), Randazzo *et al.* (1976) found all comutations were due to mutations in genes linked to the revertant locus; therefore a comutation region exists and mutation outside the comutation region is restricted. Mutations induced by NG are localized in a short segment without effecting the residual genome. Isolation of comutants in unknown loci linked to the revertant site can be done by a heterokaryon method or by the use of temperature-sensitive mutants. Comutation also appears promising as a relatively simple tool for gene linkage and mapping studies.

## K. Mutational Biosynthesis and New Metabolite Formation

Mutation of microorganisms producing secondary metabolites has resulted in the selection of strains capable of producing new modified metabolites either directly or in response to some added precursor analog. The modified metabolites usually possess structural features of the parent compound but often lack certain functional groups. Other compounds contain structurally modified functional groups which convey differing biological activities. This biogenetic approach to generate new secondary metabolites has been especially useful in the generation of new aminoglycoside antibiotics.

The phenomenon was first reported by Shier *et al.* (1969) utilizing a mutant strain of *Streptomyces fradiae*, the organism that synthesizes neomycin. Wildtype strains of *S. fradiae* normally synthesize neomycins A, B, and C, incorporating the biosynthesized diaminocyclitol subunit deoxystreptamine into all the neomycin components. They obtained a mutant strain derived from an NG-treated population which was unable to synthesize deoxystreptamine and, therefore, was dependent on an outside source of diaminocyclitols for antibiotic synthesis. When the mutant was cultured with added streptamine instead of deoxystreptamine, two new antibiotic substances were produced which were called hybrimycin $A_1$ and $A_2$. Epistreptamine, the C-2 epimer of streptamine, gave two more antibiotics, hybrimycin $B_1$ and $B_2$ (Fig. 5).

New biosynthetic analogs of butirosin, another aminoglycoside antibiotic, have been reported by Claridge et al. (1974), using mutant strains of Bacillus circulans. Other examples of new biosynthetic products produced by mutant strains are listed in Table II.

## IV. GENETIC SYSTEMS IN ECONOMICALLY IMPORTANT MICROORGANISMS

### A. Bacteria and Actinomycetes

Both eubacteria and actinomycetes are prokaryotic organisms; i.e., the nucleus is a single, naked chromosome devoid of a nuclear membrane. Bacteria have been important tools for research in molecular biology which has provided significant fundamental information for the

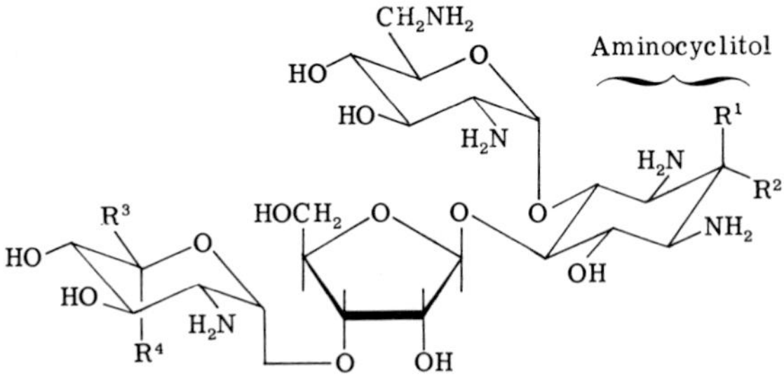

**FIGURE 5.** Structures of neomycin and new biosynthetic "hybrimycin" antibiotics. (From Shier et al., 1969.)

| Antibiotic | Aminocyclitol | $R^1$ | $R^2$ | $R^3$ | $R4^4$ |
|---|---|---|---|---|---|
| Neomycin B | Deoxystreptamine | H | H | H | $CH_2NH_2$ |
| Neomycin C | Deoxystreptamine | H | H | $CH_2NH_2$ | H |
| Hybrimycin A1 | Streptamine | H | OH | H | $CH_2NH_2$ |
| Hybrimycin A2 | Streptamine | H | OH | $CH_2NH_2$ | H |
| Hybrimycin B1 | Epistreptamine | OH | H | H | $CH_2NH_2$ |
| Hybrimycin B2 | Epistreptamine | OH | H | $CH_2NH_2$ | H |

**TABLE II.** Mutational Biosynthesis and New Product (Idiolite) Formation

| Organism | Original product | New product | Reference |
|---|---|---|---|
| *Pseudomonas aureofaciens* | Pyrrolnitrin | 4'-Fluoropyrrolnitrin | Gorman *et al.* (1968) |
| *Streptomyces peucetius* | Daunomycin | Adriamycin | Arcamone *et al.* (1969) |
| *Nocardia mediterranei* | Rifamycin B | Rifamycin W | White *et al.* (1974) |
| *Micromonospora inyoensis* | Sisomicin | Mutamicins | Testa *et al.* (1974) |
| *Streptomyces griseus* | Streptomycin | Streptomutin | Nagaoka and Demain (1975) |
| *Cephalosporium acremonium* (ATTC 20389) | Penicillin N | 6-(D)-[(2-amino-2-carboxy)-ethylthio]-acetamido-penicillanic acid | Troonen *et al.* (1976) |
| *Micromonospora purpurea* | Gentamicin | Hydroxy- and deoxy gentamicins | Rosi *et al.* (1977) |

fermentation industry during the past two decades. Among bacteria of economic importance are strains of the genus *Bacillus* (producers of antibiotics, insecticides, and enzymes), *Corynebacterium* (amino acid production), and *Pseudomonas* (hydrocarbon degradation and antibiotic production). Bacteria generally do not exhibit an alternation of haploid and diploid phases in their life cycles and were considered to be asexual organisms until Lederberg and Tatum (1946) demonstrated the existence of genetic recombination. Other mechanisms for gene transfer have since been discovered and are discussed briefly in the following sections. For a more detailed account of bacterial genetic systems, the reader is referred to the excellent text by Hayes (1968).

### 1. Conjugation

Lederberg and Tatum (1946) showed that when two strains of *E. coli* K-12 carrying different genetic markers were mixed together, some recombinants were formed. It was later found that not all *E. coli* strains could mate to give recombinants. On the basis of ability to transfer genetic material, *E. coli* K-12 can be classified into two sexes, male ($\delta$ or F⁺) and ($\mathcal{Q}$ or F⁻) (see Fig. 6). The character of "maleness" is conferred by a sex factor called F (for fertility) which is composed of DNA and determines the capability to form a conjugal union with and to transfer genetic material to $\mathcal{Q}$ bacteria.

The F factor exists in $\delta$ bacteria in one of two mutually exclusive states. In one state (F⁺), the sex factor is unassociated with the bacterial chromosome and replicates independently of it. In the other state (Hfr), the sex factor is inserted into the chromosome to form a single unit of

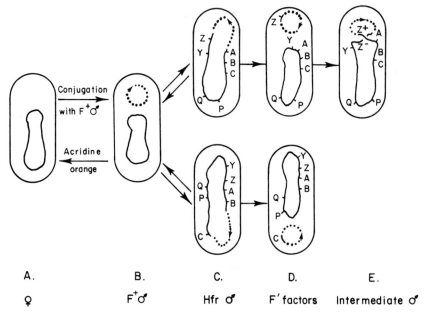

A.                  B.              C.              D.              E.

♀                   F⁺♂            Hfr ♂         F′ factors      Intermediate ♂

**FIGURE 6.** Properties and interrelationships of F⁺, *Hfr* and F′ donors. The interrupted lines represent the bacterial chromosome, and the dotted lines the sex factor; the arrow head indicates the polarity of transfer. The letters A–Z show the location of various bacterial genes. (From Clowes and Hayes, 1968.)

replication. When F⁺ bacteria come in contact with F⁻ cells, the sex factor is transferred with high efficiency to females, converting them to the F⁺ state. However, chromosome transfer is insignificant. Thus, F factor is clearly an extrachromosomal element (plasmid). On the other hand, the Hfr strain can transfer chromosomal genes at high frequency, depending on their location on the chromosome. During conjugation, the circular bacterial chromosome opens up at a specific point (origin) and enters the F⁻ cells at a constant rate in the same sequence as the arrangement on the chromosome. In this fashion, a gene can be mapped according to the time required to transfer into F⁻ bacteria during conjugation.

Some Hfr cells tend to revert to the F⁺ state under normal growth conditions, presumably due to reversal of the act of recombination which led to insertion of F factor into the chromosome. When this occurs a piece of the bacterial chromosome is often incorporated into the sex factor. Such modified sex factors are termed F-prime (F′) factors. A widely used F′ factor is one which carried a dominant allele of a gene, such as *lac*⁺ (F′ *lac*⁺). When the F′ factor is introduced into a bacterial host which carried the recessive allele of the gene, a complementing heterogenote (F' lac+/lac−) is formed. The F' factor is very useful for monitoring the presence of F factor in bacterial cells and for studying the dominant and recessive relationships of genes. Figure 6 illustrates the various types of sex in bacteria and their origin. As shown, the F factor, like many other plasmids, can be cured by acridine orange, ethidium bromide, etc.

The F factor from *E. coli* K-12 can also be transmitted to other bacteria including *Shigella, Salmonella,* and *Serratia. Salmonella* strains which acquire F factor from *E. coli* can now transfer chromosomal markers to both *Salmonella* and *E. coli* recipients. Genetic recombination via conjugation has also been reported in *Pseudomonas aeruginosa, Serratia marcescens,* and *Vibrio cholerae.*

## 2. Transduction: Phage-Mediated Gene Transfer

The phages which are capable of transduction are temperate, meaning they can establish themselves in a host bacterium without killing the cell. A bacterium harboring a temperate phage is called lysogenic. The phage in a lysogenic strain can be induced to multiply and lyse (kill) cells by a variety of treatments of which the most convenient is uv irradiation. Bacteria infected with a temperate phage may either lyse at once or become lysogenic depending on conditions. Figure 7 illustrates lysis and lysogenization in enteric bacteria.

In generalized (or unrestricted) transduction, a phage-sensitive bacterial strain is infected with transducing phage under conditions favoring lysis. The phage lysate is then freed from the bacterial cells and used to infect a recipient strain under conditions favoring lysogenization. The transductants are detected by plating on a selective medium.

In specialized transduction, a temperate phage called lambda (λ) has the special property of transducing the *gal* gene exclusively. The reason λ has this special transducing ability is because it actually integrates with the bacterial chromosome at a site close to the *gal* locus. Upon lysogenic induction, the λ is excised from the bacterial chromosome along with adjacent bacterial genes.

## 3. Transformation: DNA-Mediated Gene Transfer

Transformation is the process by which the genetic traits of bacterial strain A (donor) are transferred to strain B (recipient) by incubating the recipient cells with DNA from a donor strain (Fig. 8). The physiological characteristic of cells which can take up DNA and be transformed is called competence. Although transformation is not a common phenomenon, the demonstration of transformation in most bacterial species is probably a matter of discovering the proper conditions for competency. Genera in which species have been reported to undergo transformation

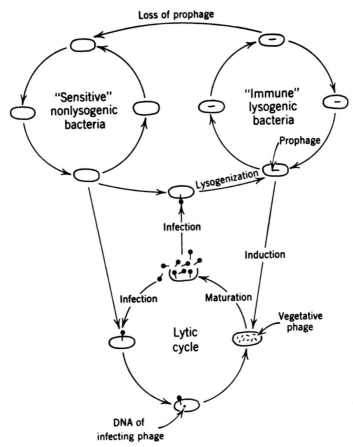

**FIGURE 7.** Lysis and lysogenization in enteric bacteria. (From Wagner and Mitchell, 1964.)

include *Bacillus, Hemophilus, Pneumococcus, Neisseria,* and *Rhizobium.* In *E. coli* the failure to demonstrate genetic transformation until recently appears to be due to the presence of hydrolytic nucleases which rapidly destroy incoming DNA. It has now been shown that when suitable nuclease-deficient mutant strains of *E. coli* are used as recipients, transformation can be readily demonstrated. Also, recently it has been possible to transform antibiotic-sensitive *E. coli* strains into antibiotic-resistant clones by incubating the recipient cells with antibiotic-resistant plasmid DNA under special conditions. (See section VIII on molecular cloning.)

Streptomycetes are distinguished from eubacteria by the formation of filamentous cells and aerial mycelia. Genetically, this implies that a longer time is required in streptomycetes to establish genetic homogeneity following gene transfer. As in *E. coli, Streptomyces*

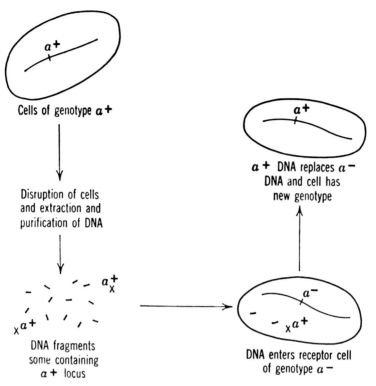

**FIGURE 8.** Generalized process of transformation in bacteria. (From Wagner and Mitchell, 1964.)

achieves its gene transfer mainly via conjugation. Transformation has been claimed in some streptomycetes but it has not been confirmed. It does occur, however, in a related species, *Thermoactinomycetes vulgaris* (Hopwood and Wright, 1972).

Hopwood (1973) suggested that the nontransformability of *Streptomyces* may be due to the excessive quantity of extracellular DNAase which readily degrades the donor DNA. Phages infecting *Streptomyces* (called actinophages) and transduction in *Streptomyces* have been reported. However, a useful transduction system in *Streptomyces* has not been demonstrated unequivocably.

Table III shows a list of actinomycetes (including *Streptomyces*) in which gene exchange via conjugation has been demonstrated. A detailed review on the genetics of *Streptomyces* has been published by Hopwood *et al.* (1973). In *S. coelicolor,* which is the most thoroughly studied *Streptomyces* species, fertility is mediated by a plasmid *scp*-1 [equivalent to F factor in *E. coli* K-12 (Hopwood *et al.*, 1973)]. As shown in Fig. 9, *scp-1*⁺ strains (a) bearing the autonomous plasmid scp 1 can

**TABLE III.**  Reported Genetic Exchange in Actinomycetes

| Organism (antibiotics produced) |
| --- |

*Streptomyces*
  *S. coelicolor* (actinorhodin and methylenomycin)
  *S. achromogenes* var. *rubradiris*[a] (rubradirin)
  *S. acrimycini*[a] (unidentified)
  *S. aureofaciens* (chlortetracycline)
  *S. bikiniesis*[a] (zorbamycin and zorbonomycin)
  *S. erythreus* (erythromycin)
  *S. fradiae* (neomycin)
  *S. glaucescens*[a]
  *S. grisoflavus*
  *S. griseus* (streptomycin)
  *S. olivaceus*[a]
  *S. rimosus*[a] (oxytetracycline)
  *S. scabies*
  *S. venezuelae*[a] (chloramphenicol)
*Nocardia*
  *N. erythropolis*[a]
  *N. mediterranei*[a] (rifamycins)
*Micromonospora*
  *M. chalcea*
  *M. echinospora (gentamicins)*
  *M. purpurea* (gentamicins)
*Mycobacterium smegmatis*
*Thermoactinomyces vulgaris*[b]

[a] A circular map with close resemblance to *S. coelicolor*.
[b] *Thermoactinomyces vulgaris* has a transformation system; in other cases a "conjugation" mechanism of recombination is implicated. (After Hopwood, 1976.)

transfer random chromosomal markers to scp 1⁻ (b) at low frequency. Strains in which scp 1 is integrated into the chromosome (c) can transfer the markers adjacent to the site of plasmid integration at very high frequency when mated with scp 1⁻ (equivalent to Hfr in *E. coli*). Scp 1′ strains are those in which a piece of chromosome has been inserted into the autonomously replicating plasmid (d) (equivalent of F′ in *E. coli*). Such strains can transfer the chromosomal markers as well as scp 1 plasmid to scp 1⁻ strains at high frequency.

Linkage maps for several species of *Streptomyces* have been shown to be circular (Hopwood *et al.*, 1973). All the maps of *Streptomyces* show an approximate equivalent location for most loci, as exemplified by the genetic map of *S. coelicolor* and *S. rimosus* (Alacevic, 1976; see Fig. 10).

An important aspect of recombination in *Streptomyces* is interspecific or interstrain recombination, particularly between two strains producing

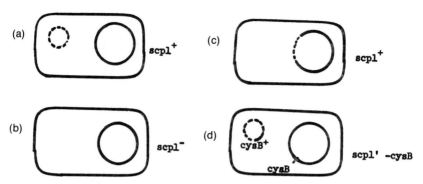

**FIGURE 9.** Various fertility types in *Streptomyces coelicolor.*

different antibiotics. Not only can interspecific crossing be used as a test for taxonomic relatedness between strains but it also could have great potential for breeding new strains for the synthesis of structurally modified antibiotics (Alacevic, 1963). Hopwood *et al.* (1977) have recently reported on genetic recombination involving protoplast fusion in streptomycetes. Their technique involves the use of polyethylene glycol (PEG)-induced protoplast fusion and regeneration with recombination frequencies so high that selectable markers were no longer necessary, thereby greatly facilitating the routine use of recombination in strain improvement programs.

## B. Filamentous Fungi and Yeasts

In contrast to eubacteria and actinomycetes, the yeasts and molds possess true nuclei and, in some species, regular sexual cycles comparable to those existing in higher plants and animals. The major difference between the life cycles of higher organisms and those of fungi is the prolonged and generally predominant haploid phase in fungi. *Neurospora crassa* has been the fungal organism which has been thoroughly exploited by geneticists. The life cycle of *N. crassa* is shown in Fig. 11. The vegetative mycelium of *N. crassa* is haploid and can be propagated almost indefinitely by serial transfer of hyphal fragments or asexual spores (conidia). Two strains of opposite mating types (*A* or *a*) are required to initiate the sexual cycle. Fertilization is effected by mixing the conidia of *a* mating type with mycelia of *A* mating type growing on appropriate media. After a period of nuclear division and nuclear migration, fusion between *A* and *a* nuclei takes place. Each fused nucleus (diploid) immediately undergoes meiosis to form four haploid nuclei which, in turn, undergo two successive mitotic divisions to form eight ascospores. The eight ascospores are contained linearly in

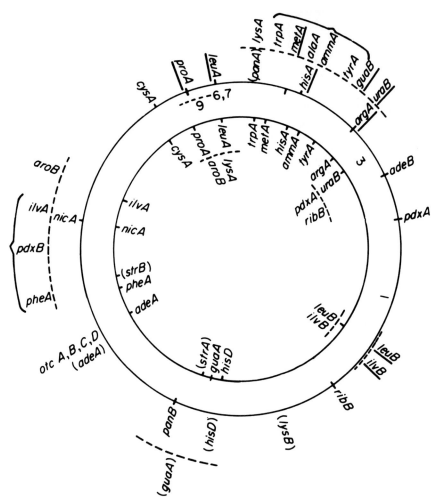

**FIGURE 10.** The map of *Streptomyces rimosus* (outer circle) compared to that of *S. coelicolor* with empty regions omitted (inner circle). (After Alacevic, 1976.)

a sac (ascus). Analysis of the linear segregation of each genetic marker as well as the recombination between markers in each ascus provide a unique way of mapping genes in *N. crassa*.

Few of the industrially important fungi, however, are endowed with such conventional sexual cycles as found in *N. crassa*. Fortunately, the filamentous fungi often form heterokaryons in which rare diploid nuclei result from the fusion of two haploid nuclei. This process is called parasexuality and is illustrated in Fig. 12. Although recombination (mitotic recombination) is much less frequent in the parasexual cycle compared

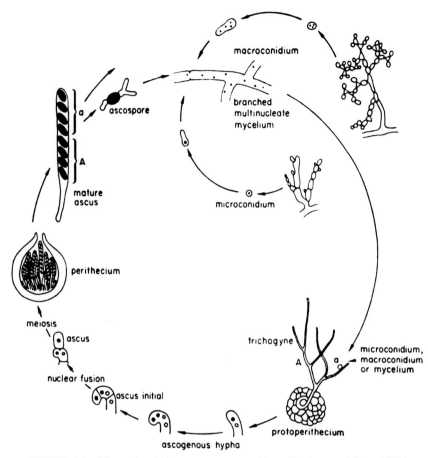

**FIGURE 11.** Life cycle of *Neurospora crassa*. (From Fincham and Day, 1973.)

to the meiotic process, it can occur by mitotic crossing over or by other mechanisms. Mitotic recombination is of great practical importance, since it makes possible genetic analysis and controlled breeding in organisms with no sexual cycle. The parasexual cycle has been demonstrated in many industrial fungi including *P. chrysogenum* (Pontecorvo and Sermonti, 1953), *C. acremonium* (Nüesch *et al.*, 1973), and *Aspergillus oryzae* (Ikeda *et al.*, 1957). A successful attempt to apply parasexual genetics to strain improvement in *P. chrysogenum* has been reported by Elander (1967). In this study, a homozygous diploid representing genome duplication of the haploid parent was an efficient producer of penicillin V.

Yeasts possess a similar life cycle to that of *N. crassa* except that they are morphologically more simple. *Saccharomyces cerevisiae* is the most

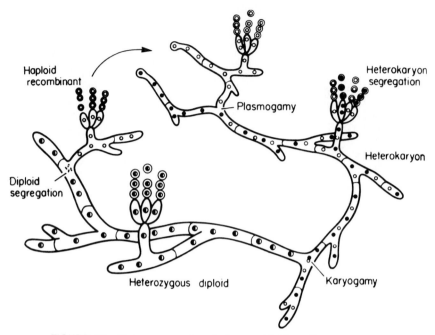

**FIGURE 12.**  The parasexual cycle in a mold. (From Sermonti, 1969.)

important economic species of yeast and is used in brewing, in baking, and for the manufacture of single-cell protein. The haploid phase of S. cerevisiae consists of free cells with single nuclei which propagate by budding (see Fig. 13). The diploid phase is initiated by fusion of free haploid cells of opposite mating type (a or α), followed by meiosis leading to the formation of four haploid cells (ascospores). Unlike the situation in Neurospora and other fungi, diploid nuclei in yeast can divide by mitosis and a diploid culture can multiply indefinitely by budding.

## C. Higher Fungi

The cultivated mushroom, Agaricus bisporus, is a member of the group of higher fungi called basidiomycetes. Many basidiomycetes have two distinct phases in their life cycle, a haploid homokaryotic phase and a dikaryotic phase in which each cell contains two nuclei of different mating types. During the dikaryotic phase, special cells called basidia are formed on which the products of meiosis (basidiospores) are born externally. The number of basidiospores on each basidium can be either two or four in A. bisporus. The life cycle of a representative

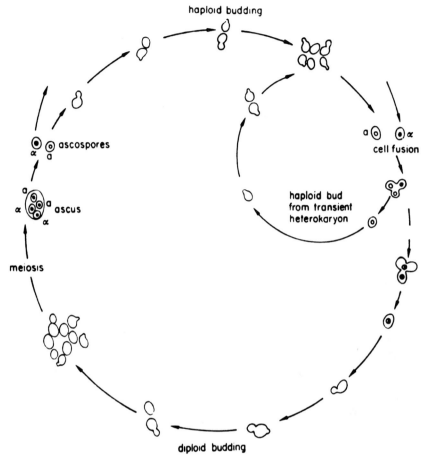

**FIGURE 13.** Life cycle of *Saccharomyces cerevisiae*. (From Fincham and Day, 1973.)

basidiomycete, *Coprinus lagopus,* is shown in Fig. 14. Unlike *N. crassa,* in which the mating type is determined by a single locus (*A/a*), the mating type of most basidiomycetes is determined by two genetic loci ($A_1/B_2$ and $B_1/B_2$). This means that there are four possible mating types determined by the two loci—$A_1B_1$, $A_1B_2$, $A_2B_1$, and $A_2B_2$. For two strains to be genetically compatible (to form dikaryons and subsequent fruiting), each of the two nuclei should possess different alleles on each locus, i.e., $A_1B_1$ versus $A_2B_2$, $A_1B_2$ versus $A_2B_1$.

In mushroom cultivation, the most serious problem has been the occurrence of spontaneous mutations leading to strain variability and instability. Thus, heterokaryotic cells resulting from spontaneous mutations appear to be a common phenomenon. Accordingly spawn producers have to exercise constant selection to improve the quality of mushroom spawn by testing its ability to produce mushrooms with the appropriate attributes. Recent work by Raper *et al.* (1972) showed, contrary to common belief, that *A. bisporus* has only two mating types ($A_1$ and $A_2$) and that it occasionally incorporates a pair of compatible nuclei ($A_1$ + $A_2$) into a single basidiospore to produce a self-fertile monosporic culture. As our understanding of the life cycle and mating system increases, the prospect for combining desirable characteristics through interstrain breeding should become a reality.

## V. GENETICS OF DIFFERENTIATION IN RELATION TO SECONDARY METABOLITE FORMATION

With the exception of biomass production, most industrial fermentations deal with secondary metabolites. Although citric acid and various amino acids are primary metabolites, their production involves certain characteristics in common with secondary metabolites; i.e., overproduction is not essential for growth or survival of the producer organisms. In a typical secondary metabolite fermentation, there is an initial phase of cell growth with little product formation (trophophase), followed by a period, often after the cessation of growth, in which product is generated (idiophase). Inevitably, many morphological changes take place during the transition from trophophase to idiophase. This is especially true for mycelial organisms such as fungi and streptomycetes.

In the cephalosporin C fermentation, submerged cultures of *C. acremonium* exist as four morphological forms: conidia, hyphae, germlings, and arthrospores. Many high-yielding mutants of *C. acremonium* were found to have an increased capacity to form arthrospores (Nash and Huber, 1971). Supplementation of methionine or its analog, norleucine, stimulated cephalosporin C production as well as arthrospore formation (Drew *et al.*, 1976).

Many high-producing strains of *P. chrysogenum* and *C. acremonium* exhibit decreased vegetative vigor and/or reduced sporulation on plate culture. Figure 15 illustrates the progressive reduction in colony diameter and sporulation ability in the improved strains of *P. chrysogenum* within the Wisconsin family of strains (Sermonti, 1969). However, it is possible to derive a well-sporulating isolate from a poorly sporulating parent without sacrificing the productivity of the parent. For example, strain Wis 51-20 was a good producer of penicillin but showed poor growth and sporulating ability. After repeated mutations and reisolation, strains with superior sporulation capacity and high productivity were derived from strain Wis 51-20 (Backus and Stauffer, 1955). Similar re-

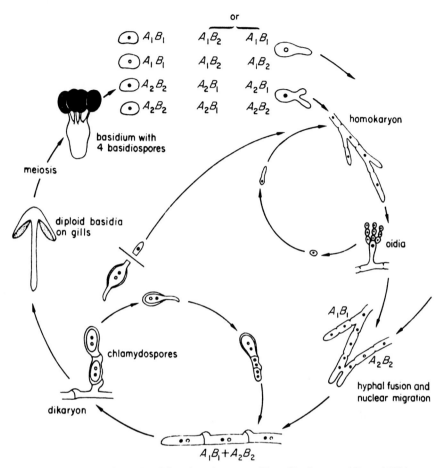

**FIGURE 14.** Life cycle of *Coprinus lagopus*. (From Fincham and Day, 1973.)

sults were obtained with a rifamycin-producing asporogenous mutant of *Streptomyces mediterranei* (Esposito *et al.*, 1972).

Although most morphological mutants are inferior producers of secondary metabolites, frequently the most outstanding higher producing mutants differ in morphology from their less productive parent. As seen from a penicillin strain improvement program (Backus and Stauffer, 1955), the types of morphological mutants which exhibited high productivity were those which gave restricted growth (colonial mutants) on plates (Fig. 10). In *Neurospora,* the biochemical lesion(s) for colonial mutations has been traced to altered enzymes in the pentose phosphate pathway, such as glucose-6-phosphate dehydrogenase, 6-phosphogluconate dehydrogenase, and phosphoglucomutase (Brody

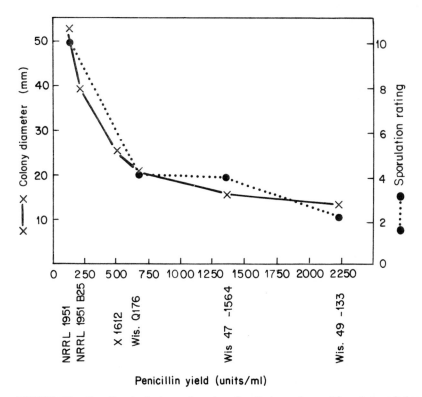

**FIGURE 15.** Growth rate (colony diameter after 7 days of growth) and sporulating ability (arbitrary scale) of colonies of different strains of *Penicillium chrysogenum* (Wisconsin family) plotted against the productivity of each strain. (The code of the strains corresponding to the various points is indicated below.) (From Sermonti, 1969.)

and Tatum, 1966). Since these enzymes regulate the levels of various sugar phosphates and coenzyme $NADPH_2$, it is not surprising that the production of antibiotics, many of which require sugar phosphates and $NADPH_2$ for their biosynthesis, are affected by colonial mutations (Demain, 1972).

In spore-forming bacteria such as bacilli, the production of enzymes (proteases and amylases) and antibiotics (bacitracin, gramicidin, polymyxin, circulin) are often associated with onset of sporulation (Schaeffer, 1969). In *Penicillium griseofulvum,* the conditions determining maximum sporulation are quite similar to those promoting griseofulvin production (Morton, 1961). Although mutations resulting in simultaneous loss in sporulation and ability to synthesize enzymes or antibiotics (pleiotropic mutations) are fairly common, there appears to be no causal relationship between these two processes. In bacilli, many asporogenous mutants ($sp^-$) still produced copious amounts of amylase, protease, or antibiotics and, conversely, many mutants selected as non-producers of amylase, protease, or antibiotics (bacitracin or polymyxin) sporulated normally (Schaeffer, 1969).

The formation of aerial hyphae and antibiotics, and other characteristics, in many strains of *Streptomyces,* appear to be regulated by plasmid genes (Okanishi *et al.,* 1970; Redshaw *et al.,* 1976; Pogell *et al.,* 1976; Wright and Hopwood, 1976). Such strains can be cured of ability to form aerial mycelium as well as their ability to produce antibiotics by treatment with heat, acridine orange, or ethidium bromide. A chromosomal mutation called *bld* which eliminated aerial mycelium also had the pleiotropic effect of eliminating methylenomycin A production, which is plasmid determined (Pogell *et al.,* 1976).

In submerged fungal fermentations, mold mycelia can grow in either filamentous or pellet forms depending on physiological conditions, such as medium, aeration, and inoculum size (Whitaker and Long, 1973). The filamentous form is more difficult to aerate efficiently because of its higher viscosity. Thus, it is not surprising that pellet forms with diameters of 0.1–0.2 mm are more desirable for citric acid production than filamentous forms in *A. niger.* In the case of the penicillin fermentation, there appears to be no consensus as to which of the two forms is more desirable. In an experiment employing two morphologically distinct strains of *P. chrysogenum* fermented with excess oxygenation under identical conditions, the pellet variant produced twice the penicillin titer of the filamentous strain (Carilli *et al.,* 1961).

## VI. THE ROLE OF PLASMIDS IN MORPHOLOGICAL DIFFERENTIATION AND SECONDARY METABOLITE FORMATION

Plasmids are extrachromosomal genes which replicate and segregate independently of chromosomal (nuclear) genes. Besides the well-known plasmids which confer drug resistance (R factors), fertility (F and *scp*-1 factors), and toxin production (colicin $E_1$, $E_2$) in eubacteria, there are many other characteristics in prokaryotes and eukaryotes which are determined by plasmids. In the eubacteria, covalently closed circular (ccc) DNA can be isolated from strains suspected of plasmid gene involvement. Detailed accounts of bacterial plasmids, which are beyond the scope of this report, can be found in recent reviews on the subject (Helinski, 1973; Falkow, 1975). In *Streptomyces,* plasmid genes have been reported to be involved in the formation of aerial hyphae, fertility, enzymes, and antibiotics. Table IV lists various characteristics and

**TABLE IV.**   Characters and Products of *Streptomyces* Determined by Plasmid Genes

| Character or product | Species | Reference |
|---|---|---|
| Fertility | *S. coelicolor* | Hopwood *et al.* (1973) |
| Aerial hyphae | *S. alboniger* | Redshaw *et al.* (1976) |
| | *S. venezuelae* | Okanishi *et al.* (1970) |
| Resistance to | *S. rimosus* | Boronin and Sadovnikova (1976) |
| antibiotics | *S. coelicolor* | Kirby *et al.* (1975) |
| | *S. bikiniensis* | Shaw and Piwowarski (1977) |
| Melanin excretion | *S. glaucescens* | Baumann and Kocher (1976) |
| Tyrosinase (melanin | *S. scabies* | Gregory and Huang (1964a,b) |
| formation) | *S. venezuelae* | Okanishi *et al.* (1970 |
| | *S. kasugaensis* | Okanishi *et al.* (1970) |
| β-Lactamase | *Streptomyces* spp. | Oganwara and Nozaki (1977) |
| Kasugamycin | *S. kasugaensis* | Okanishi *et al.* (1970) |
| Aureothricin | *S. kasugaensis* | Okanishi *et al.* (1970) |
| Chloramphenicol | *S. venezuelae* | Akagawa *et al.* (1975) |
| Methylenomycin A | *S. coelicolor* | Kirby *et al.* (1975) |
| Turimycin | *S. hygroscopicus* | Kahler and Noack (1974) |
| Streptomycin | *S. bikiniensis* | Shaw and Piwowarski (1977) |

products in species of *Streptomyces* which have been reported to be determined by plasmid genes.

In *S. coelicolor,* plasmid scp 1 not only confers fertility to strains that harbor this plasmid but also determines the production of and resistance to methylenomycin A (Kirby *et al.*, 1975) (see Section IV for fertility determined by scp 1). Attempts to isolate scp 1 plasmid DNA have not been successful to date. However, a 10–20 × $10^6$ dalton ccc DNA has been isolated from both scp $1^+$ and scp $1^-$ strains (Schrempf *et al.*, 1975). This ccc DNA appears to correspond to a new sex factor, scp 2, in *S. coelicolor.*

Most of the evidence for plasmid involvement in genetic control has been based on curing experiments, i.e., demonstration of increased frequency of nonproducing strains (as compared to mutation frequency) after treatment with agents such as acridine orange, ethidium bromide, or high temperature (Okanishi *et al.*, 1970; Redshaw *et al.*, 1976). In the case of *S. scabies* (tyrosinase), *Streptomyces venezuelae* (chloramphenicol), and *S. coelicolor* (scp 1), crosses provided additional evidence that extrachromosomal genes were involved. In *S. glaucescens,* genetic analysis showed that the tyrosinase gene was located on the chromosome, whereas the excretion character of melanin was determined by an extrachromosomal gene (Baumann and Kocher, 1976).

Among other plasmids of potential economic importance are the plasmids involved in the biodegradation of organic compounds in

*Pseudomonas* (Chakrabarty, 1976). Many of these plasmids are transmissible and most are compatible with one another. Thus, it is possible to maintain several hydrocarbon-degrading plasmids in a single bacterial strain. Whether such a strain will have an enhanced capability to utilize crude oil remains to be seen. Also noteworthy is the recent discovery that the ability of *Agrobacterium tumefaciens* to induce crown gall tumors in plants can be attributed to the presence of a large plasmid (1.2 × 10⁸ daltons). The plasmid has also been detected in crown gall tumors. The tumor-inducing ability of the bacterium can be eliminated irreversibly by propagation at high temperature and can be transferred from a virulent to an avirulent strain by means of plasmid transfer (Watson *et al.*, 1975).

Plasmids or plasmidlike genes have also been reported in eukaryotic organisms. Some strains of *S. cerevisiae* contain copies of 2-$\mu$m ccc DNA (Livingston and Klein, 1977). A single-step spontaneous mutant of *S. cerevisiae,* resistant simultaneously to oligomycin, venturicidin, chloramphenicol, cycloheximide, and triethyltin, was also found to contain 2-$\mu$m DNA (Guérineau *et al.*, 1974). Oligomycin resistance appears to be controlled by a chromosomal gene, even though its expression was correlated with the presence of 2-$\mu$m DNA within the cells. The resistance to other drugs can be eliminated by ethidium bromide, suggesting their extrachromosomal nature (although unrelated to 2-$\mu$m DNA). A plasmidlike behavior has been recently demonstrated in transformed strains of *N. crassa* (Mishra, 1976). Two classes of inositol-independent (*inl*$^+$) transformants were observed; one class showed the Mendelian transmission of *inl*$^+$ character in meiotic crosses, while the other class showed non-Mendelian transmission. The latter class can be cured of the *inl*$^+$ character by treatment with ethidium bromide.

## VII. INDUSTRIAL STRAIN IMPROVEMENT PROGRAMS IN PENICILLIN- AND CEPHALOSPORIN-PRODUCING FUNGI

### A. The Penicillin Fermentation

Large-scale programs concerned with the induction, selection, and utilization of superior penicillin-producing variants of *P. chrysogenum* have been carried out in government, university, and industrial laboratories for over 30 years. From the cumulative screening of hundreds of thousands of strains, a series of very superior penicillin producers has been developed, certain of which are utilized today for the commercial manufacture of penicillin (Elander, 1967; Elander *et al.*, 1973). Over the years, "mutation breeding" has been undertaken with strains of *P. chrysogenum* more than with any other industrially important microorganism. The genealogy of the Wisconsin family of strains of *P. chrysogenum* and two important modern industrial lineages is presented in Fig. 16. Although the initial screening was based on natural (nonmutated) colony populations, subsequent natural populations failed to lead to major yield increases. For example, at Stanford University some 60,000 cultures were screened for improved productivity with no success (Sermonti, 1969). Key strains in the early ancestry of the Wisconsin series were the famous Q-176 culture, which had significantly improved antibiotic titers, and strain BL3-D10, which did not produce the characteristic and troublesome chrysogenin pigment (Backus and Stauffer, 1955). All further mutant selections were derived from the Q-176 culture over the next decade.

Three distinct selection lines were established from the pigmentless mutant. One line was based on the selection of natural variants, another based on uv-radiation survivors, and a third highly successful line based on treatment with a nitrogen mustard mutagen. Industrial lines were later derived from the 51-20 strain utilizing a variety of different mutagens, including NG, diepoxybutane, and nitrous acid (Elander *et al.*, 1973).

Throughout the Wisconsin, Lilly, and Wyeth programs, it was noted that the highest morphological mutation rate with uv radiation did not necessarily coincide with the highest rate of kill. The Wisconsin group reported that lower doses, i.e., treatments yielding 25–30% survivors, were often effective in inducing higher productivity mutants than a dose which would effect a maximum kill (Backus and Stauffer, 1955).

During the period of intensive selection work, attention was focused on searching for strain characteristics which appeared to correlate with high yield. In general, the Wisconsin group demonstrated a correlation between increased productivity and reduced sporulation and growth, characteristics which also appear to correlate with improved cephalosporin C variants of *C. acremonium* (Elander, 1975; Queener *et al.*, 1975). In the Wisconsin series, the greatest change was observed in the early ancestry. Between the NRRL-1951 and Q-176 strains, growth and sporulation were reduced by 60%, with a concomitant yield increase of nearly sixfold. Later, an additional threefold increase in antibiotic titer was associated with only an additional 10% reduction in mycelial vigor. These changes may represent a correlated response due to linkage of loci determining growth with those influencing penicillin titer. Selection can also be utilized to isolate strains with improved growth and sporulation characteristics, the latter being most important for long-term preservation and providing for vigorous vegetative seed development for large-tank fermentations (Elander *et al.*, 1973). The correlation between strain productivity, decreased sporulation, and weak vegetative de-

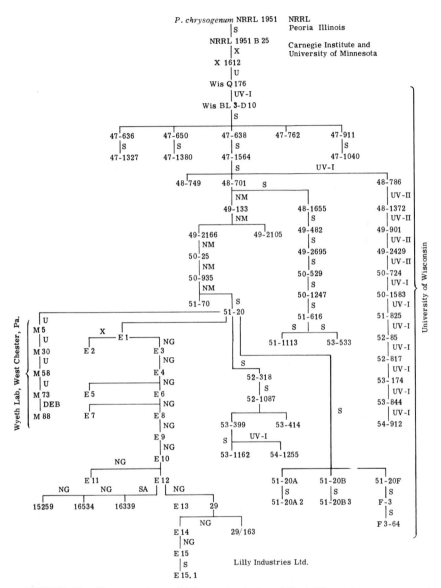

**FIGURE 16.** The genealogy of Wisconsin strains of *Penicillium chrysogenum* and subsequent development of two industrial lines. Key: Mutations brought about spontaneously, S; by X irradiation, x; by uv irradiation, wavelength unspecified, U; 275 nm, UV-I; 253 nm, UV-II; nitrogen mustard, NM; nitrosoguanidine, NG; diepoxybutane, DEB. (Based on Stauffer, 1961; and Elander, 1967.)

velopment may reflect conflicting physiological or metabolic balances of pleiotropic effects of high-yield determining genes. However, it is not possible at this time to assess the basis of genetic and environmental interactions as related to improved yield because of the lack of relevant information. There are other correlations which include strain tolerance to phenylacetic acid (Fuska and Welwardova, 1969), ability to accumulate intracellular sulfate (Segel and Johnson, 1961), ability to assimilate carbohydrate and side-chain precursor (Pan *et al.*, 1972), sensitivity to iron (Pan *et al.*, 1975), penicillin acylase activity (Erickson and Dean, 1966), levels and valine sensitivity of acetohydroxy acid synthetase (Goulden and Chattaway, 1969), and penicillin acyltransferase activity (Preuss and Johnson, 1967).

Relatively little is known about the genetic basis of penicillin production. Table V illustrates the improvement in penicillin productivity culminating in strains developed at Eli Lilly and Company. These data indicate that the yield increase at each step is relatively small so that it could often be accounted for by the accumulation of nuclear mutants having relatively small effects. However, when a comparison was made of the increased productivity of mutant E15-1 to the NRRL 1951 culture, a significant 55-fold improvement in productivity over the Fleming strain was noted!

Sermonti (1959) carried out heterokaryon experiments between a low-producing auxotrophic strain NRRL 1951 *pro* and a high-producing

**TABLE V.** Improvement in Penicillin Production in a Family of Mutant Strains of *Penicillium chrysogenum* Developed from Strain NRRL 1951[a]

|  | Improvement in production over: | |
| --- | --- | --- |
| Strain[b] | Previous strain | Fleming strain[c] |
| NRRL 1951 | 0.50 | 0.50 |
| NRRL 1951 B25 | 2.70 | 4.05 |
| X1612 | 0.06 | 4.90 |
| Q176 | 0.52 | 9.00 |
| 47-1564 | −0.04 | 8.55 |
| 48-701 | −0.14 | 8.15 |
| 49-133 | 0.08 | 12.70 |
| 51-20 | 0.73 | 22.60 |
| E15 | 1.05 | 50.64 |
| E15-1 | 0.09 | 55.00 |
| Cumulative total | 5.55 | 176.09 |

[a] From Elander (1967).

[b] The strains concerned can be identified in the family lineage in Fig. 16.

[c] The penicillin titers are from shake-flask fermentations using modern fermentation conditions.

strain Wis 49-133 *nic*. Homokaryotic segregants from the heterokaryon showed a clear association between the nuclear marker and yield; i.e., *pro* segregants were low-producers and *nic* segregants were high-yielding. Thus, penicillin yield is determined by nuclear genes and there is evidence from both heterokaryons and parasexual recombinants that several different genes are involved. Both the *y met* and *w ade* hetero-karyons of Elander (1967) (also see Fig. 17) and the "New Hybrid" strain of Alikhanian and Borisova (1956) show significantly higher yields than their parent homokaryons, suggesting complementation between nonal-lelic genes.

Caglioti and Sermonti (1956) attempted to map *pen*-1, a determinant

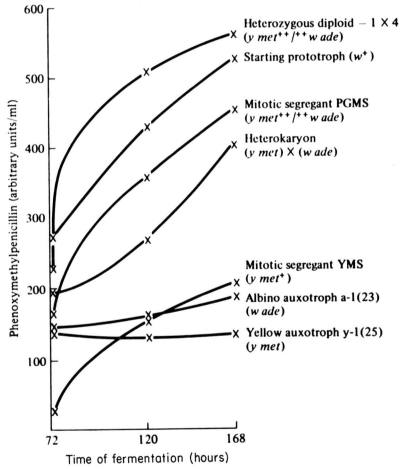

**FIGURE 17.**   Penicillin production by recombinant strains of *Penicillium chrysogenum* E15. (From Elander, 1967.)

for increased yield by mitotic recombination. They showed its locus was distal to *pro*-1 and *met*-1, two auxotrophic markers in one of three linkage groups they tentatively identified. More recently Ball (1971, 1973a,b) confirmed by haploidization analysis the existence of three linkage groups in a Glaxo strain. Ball induced increased yield in five separate strains, each of which carried conidial color or auxotrophic markers (Fig. 18).

Two phenomena have been described which suggest that genes determining increased yield of penicillin are recessive and that independently induced mutations could be allelic. In most cases when a diploid strain is derived via the parasexual cycle between a low-yielding and a high-yielding strain, its yield is comparable to that of the former. Diploids obtained between strains from a common ancestor, but carrying inde-

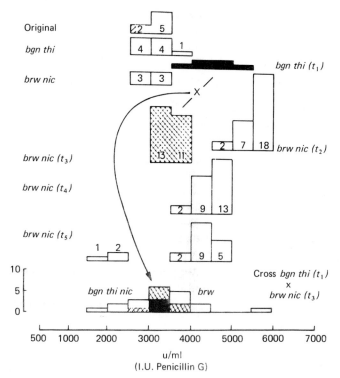

**FIGURE 18.** Penicillin production of some genetically marked strains of *Penicillium chrysogenum* and the yield of a cross between two of them (*bgn thi* $t_1$ × *brw nic* $t_3$). Key: Number in histograms indicate number of strains. Symbols: *bgn,* bright green and *brw,* brown colonies; *nic thi,* auxotrophic for nicotinic acid and thiamin, respectively; $t_1$–$t_5$, determinants of penicillin yield. Solid histogram, *bgn thi* $t_1$ phenotype; crosshatched histogram, *brw nic* $t_3$ phenotype, others indicated. (Based on data of Ball, 1973a, as outlined by Burnett, 1975.)

pendently induced mutations for higher yield, gave yields equivalent to or not much lower than their parental strain, but notably higher than the yield of the common ancestor (Macdonald *et al.*, 1964; Elander, 1967; Elander *et al.*, 1973). An explanation for these observations is that the strains carried a number of allelic mutant sites in common, even though the sites had been independently mutated. It appears there are a number of genetic determinants, recessive in their expression and located on more than one chromosome, which determine increased penicillin yield in *P. chrysogenum*. Whether these determinants are oligogenes, polygenes, or a combination of the two has not been ascertained. Their expression can be modified by genetic background, for most conidial color and auxotrophin markers reduced yield drastically.

To date, the technique of mutation and selection is the most reliable procedure for improving penicillin titers. Breeding involves the parasexual cycle and, therefore, the peculiarities of the cycle in *P. chrysogenum* present difficulties in achieving recombination and subsequent unimpaired segregation. Strain variability and instability cause frequent problems. It has proved difficult to initiate the parasexual cycle because heterokaryons are not produced easily with the Wisconsin strains of *P. chrysogenum*. Even when heterokaryons are produced, numerous difficulties have been encountered with the instability of diploids and their segregants. Elander (1967) described a highly stable diploid strain at Eli Lilly derived from the haploid production strain E-15 (Fig. 17). Spontaneous variation in the diploid and haploid was 9.2 and 31.9%, respectively, as assessed by the color types which segregated. The difference was even more striking after exposure to uv irradiation, where the diploid : haploid derived variants were 11.6 and 41.1%, respectively. In this case, the diploid strain was far more stable than its haploid progenitor.

In contrast, Ball (1971, 1973b) reported that diploids show greater instability than their haploid progenitors but within the same range. The diploid showed twice as many poor sporulating types as the haploid, while densely sporulating types were increased tenfold. Ball (1973a) and Roper (1973) have compared this phenomenon with that of "mitotic nonconformity" described by Nga and Roper (1969) in *A. nidulans*. Mitotic nonconformity stems from the existence of duplicate segments of small fractions of the genome which have probably originated through translocations. If the difficulties described in producing diploids can be overcome, they offer many attractions for the fungal geneticist. First, there is the possibility of a heterotic effect on yield (Sermonti, 1959; see Fig. 17). Even if this does not occur, the possibility exists of selecting productive segregants from diploid strains. If the latter approach is to be used, it is recommended to start initially with strains lacking chromosomal rearrangements or to induce variants in increased yield from an existing strain without inducing chromosomal aberrations. Since spontaneous dediploidization can always occur, systems have been proposed to reduce such effects. Macdonald (1964) has suggested the use of parental haploids which grow poorly in complex fermentation media so that the parental segregants arising during fermentation would be selected against. Azevedo and Roper (1967) suggested the induction of recessive lethals in the diploid so that haploid segregants would be eliminated as they arose. It has been suggested that the Lilly diploid (Fig. 17) may be a balanced lethal diploid in that it was extremely stable but, even after treatment with uv irradiation, the proportion of viable segregants it produced was hardly increased. Provided that diploid stability and unrestricted parasexual recombination can be achieved, it should be possible to make further selection progress. Elander (1967) described selection No. 2, a spontaneous segregant from the Lilly diploid, which produced nearly 25% more antibiotic than its parent, which itself yielded better than the production haploid E-15. Ball (1973a,b) has also described recombinant segregants with improved yield (see Fig. 18).

The first 30 years of mutation selection and scientific breeding have demonstrated the potentials of mutation breeding and the dangers present in utilizing mutagens which induce chromosomal aberrations (Ball, 1973b). The next 25 years should exhibit more extensive application of the parasexual cycle to breeding (Merrick, 1976; Hopwood, 1977).

## B. The Cephalosporin Fermentation

An intensive strain improvement program was initiated in England in 1957 to improve, through mutagenesis and fermentation development, the poor production of cephalosporin C in Brotzu's Sardinian strain of *C. acremonium* (Elander, 1975). Mutagenesis of the Brotzu isolate resulted in the eventual selection of the improved mutant, M-8650, which served as the progenitor strain for large-scale industrial programs (Fig. 19). The synthesis of cephalosporin C in laboratory fermentations using the early British fermentation conditions for a series of improved uv variants is shown in Fig. 20.

Strain CW-19 produced threefold more antibiotic than the Brotzu culture and, when fermented under more optimal fermentation conditions, the culture was capable of synthesizing nearly 15-fold more antibiotic than the progenitor strain. The CW-19 variant also had a significantly improved cephalosporin C/penicillin N ratio (Elander *et al.*, 1976).

Biometric considerations of the overall data concerning "normal" and "mutant" (morphological and biochemical) populations of uv-survivor strains of *C. acremonium* and their potential to synthesize cephalosporin

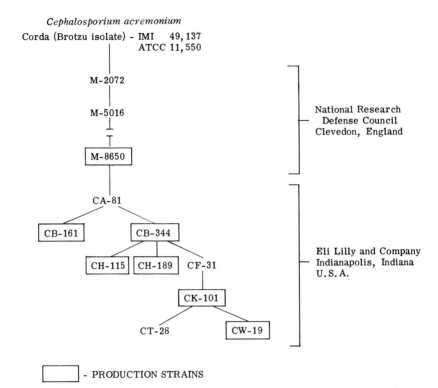

*Cephalosporium acremonium*
Corda (Brotzu isolate) - IMI   49,137
                         ATCC 11,550

M-2072

M-5016

M-8650

National Research
Defense Council
Clevedon, England

CA-81

CB-161     CB-344

CH-115  CH-189   CF-31

CK-101

CT-28            CW-19

Eli Lilly and Company
Indianapolis, Indiana
U.S.A.

☐ - PRODUCTION STRAINS

**FIGURE 19.** Lineage of early Lilly uv (275 nm) variants of *Cephalosporium acremonium.* (After Elander, 1975.)

C showed an approximate 11 to 1 advantage for "normal" clones versus "mutant" clones in searching for mutants possessing the capacity to produce 20% more antibiotic. The above probability statements were calculated for many survivor populations where suitably large numbers of "normal" and "mutant" populations were evaluated on a statistical basis (Brown and Elander, 1966).

The improved uv variants differed markedly from the progenitor strains in cultural characteristics and biochemical properties. Untreated colony populations of the improved variants showed a progressive reduction in colony diameter and decreased vegetative development. The improved mutants were also characterized by decreased sporulation vigor, features which were also characteristic of the improved penicillin variants (Elander *et al.*, 1976). Other mutagenic agents have also been reported to elicit improved antibiotic-producing variants.

Recent patent literature from Japan has reported that strains of *C. acremonium* or *C. polyaleurum* resistant to the polyene antibiotics nystatin, kabicidin, or trichomycin produced greater than 10 gm/liter of

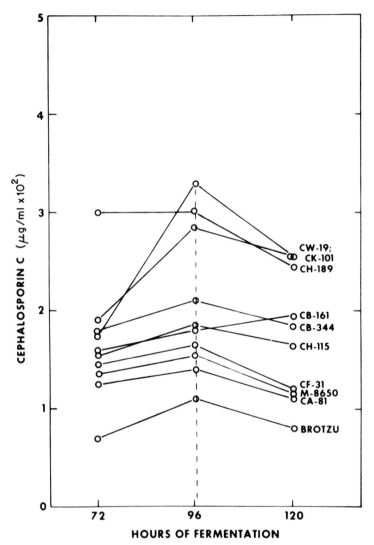

**FIGURE 20.** Cephalosporin production by Brotzu strain and early improved Lilly variants of *Cephalosporium acremonium*. (After Elander, 1975.)

cephalosporin C (Japanese Patent JA-110723, issued 1/16/75 to Takeda Chem. Ind. Ltd.). In another recent patent issued to the same company, certain polyploid forms of *C. acremonium* or *C. polyaleurum* were reported to be potent cephalosporin C producers. The higher ploidy clones were induced following exposure to camphor and acenaphthene and the subsequent selection of giant cells (Japanese Patent JS-109680, issued 1/16/75 to Takeda Chem. Ind. Ltd.).

**TABLE VI** Comparison of Random Selection versus Directed Selection Procedure in Strains of *Penicillium chrysogenum* and *Acremonium chrysogenum*[a]

| Type | Treated | Organisms Examined | Selection procedure | No. Tested | No. Retained | % Retained[b] |
|---|---|---|---|---|---|---|
| Random selection | uv, NG | *P. chrysogenum* and *C. acremonium* | None | 860 | 7 | 0.81 |
| Directed selection | uv, NG X-ray | *P. chrysogenum* and *C. acremonium* | 1. Colony-plate | 438 | 6 | 1.36 |
| | | | 2. Auxotrophs | 35 | 2 | 5.71 |
| | | | 3. Haploidization-inducing agents | 503 | 8 | 1.59 |
| | | | 4. Mitotic inhibitors | 225 | 3 | 1.33 |
| | | | 5. Mercury | 162 | 2 | 1.23 |
| | | | 6. Amino acid analogs | 567 | 8 | 1.41 |
| | | | 7. Sulfur analogs | 452 | 22 | 3.98 |
| | | *C. acremonium* (only) | 1. Methionine analogs | 605 | 22 | 3.64 |
| | | | 2. Increased sensitivity to methionine | | | |
| | | | a. Growth | 247 | 4 | 1.62 |
| | | | b. β-lactam synthesis | 52 | 2 | 3.85 |

[a] Adapted from Chang and Elander (1979).
[b] Superior on both primary and secondary screening tests.

Nash and Huber (1971) reported that submerged cultures of *C. acremonium* differentiate into small swollen fragments termed arthrospores and the differentiation coincides generally with the maximal rate of cephalosporin C synthesis. In their improved variants, arthrospore formation was proportional to increased antibiotic formation. Methionine supplementation enhanced the onset of differentiation and the requirement for methionine was increased for higher yielding mutants. Methionine and sulfur metabolism are very important for cephalosporin C synthesis and the metabolism of methionine, its analog norleucine (Drew *et al.*, 1976), and sulfate may trigger antibiotic synthesis through cellular differentiation. Dennen and Carver (1969) have reported that strains with increased capacity to synthesize cephalosporin C have decreased capacity to degrade methionine for cellular requirements and their arylsulfatase derepression is increased proportionately.

The probable pathway for sulfur metabolism in *C. acremonium* has been described by Drew and Demain (1975). Cysteine appears to be the immediate donor of sulfur to cephalosporin C, but the amino acid is not highly stimulatory for cephalosporin C formation in media containing sulfate (Demain, 1973, 1974). Methionine shows excellent stimulation of antibiotic synthesis and its stimulation is not due to sulfur donation but to some unresolved role in antibiotic regulation. Nüesch *et al.* (1973) and Drew and Demain (1975) have obtained mutants with genetic blocks between sulfate and cysteine and cystathionine and homocysteine. The mutants (designated 274-1 and 8650 *slp*) grow on cysteine, cystathionine, and homocysteine (but not on sulfate) but these compounds do not replace methionine with respect to cephalosporin C stimulation. Nüesch *et al.* (1973) reported on a Ciba *slp* mutant blocked in the sulfate reduction pathway prior to sulfide formation. The mutant was able to assimilate more exogenous methionine and had potential to synthesize nearly fourfold more antibiotic than its sulfide-proficient parent. Revertant strains of the mutant assimilated less methionine and synthesized only low levels of cephalosporin C. Drew *et al.* (1976) showed a similar result for the 274-1 mutant. It produced over 1100 $\mu$g/ml of cephalosporin C, whereas its prototrophic parent strain CW-19 formed only 400 $\mu$g/ml under identical fermentation conditions. Nüesch and co-workers have proposed that higher levels of cephalosporin C obtained with nonsulfate-utilizing mutants are due to their inability to synthesize cysteine, an amino acid which acts as a repressor and inhibitor of methionine permease.

In a study on a series of improved cephalosporin C mutants, Queener *et al.* (1975) reported that the specific activity of glutamate dehydrogenase (GDH) was derepressed upon entry into the stationary growth phase, whereas two mutants in a low-yielding series had repressed levels of

GDH at the same growth phase. The altered regulation pattern for GDH may have removed a nitrogen limitation for cephalosporin C synthesis. In this study and in a previous report by Elander *et al.* (1976), there appears to be an inverse relationship between vegetative development and enhanced cephalosporin C synthesis.

Komatsu *et al.* (1975) isolated an improved mutant of *C. acremonium* which synthesized copious amounts of cephalosporin C in the presence of sulfate. The mutant was also reported to be sensitive to methionine in that it displayed potency levels more than twofold that of the parent in the presence of sulfate, but its productivity was severely inhibited by more than 0.5% of methionine which gave high cephalosporin C production with the parent strain. Recent studies by Komatsu and Kodaira (1977) showed that the improved mutant had enhanced L-serine sulfhydrylase activity compared to the parent. They hypothesize that the mutant is superior to its parent in cephalosporin synthesis in that it has an improved ability to maintain a high level of the cysteine pool. Although there is no evidence to show that L-serine sulfhydrylase actually functions in the biosynthesis of cysteine in *C. acremonium,* they maintain that a second enzyme in the cysteine pathway, cystathionine $\beta$-synthase, may facilitate the conversion of methionine to cysteine, thereby enabling the mutant to maintain a high pool of cysteine, which appears to be essential for high cephalosporin synthesis in the presence of sulfate.

Rational selection procedures have been recently reported to be more efficient in selecting superior penicillin V- and cephalosporin C-producing mutants in large-scale selection programs at the Bristol-Myers Company (Elander, 1978; Chang and Elander, 1979). The rational selection procedures are based on known or probable biochemical mechanisms, thereby removing much of the empiricism commonly associated with labor-intensive random screening and selection. Table VI shows a comparison of rational *vs* random selection for the selection of superior fermentation variations of *P. chrysogenum* and *C. acremonium* (also classified as *Acremonium chrysogenum*).

## VIII. POTENTIAL ROLES FOR RECOMBINANT DNA, CLONING, AND GENE AMPLIFICATION IN MICROBIAL SELECTION TECHNOLOGY

A series of recent major technological breakthroughs in molecular biology has made it possible now to introduce foreign DNA into *E. coli* or other bacteria using plasmids or phages as cloning vehicles (Cohen, 1975). Two key developments have contributed more than anything else to this new technology. First, the observation that treatment of *E. coli* cells with $CaCl_2$ renders them competent for transformation by plasmid DNA (Cohen and Chang, 1975), and second, the discovery of a battery of restriction endonucleases, which cleave DNA at specific sites (Roulland-Dussoix *et al.*, 1975). A new field of genetic engineering aiming at introducing foreign DNA into bacteria (also termed "gene splicing," "molecular cloning," or "recombinant DNA") has evolved over the past 4 years.

A typical recombinant DNA experiment is diagrammed in Fig. 21. Briefly, the experiment consists of the following steps: (1) treatment of the plasmid DNA (vector) and foreign DNA, respectively, with endonuclease to generate fragments with complementary, overlapping single-strand ends ("cohesive ends"); (2) enzymatic ligation of fragments of plasmid DNA's with those of foreign DNA to form recombinant DNA's; and (3) tranformation of *E. coli* cells with recombinant DNA's. The use of drug-resistant plasmid provides a convenient means of selecting for clones containing recombinant DNA (Cohen *et al.*, 1973).

Since many plasmids, such as Col $E_1$ (determines colicin $E_1$ production and resistance), are present as multiple copies in cells (25–30 copies for Col $E_1$ plasmid), a gene can be greatly amplified in cells after being cloned in *E. coli* cells. Such gene amplification has been observed with *E. coli* tryptophan operon genes cloned into Col $E_1$ plasmid (Hershfield *et al.*, 1974). When this plasmid was introduced into an *E.*

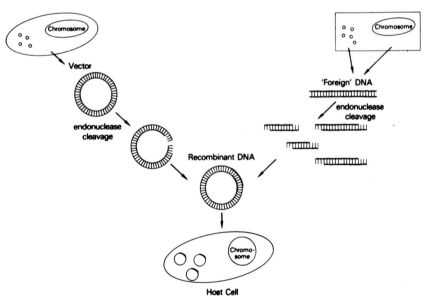

**FIGURE 21.** A typical recombinant DNA experiment. (From Singer, 1977.)

*coli* strain and the tryptophan operon induced by addition of inducer, the specific activity of the tryptophan operon enzymes increased more than 10-fold. Additional gene amplification through increase in plasmid copy number was achieved when plasmid-containing bacteria were grown in the presence of chloramphenicol. Since chloramphenicol inhibits protein synthesis, no amplification of gene products (proteins) was obtained. Future research may provide a means for amplifying plasmid gene products (proteins) as well as gene copies.

To date, many plasmid cloning vehicles as well as restriction endonucleases are readily available for cloning experiments (Helinski *et al.*, 1975). Genes from totally different biological classes, such as *Xenopus* and *Drosophila,* have been cloned in *E. coli* (Cohen *et al.*, 1973). In many instances, the foreign genes are able to replicate and are even transcribed, but not translated. Efforts are also being made to introduce into *E. coli* genes specifying such synthetic functions as nitrogen fixation, antibiotic and enzyme production, and antibody and hormone synthesis (Wade, 1977). Hormones such as insulin, somatostatin, and human growth hormone are particularly amenable to cloning since they are all relatively small peptides whose sequences are known.

Recombinant bacterial plasmids have been constructed that contain various regions of the rat insulin gene sequence (Ullrich *et al.*, 1977). In this instance the foreign DNA (rat insulin genes) was prepared enzymatically using reverse transcriptase (RNA-dependent DNA polymerase) with rat insulin messenger RNA (mRNA) as template. It remains to be seen whether the *E. coli* clones containing insulin gene segments are capable of producing insulin or insulin mRNA. Another recent development in recombinant DNA research is the successful transfer of nitrogen fixing gene (*nif*) from *Klebsiella pneumoniae* to *E. coli* by means of a recombinant plasmid (Dixon and Postgate, 1972). Even though the new strain of *E. coli* was able to synthesize nitrogenase, it could not fix nitrogen unless it was protected from oxygen. It appears that the technology for creating a higher plant capable of efficient nitrogen fixation is still in the distant future.

As can be seen from the above examples, industrial microbiologists are now faced with enormous problems in their attempts to apply recombinant DNA techniques to "tailor make" microorganisms. Some of the more obvious problems are as follows:

1. Even if one discounts the evolutionary barriers between species, the "spliced" genes in the new host cell have to overcome various regulatory limitations in order to be expressed.

2. The clones containing foreign DNA may not be genetically stable; the foreign DNA may be eliminated by segregation unless certain selec-

tion pressure is applied. An example of this situation is the mouse–human hybrid cells which tend to lose human chromosomes during propagation (Weiss and Green, 1967).

3. To clone a whole array of antibiotic pathway genes (e.g., more than ten genes are required for cephalosporin C biosynthesis) onto a plasmid could be an awesome task unless the genes are clustered. However, it should be less difficult to clone the structural genes for hormones or enzymes (amylase or protease) or genes with regulatory functions.

4. Proper cloning vehicles have not been developed for industrial microorganisms (mostly actinomycetes and fungi). Although plasmids have been found in streptomycetes, none has been reported to be useful as a cloning vehicle. A recent report that *Staphylococcus aureus* plasmids coding for tetracycline and chloramphenicol resistance can be introduced by transformation into *B. subtilis* (Ehrlich, 1977) could mean that some bacterial plasmids can be transferred to *Streptomyces* and be used as cloning vehicles. Recently, the *E. coli* gene coding for leu-2 has been introduced and expressed in *Saccharomyces cerevisiae* (Hinnen *et al.*, 1978).

The commercial manufacture of somatostatin and of human insulin using recombinant DNA technology may soon be possible as scientists at the University of California at Los Angeles and Genentech Inc. have inserted chemically synthesized genes into a bacterial host. The human gene coding for somatostatin is a small protein consisting of 14 amino acids. The artificial gene is then attached to a gene segment (operon) coding for $\beta$-galactosidase which is transferred into the *E. coli* cell via a plasmid known as pBR322 (Itakura *et al.*, 1977).

The chemical synthesis of the gene sequence coding for human insulin posed a more formidable task in that the gene coding for the A chain alone consists of 21 amino acids. The B chain gene sequence consists of some 30 amino acids. The synthetic A and B gene sequences were then linked to the $\beta$-galactosidase operon and inserted separately into different *E. coli* host cells (see Fig. 22). The *lac* operon was then "switched on" by the addition of an appropriate inducer substrate and the message was translated into the appropriate human insulin protein fragment. The A and B insulin chains were next purified and rejoined chemically to generate active human insulin (Anon., 1978a). Other areas currently under development using recombinant DNA technology include the eventual commercial synthesis of L-asparaginase, human interferon and other hormones, vaccines, and antiviral products (Anon., 1978b).

Finally, although the potential benefits provided by applying recombinant DNA techniques may be substantial, the potential biohazards

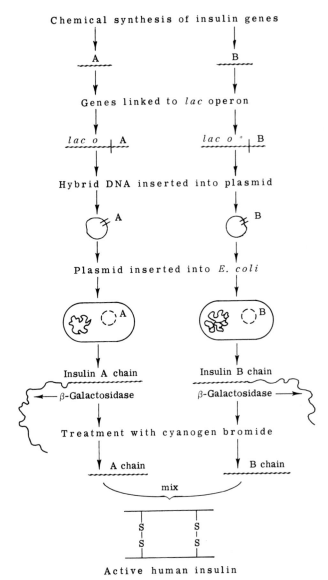

**FIGURE 22.** Recombinant DNA technology used for synthesizing human insulin. (From Anon., 1978a.)

inherent in this type of research probably have been overly exaggerated; however, they should not be overlooked (Szybalski, 1977). Undoubtedly, both physical as well as biological containment is needed so as to minimize any chance of external biohazard. In response to this need, governments in many developed countries, notably the United States, Germany, Britain, and Japan, have issued safety guidelines, or are in the process of formulating policy, to regulate recombinant DNA research (Singer, 1977).

## REFERENCES

Anonymous (1978a). *New Sci.* **79,** 926.
Anonymous (1978b). *Sci. News J. Phys. Chem.* **114,** 195.
Abe, S. (1972). *In* "The Microbial Production of Amino Acids" (K. Yamada, ed.), pp. 39–66. Wiley (Halsted), New York.
Akagawa, H., Okanishi, M., and Umezawa, H. (1975). *J. Gen. Microbiol.* **90,** 336–346.
Akiyama, S., Suzuki, T., Sumino, Y., Nakao, Y. and Fukuda, Y. (1972). *Ferment. Technol. Today, Proc. Int. Ferment. Symp., 4th, 1972,* pp. 613–620.
Alacevic, M. (1963). *Nature (London)* **197,** 1323–1324.
Alacevic, M. (1976). *Proc. Int. Symp. Genet. In. Microorg., 2nd, 1974,* pp. 513–519.
Alikhanian, S. I. (1962). *Adv. Appl. Microbiol.* **4,** 1–50.
Alikhanian, S. I., and Borisova, L. N. (1956). *Izv. Akad. Nauk SSSR* **2,** 74–79.
Alikhanian, S. I., and Iljina, (1958). *Nature (London)* **181,** 1476–1477.
Alikhanian, S. I., Orlova, N. U., Mindlin, S. Z., and Zaitseva, Z. M. (1961). *Nature (London)* **189,** 939–940.
Arcamone, F., Cassinelli, G., Fantini, G., Grein, A., Onezzi, P., Pol, C., and Spalla, C. (1969). *Biotechnol. Bioeng.* **11,** 1101–1110.
Arima, K. (1977). *Dev. Ind. Microbiol.* **18,** 79–117.
Azevedo, J. L., and Roper, J. A. (1967). *J. Gen. Microbiol.* **49,** 149–155.
Backus, M. P., and Stauffer, J. F. (1955). *Mycologia* **47,** 429–463.
Ball, C. (1971). *J. Gen. Microbiol.* **66,** 63–69.
Ball, C. (1973a). *In* "Genetics of Industrial Microorganisms" (Z. Vaněk, Z. Hošťálek, and J. Cudlín, eds.), Vol. 2, pp. 227–237. Elsevier, Amsterdam.
Ball, C. (1973b). *Prog. Ind. Microbiol.* **12,** 47–72.
Barchielli, R., Baretti, G., DiMarco, A., Julita, P., Migliacci, A., Munghetti, A., and Spalla, C. (1960). *Biochem. J.* **74,** 382–387.
Bauman, R. and Kocher, H. P. (1976). *Proc. Int. Symp. Genet. Ind. Microorg., 2nd, 1974* pp. 535–552.
Boronin, A. M., and Sadovnikova, N. (1972). *Genetika* **11,** 174–176.
Brody, S., and Tatum, E. L. (1966). *Proc. Natl. Acad. Sci. U.S.A.* **56,** 1290–1297.
Brown, W. F., and Elander, R. P. (1966). *Dev. Ind. Microbiol.* **7,** 114–123.
Burnett, J. H. (1975). *In* "Mycogenetics: An Introduction to the General Genetics of Fungi," pp. 235–258. Wiley, New York.
Caglioti, M. T., and Sermonti, G. (1956). *J. Gen. Microbiol.* **14,** 38–46.
Carilli, A., Chain, E. B., Ghulandi, G., and Morisi, G. (1961). *Sci. Rep. Ist. Super. Sanita* **1,** 177–189.
Carvajal, F. (1952). *Mycologia* **45,** 209–234.

Chakrabarty, A. M. (1976). *Annu. Rev. Genet.* **10,** 7–30.

Chang, L. T., and Elander, R. P. (1979). *Dev. Ind. Microbiol.* **20** (*in press*).

Čižmek, S. (1976). *Abstr. Int. Ferment. Symp., 5th, 1976,* p. 216.

Claridge, C. A., Bush, J. A., DeFuria, M. D., and Price, K. E. (1974). *Dev. Ind. Microbiol.* **15,** 101–113.

Clarke, P. H. (1976). *Proc.—Int. Symp. Genet. In. Microorg., 2nd, 1974* pp. 15–28.

Clowes, R. C., and Hayes, W. (1968). "Experiments in Microbial Genetics." Wiley, New York.

Cohen, S. N. (1975). *Sci. Am.* **233,** 24–33.

Cohen, S. N., and Chang, A. C. Y. (1975). *In* "Microbiology—1974" (D. Schlessinger, ed.), pp. 187–198. Am. Soc. Microbiol., Washington, D.C.

Cohen, S. N., Chang, A. C. Y., Boyer, H. W., and Hellwig, R. B. (1973). *Proc. Natl. Acad. Sci. U.S.A.* **70,** 3240–3244.

Conover, L. H. (1971). *Adv. Chem. Ser.* **108,** 33–80.

Davies, O. L. (1964). *Biometrics* **20,** 576–591.

Delic, V., Pigac, J., and Sermonti, G. (1969). *J. Gen. Microbiol.* **55,** 103–110.

Demain, A. L. (1972). *Ferment. Technol. Today, Proc. Int. Ferment. Symp., 4th, 1972* pp. 239–245.

Demain, A. L. (1973). *Adv. Appl. Microbiol.* **16,** 177–202.

Demain, A. L. (1974). *Lloydia* **37,** 147–163.

Dennen, D. W., and Carver, D. D. (1969). *Can. J. Microbiol.* **15,** 175–181.

Dixon, R. A., and Postgate, J. R. (1972). *Nature (London)* **237,** 102–103.

Drake, J. W., and Baltz, R. H. (1976). *Annu. Rev. Biochem.* **45,** 11–37.

Drew, S. W., and Demain, A. L. (1975). *Antimicrob. Agents Chemother.* **8,** 5–10.

Drew, S. W., Winstanley, D. J., and Demain, A. L. (1976). *Appl. Microbiol.* **31,** 143–145.

Dulaney, E. L., and Dulaney, D. D. (1967). *Trans. N. Y. Acad. Sci.* [2] **29,** 782–799.

Ehrlich, S. D. (1977). *Proc. Natl. Acad. Sci. U.S.A.* **74,** 1680–1682.

Elander, R. P. (1967). *In* "Induced Mutations and their Utilization" (H. Stubbe, ed.), pp. 403–423. Akademie-Verlag, Berlin.

Elander, R. P. (1975). *Dev. Ind. Microbiol.* **16,** 356–374.

Elander, R. P. (1979). *In* "Genetics of Industrial Microorganisms." (O. K. Sebek, and A. I. Laskin, eds.), pp. 21–35. American Soc. Microbiol., Wash., D.C.

Elander, R. P., and Espenshade, M. A. (1976). *In* "Industrial Microbiology" (B. M. Miller and W. Litsky, eds.), pp. 192–256. McGraw-Hill, New York.

Elander, R. P., Mabe, J. A., Hamill, R. L., and Gorman, M. (1971). *Folia Microbiol. (Prague)* **16,** 156–165.

Elander, R. P., Espenshade, M. A., Pathak, S. G., and Pan, C. H. (1973). *In* "Genetics of Industrial Microorganisms" (Z. Vaněk, Z. Hošťálek, and J. Cudlín, eds.). Vol. 2, pp. 239–253. Elsevier, Amsterdam.

Elander, R. P., Corum, C. J., DeValeria, H., and Wilgus, R. M. (1976). *Proc. Int. Symp. Genet. Ind. Microorg., 2nd, 1974,* pp. 253–271. Academic Press, New York.

Erickson, R. C., and Dean, L. D. (1966). *Appl. Microbiol.* **14,** 1407–1448.

Esposito, A., Licciardello, G., Murthy, Y. K. S., Sacerdoti, S. A., and Sparapani, P. (1972). *Ferment. Technol. Today, Proc. Int. Ferment. Symp., 4th, 1972* Abstract, p. 217.

Falkow, S. (1975). "Infectious Multiple Drug Resistance." Pion Ltd., London.

Fantini, A. A. (1976). *In* "Methods in Enzymology" (J. H. Hash, ed.), Vol. 43, pp. 24–41. Academic Press, New York.

Fincham, J. R. S., and Day, P. R. (1973). "Fungal Genetics," 3rd ed. Blackwell, Oxford.

Fuska, J., and Welwardova, F. (1969). *Biologia (Bratislava)* **24,** 691–698.

Godfrey, O. (1974). *Can. J. Microbiol.* **20,** 1479–1486.

Gorman, M., Hamill, R. L., Elander, R. P., and Mabe, J. (1968). *Biochem. Biophys. Res. Commun.* **31,** 294–297.

Goulden, S. A., and Chattaway, F. W. (1969). *J. Gen. Microbiol.* **59,** 111–118.

Gregory, F. K., and Huang, J. C. C. (1964a). *J. Bacteriol.* **87,** 1281–1286.

Gregory, F. K., and Huang, J. C. C. (1964b). *J. Bacteriol.* **87,** 1287–1294.

Guérineau, M., Slominski, P. P., and Auner, P. R. (1974). *Biochem. Biophys. Res. Commun.* **61,** 462–469.

Guerola, N., Ingraham, J. C., and Cerda-Olmedo, E. (1971). *Nature (London), New Biol.* **230,** 122–124.

Halsall, D. M. (1975). *Biochem. Genet.* **3,** 109–116.

Hayes, W. (1968). "The Genetics of Bacteria and their Viruses," 2nd ed. Wiley, New York.

Helinski, D. R. (1973). *Annu. Rev. Microbiol.* **27,** 437–470.

Helinski, D. R., Lovett, M. A., Williams, P. H., Katz, L., Kupersztock-Portnay, Y. M., Guiney, D. G., and Blair, D. G. (1975). *In* "Microbiology—1974" (D. Schlessinger, ed.), pp. 104–114. Am. Soc. Microbiol., Washington D.C.

Hershfield, V., Boyer, H. W., Yanovsky, C., Lovett, M. A., and Helinski, D. R. (1974). *Proc. Natl. Acad. Sci. U.S.A.* **71,** 3455–3459.

Hinnen, A., Hicks, J. B., and Fink, G. R. (1978). *Proc. Natl. Acad. Sci., U.S.A.* **75,** 1929–1933.

Hohfeld, R., and Vielmutter, W. (1973). *Nature (London), New Biol.* **242,** 130–133.

Holliday, R. (1956). *Nature (London)* **178,** 987–990.

Hopwood, D. A. (1970). *Methods Microbiol.* **3A,** 363–433.

Hopwood, D. A. (1973). *In* "Actinomycetales" (G. Sykes and F. A. Skinner, eds.), pp. 131–153. Academic Press, New York.

Hopwood, D. A. (1976). *In* "Microbiology—1976" (D. Schlessinger, ed.), pp. 558–562. American Society for Microbiology, Washington, D.C.

Hopwood, D. A. (1977). *Dev. Ind. Microbiol.* **18,** 9–21.

Hopwood, D. A., and Merrick, M. J. (1977). *Bacteriol. Rev.* **41,** 595–635.

Hopwood, D. A., and Wright, A. M. (1972). *J. Gen. Microbiol.* **71,** 383–398.

Hopwood, D. A., Chater, K. F., Dowding, J. E., and Vivian, A. (1973). *Bacteriol. Rev.* **37,** 371–405.

Hopwood, D. A., Wright, H. M., Bibb, M. J., and Cohen, S. N. (1977). *Nature (London)* **268,** 171–174.

Hošťálek, Z., and Vaněk, Z. (1973). *In* "Genetics of Industrial Microorganisms" (Z. Vaněk, Z. Hošťálek, and J. Cudlín, eds.), Vol. 2, pp. 353–371. Elsevier, Amsterdam.

Ibuki, A., and Takinami, K. (1975). *Abstr. Meet., Agricl. Soc. Jpn., 1975,* p. 341.

Ikeda, Y., Nakamura, K., Uchida, K., and Ishitani, C. (1957). *J. Gen. Appl. Microbiol.* **3,** 93–126.

Ilczuk, Z. (1971). *Nahrung* **15,** 381–386.

Itakura, K., Hirose, T., Crea, R., Riggs, A., Heyneker, H., Bolivar, F., and Boyer, H. W. (1977). *Science* **198,** 1056–1063.

James, L. V., Rubbo, S. D., and Gardener, J. F. (1956). *J. Gen. Microbiol.* **14,** 223–277.

Kahler, R., and Noack, D. (1974). *Z. Allg. Mikrobiol.* **14,** 529–535.

Katagiri, I. (1954). *J. Antibiot.* **7,** 45–52.

Kelner, A. (1949). *J. Bacteriol.* **57,** 73–92.

Kirby, R., Wright, L. F., and Hopwood, D. A. (1975). *Nature (London)* **254,** 265–267.

Komatsu, K., and Kodaira, R. (1977). *J. Antibiot.* **30,** 226–233.

Komatsu, K., Mizumo, M., and Kodaira, R. (1975). *J. Antibiot.* **28,** 881–888.

Lederberg, J., and Tatum, E. L. (1946). *Nature (London)* **158,** 558–560.

Livingston, D. M., and Klein, D. M. (1977). *J. Bacteriol.* **129,** 472–481.

McCormick, J. R. D. (1967). *In* "Antibiotics" (D. Gottlieb and P. D. Shaw, eds.), Vol. 2, pp. 113–121. Springer-Verlag, Berlin and New York.

McCormick, J. R. D., Sjölander, N. L., Hirsch, U., Jensen, E., and Doershuk, A. P. (1957). *J. Amer. Chem. Soc.* **79,** 4561–4563.

Macdonald, K. D. (1964). *Nature (London)* **204,** 404–405.

Macdonald, K. D., Hutchinson, J. M., and Gillett, W. A. (1964). *Antonie van Leeuwenhoek* **30,** 209–224.

Matselyukh, D. P., and Mukvic, N. J. (1973). *Mikrobiol. Zh.* **35,** 411–417.

Merrick, M. J. (1976). *Proc.—Int. Symp. Genet. Ind. Microorg., 2nd, 1974* pp. 229–242.

Miller, P. A., Hash, J. H., Lincks, M., and Bohonos, N. (1965). *Biochem. Biophys. Res. Commun.* **3,** 575–577.

Mishra, N. C. (1976). *Nature (London)* **264,** 251–253.

Mitscher, L. A. (1968). *J. Pharm. Sci.* **57,** 1633–1638.

Morton, A. G. (1961). *Proc. R. Soc. London, Ser. B* **153,** 548–569.

Nagaoka, K., and Demain, A. L. (1975). *J. Antibiot.* **28,** 627–635.

Nakayama, K. (1976). *Process Biochem.* **11,** 4–9.

Nash, C. H., and Huber, F. M. (1971). *Appl. Microbiol.* **22,** 6–10.

Nga, B. H., and Roper, J. A. (1969). *Genet. Res.* **14,** 63–70.

Nüesch, J., Treichler, H. J., and Liersch, M. (1973). *In* "Genetics of Industrial Microorganisms" (Z. Vaněk, Z. Hošťálek, and J. Cudlín, eds.), Vol. 2, pp. 309–334. Elsevier, Amsterdam.

Nüesch, J., Hinnen, A., Liersch, M., and Treichler, H. J. (1976). *Proc.—Int. Symp. Genet. Microorg., 2nd, 1974* pp. 451–472.

Oeschger, M. P., and Berlyn, M. K. B. (1974). *Mol. Gen. Genet.* **134,** 77–83.

Ogawara, H., and Nozaki, S. (1977). *J. Antibiot.* **30,** 337–339.

Okanishi, M., and Gregory, K. F. (1970). *Can. J. Microbiol.* **16,** 1139–1143.

Okanishi, M., Ohta, T., and Umezawa, H. (1970). *J. Antibiot.* **23,** 45–47.

Orgel, L. E. (1965). *Adv. Enzymol.* **27,** 289–346.

Ostroukov, A. A., and Kuznetsov, V. P. (1963). *Antibiotiki* **8,** 33–35.

Ostrowska-Krysiak, B. (1971). *Acta Microbiol. Pol., Ser. B* **3,** 23–29.

Pan, C. H., Hepler, L., and Elander, R. P. (1972). *Dev. Ind. Microbiol.* **13,** 103–112.

Pan, C. H., Hepler, L., and Elander, R. P. (1975). *J. Ferment. Technol.* **53,** 854–861.

Pathak, S. G., and Elander, R. P. (1971). *Appl. Microbiol.* **30,** 750–758.

Perlman, D., and Ogawa, Y. (1976). *J. Antibiot.* **29,** 1112–1113.

Pogell, B., Sankaran, L., Redshaw, P. A., and McCann, P. A. (1976). *In* "Microbiology— 1976" (D. Schlessinger, ed.), pp. 543–547. Am. Soc. Microbiol., Washington, D.C.

Pontecorvo, G., and Sermonti, G. (1953). *Nature (London)* **172,** 126–127.

Preuss, D. L., and Johnson, M. J. (1967). *J. Bacteriol.* **94,** 1502–1508.

Queener, S. W., Capone, J. J., Radue, A. B., and Nagarajan, R. (1974). *Antimicrob. Agents & Chemother.* **6,** 334–337.

Queener, S. W., McDermott, J., and Radue, A. B. (1975). *Antimicrob. Agents & Chemother.* **7,** 646–651.

Randazzo, R., Sermont, G., Carere, A., and Bignami, M. (1973). *J. Bacteriol.* **113,** 500–509.

Randazzo, R., Sciandrello, G., Carere, A., Bignami, M., Velcich, A., and Sermonti, G. (1976). *Mutat. Res.* **36,** 291–297.

Raper, J. R., Raper, C. A., and Miller, R. E. (1972). *Mycologia* **64,** 1088–1117.

Redshaw, P. A., McCann, P. A., Sankaran, L., and Pogell, B. M. (1976). *J. Bacteriol* **125,** 698–705.

Roper, J. A. (1973). *In* "Genetics of Industrial Microorganisms" (Z. Vaněk, Z. Hošťálek, and J. Cudlín, eds.), Vol. 2, pp. 81–88. Elsevier, Amsterdam.

Rosi, D., Goss, W. A., and Daum, S. J. (1977). *J. Antibiot.* **30,** 88–97.

Roulland-Dussoix, D., Yoshimori, R., Green, P., Betlack, M., Goodman, H. M., and Boyer, H. W. (1975). *In* "Microbiology—1974" (D. Schlessinger, ed.), pp. 187–198. Am. Soc. Microbiol., Washington, D.C.

Sanchez, S., and Quinto, C. M. (1975). *Appl. Microbiol.* **30,** 750–758.

Schaeffer, P. (1969). *Bacteriol. Rev.* **33,** 48–71.

Schrempf, H., Bujard, H., Hopwood, D. A., and Goebel, W. (1975). *J. Bacteriol.* **121,** 416–421.

Segal, I. H., and Johnson, M. J. (1961). *J. Bacteriol.* **81,** 81–98.

Sermonti, G. (1959). *Ann. N.Y. Acad. Sci.* **81,** 950–956.

Sermonti, G. (1969). "Genetics of Antibiotic-Producing Microorganisms." Wiley, New York.

Shaw, P. D., and Piwowarski, J. (1977). *J. Antibiot.* **30,** 404–408.

Shier, W. T., Rinehart, K. L., Jr., and Gottlieb, D. (1969). *Proc. Natl. Acad. Sci. U.S.A.* **63,** 198–204.

Singer, M. (1977). *Gene* **1,** 123–139.

Spizek, J., Malek, I., Dalezilova, L., Vondracek, M., and Vaněk, Z. (1965). *Foli Microbiol.* **10,** 259–262.

Stark, W. M., Knox, N. G., and Wilgus, R. M. (1971). *Folia Microbiol. (Prague)* **16,** 205–217.

Stauffer, J. F. (1961). *Sci. Reps. Ist. Super. Sanita* **1,** 441–447.

Szybalski, W. (1977). *Gene* **1,** 181–183.

Testa, R. T., Wagman, G. H., Daniel, P. J. L., and Weinstein, M. (1974). *J. Antibiot.* **27,** 917–921.

Troonen, H., Roelants, P., and Boon, B. (1976). *J. Antibiot.* **29,** 1258–1267.

Ullrich, A., Shine, J., Chirgwin, J., Pictet, R., Tischer, E., Rutter, W. J., and Goodman, H. M. (1977). *Science* **196,** 1313–1319.

Vezina, C., Bolduc, C., Kudelski, A., and Sehgal, S. N. (1976). *J. Antibiot.* **29,** 248–264.

Wade, N. (1977). *Science* **195,** 558–560.

Wagner, R. P., and Mitchell, H. K. (1964). "Genetics and Metabolism," 2nd ed. Wiley, New York.

Watson, B., Currier, T. C., Gordon, M. P., Chilton, M., and Nester, E. W. (1975). *J. Bacteriol.* **123,** 255–264.

Weiss, M., and Green, H. (1967). *Proc. Natl. Acad. Sci. U.S.A.* **58,** 1104–1111.

Whitaker, A., and Long, P. A. (1973). *Process Biochem.* **8,** 27–31.

White, R. J., Martinelli, E., and Lancini, G. (1974). *Proc. Natl. Acad. Sci. U.S.A.* **71,** 3260–3264.

Wright, L. F., and Hopwood, D. A. (1976). *J. Gen. Microbiol.* **95,** 96–106.

Chapter 13

# Methods for Laboratory Fermentations

N. D. DAVIS
W. T. BLEVINS

## I. INTRODUCTION

Fermentation processes applicable to industrial use are first investigated in the laboratory prior to scale-up and development of production

**303**

MICROBIAL TECHNOLOGY, 2nd ed., VOL. II
Copyright © 1979 by Academic Press, Inc.
All rights of reproduction in any form reserved. ISBN 0-12-551502-2

processes. Although methods used in the fermentation laboratory for these small- to intermediate-scale fermentations are not necessarily unique to this specialty of microbiology, certain equipment and philosophies are routinely applied. These constitute the subject matter of this chapter.

## II. EXPERIMENTAL DESIGN

### A. The Hypothesis

The fermentation technologist is an experimental scientist who must design each experiment with great attention to detail. Carelessly designed experiments are a waste of resources. An experiment tests a hypothesis, and the hypothesis should be precisely written and its assumptions and implications thoroughly examined and understood in planning the experiment.

### B. Fermentation Experiments

An experiment should be designed in such a manner that it unequivocally demonstrates that the hypothesis is either true or false; such an experiment is quite rare. In practice it is usually a matter of to what relative degree the data support or contradict the hypothesis.

In designing an experiment, one must decide in advance which measurements will be made, how they will be made, and what data format will be required. The hypothesis dictates the general form the experiment must take. The nature of the data to be accumulated is the dominant factor in making such decisions as the number of replications, method of randomization, and number of repetitions. Inherent biological variation and precision of measurements and analyses are always factors to be considered. The experimenter should be acquainted with range, variance, standard deviation, significant differences among measurements, and confidence limits for interpreting data. Thus, the fermentation scientist needs a thorough grasp of the principles of biostatistics (Campbell, 1974) and should consult with a statistician during the design of an experiment.

### C. Record Keeping and Data Processing

Complete, accurate, and detailed record keeping is essential. Recording data directly into bound record books, dating and initialing entries, and recording all data, observations, and conclusions should be routine.

Experiments should be numbered and data organized in a manner that allows easy access to original data, analyses, hypotheses, and conclusions years later. Information control, record keeping, and methods of good laboratory management are detailed by Hockenhull (1977).

Computer facilities for data processing are available in many laboratories and are considered a necessity rather than a luxury. Also, the age of "smart" instruments coupled to mini- and midicomputers is here and growing rapidly; computer analysis and control of fermentations is becoming widespread. A basic knowledge of computer technology is becoming increasingly important to the fermentation technologist. Computer control of fermentations is discussed in this volume, Chapter 15. Computer technology applicable to laboratory fermentations has recently been discussed by Humphrey (1976) and Isenberg and Mac-Lowry (1976).

## III. SCREENING TECHNIQUES

### A. Primary Screening

#### 1. General Considerations

A mixed population of microorganisms may contain a number of species which have potential for industrial applications. Primary screening programs involve highly selective procedures for detecting and isolating individual species from a mixed population. The most frequently used screening procedure is to isolate microorganisms from soil dilutions. Soil is useful since it is a habitat where organisms are involved in decomposition and resynthesis of complex organic materials and in various transformations of inorganic materials. The spread-plate or crowded-plate technique commonly is used; these methods involve use of various differential and selective media. General principles for microbial culture selection are reviewed in this volume, Chapter 12. Hopwood (1970) has discussed the isolation of mutants.

#### 2. Biologically Active Substances

**a. General Techniques.** During growth of a mixed population of microorganisms, a chemically defined liquid culture medium may be utilized so that a given species will grow at a faster rate than the other organisms. Successive transfers to this specialized medium permit the desired organisms to become the predominant species. Selection may also develop if unwanted organisms are inhibited, while organisms of interest are permitted to grow. One can, therefore, "enrich" a medium in a manner that favors the desired species or physiological type, and these species can be isolated to pure culture by plating on solid medium of the same composition (Stanier et al., 1976). For example, an organism capable of using atmospheric nitrogen ($N_2$) as the sole source of nitrogen may be isolated by using a medium devoid of fixed nitrogen. If $N_2$ is available, only those organisms that can fix atmospheric nitrogen can grow. Enrichment culture experiments utilize such factors as the carbon source, energy supply, temperature, oxygen level, and hydrogen ion concentration. Microorganisms capable of specific biotransformations may be isolated by this method, as may degraders of cellulose or other polymers, hydrocarbons, and other compounds of interest.

Selective pressures can also be exerted by using osmotic pressure, an antibiotic, various inhibitors, or other factors as desired. For direct isolation, a mixed inoculum containing several species is spread directly on the surface of a selective solid medium; all microorganisms which can grow will produce colonies. The maximum possible number of types of microbes that can grow in a particular environment can be isolated by this method.

Once an organism capable of producing some amino acid, vitamin, or enzyme is isolated and purified, strain improvement techniques are usually employed to improve the growth of the organism and/or the production of the desired fermentation product. These methods are discussed by Fantini (1976).

**b. Antibiotics and Other Growth-Inhibiting Agents.** The discovery of penicillin by Alexander Fleming in 1929 and the demands for antibiotics during World War II stimulated technological advances for the detection and large-scale production of antibiotics and other antagonistic compounds. Intensive screening programs were developed to find microorganisms capable of producing chemotherapeutically useful antibiotics (Spooner and Sykes, 1972).

Antibiotics generally are defined as those organic compounds produced by microorganisms that are antagonistic in low concentrations to the growth or survival of microorganisms (Casida, 1968). Other metabolic products, such as ketones, acids, and amines, may be inhibitory to growth but are not considered to be antibiotics. Antibiotics are produced primarily by bacteria belonging to the genera *Streptomyces, Nocardia,* and *Micromonospora* (the actinomycetes); by *Bacillus* (a eubacterium); and by fungi, notably *Penicillium.* Of over 3000 known antibiotics, approximately 70% are produced by actinomycetes, 20% by fungi, and 10% by eubacteria.

A detailed survey of laboratory techniques for isolation of antibiotic producers and the detection and identification of antibiotics has recently been published (Hash, 1976). In a screening program, several general items are of importance. Although there are a few thermophilic and psychrophilic antibiotic producers, most prokaryotic organisms are mesophilic, growing best over a temperature range of 23°–37°C. Antibiotic-producing fungi grow best at 20°–22°C; actinomycetes grow optimally around 28°C. Fungi can grow at a lower pH than most of the bacteria, so acidity can be used to inhibit bacterial growth during the screening process. Furthermore, the pH for optimal growth may be different from the pH favoring optimal production of the antibiotic, and the pH of assay media can drastically affect the activity of the antibiotic. Also, most of the antibiotic producers are highly aerobic, so appropriate aeration techniques are employed.

The nutritional conditions used for growth of antibiotic producers are also important. The isolation media for actinomycetes are usually enriched with carbon sources, such as glycerol, chitin, or starch, with organic nitrogen sources such as casein, arginine, and asparagine. Nitrate or ammonia also may serve as nitrogen sources.

In screening soil or mud, samples may be pretreated in several ways to increase the proportion of actinomycetes. Air-drying the sample will reduce the number of vegetative bacterial cells while allowing actinomycete spores to survive. Mixing soil with $CaCO_3$ for several days also may be effective. Overgrowth by fungi can be suppressed by using antifungal agents in the growth medium. For isolation, appropriate dilutions of the sample are plated onto enriched agar by the spread-plate method. Agar plates should be relatively dry on the surface to prevent the spread of motile bacteria. Following incubation for 5–7 days at 25°–28°C, actinomycetes appear as tough, chalky, white to grey colonies. Those that appear to be antagonistic to surrounding bacteria should be removed for further testing. Antagonism is best observed using a crowded-plate technique, in which plates having 400 or more colonies are examined. Antibiotic-producing colonies are subcultured to a similar medium and purified by streaking.

Nonantibiotic compounds that are inhibitory to growth of test bacteria may be produced by various organisms. Organic acids or amines, for example, may antagonize other organisms growing around colonies producing these compounds. These organisms initially might be presumed to produce antibiotics, so further testing is always necessary. These basic or acidic compounds can be detected by incorporating various pH indicators into poorly buffered agar media. Colonies producing these compounds may cause a color change in the surrounding agar due to changes in acidity or alkalinity.

**c. Vitamins and Other Growth Factors.** Primary screening for organisms that produce vitamins or other growth factors requires a bioassay organism that has an absolute requirement for the particular growth factor in question. If an isolate produces the growth factor extracellularly, its metabolic activities will permit growth of the bioassay organism. The detection of organisms that produce growth factors needed in human nutrition relies on microbiological assays using auxotrophs that have a specific requirement for a particular growth factor in order to grow. This can encompass requirements for vitamins, amino acids, purines, pyrimidines, etc.

Vitamins serve as precursors for coenzymes that function as catalysts in various enzyme systems; thus, deficiencies can lead to metabolic blocks reversible only when the vitamin is present in the nutrient supply (Hanson, 1967; Skeggs, 1976). These vitamin-requiring organisms may occur naturally (for example, lactic acid bacteria) or may be auxotrophic mutants isolated in the laboratory (Hopwood, 1970). Various factors must be considered in choosing an assay system for specific growth factors. The assay organism may be able to grow if only a precursor of the growth factor is present. For example, the thiazole or pyrimidine moiety of thiamine can be used by *Saccharomyces cerevisiae* during growth; thus, this presumptive thiamine-requiring organism can grow without the vitamin actually being supplied. Furthermore, end products of enzymatic reactions catalyzed by the growth factor might support growth if they are present in the medium. This occurs with some assay organisms for vitamin $B_{12}$, wherein deoxyribosides may support growth (Skeggs, 1976).

The medium for assaying for a particular growth factor should contain all nutrients necessary for optimal growth except for the one factor being measured. The medium should ideally support no growth in the absence of the growth factor, so that no turbidity develops in culture tubes lacking the factor. Most vitamin assays can be done using commercially prepared dehydrated media.

A screening approach similar to that used for antibiotics can be employed to isolate microorganisms that can synthesize extracellular vitamins, amino acids, or other growth factors from natural sources. For example, glutamic acid-producing cultures were first isolated in Japan by screening feces of various birds and mammals. Cross-feeding of the growth factor-requiring auxotroph by the growth factor producer will occur on a medium deficient in the particular required growth factor. Large numbers of isolates can be screened using the appropriate microbiological assay organism, and zones of growth stimulation by streaks or colonies of the isolate is indicative of growth factor synthesis.

Plates of a medium favoring production of a particular growth factor (if this is known) are allowed to incubate for 2–3 days at 28°C following inoculation of an appropriate dilution of some natural source. These plates may be replicated to a medium of the same composition using the replica-plate method of Lederberg and Lederberg (1952); colonies on newly inoculated plates will arise at the same site as on the original plate. After 2–3 days growth, the microorganisms on the replicated plates can be killed by exposure to lethal doses of ultraviolet (uv) light. The plates are then overlaid with a melted medium containing the specific assay organism. After incubation for 24 hours, the assay organism will grow in the immediate areas where any isolated colonies had produced the desired growth factor. By comparing the locations of the turbid areas with those on the original plate, the viable organisms which were responsible for production of the growth factor can be isolated (Miller and Litsky, 1976). These strains can be purified by the streak-plate technique and grown in a liquid medium; subsequent isolation and identification of the growth factor can be accomplished.

Amino acid producers have come primarily from species of bacteria; thus, fungal inhibitors can be added to the medium to prevent overgrowth by these eukaryotes (Dulaney, 1967). On the other hand, vitamins have been produced mostly by eukaryotes such as yeasts, fungi, and algae, especially for large-scale productions.

**d. Enzymes.** Enzyme-catalyzed reactions using intact growing cells present problems that have stimulated studies involving the use of cell-free enzyme preparations in fermentations. Enzymes are widely used in pharmaceutical, brewing, textile, cheese, leather, meat, corn syrup, wine, baking, and paper industries. Plants, animals, and microorganisms may serve as sources of these enzymes, but the convenience and economy of using microorganisms warrants special emphasis on this aspect of enzyme production (Underkofler, 1976).

Species of bacteria, molds, and yeasts have been useful in large-scale production of enzymes. As outlined by Beckhorn (1967), the organism must be nonpathogenic, produce no toxic products, and be genetically stable with respect to the production of the enzyme in question. Cultures may be screened from stock collections or from natural sources. A useful technique is the enrichment culture method discussed previously; another technique is one in which the potential substrate is incorporated into the growth medium. For instance, production of amylase and protease by various isolates can be detected by incorporating soluble starch or casein, respectively, into the agar medium. Clear zones of substrate degradation are observed around the isolates that are synthesizing the extracellular enzymes. Lack of a detection system on agar requires enzyme analysis of the supernatant fraction following growth in liquid media. Isolates that show potential can be mutated to try to improve production (Hopwood, 1970).

### 3. Biomass Production

Aspects of biomass production procedures have been treated extensively in earlier chapters and also in the 1975 International Symposium on Single Cell Protein (Tannenbaum and Wang, 1975). Filamentous fungi, yeasts, bacteria, and algae all have been used for production of single-cell protein (SCP). Bacteria and yeast have the briefest generation times, thus producing more cells faster. Bacteria have a comparatively high nucleic acid content that contributes to toxicity of bacterial SCP intended for human consumption. However, slower growth rates can limit nucleic acid content, thus minimizing this disadvantage. Yeasts and algae are rich sources of B vitamins, while fungal mycelia generally contain lesser amounts. Also, bacteria are higher in certain vitamins and in sulfur-containing amino acids than are yeasts. The ideal organism for biomass production should be thermoduric and have a high protein content, relatively low nucleic acid content, a rapid growth rate, and high survivability or robustness. Toxic and pathogenic properties of the organism must be determined by feeding studies when evaluating its potential as a food supplement.

### 4. Fermentation Products

The products of a fermentation process may be tremendously diverse, requiring separation, identification, and quantitation of many compounds. A complete analysis of fermentation products can be accomplished by methods discussed by Dawes et al. (1971). Gaseous products are best analyzed by gas chromatography–mass spectrometry (GC–MS). Headspace vapors of fermentation vessels or fermentation tubes may be sampled and analyzed directly for methane, $CO_2$, etc. Separation of volatile products from cleared fermentation media can be achieved by distillation after adjustment to the desired pH (7.0–7.5 for neutral volatiles). Volatiles that may be present are methanol, ethanol, isopropanol, butanol, acetoin, diacetyl, acetone, acetaldehyde, or volatile fatty acids (steam distilled at pH 2.8–3.0) such as formic, acetic, propionic, and butyric acids. Also, various volatile organic compounds can be trapped on a chromatography column (or freezing trap) connected in tandem to a closed fermentation train. These compounds can subsequently be eluted and analyzed by GC–MS.

Ether extraction of a neutral, cleared solution will leave sugars, sugar alcohols, and organic acids in the aqueous phase; these can be acidified and subjected to continuous ether extraction. Organic acids will be removed in the ether phase with sugars and sugar alcohols remaining in the residual aqueous phase. The nonvolatile components may be separated and quantitated by various techniques (Dawes et al., 1971).

### 5. Biological Techniques

A biological assay may be necessary for detection or assay of some fermentation products (Roberts and Boyce, 1972). A sensitive test organism may be observed for effects of the compound on growth, or the concentration of the compound may be determined by enzymatic analysis. Use of auxotrophic test organisms that require specific compounds for growth has been discussed earlier. The amount of growth observed is directly proportional to the concentration of the required compound, as long as that compound is the limiting growth factor. The amount of growth can be measured turbidimetrically. Bacteria, yeast, and fungi have been used as test organisms in the analysis of amino acids, antibiotics, vitamins, etc.

Employing a solid medium inoculated with a test organism, diffusion assays are carried out in which zones of growth or inhibition will occur as a particular compound diffuses from a cup or disk into the surrounding agar. The diameter of the zone can be compared with a standard concentration (Cooper, 1963).

A turbidimetric end-point determination can be made on inhibitory substances by observing the inhibitory effect on growth of twofold serial dilutions of the agent. The minimal dilution at which growth of the test organism is still inhibited can be used to determine the relative amount of agent in the undiluted sample (Kavanaugh, 1963).

Metabolic responses to the presence of a fermentation product may be diverse, and various detection methods may be used. Among these are titration of acid produced, measurement of oxygen absorbed or $CO_2$ evolved, or measurement of a variety of enzyme activities. These techniques may require use of respirometers, radioisotopes, spectrophotometers, chromatography or other methods.

## B. Secondary Screening

### 1. General Considerations

Secondary screening qualitatively and quantitatively expands information gained from the primary screening program. Secondary screening represents a major commitment of time and resources to determine whether a given microorganism possesses the potential for use in a new or improved industrial fermentation. As such, secondary screening must serve to eliminate worthless microorganisms and fermentation procedures from further consideration as quickly and as inexpensively as possible.

Secondary screening involves many of the same techniques used in primary screening. Usually small flasks are utilized during early stages, small fermentors during intermediate stages, and pilot-plant fermentors for scaleup during later stages. While pH indicators or calcium carbonate may have been used in primary screening to reveal microorganisms that have produced acids, secondary screening may be used to determine precisely which acids are produced and the quantity of each acid accumulated. Effects of various environmental parameters on total and individual acids produced may be studied in order to develop a procedure that maximizes the production of a specific acid. Thus, much more elaborate experimentation is required for secondary than for primary screening.

## 2. Product Recovery

Standard techniques for biomass recovery include centrifugation and filtration. These and other methods for harvesting biomass are detailed by Thomson and Foster (1970). Recovery of a fermentation product may be a simple procedure, such as adjusting the acidity of the medium to bring about flocculation or precipitation; the desired product then can be recovered by centrifugation or filtration. In some cases the product can be precipitated as water-insoluble calcium or barium salts. Precipitation of proteins from saturated salt solutions is also a useful technique. Concentration and precipitation in an organic solvent, in which the product of interest is insoluble, is best for certain metabolites, i.e., addition of a concentrated chloroform solution of aflatoxin into hexane, whereupon the aflatoxin is precipitated.

Recovery techniques also may be based on vacuum distillation, steam distillation, and fractional distillations that are much used in the well-equipped fermentation laboratory.

Solvent extraction following pH adjustment is another fundamental technique. Adjustment of pH can suppress ionization of acids or bases in aqueous solutions. Many nonionized organic molecules may then be extracted into nonmiscible organic solvents. Evaporation of the solvent then concentrates the metabolites. Product recovery methods are ideally designed not only to harvest the desired product from the fermentation medium, but also to concentrate and purify it if possible.

## 3. Product Detection

A product detection method is designed to take advantage of known chemical, physical, or biological properties of the fermentation product of interest. Product accumulation can be monitored indirectly by taking advantage of any correlation that exists between metabolite formation and some other variable, such as medium pH, optical density, color, and titratable acidity. Growth can be monitored or measured by a large variety of methods that are discussed in detail by Humphrey (1976), Isenberg and MacLowry (1976), and Kubitschek (1969). Biomass estimations may involve use of centrifugation to determine packed cell volume or weight, nephelometry to determine cell numbers via standard curves, cell counting chambers of several types, and electronic instruments. Other methods are based on dry weight measurements, nitrogen determinations, radionuclide uptake, and on-stream monitoring of optical density (OD) using a micro-OD unit coupled to a digital display device or a recorder via fiber optics. Evaluation of growth by physical and chemical means is discussed by Mallette (1969). The pH and other environmental parameters can be measured on-stream by special electrodes connected to appropriate meters or other devices. Oxygen electrodes are discussed by Beechey (1972). Electrode measurements of carbon dioxide are reviewed by Nicholls (1972). Samples may be withdrawn periodically and the pH or other variables measured. Titratable acidity of samples is more informative than simple pH measurements and also can form the basis of a method of quantitative analysis. Volatile titratable acidity, due to acetic acid, can be determined separately from nonvolatile acidity due to other acids following steam distillation.

Visible and uv spectrophotometric monitoring methods are very sensitive and convenient where applicable. Compounds that absorb visible or uv radiation at specific energy levels can be monitored using appropriately equipped spectrophotometers. Fluorometry is a sensitive method that can be used to follow the accumulation of fluorescing metabolites such as riboflavin. Biologically active substances can be monitored using various biological assays, as previously discussed.

Spot testing of culture media is sometimes convenient for monitoring fermentation products (Feigl, 1977), although test sensitivity and small concentrations of the product of interest often prevent development of a useful procedure. For example, kojic acid forms a deep red color in the presence of alcoholic ferric chloride. This test is very sensitive and the intensity of the color produced can be used as a quantitative analysis for kojate. Obviously, sensitivity of such tests can be increased by extraction and concentration of the product. Most end products can be monitored by various combinations of chromatography and qualitative chemical analyses. Instrumental methods are useful for monitoring as well as for analyses (Gouw, 1972; Weber, 1973). It is a challenge to the fermentation technologist to select from among the many techniques available and to develop a monitoring system for a specific microbial end product.

## 4. General Analytical Methods

Qualitative analyses of fermentation products generally consist of isolation, concentration, purification, and identification or characterization of the product. Dawes et al. (1971) comprehensively discuss methods of analysis of fermentation products; Herbert et al. (1971) consider chemical analyses of microbial cells. Concentration of metabolites is often essential and can be accomplished by standard separation methods (Miller, 1975), such as precipitation and filtration or centrifugation, as previously discussed. Other techniques include extraction (Sutherland and Wilkinson, 1971) or removal of water through evaporation, or by specialized techniques such as freeze drying, dialysis, hollow fibers, absorption onto silica, charcoal, cellulose, etc. Liquid–liquid or soxhlet extractions are also common techniques for concentration and partial purification of various classes of compounds. After extraction, organic solvents are easily evaporated , and the residue containing the fermentation product(s) is further processed.

Identification of components in concentrated extracts is frequently accomplished following some form of chromatography (Perry et al., 1972). These include gas, paper, thin-layer (TLC), and column chromatography procedures, basic types of which are adsorption, ion-exchange, partition, and exclusion chromatography. High-performance liquid chromatography (Brown, 1973) is also a standard qualitative and quantitative tool in many analytical laboratories.

Thin-layer chromatography is a particularly useful and versatile tool for the fermentation specialist who must isolate and identify small quantities of individual products from a mixture of microbial metabolites (Stahl, 1969). Preparatory-scale TLC generally can quickly provide sufficient quantities of purified materials to achieve identification. Also, spot testing directly on TLC plates may reveal the class of organic compounds to which a particular metabolite belongs, thus facilitating other qualitative analyses (Krebs et al., 1969; Feigl, 1977). Many products can be easily and quickly isolated by preparatory-scale TLC and identified by comparing uv or visible absorption spectra with published spectra or with spectra of authentic samples of suspected metabolites. Infrared (ir) spectrophotometry also is convenient on quantities of material isolated from preparatory TLC plates. Published spectra and computer or manual searches of published ir indices are convenient for identifying a great many organic compounds. Mass spectrometry (MS) is another powerful tool for identification of organic compounds. Low- and high-resolution MS analyses may be purchased from commercial or university laboratories. Molecular weight and elemental analyses coupled with uv, visible, and ir spectra are frequently sufficient for positive identification of a fermentation product. Additional data, such as melting points, TLC $R_f$ values in different solvent systems, spot chemical tests, and fluorescence analyses usually provide more than sufficient data for identification except for previously unidentified metabolites.

## IV. MEDIA SELECTION

### A. General Considerations

The screening and selection of an organism for fermentation processes require selective and differential media for detection and isolation of the desired microorganism. The chemical composition of the medium to be used for the fermentation and the environmental conditions to be employed are very important. The medium must contain nutrients that supply carbon and energy, necessary growth factors such as vitamins and amino acids, inorganic salts, and any precursors essential to product formation. The choice of a good medium is of interest from the standpoint of utility and economy. Booth (1971) summarized information pertaining to culture media for fungi. Lapage et al. (1970) have compiled information on media for growth of microorganisms in general. Calam (1969), Bridson and Brecker (1970), and Pirt (1975) can be consulted for detailed information on most aspects of the culture of microorganisms in liquid media. Handling and storage of media and equipment are discussed by Elliott and Georgala (1969).

### B. General Purpose Media

#### 1. Solid or Semisolid Media

**a. Solidifying Agents.** Primary screening for potentially useful microorganisms requires the use of solid media in many instances (Codner, 1969). Some agent is added to a liquid medium, heated to the appropriate temperature for solubilizing the agent, mixed thoroughly, and allowed to solidify. The ideal solidifying agent is one which is not degraded by the microorganism(s) to be studied, so that growth is totally at the expense of added nutrients.

The most commonly used solidifying agent is agar, a polysaccharide not degraded by most bacteria. Normally a 1.5–2.0% concentration of the dehydrated agent is added to a liquid, heated to boiling, then allowed to solidify at approximately 42°C. For semisolid media, a concentration of 0.7% or less can be employed.

Other semisolid culture media contain a high concentration of bran or some cereal adjusted to the desired moisture content. The microorganisms are grown either in stationary culture on the surface of the medium in trays or while being agitated in rotating drums.

**b. Synthetic, Semisynthetic, and Complex Media.** Fermentation media can be classified as synthetic, semisynthetic, or complex. A synthetic medium is one in which the exact composition and quantity of ingredients is known; a semisynthetic medium is one which has very few unknown ingredients. A synthetic or semisynthetic medium is very useful in determining the specific capabilities of an organism. Such media can be manipulated by varying levels of known compounds to observe the effects of such alteration on growth and/or product formation. A medium solidified with agar technically is a semisynthetic medium, although the exact chemical nature of other ingredients is known, since agar is not a pure polymer as normally employed.

A complex medium is one in which specific components are unknown because of the addition of crude ingredients such as molasses, corn syrup, wheat bran, corn steep liquor, yeast extract, or protein digests. In later stages of screening, these crude ingredients may be valuable in promoting growth and/or product formation and would thus be important in large-scale operations. These complex media may stimulate growth and product formation because of unknown growth factors. Conversely, inhibitory effects may be observed when certain crude substrates are used due to feedback control mechanisms operating in the microorganism. Another disadvantage is that addition of proteins or other crude ingredients may increase foaming in liquid culture; this is not a problem when solid or semisolid media are used. Usually, addition of specific inorganic nutrients is unnecessary when complex media are used, since such media contain sufficient quantities of anions and cations to sustain growth. If a synthetic medium is to be used, these ions must be added. If toxic ions are present in some crude ingredient, chelating agents to bind these ions may be added to decrease inhibitory effects.

The utility of solid or semisolid media in primary screening is obvious; however, in large-scale operations submerged liquid culture is often preferred because it is more flexible and easier to control.

### 2. Liquid Media

**a. Complex and Semisynthetic Media.** During screening, the capabilities of an organism initially may be determined using a synthetic medium, the composition of which can be easily altered. In intermediate scale-up studies, submerged cultures with liquid media must be employed, since this probably will be the method of choice in a large-scale operation. Use of complex media will probably be cheaper. Other advantages and disadvantages have already been discussed. The complex medium must support sufficient growth and also promote enzyme or product formation. Crude sources of carbohydrates (starch wastes, cellulosic wastes, molasses, corn syrup, sulfite waste liquor), nitrogenous compounds (casein, gelatin, protein digests), and growth factors (yeast extract, corn steep liquor, etc.) may be added and the effects on growth and/or product formation observed. On a laboratory scale, liquid cultures can be incubated in shake flasks or small fermentors following sterilization of the media and inoculation with the organism. Solubility of the crude additives is not a great problem since extensive agitation probably will be employed. However, shallow layers of a liquid medium may be necessary for growing aerobic organisms in stationary culture.

**b. Synthetic Media.** As previously discussed, media for which the exact composition and quantities of added compounds are known are called synthetic media. Such media often are required for studies on growth, metabolism, and product formation. Liquid synthetic media are employed for substrate specificity and growth factor studies. Also, the effect of precursors of some product are best observed in a synthetic medium in which other interfering compounds have been eliminated. Results of any study are more often reproducible when a synthetic medium is used. A synthetic medium is more convenient when detecting intermediates or end products, since extracted samples will be cleaner than if a complex medium were used. Also, results from using radioisotopes in metabolic studies are more meaningful if synthetic media are employed.

### 3. Miscellaneous Problems

**a. pH and Buffers.** During fermentation processes, microorganisms synthesize metabolic products which may drastically change the pH of the media. Excessive production of acid end products, ammonia, $CO_2$, or amines may cause pH changes. Altered pH may be unfavorable for optimal growth and/or product formation; thus, it is often critical to maintain the medium at a certain pH. Agents that possess buffering capacity may be added to retard drastic pH changes. The addition to media of $CaCO_3$ to minimize effects of acid production has already been mentioned. Complex media with high levels of proteins may possess good buffering capacity near neutral pH. Various inorganic compounds, such as mono- and dihydrogen potassium phosphates, are commonly used for buffering, as are organic acids or amino acids.

**b. Oxidation–Reduction Potentials.** The oxygen tension required for a fermentation process determines the importance of adding components that help maintain a very low oxygen tension as in anaerobic fermentations. In aerobic fermentations, the oxidation–reduction potential presents little problem. However, large-scale operations often experience problems with insufficient aeration. In larger fermentors, adequate aeration can be achieved by forced air or mechanical agitation. Methods for supplying oxygen demands are detailed by Weinshank and Garver (1967).

**c. Precipitates and Other Problems.** Combinations of certain salts sometimes cause precipitations to occur which reduce the availability of an essential ion in growth media. This often can be alleviated by separately adding pure solutions of the presterilized salts following autoclaving of the salt-free medium. The pH of the medium may also be important in preventing formation of precipitates. In high-protein media, low pH may cause denaturation and precipitation of some proteins. Precipitation may not be a significant problem if the fermentation requires extensive agitation or aeration.

## V. EQUIPMENT

## A. Static Cultures

### 1. Culture Vessels

The choice of containers for growth of cultures, sterilization and storage of media, inoculum preparation, etc., varies with the type of organism and fermentation (aerobic or anaerobic). In primary screening, hundreds of samples may be screened using small volumes of a medium in tubes, bottles, or flasks. Significant cell mass in static culture is obtained with larger volumes incubated in 1- to 3-liter flasks, 5-gal carboys, or in other large vessels with shallow layers of medium to provide adequate oxygen exchange. Flat bottles incubated on their sides also are commonly used to provide a large surface area for growth. Special side arm flasks are used for gaseous substrates.

Fungal fermentations may be conducted by growing cultures as surface mats, although submerged fermentations that are agitated or aerated are more common. Intermediate-scale, static cultures generally are incubated in trays or shallow pans that have been inoculated with a spore suspension. Calam (1969) has reviewed the culture of microorganisms in liquid medium, including special techniques, i.e., anaerobic and hydrocarbon fermentations.

## 2. Incubators and Environmental Cabinets

Investigations may require static cultures using solid or semisolid substrates and precise control of atmospheric conditions. Where temperature is the important parameter, refrigerator-style incubators may be used. However, many studies dealing with fermented foods, tropical deterioration, mycotoxin production, etc., require maximal aeration coupled with maximal nutrient concentration and optimal moisture. Control of relative humidity, temperature, oxygen, carbon dioxide, etc., can best be obtained in environmental cabinets, which are capable of controlling temperatures from 5° to 90°C and relative humidity from 40 to 99%. Some environmental cabinets may provide a means for monitoring and regulating gaseous composition, i.e., $O_2$, $N_2$, and $CO_2$ content. Grain on trays in closed cabinets can be pasteurized or sterilized in place, or grain in flasks with cotton plugs can be sterilized separately in an autoclave. Recorders provide a permanent record of environmental parameters.

## B. Agitated Cultures

### 1. Shakers and Incubator Shakers

Bench-top and floor-model variable-speed gyratory or reciprocating incubator shakers with interchangeable platforms are available commercially for aerobic culture. Low-temperature incubators, water bath shakers, and incubators with light sources for photosynthesis investigations also can be purchased. Wrist action or twist action shakers may be more desirable in some studies. For incubation of large numbers of flasks, multitier shakers with changeable holders or platforms for various sizes of flasks also are available.

### 2. Rotating Drums

A rotating-drum method serves to minimize limiting effects of various diffusion gradients. Substrate moisture, nutrient concentration, and substrate–organism contact all can be maximized. Also, effects of aeration and agitation can be independently adjusted and investigated, as can the effects of temperature. This equipment consists of a horizontally mounted drum which rotates constantly during fermentation. Baffles on the inner wall facilitate agitation. Sterile air of predetermined relative humidity and temperature may be introduced into the system and effluent gases may be scrubbed of toxic particles. Also, back pressure may be regulated if desired. Various sizes of bottles and carboys also may be continuously rolled at varied speeds by roller platforms or cabinets to provide highly aerated thin films of media suitable for culturing many microbial and eukaryotic cells. This equipment may be relatively simple, open systems or may be integrated into controlled environmental cabinets.

The rotating-drum method has been used successfully for such processes as gluconic acid production, mold amylase production on wheat bran, and production of bacterial or eukaryotic cells. A chief limitation of the method is the difficulty of scale-up to pilot-plant or larger sized industrial fermentations.

## C. Submerged Cultures

### 1. Flasks and Carboys

If a fermentor is not available for growth of submerged cultures, large flasks or carboys can be used. Aeration is facilitated if flasks have baffles extending from the bottom or sides which increase aeration during shaking or stirring. Aeration is best achieved using spargers and filtered sterilized air delivered under pressure to produce fine bubbles. Brown (1970) discusses details of aeration in submerged culture. The size of holes in the sparger may be varied to avoid clogging by mycelia or bacterial clumps. Aeration via magnetic stirring or mechanical agitation of carboys also can be accomplished with proper equipment.

### 2. Fermentors

Fermentors up to about 16 liters typically are used during final stages of secondary screening and for subsequent investigations to determine whether satisfactory scale-up of fermentations is possible. This is essential because microorganisms do not always perform in submerged culture as they do in stationary culture. Yields of desired metabolites may be drastically reduced and in many cases the entire spectrum of metabolites is changed. Small to intermediate fermentors (Figs. 1 and 3) also are required to determine the effects of aeration, temperature, medium composition, pH, time, etc., on the progress and rate of a submerged culture fermentation and on yields of particular products. Precise effects of various factors influencing a fermentation are best determined using a chemostat (Fig. 2) as discussed below. Design and use of fermentors are discussed by Wang and Fewkes (1976). One-liter fermentors are mostly useful for preparation and adaptation of inoculum, whereas 4- to 28-liter fermentors (Fig. 1) or chemostats are convenient for fundamental inves-

**FIGURE 1.** A steam sterilizable in-place fermentor. (Courtesy New Brunswick Scientific Co. Inc., Edison, New Jersey.)

tigations. Intermediate scale-up utilizes larger fermentors generally of from 30 to 150 liters or more (Figs. 3 and 4). These intermediate and pilot-plant fermentors bridge the gap between the laboratory and the production-size fermentors.

### 3. Continuous Culture

Precise control of fermentations can be achieved in a chemostat (Fig. 2) using continuous culture techniques. Fresh medium is continuously added to a fermentor or carboy, while the culture is withdrawn at the same rate. The flow rate can be adjusted so as to maintain the desired steady state conditions within the fermentor. Theory, construction, and use of chemostats for continuous culture are discussed by Tempest (1970), Evans et al. (1970), and Pirt (1976). The effects of environmental conditions and growth-limiting parameters may be studied using this technique and the culture can be maintained in a constant or steady physiological state.

Most of this equipment is commercially available and, except for the larger pilot-plant equipment, is exhibited annually at the national meetings of the American Society for Microbiology and other scientific societies. These exhibits afford opportunities to become familiar with the state of the art in fermentation equipment.

## VI. INTERMEDIATE SCALE-UP

### A. General Considerations

Scale-up is the process of increasing the size of a fermentation system to arrive at a full-scale plant fermentation in the most efficient and economical manner. Problems will be encountered and resolved at the small-scale level so that only minor adjustments are necessary at the production-scale level. Obviously, mistakes made at the larger level are more expensive. Also, additional information is gathered during scale-up so that the role of each factor affecting the fermentation is clearly understood and more amenable to control. Factors to be considered during scale-up include those of the small fermentor stage plus new problems that must often be resolved by a chemical engineer rather than the fermentation technologist. Many of these problems are considered in other sections of this text.

### B. Media

Prior to scale-up of a fermentation one must determine the exact composition of the nutrient solution. This is usually determined in small-scale fermentors. Major changes should not be necessary during scale-up, although minor changes undoubtedly will be required. One or more alternatives in the basic medium composition should be developed in the event that ingredient shortages or other unforeseen economic factors might force changes in composition.

Convenience of medium preparation and prevention of contamination

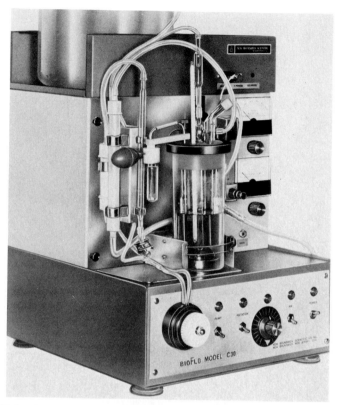

**FIGURE 2.** A continuous culture bench-top chemostat with a 350-ml working capacity. (Courtesy New Brunswick Scientific Co. Inc., Edison, New Jersey.)

are important economic factors to be considered during or before scale-up. Medium composition should be simple and should require the fewest manipulations and the least possible amount of preparation time. Steam sterilization is costly and troublesome and is not always necessary. Inoculation techniques are discussed by Meyrath and Suchanek (1972). The use of mass inoculum solves many problems during scale-up. A 1–5% inoculum of a fast growing organism is added insurance against contamination by unadapted wild strain microorganisms. In some cases, it is possible to recycle biomass one or more times reducing the time and expense of inoculum buildup and the total time and cost of the fermentation. Shortest fermentation lag times result when the inoculum is grown in and adapted to the same medium as that used for production. However, some fermentations require different media for growth and production, and it is not possible to use an adapted inoculum.

**FIGURE 3.** A small to intermediate-sized pilot-plant fermentor with control units (Courtesy Fermentation Design, Inc., Edison, New Jersey.)

## C. Equipment

Scale-up of a fermentation is usually accomplished via submerged culture, pilot-plant fermentors of from 30 to 150 liters capacity (Figs. 3 and 4). Pilot-plant fermentation units generally provide monitoring equipment and controls that allow sufficient control of fermentation parameters to maximize production under controlled and reproducible

**FIGURE 4.** A 150-liter capacity pilot-plant fermentor with control unit. (Courtesy Fermentation Design, Inc., Edison, New Jersey.)

conditions. Various aspects of the fundamentals of fermentor design are discussed by Blakebrough (1969) and in this volume, Chapters 11 and 14.

## D. Sterilization

Laboratory fermentation media and equipment are generally sterilized using steam under pressure to attain a temperature of 121°C for 15 minutes or more; other temperatures and times also may be used. Principles and practices for sterilization of laboratory apparatus and media are discussed in detail by Stumbo (1976). Nonpressurized steam (100°C and atmospheric pressure) may be used for from 2 minutes to several hours to sterilize equipment depending on the organisms to be killed.

Dry heat also may be used to kill most vegetative cells, i.e., 120°–180°C for 12–24 hours or more. Infrared heat has also been used successfully, as has steam combined with chemicals such as formaldehyde. Gases such as ethylene oxide can be used to sterilize equipment and media, although noxious or toxic fumes may be hazardous to personnel. Chemical sterilization with sodium or potassium metabisulfite or $SO_2$ gas is used in the wine and other industries. Sterilization by ultrafiltration is the method of choice where heat-sensitive nutrients must be separately sterilized.

It is not always essential to sterilize or even to pasteurize nutrient solutions. For example, where the pH of a medium can be adjusted to 1 or 2, and a mass inoculum of an adapted organism is used, chance of contamination is greatly reduced. Where expense is not prohibitive, various antibiotics also may be added to the medium for protection against contamination. Complex natural media such as agricultural commodities or foods may require higher temperatures or longer than customary time periods for sterilization. A technique often used for semisolid or solid media is to sterilize them two times during a 24-hour period.

### E. Monitoring and Control

Oxygen, pH, temperature, electromotive force, $CO_2$, and other environmental parameters may be monitored and controlled in small fermentors manually or by automatic devices, as discussed previously in this chapter. Instrumentation for fermentation units is discussed in detail in Chapter 14. Small aliquots of medium, headspace gas, etc. may also be withdrawn manually and analyzed, and adjustments may be made as desired by the operator. Although expensive monitors and controllers generally are necessary for chemostats, pilot-plant fermentors, and production-size fermentors, manual monitoring and control is often sufficient for small units. Foaming is a common problem in most fermentors, especially where the medium contains ingredients such as proteins, casein, peptone, beef extract, corn steep liquor, whey, and yeast extract. Foam control is discussed in detail by Bryant (1970). Foaming problems are best controlled automatically by use of appropriate probes, detectors, pumps, and tubing. They are often only required during the first 24–48 hours of a fermentation. A variety of silicone-based control agents is available commercially. Also, various combinations of natural vegetable oils or long-chain fatty acids can be used as foam control agents. Occasionally foam control agents are toxic to a microorganism, especially if used too liberally.

## F. Product Recovery

Problems of harvesting the biomass and clarifying medium generally become more numerous and difficult to deal with on scale-up except where special equipment is available. Biomass is commonly harvested by centrifugation, by spraying onto heated drums, or by filtering with use of filter aids such as Hyflo Supercel, kieselguhr, or various forms of silica, diatomaceous earth, asbestos, or other materials. Filter presses using cloth or asbestos–paper mixtures are widely used. Newer technology also makes use of stacks of membrane filters or cartridges. Large-scale continuous and noncontinuous centrifuges are generally used for clarifying cultures of pathogenic bacteria. A disadvantage of some filter systems is that the biomass cannot subsequently be separated from filter aid materials. The centrifuge does not have this disadvantage. Various techniques for biomass recovery are discussed in detail by Thomson and Foster (1970), Mulvany (1969), and Belter (this volume, Chapter 16).

Scale-up of product recovery methods is generally a matter of selection of the most suitable alternative from among those investigated during development of the laboratory fermentation. Plant engineers must be consulted at this stage of development. Important ancillary factors to be considered are those pertaining to economics, efficiency, safety, and convenience. Also, methods for safe disposal of waste materials and recovery of valuable by-products should be considered. Again, attention should be paid to the general principle that at each stage of scale-up, only minor adjustments should be necessary. All aspects of a fermentation should be planned in advance with the final plant process in mind in order to minimize the problems that invariably arise on scale-up.

## REFERENCES

Beckhorn, E. J. (1967). In "Microbial Technology" (H. J. Peppler, ed.), pp. 366–380. Van Nostrand-Reinhold, Princeton, New Jersey.
Beechey, R. B. (1972). Methods Microbiol. **6B**, 25–53.
Blakebrough, N. (1969). Methods Microbiol. **1**, 473–504.
Booth, C. (1971). Methods Microbiol. **4**, 1–47.
Bridson, E. Y., and Brecker, A. (1970). Methods Microbiol. **3A**, 229–295.
Brown, D. E. (1970). Methods Microbiol. **2**, 125–174.
Brown, P. R. (1973). "High Pressure Liquid Chromatography: Biochemical and Biomedical Applications." Academic Press, New York.
Bryant, J. (1970). Methods Microbiol. **2**, 188–212.
Calam, C. T. (1969). Methods Microbiol. **1**, 255–326.
Campbell, R. C. (1974). "Statistics for Biologists." Cambridge Univ. Press, London and New York.
Casida, L. E., Jr. (1968). "Industrial Microbiology." Wiley, New York.

Codner, R. C. (1969). *Methods Microbiol.* **1,** 427–454.

Cooper, K. E. (1963). *In* "Analytical Microbiology" (F. Kavanaugh, ed.), pp. 1–86. Academic Press, New York.

Dawes, E. A., McGill, D. J., and Midgley, M. (1971). *Methods Microbiol.* **6A,** 53–215.

Dulaney, E. L. (1967). *In* "Microbial Technology" (H. J. Peppler, ed.) pp. 308–343. Von Nostrand-Reinhold, Princeton, New Jersey.

Elliott, E. C., and Georgala, D. L. (1969). *Methods Microbiol.* **1,** 1–20.

Evans, C. G. T., Herbert, D., and Tempest, D. W. (1970). *Methods Microbiol.* **2,** 277–327.

Fantini, A. A. (1976). *In* "Methods in Enzymology" (J. H. Hash, ed.), Vol. 43, pp. 24–41. Academic Press, New York.

Feigl, F. (1977). "Spot Tests in Organic Analysis." Elsevier, Amsterdam.

Gouw, T. H. (1972). "Guide to Modern Methods of Instrumental Analysis." Wiley, New York.

Hanson, A. M. (1967). *In* "Microbial Technology" (H. J. Peppler, ed.), pp. 222–250. Von Nostrand-Reinhold, Princeton, New Jersey.

Hash, J. H., ed. (1976). "Methods in Enzymology," Vol. 43. Academic Press, New York.

Herbert, D., Phipps, P. J., and Strange, R. E. (1971). *Methods Microbiol.* **5B,** 210–344.

Hockenhull, D. J. D. (1977). *Adv. Appl. Microbiol.* **21,** 125–159.

Hopwood, D. A. (1970). *Methods Microbiol.* **3A,** 363–433.

Humphrey, A. E. (1976). *Dev. Ind. Microbiol.* **18,** 58–70.

Isenberg, H. D., and MacLowry, J. D. (1976). *Annu. Rev. Microbiol.* **30,** 483–505.

Kavanaugh, F. ed. (1963). "Analytical Microbiology." Academic Press, New York.

Krebs, K. G., Heusser, D., and Wimmer, H. (1969). *In* "Thin-Layer Chromatography" (E. Stahl, ed.), 2nd ed., pp. 854–909. Springer-Verlag, Berlin and New York.

Kubitschek, H. E. (1969). *Methods Microbiol.* **1,** 594–610.

Lapage, S. P., Shelton, J. E., and Mitchell, T. G. (1970). *Methods Microbiol.* **3A,** 1–33.

Lederberg, J., and Lederberg, E. M. (1952). *J. Bacteriol.* **63,** 399–406.

Mallette, M. F. (1969). *Methods Microbiol.* **1,** 522–566.

Meyrath, J., and Suchanek, G. (1972). *Methods Microbiol.* **7B,** 159–209.

Miller, B. M., and Litsky, W. (1976). "Industrial Microbiology." McGraw-Hill, New York.

Miller, J. M. (1975). "Separation Methods in Chemical Analysis." Wiley, New York.

Mulvany, J. G. (1969). *Methods Microbiol.* **1,** 205–253.

Nicholls, D. G. (1972). *Methods Microbiol.* **6B,** 55–63.

Perry, S. G., Amos, R., and Brewer, P. I. (1972). "Practical Liquid Chromatography." Plenum, New York.

Pirt, S. J. (1975). "Principles of Microbe and Cell Cultivation." Wiley, New York.

Pirt, S. J. (1976). *Dev. Ind. Microbiol.* **17,** 1–10.

Roberts, M., and Boyce, C. B. C. (1972). *Methods Microbiol.* **7A,** 153–190.

Skeggs, H. R. (1976). *In* "Industrial Microbiology" (B. M. Miller and W. Litsky, eds.), pp. 47–59. McGraw-Hill, New York.

Spooner, D. F., and Sykes, G. (1972). *Methods Microbiol.* **7B,** 211–276.

Stahl, E., ed. (1969). "Thin-Layer Chromatography," 2nd ed. Springer-Verlag, Berlin and New York.

Stanier, R. Y., Adelberg, E. A., and Ingraham, J. (1976). "The Microbial World." Prentice-Hall, Englewood Cliffs, New Jersey.

Stumbo, C. R. (1976). *In* "Industrial Microbiology" (B. M. Miller and W. Litsky, eds.), pp. 412–450. McGraw-Hill, New York.

Sutherland, I. W., and Wilkinson, J. F. (1971). *Methods Microbiol.* **5B,** 345–383.

Tannenbaum, S. R., and Wang, D. I. C. (1975). "Single Cell Protein II." MIT Press, Cambridge, Massachusetts.

Tempest, D. W. (1970). *Methods Microbiol.* **2,** 260–276.

Thomson, R. O., and Foster, W. H. (1970). *Methods Microbiol.* **2,** 377–405.

Underkofler, L. A. (1976). *In* "Industrial Microbiology" (B. M. Miller and W. Litsky, eds.), pp. 128–164. McGraw-Hill, New York.

Wang, D. I. C., and Fewkes, R. C. J. (1976). *Dev. Ind. Microbiol.* **18,** 39–56.

Weber, D. J. (1973). "Principles and Application of Instrumentation in The Biological Sciences." Brigham Young Univ. Printing Service, Provo, Utah.

Weinshank, D. J., and Garver, J. D. (1967). *In* "Microbial Technology" (H. J. Peppler, ed.), pp. 417–449. Van Nostrand-Reinhold, Princeton, New Jersey.

# Chapter 14

# Instrumentation of Fermentation Systems

L. P. TANNEN
L. K. NYIRI

## I. CONCEPTS OF BIOLOGICAL PROCESS ANALYSIS AND CONTROL

### A. Introduction

The objectives of applying instrumentation to fermentors and other types of biochemical reactors are: (1) to analyze the process status and

MICROBIAL TECHNOLOGY, 2nd ed., VOL.II
Copyright © 1979 by Academic Press, Inc.
All rights of reproduction in any form reserved. ISBN 0-12-551502-2

(2) to establish an optimum set of environmental conditions in order to maximize specific cellular or enzymatic activities. For the most part the information which follows directs its attention to pilot- and semiworks-scale equipment requirements but the data can be equally applied to large commerical production equipment. In many instances the problems noted for small-scale equipment would not be a significant factor at the larger scale. Our rationale in concentrating on the smaller scale in this chapter is due to the fact that we are still, unfortunately, in an embryonic stage of instrumentation development in fermentation and it is on the pilot scale that much of the systems work must be tested and proven before commercial application can be seriously considered.

From an electronic systems engineering point of view, fermentation does not differ significantly from typical chemical processes (Considine, 1957). The difference between chemical and biological process analysis and control is *what* and *how* to measure in order to obtain an adequate picture of the biological process and its dynamics on the basis of which a process-control scheme can be implemented. In this review, biological system is defined as any *in vitro* culture of microbial, plant, or animal cells. Because of the characteristics of biological systems, we are faced with problems where very complex processes are analyzed and controlled by electronic means which were developed for relatively simple (e.g., chemical) processes. This general discrepancy, despite developments over the past decade, has not yet been satisfactorily resolved for the fermentation industry.

## B. Terminology

The relationship between process status and controlled variables makes the definition of certain terms necessary. There have been attempts to adapt standard terminologies generally used in control engineering to biochemical engineering. Constantinides *et al.* (1970a) used the expression *state variable* by which the state of the system is described and the variable is not under direct control. Further, *control variables* are those variables which can be directly controlled by the experimenter and which influence the way the system changes from its initial state to that at any later time, e.g., temperature, DO (dissolved oxygen) concentration. Nyiri *et al.* (1975a) explored this concept further by defining state variables for fermentation processes as *process status indicators* incorporating all off-line or on-line obtainable signals which directly or indirectly describe the conditions of the biological process. Process status indicators are then sugar concentration, cell mass, oxygen uptake rate, respiratory quotient, etc., values of which either can be directly measured in the broth (e.g., cell mass with an on-line turbidometer), derived from combination of available on-line signals (e.g., combining off-gas $O_2$ and $CO_2$ measurements to obtain respiratory quotient), or determined by wet chemical analysis of samples. The expression for control (or controlled) variable remains unchanged with the understanding that these control variables individually, or in combination, create the environmental conditions in which the cells express a certain metabolic activity. Therefore, in this context, the terms *control variable* and *environmental variable* will be synonymous.

## II. TYPES OF INSTRUMENTATION USED IN BIOLOGICAL PROCESS ANALYSIS

Analysis of conditions in a biological process is performed by two general methods; namely, off-line operating instruments and on-line operating instruments.

### A. Off-Line Operating Instruments

Off-line operating instruments are not connected directly to the fermentor or biochemical processor; consequently, sampling, sample preparation or treatment, and sample analysis are performed and the analysis results provide information on the process status. However, the elapsed time between the sampling and the assay results can be significant. Information obtained by this method is for the most part useful for postexperimental evaluation of the process but cannot normally be used for process-control purposes. Current examples of off-line operating instruments include spectrophotometers, mass spectrometers, gas chromatographs, enzyme activity analyzers, and electron microscopes, as well as specialized wet chemical and microbiological procedures for product analyses (e.g., reducing sugar or dry weight measurements). Recently, there has been a trend toward connecting certain of these instruments to fermentor units by means of automatic samplers and in-line sample treatment devices, using electronic conversion of the analyzer signals for data logging. A typical example of such an arrangement is the application of a Technicon Auto Analyzer for measuring in-process sugar levels where the elapsed time between sampling and results (approximately 20 minutes) may allow the data to be used for process-control action in certain biological processes.

### B. On-Line Operating Instruments

On-line operating instruments are directly connected to the fermentor or biochemical processor unit and have the capability to sense a specific variable which directly or indirectly reflects a process condition and/or the performance characteristics of a controller. It is noted here that there are relatively few on-line direct sensing devices which give indication of the process status at the cellular level (e.g., NAD–NADH$_2$ measurement belongs to this category). The predominant types of on-line operating instruments indirectly reflect a process condition in the whole broth culture; e.g., a dissolved oxygen sensor connected to an analyzer indicates the dissolved oxygen level in the broth, which is a consequence of the metabolic activity of the culture and the oxygen transfer capacity of the fermentor system. Most of these instruments therefore are reflecting the performance characteristics of a controller which interacts with a process variable. A dissolved oxygen sensor connected to a dissolved oxygen controller reflects the performance characteristics of the method by which the transfer of oxygen from the gas phase into the liquid phase is performed while the culture consumes the available oxygen.

### III. MEASUREMENT AND CONTROL OF ENVIRONMENTAL VARIABLES

Biological processes are essentially under the control of two major factors: genetics and environmental conditions. These two factors do interact, however, in the everyday practice of fermentation technology during *in vitro* culture of a given mutant where only the environmental conditions are controlled. The environmental variables include temperature, hydrogen ion concentration (pH), partial pressure of dissolved oxygen (DO), and concentration and proportion of certain nutrients (carbon and nitrogen sources as well as salts). The composite condition of these individual variables creates an environment which influences the cells' metabolic activity. A given value of an environmental variable is usually maintained by conventional controllers such as are used in typical chemical engineering practice. An example of such a control scheme is shown in Fig. 1, which describes the control of temperature in a reactor. This is considered the most elementary system with which essentially every fermentor is equipped. The example shows that, in fact, the fermentor's temperature is controlled by measuring the temperature (which is the result of metabolic heat production and removal) via an electronic signal-producing thermometer, by comparing the deviation between the actual and desired temperature, and by manipulating a control valve, the position of which controls the flow rate of cooling water to a cooling surface. In a complex situation more than one manipulated variable performs the control of a process variable. This is exemplified in Fig. 2, where the schematics of control of dissolved oxygen are demon-

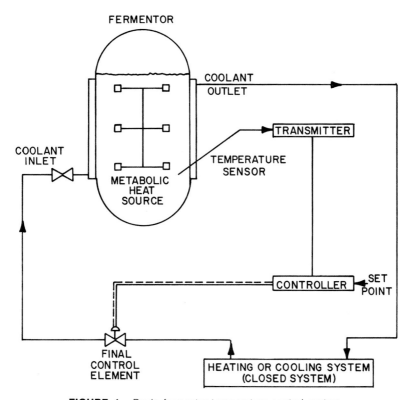

**FIGURE 1.** Basic fermentor temperature-control system.

strated. In this case the controller, based on a DO level measurement and comparison with a set point, governs both the inlet air ($O_2$) flow rate control valve and the impeller motor speed controller in an interactive way to maintain a dissolved oxygen level under dynamic conditions where the oxygen consumption and oxygen input rates must be in balance. A close examination of the process, however, reveals that, in fact, the air ($O_2$ gas) flow rate and the agitation speed interactively influence the oxygen mass transfer coefficient ($K_L a$), which results in the dissolution of gaseous $O_2$ into the culture broth (Aiba *et al.*, 1973).

In view of these facts, fermentation or biochemical process control can be described as adaptation of environmental conditions to genetically controlled physiological conditions to influence cellular or enzymatic activities. It is a series of events which ultimately depends on the proper on-line analysis of the existing process status and the manipulation of final control elements (valves, meters, pumps).

During the culture of cells there are certain process conditions which

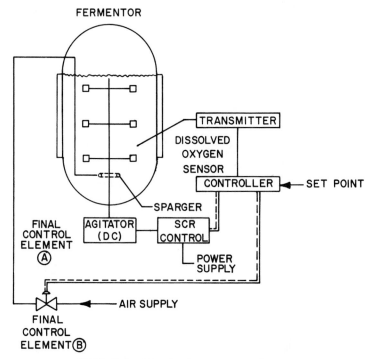

**FIGURE 2.**   Dissolved oxygen-control system.

undergo change. These conditons can be grouped into four major categories: (1) physical conditions, (2) physicochemical conditions, (3) cellular physiology conditions, and (4) cell biochemistry conditions. All of them require analysis and control in order to achieve optimum productivity of a cell culture. Examples of some typical systems for measurement and control of functions within these categories will now be presented.

## A. Physical Conditions

### 1. Volume Measurement

The first important physical condition of a liquid culture is its volume or working volume. Information on liquid volume (or level) is essential in liquid flow control (e.g., filling up vessels, continuous culture feeds, or introduction of liquid nutrients) which ultimately affect metabolic activity. There are two systems which are generally used for on-line volume determinations: (1) liquid level sensors, and (2) weight or mass measurement devices.

For sensing of liquid level experience indicates fairly reliable performance can be obtained in small-scale equipment with capacitance probes. As an example, the principle of operation of the Robertshaw capacitance level probe is described.

The sensor uses an electronic circuit which compares the rapid cyclical rate of charge of two capacitor systems, one reference capacitance internal to the electronic system, and the other, the external sensor capacitance probe itself, located in the fermentor. When the liquid level in the fermentor changes, the sensor probes capacitance and hence rate of charge changes. This rate difference is measured and provides an output signal proportional to the change in liquid level. Experience has shown that capacitance probe systems will provide level measurement with an accuracy of $\pm 2-4\%$ (full scale) depending on the liquid mixing surface turbulence and foaming conditions of the fermentor.

For larger scale systems (5000 liters and above) the $\Delta P$ measurement method is usually used, where a flush mounted diaphragm pressure cell is located in the base of the vessel for liquid height, and hence volume, measurement. Among suppliers of volume measurement systems are Ranco Controls, Drexelbrook, and Robertshaw Controls.

### 2. Weight/Mass Measurement

Weight or mass of the liquid can be determined by scales of various types. In this method the vessel is suspended on a scale and the combined weight of the vessel and liquid is electronically measured, usually by strain-type gauges. Several methods for electronically or mechanically eliminating the tare weight of the system can be used. Advantages of this method are that (a) the system is external to the vessel and need not be steam sterilizable, (b) its calibration is relatively simple, and (c) its accuracy is within $\pm 1-2\%$ (based on field experience). Among the disadvantages are the following: (a) Special engineering is needed for suspension of the vessel on the scale; consequently, this approach cannot be readily applied to existing installations. (b) There is a lower volume limit (approximately 250 liters) below which measurement accuracy falls off rapidly as the dynamic load becomes less than the tare weight. (c) Certain systems are sensitive to vibration and ambient temperature changes and compensation must be allowed for this in design. (d) The system is more costly than the simpler level probe method by a factor of four to five times. Suppliers of strain gauge systems are BLH Electronics, Transducers, Inc., and Micro-Strain, Inc.

### 3. Liquid Metering

Liquid metering is an important physical process variable since it serves as the basis for material balance analysis when fresh nutrient feed is added during the culture. There are several devices from industrial and biological waste treatment applications which have been adapted to fermentation processes. Solomons (1969) treated this subject previously and unfortunately there have been no major developments reported during the past 10 years.

Difficulties in measurement and control of liquid flows to fermentation units stem from the following factors: (1) There is significant difference in sensing flow values on laboratory, (small) pilot, and semicommercial scales. Consequently, accuracy of flow measurement is scale-dependent. (2) There are significant differences between handling (pumping) solutes of low or high viscosity and those with or without suspended solid particles.

Neither the problem of scale nor the rheological dependency of the flowmeter measurement and subsequent control has been systematically studied; consequently, no conclusions can be drawn. In Table I some representative examples of available liquid flow measurement devices are shown with their claimed accuracy. It should be cautioned that attempts to use diaphragm-type steam sterilizable metering pumps for very low liquid flow rates (milliliters to several liters per hour) can be highly inaccurate. Care should be exercised when such pumps are used and sufficient performance testing carried out with the fluid to be pumped to develop a measure of true accuracy.

### 4. Gas Flow Measurement

Introduction of air into aerobic biological processes serves essentially two purposes: (1) It supplies the necessary oxygen (Aiba et al., 1973); and (2) It assures the removal of carbon dioxide (ventilation) (Lengyel and Nyiri, 1966) and other volatile gaseous products of metabolism.

In certain cases introduction of pure oxygen (generally up to 40% of total gas volume) or $CO_2$ (generally up to 5% of total gas volume) (Kruse and Patterson, 1973), is necessary, which requires specially calibrated measuring devices.

On-line monitoring of gas flow can be performed by means of a simple variable area meter. This is usually a glass tube with a tapered bore resulting in a variable cross-sectional area and containing a float (a small sphere or a hollow thimble). This is the simplest system and is generally used for manual airflow control. The flowmeter device is equipped with a throttling valve to adjust the gasflow rate. Field experience indicate ±2% (full-scale) accuracy, depending on the inlet gas pressure. Such gas flowmeters are available in unit ranges from as low as 10 $cm^3$/min and with turndown ratios of 10 : 1. These simpler units do not provide any outputs for use with data logging or electronic computa-

**TABLE I.** Liquid Metering Systems

| Manufacturer | Model | Type | Range | Accuracy[a] |
|---|---|---|---|---|
| A. Flow monitors (meters) | | | | |
| Brooks | Series 22 | Propeller | 0–17,610 gal/min | ± 2% |
| Fisher & Porter | 10A Series | Rotameter | 0.1 cm³/min– 4400 gal/min | ± 2% |
| Foxboro | Series 81 and 82 | Turbine | Variable | ± 0.2% |
| Foxboro | Series 2000 | Electro- magnetic | 0.07– 115,000 gal/min | ± 1% |
| B. Metering pumps (steam sterilizable in-place) | | | | |
| Interpace Corp. | Lapp | Diaphragm | 0.5–5500 liter/hour | ± 1% of set point |
| Chemcon, Inc. | Series A52 | Diaphragm | 0–570 liter/hour | ± 1.5% of set point |
| | Series 1100 | Diaphragm | 0–1700 liter/hour | ± 0.5% of set point |
| Milton Roy Co. | Milroyal Diaphragm simplex | Diaphragm | 0.7–900 liter/hour | ± 1% of set point |
| | Milroyal Diaphragm duplex | Diaphragm | 13.5–4500 liter/hour | ± 1% of set point |

[a] Manufacturer's published claim over 10–100% scale range.

tion systems. A large number of companies supply such devices, including Fisher-Porter, Brooks, and Foxboro.

Gas flowmeters with electrical transducers which can provide output signals for use in data logging and control are available. These include devices such as (a) magnetic flow rate sensing elements, (b) thermal mass flow rate sensing elements, and (c) laminar flow rate sensing devices. A typical thermal mass gas flowmeter is shown in Fig. 3.

The thermal mass flowmeter has a ±1% full-scale (FS) accuracy and is based on the principal of detecting a temperature differential across a heater device placed in the path of the gas flow. Two temperature sensors, one upstream and one downstream of the heat source are made part of a bridge circuit which becomes unbalanced when there is flow past the heat source. This type of device is most suited to flow rates of under 500 liters/minute. The output of the transducer is in the range of 0–5 V.

Another type of gas flow metering device which can provide similar signal outputs to the thermal mass type are laminar gas flow elements (Fig. 3). Laminar flow measurement devices operate on capillary flow principles. The sensors flow versus differential pressure curves approach linearity and this makes them applicable over wide ranges of flow measurement. Mechanically, a matrix device inserted in the line breaks the flow into multicapillary paths. As the gas flows through these "capillaries" the friction created causes a pressure drop which is measured across the entire matrix device. Design of most devices is based on a 4- to 8-inch permanent water differential pressure loss. Units are available to measure flow over the range 0.000092 to 2250 Standard cubic feet per minute (SCFM) and are accurate to within ±0.5% error, FS. Sources of thermal mass and laminar flow measuring devices are Brooks Instrument, Fisher-Porter, and Meriam Instruments.

### 5. Temperature Measurement

Heat energy is considered to be the only environmental variable which freely penetrates the cell wall and directly influences intracellular metabolism. Therefore, temperature measurement (and control) is considered to be the most important direct process variable. Because of the hundreds of available variations, proper temperature sensor selection for biological processes depends on the specific use of the signal.

Typical sensor devices which can be used for fermentation application are described below.

**a. Platinum Resistance Sensor (RTD).** This device has a range capability of −150°C to 600°C with a linearity error of approximately 1% over the range 0°–300°C. Because current is passed through this type of

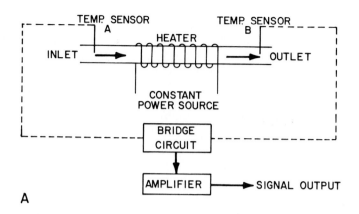

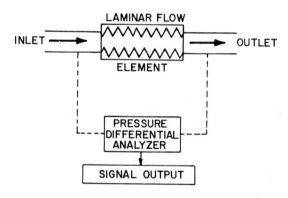

**FIGURE 3.** Gas flowmeter devices. (a) Thermal mass flow device; (b) laminar flow device.

device (creating a heat "load" at the probe) it should not be employed for nonagitated systems, e.g., typically anaerobic systems. The self-heating error in turbulent systems usually is less than a ±0.1°C error. The output signal of a typical platinum resistance sensor is reported as 2 mV/°C at 200°C.

**b. Thermistors.** These semiconductor devices are relatively inexpensive and quite sensitive to temperature changes. However, their output (resistance–temperature relationship) is highly nonlinear over large temperature ranges and numerous calibration points are required. Because of their large outputs they can be employed for temperature control systems over very narrow ranges, such as those found in typical fermentaton applications (25°–45°C) with extremely good accuracy. Again, care must be taken with the use of thermistors to calibrate them against a standard mercury thermometer over the desired temperature range of application.

**c. Thermocouples.** Thermocouple devices have been used in a sufficient number of situations so that a detailed description of their operation is felt unnecessary. Their most serious drawback, however, is the need to provide a "cold junction" reference point in the system (either physical or electronic) and hence the system is a two-point measurement arrangement rather than single point. Also, the circuitry required to monitor the thermocouple output is more elaborate (and costly) than that required for either of the other two types of sensors; hence, it is not always selected.

**d. Filled Bulbs.** Both gas- and liquid-filled bulbs are available for direct temperature sensing and cover most equipment with suitable capillary cable connections. Where the distance between sensor and analyzer or readout device is substantial (greater than 30 m), the use of these capillary connected bulbs is not usually recommended as their response time suffers and the long-term mechanical problems of installation and maintenance can present chronic headaches for their user. However, because of their fundamental simplicity and low cost they still are the predominant sensor found in many chemical process systems.

As might be expected, accuracy and cost of the thermometers are directly proportional. However, since the accuracy of temperature controllers ordinarily do not surpass ±0.1°C there is no reason to select thermometers with better resolution. Good operating practice is to use mercury bulb thermometers located in a vessel thermowell along with the selected sensor device for control purposes.

For computation of process heat balances, additional measurement of the ambient temperature and the inlet and outlet coolant temperature are also required.

Sources of vendors who can provide various electronic-type temperature sensors are Gulton Industries, LFE Corp.; and Rosemount, Inc.

### 6. Pressure Measurement

Static head pressure inside the bioreactor influences the partial pressure conditions and, hence, solubility of gases ($O_2$, $CO_2$, in particular) and volatile compounds. A positive head pressure (about 1.2 atm absolute) can also contribute to the maintenance of sterility inside bioreactors operating under aseptic conditions. In addition it is important to measure internal vessel pressure conditions during sterilization as well as the pressure condition of the main supply services (steam, water, and air). Depending on the mode of application, there are two pressure measuring systems generally accepted for biochemical reactors: (1) pressure devices without electrical signal outputs, and (2) pressure transducers.

In their applications there are several points to note.

1. At nonsterile points, generally Bourdon gauges are used. For steam pressure measurement a gauge protector is usually employed.

2. For determination of vessel pressure under aseptic conditions a diaphragm is placed between the sterile environment and the gauge.

3. For fermentor head pressure measurement, which also provides an electronic signal useful for on-line analysis as well as for automatic control of a given head pressure when gas flow rates are changed, pressure transducers are required.

An example of a typical pressure transducer with an electronic output is shown in Fig. 4. Pressure entering the port moves the diaphragm, causing a frictionless floating cone to be repositioned within the magnetic field of a linear variable differential transformer (LVDT). Movement of the transformer cone alters the inductive coupling between the primary and the secondary coils, resulting in a voltage change proportional to the displacement caused by variations in the input pressure. Typical specification for such a device are (a) a range of 3–15 to 0–3000 psi, (b) a linearity of $\pm 0.25\%$, (c) a stability of $0.01\%/°F$ (from $+10°F$ to $140°F$), and (d) a response time of 100 msec. One source of pressure transducer devices is Schaevitz Engineering.

### 7. Agitation Speed Measurement

The measurement of agitation speed can be by either direct or indirect monitoring devices. Indirect monitoring devices include both magnetic and optical methods. Magnetic sensors are noncontact transducers whose output voltage reflects the rate of change that is taking place in their magnetic field. The sensors are easily mountable adjacent to the external drive shaft (in a nonferrous housing) and opposite special gear teeth devices mounted directly to the drive shaft. These devices come in both analog and digital outputs. A typical supplier of such magnetic sensor systems is Electro Corporation, Inc.

Direct gear or belt-driven tachometers are also available for monitoring of agitator speed but are only advisable for use at low rpm (<250). For higher speed monitoring, indirect magnetic devices are preferred. Note that output meters generally read in revolution per minute (rpm). Caution must be exercised in comparing agitation speed literature data from different experimenters in view of the fact that the diameter of the

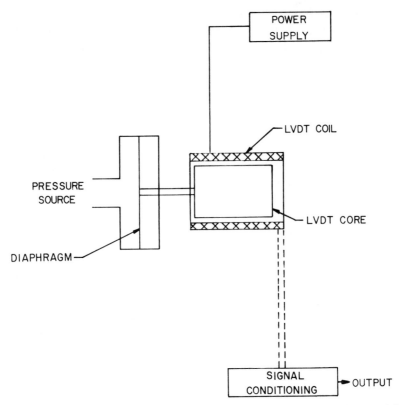

**FIGURE 4.** Schematic of a diaphragm pressure transducer. LVDT, linear variable differential transformer.

impeller will determine the absolute rotation speed of the tip of the impeller, which in turn defines the liquid shear rate and the liquid turnover rate in the culture vessel (Metzner and Taylor, 1962). In order to develop scale-up correlations between agitation and energy uptake by this method, it is necessary to compare tipspeed (*TS*) values derived from Eq. (1):

$$TS = ND_i \qquad\qquad (1)$$

where $N$ = rotation speed of impeller (per minute)
$D_i$ = diameter of impeller (linear units: ft, cm)

Accuracy of agitation measurement devices ranges between +2% (FS).

### 8. Motor Power Uptake

Total energy consumed in operation of the agitator is generally measured at the armature of the drive motor in units of watts consumed per unit of time. For this measurement a Hall Effect wattmeter can be used. The operating principle of such a device, which measures power by use of the Hall Effect, first discovered by E. H. Hall in 1879, is that a conducting material placed in a magnetic field perpendicular to the direction of current flow produces a voltage in a direction perpendicular to both the initial current direction and the magnetic field.

In the early 1950s materials became available to utilize this principle and wattmeters employing this principle provide very reliable power measurements for pilot-scale fermentors.

Accuracy of a Hall Effect wattmeter is $\pm 1\%$ FS. The information obtained, however, represents only the total motor power uptake and does not necessarily reflect the actual amount of energy which is transferred into the liquid via the impellers. The design and scale of the vessel agitator system (i.e., size of shaft, bearings, seals, gears, pulleys, and/or belts) will significantly influence the loss of power between the motor armature and the impeller. A differential measurement is needed in which both the motor power uptake and the definition of power input into the liquid by the impeller can be determined. One source of Hall Effect power measuring devices is Ohio Semitronics Co.

## 9. Power Input to the Fluid

During the past 10 years there have been reports involving pilot-plant scale fermentors where both dynamometers and strain gauges have been used for the determination of the actual power input to the fluid. Various types of dynamometers and strain gauges were reviewed by Solomons (1969a) and Aiba et al. (1973). A detector device utilizing four strain gauges applied directly to the agitator shaft of a 270-liter fermentor was reported by Harmes (1972). The author discusses the degree of care which must be exercised in calibration, assembly, and operation of such a device in order to obtain acceptable process data. Based on such measurements, data were obtained which showed the correlation between the agitation speed and the efficiency of power input to the fluid ($\phi$) ($\phi$ = power measured at the shaft divided by the power measured at the motor armature) in a 270-liter fermentor with 200 liters of working volume. The reported value of $\phi$ increased with increasing agitation speed. At maximum rpm, almost 30% of the energy supplied to the motor was still lost between the motor and the impeller shaft for the agitator system design used with this unit.

A compromise solution to the problem of obtaining power measurement of the energy imparted to the fermentation fluid is the use of an in-line torque measuring device located external to the vessel. The losses due to bearings and seals will be included in this type of measurement, but by suitable no-load testing it can be effectively eliminated from the reading and a reasonable approximation of power imparted to the fluid can be made. One source of such a torque instrument is Vibrac Corporation.

## B. Physicochemical Conditions

The physicochemistry of the culture is related to the culture broth rheology which is a continuously changing condition during the process (Solomons, 1969a; Tuffile and Pinho, 1970). Despite the recognition of the importance of culture broth rheology (Aiba et al., 1973) no systematic work has been performed with respect to development of an on-line measuring device by which the rheology characteristics of the culture broth can be identified. An instrument marketed by Automation Products, Inc. (Dynatrol Viscosity Control Device) might possibly be applied to the fermentation broth rheology if the problem of steam sterilization can be overcome.

Attempts to determine, on-line, the liquid rheology using computed correlations between agitation speed (shear rate) and power uptake (shear stress) in liquids of increasing viscosity (Charles and Toth, 1974) served as an example of utilizing the entire fermentor system as a viscometer. However, this study was confined only to xanthan solutions. Further systematic research is required in this direction to develop these early results into a practical engineering tool.

## C. Cellular Physiology

Under the category cellular physiology, the gas-exchange conditions of a culture are of paramount concern. Gas exchange relates to the activity of the aerobically growing cells to consume oxygen and release carbon dioxide. Rates of these activities reflect the cell's overall physiological condition, which is related to the intracellular metabolic activities. Gas exchange or respiratory activity of cells has been investigated in an off-line mode using Warburg respirometers (Umbreit et al., 1957) and washed cells obtained from culture samples. Recently, development of on-line operating gas analyzers has made it possible to analyze respiratory activity in a continuous, real-time mode. In view of the literature which has appeared in this area, one can consider it to be one of the most advanced and practical on-line monitoring facilities for biological process status indication. Instruments which are used to obtain on-line information on gas-exchange conditions are classed in two major categories: (1) liquid phase (dissolved) gas analyzers, and (2) gas phase gas analyzers.

## 1. Liquid Phase (Dissolved) Gas Analyzer

**a. Determination of Dissolved Oxygen.** Dissolved oxygen analyzers consist of at least two essential elements: (1) a dissolved oxygen sensor, and (2) a signal amplifier and indicator. Dissolved oxygen sensors of various kinds are described by Solomons (1969b), Aiba et al. (1973), and Vincent and Priestley (1975) quoting the relevant original publications. Table II presents several of the currently available commercial dissolved oxygen sensors (probes) usable in aseptic and nonsterile fermentations.

Reviewing the technical background and principles of operation of dissolved oxygen probes, the following factors should be noted:

1. Essentially all of the dissolved oxygen sensors operate on the principle of reduction of oxygen at the surface of a metal electrode as shown by Mackereth (1964):

$$\text{Metal} + \tfrac{1}{2}O_2 + H_2O + 2\bar{e} \rightarrow 2\,OH^-$$

At the anode the loss of electrons results in production of ions:

$$\text{Metal} \rightarrow \text{Metal}^2 + 2\,\bar{e}$$

The metal ions combine with the hydroxyl ions to form a metal hydroxide [$ME(OH)_2$]. This interchange occurs between the compounds in the electrolyte surrounding the cathode and anode.

In summary, metal salts will be removed from the anode surface and accumulate in the electrolyte. Consequently, the expendable items in the dissolved oxygen sensor are the metal anode and the electrolyye. Electrons generated by the interaction of oxygen and metals offset the cathode–anode current balance and electric current is generated which is proportional to the amount of $O_2$ reacted with the metals. Cathodes are generally made of noble metals (most frequently used are platinum or silver). Anodes are generally made of lead.

2. Field experience indicates that oxygen-permeable, membrane-covered sensors have greater stability than a nonmembrane-covered sensor, although the membrane does affect the probe's response time and can be fouled or perforated.

3. Several researchers (Siegel and Gaden, 1962; Taylor et al., 1971) had proved that the current which is produced by the interaction of oxygen and the electrode is a measure of oxygen partial pressure ($pO_2$) and not the dissolved oxygen concentration. The value of the current ($I$) from the sensor (at a given temperature and pressure) depends on the electrons transferred ($M$), the Faraday constant ($F$), the area of the cathode ($A$), the permeability coefficient of membrane ($P_m$), the thickness of the membrane ($b$) and the partial pressure of the oxygen ($pO_2$):

**TABLE II.** Dissolved Oxygen Sensors

| Manufacturer | Type of sensor | DO measurement range | Temperature compensation | Pressure compensation | Response time (sec)[b] |
|---|---|---|---|---|---|
| Beckman Instruments | Polarographic | 0–9.99 ppm | Yes | No | 30 |
| Instrumentation | Polarographic | 0–20 ppm | Yes | Yes | 20 |
| Laboratory, Inc.[a] | | | | | |
| New Brunswick | Galvanic | 0–8 ppm | No | Yes (vented) | 45 |
| Scientific Co., Inc.[a] | | | | | |
| Transidine General Corp. | Polarographic | 0–1999 mm Hg | No | No | — |
| Yellow Springs | Galvanic | 0–8 ppm | No | Yes (vented) | 60–120 |
| Instruments[a] | | | | | |

[a] Steam sterilizable (min. 121°C).
[b] 90% of reading at 20°C.

$$I = nFA(Pm/b)\,pO_2 \qquad (2)$$

4. The physiological significance of the dissolved oxygen measurement in aerobic cultures can be depicted by the following equation (Aiba et al. 1973):

$$d\overline{C}/dt = k_La\,(C^* - \overline{C}) - Q_{O_2}X \qquad (3)$$

where $C^*$ = concentration of dissolved oxygen which is in equilibrium with $pO_2$ in the bulk gas phase ($mM\ O_2$)

$\overline{C}$ = the actual dissolved oxygen concentration in the bulk of the liquid ($mM\ O_2$)

$k_La$ = the volumetric oxygen transfer coefficient ($1/t$)

$Q_{O_2}$ = specific rate of oxygen uptake ($mM\ O_2/t$)

$X$ = cell mass concentration (weight)

On this basis the actual dissolved oxygen level or its concentration [at a given gas ($O_2$) flow rate, impeller tipspeed, broth temperature, and vessel pressure] reflects conditions which are related both to rheology-dependent $O_2$ gas transfer rate ($k_La$), and the oxygen uptake rate ($Q_{O_2}$) of the culture.

Dissolved oxygen probes have been used to determine $k_La$ and $Q_{O_2}X$ (Lengyel and Nyiri, 1965; Taguchi and Humphrey, 1966; Bandyopadhyay et al., 1967) as well as to control the broth dissolved oxygen concentration (Siegel and Gaden, 1962; Lengyel and Nyiri, 1965; Aiba et al., 1973). Dissolved oxygen control is based on on-line continuous measurement of $pO_2$. For determination of $k_La$ and/or $Q_{O_2}X$, techniques employing step changes in the air-flow ($O_2$) rate are used. From the recorded pattern of the changes in dissolved oxygen concentration (as the result of step changes in the air-flow rate) the values of $k_La$ and $Q_{O_2}X$ can be computed (Linek et al., 1973).

**b. Determination of Dissolved Carbon Dioxide.**   One of the end products of complete oxidation of carbon-containing compounds is carbon dioxide ($CO_2$). The effect of $CO_2$ on cellular metabolism has been considered as important as that of oxygen (Wimpenny, 1968; Nyiri, 1979). The regulatory effect of $CO_2$ was successfully applied to the industrial production of $\alpha$-amylase and protease (Zajic and Liu, 1970; Gandhi and Kjaergaard, 1975), penicillin (Lengyel and Nyiri, 1968), and inosine (Shibai et al., 1973).

Despite its recognized process significance, development of techniques for on-line sensing of dissolved carbon dioxide ($HCO^-_3$) are far behind the techniques used for the analysis of dissolved oxygen concentration. On-line dissolved carbon dioxide measurement can be performed using a sensor developed by Severinghaus and Bradley (1958) for human physiology studies.

The operating principle of the instrument is the measurement of the pH change of a $HCO_3^- + H^+$ solution influenced by $HCO_3^-$ ions. The sensor itself is a pH probe surrounded by a $HCO_3^- + H^+$ solution which is separated from the medium with a $CO_2$ ($HCO_3^-$)-permeable membrane. Presence of $HCO_3^-$ in the system makes the sensor sterilizable only by chemical means. In the fermentation industry, this fact has handicapped the widespread application of this otherwise accurately operating sensor. Recently the sensor was adapted by Japanese scientists to fermentors (Ishizaki et al., 1973; Ishizaki and Hirose, 1973) for studies of the effect of $CO_2$ ($HCO_3^-$) on various microbial fermentors and the bioengineering aspects of ventilation. Another $HCO_3^-$ sensor is marketed by Orion Research, Inc. However, its operational principle requires the liberation of $HCO_3^-$ into $CO_2$ gas which makes pH adjustment (to pH 5) necessary. This fact makes it difficult to adapt the sensor for on-line analysis of dissolved carbon dioxide.

### 2. Gas Phase Gas Analyzers

**a. Determination of Exhaust Gas Oxygen.** Instruments which measure $O_2$ gas concentration in the exhaust gases generally are of the paramagnetic type. Solomons (1969b) discussed in detail their operational principles, which essentially have not changed during the past 10 years. Table III presents a list of several vendors who currently manufacture $O_2$ gas analyzers which can be utilized for fermentation purposes.

Researchers have also constructed $O_2$ gas analyzers. Utilizing a nonsterilizable $O_2$ probe to the exit gas line of a fermentor and connecting the $O_2$ concentration signal with a gas flowmeter signal, it was possible to electronically calculate the oxygen uptake rate of a yeast culture of a given volume with an accuracy of $\pm 2$ m$M$ $O_2$/liter/hour (Nyiri and Jefferis, 1974). This was among the first "net-effect" sensors which combined more than one on-line operating sensor signal to obtain a third, nondirectly measurable process status indicator, in this particular case the oxygen uptake rate (OXUP or OUR) of the fermentation culture.

**b. Determination of Exhaust Gas $CO_2$.** Since $CO_2$ absorbs in the infrared range, most gaseous $CO_2$ analyzers commercially developed use this property for analyses. Table IV presents a list of such gas analyzers and their general specifications. Field experience has indicated excellent applicability of infrared devices for $CO_2$ off-gas analysis. One manufacturer (Infrared) provides an optional "dual channel" design by which another low molecular weight volatile compound which ab-

**TABLE III.** Oxygen Gas Analyzers

| Manufacturer | Model | Ranges ($O_2\%$) | Response time (sec)[a] | Accuracy (% FS) | Sample temperature range (°F) | Measurement Method |
|---|---|---|---|---|---|---|
| Beckman Instruments | 755 | 0–1 to 0–100 | 7 | ± 2 | − 20–120 | Paramagnetic |
| Mine Safety Appliances | 802 | 0–0.5 to 0–100 | 80 | ± 2 | 32–118 | Paramagnetic |
| Taylor Instrument (Sybron) | OA 184 | 0–25 to 0–100 | 7 | ± 1 | 14–122 | Paramagnetic (both inlet and outlet) |

[a] 90% for step change.

**TABLE IV.** Carbon Dioxide Gas Analyzers

| Manufacturer | Model | Range (% $CO_2$) | Response time (sec)[a] | Accuracy | Drift (zero, span) (per 24 hours) |
|---|---|---|---|---|---|
| Anarad, Inc. | AR500 | 0–30 | 5 | ± 1% | ± 1% |
| Cavitron Corp. | | | | | |
| Beckman Instruments | 865 | 0–5 | 0.5 | ± 1% FS | ± 1% |
| | | 0–15 | | | |
| Infrared Industries | IR703 | 0–1 | 5 | ± 1% C | 1°C |
| | IR702 | 0–30 | | | |
| Mine Safety | 200 | 0–30 | 5 | ± 1% | ± 1% |
| (M.S.A.) | | | | | |

[a] For 90% response time to step change.

sorbs in the infrared region can also be measured (e.g., $CH_4$, $NH_3$, acetone, and ethyl alcohol).

Exhaust gas analyzers (either $O_2$, $CO_2$, or both) obtain their samples from the exit gas line (after the exit gas filter, foam trap, or dehumidifier). The $O_2$ and $CO_2$ concentration of the incoming air is assumed constant at 20.91 and 0.03%, respectively, in order to develop $O_2$ and $CO_2$ balances around the fermentor (Aiba et al., 1973). If the incoming air is enriched with $O_2$ and/or $CO_2$ the use of an additional gas analyzer in the inlet gas stream is mandatory. The signal output of most commerically available gas analyzers provides data as a percentage of gas concentration. Additional computations are necessary to convert these readings into bioengineering units (e.g., moles) (Fiechter and VonMetenburg, 1963).

Other exhaust gases of potential interest include $N_2$ ($N_2$ fixing microbe physiology) and $H_2$ (*Hydrogenomonas* sp. cultures). $N_2$ can be determined in the off-gas using gas chromatographs, whereas $H_2$ in the exit gas can be determined using a thermal conductivity analyzer (Solomons, 1969b).

Though there has been little published to date regarding mass spectroscopic techniques for off-gas analyses, the authors are aware of several laboratories experimenting with this method. As the cost of mass spectrometer instruments has been reduced their competitive position with respect to multiple analyses, such as high-speed gas chromatographs, has improved. Gas chromatographs have been sucessfully applied in several laboratories for the measurement of off-gas $O_2$, $CO_2$, $H_2O$, and other low molecular weight volatile compounds and should be given serious consideration where multiple vessel application for these measurements are needed. Further discussion of the use of both of these methods for off-line analyses will be found in Section III, D, 2, a.

## D. Cell Biochemistry

Biochemical measurement of cellular activity is, in fact, measurement of enzymatic activity which, based on our current state of the art, cannot be directly analyzed. On the other hand, the disappearance or appearance of certain compounds (substrates, intermediate products) can indirectly tell the researcher about the biochemical events taking place within the cell. The compounds to be measured must have some physical or chemical characteristics on the basis of which their quantitative determination becomes possible. Typical compound properties used are the ionized forms ($H^+$, $OH^-$), molecular weight, volatility (ethanol), variation of functional groups (glucose), or an ability to absorb/scatter light ($NAD-NADH_2$). Based on these physical or chemical characteristics, instruments have been developed for both off-line and on-line determinations of their concentrations. Currently, there is a trend toward development of automatic sampling devices by which the fermentation unit can be connected with an otherwise off-line analytical instrument resulting in "on-line" operation.

### 1. Direct Reading Sensors

**a. pH Measurement.** The only steam sterilizable truly direct reading sensor with high selectivity is the pH probe. Of the varieties described by Solomons (1969a), Aiba *et al.* (1973), Vincent and Priestley (1975), we will comment on one sensor which, according to extensive field experience has become a commonly used device in the fermentation industry. This is the combination Ingold pH probe. Advantages of this pH probe can be listed as follows: (1) Combination of the $H^+$ sensor (half cell) and the reference sensor (half-cell) results in one probe which, therefore, requires only one penetration in the fermentor (bioreactor). (2) Based on widespread field experience, the probe is repeatedly steam sterilizable (more than 100 times at 121°C). (3) Design of the sensor makes it possible to sterilize (one or more models) either in an autoclave or by in-place sterilization techniques. (4) One model (Model 768-35) is designed to be removed under pressure, resterilized, and replaced during process operation. Arrangement of this retractable system is shown in Fig. 5. Connectors and cables make it compatible with most pH analyzers. Recently, another manufacturer, L&N, has come out with a single probe which appears to be comparable to the Ingold unit but there is, as of now, limited market experience with the probe. Table V lists several sources of sterilizable in-place pH electrodes.

**b. Specific Ion Measurements.** Orion Research and Radiometer have developed on-line sensors for determination of various ions. A list

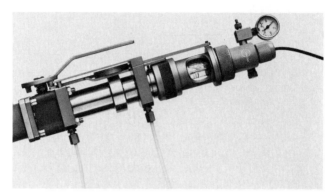

**FIGURE 5.** Ingold Model 768-35 steam sterilizable retractable electrode housing for use with Ingold combination pH electrodes.

of specific ion electrodes of interest in biochemical processes is shown in Table VI. It should be noted that none of these probes is steam sterilizable and the two most critical sensors [$CO_2$ ($HCO^-_3$) and $NH_3$ ($NH^+_4$)] operate in pH ranges (pH 5 for $CO_2$ and pH 11–13 for $NH_3$) usually well outside a majority of biological processes. It is felt that the possibilities for measurement of the effects of certain ions on the cell's metabolic activity have not been significantly exploited in either microbial, plant, or animal cell cultures (Kruse and Paterson, 1973; Schroder and Weide, 1974).

**c. Redox Potential Measurement.** Based on Nernst and other pioneering investigations, several review articles have appeared dealing with the theoretical background and practical application of redox potential in microbiology and biochemical engineering (Hewitt, 1950; Rabotnova, 1963; Kjaergaard, 1977). In this discussion extracellular and intracellular redox potential conditions should be distinguished. Extracellular redox potential (synonyms: $rH_2$, $E_{h_1}$, ORP; oxidation–reduction

**TABLE V.** Steam Sterilizable pH Sensors

| Manufacturer | Type | Model | Temperature Range (°C) | Pressure range |
|---|---|---|---|---|
| Activion Glass, Ltd. | Combination | 11-801 | 10–140 | N/A |
|  |  | 11-902 | 10–140 |  |
| Ingold | Combination | 761/764 Series | – 10–130 | 6 bar |
| Leeds & Northup | Combination | 117494 | 20–130 | 3.5 bar |
|  |  | 117495 |  |  |

**TABLE VI.** Commercially Available Ion Sensors for Biological Processes[a]

| Ion species | pH range for direct measurement | Concentration range (moles) |
|---|---|---|
| Ammonia | 11–13 | $1–10^{-6}$ |
| Calcium | 6–8 | $1–10^{-5}$ |
| Carbon dioxide | 5 | $10^{-2}–10^{-4}$ |
| Chloride | 2–11 | $10°–8 \times 10^{-6}$ |
| Magnesium | — | $10^{-4}$ (min) |
| Nitrate | 3–10 | $10°–6 \times 10^{-6}$ |
| Nitrate | 0–2 | $10^{-2}–5 \times 10^{-7}$ |
| Potassium | 3–10 | $10°–10^{-5}$ |
| Sodium | 9–10 | Saturation to $10^{-6}$ |
| Phosphate | — | $5 \times 10^{-3}$ (min) |
| Sulfate | — | $5 \times 10^{-3}$ (min) |

[a] Based on Orion Research Literature, 8th ed. May, 1977.

potential) is generally measured with two half-cells which compare the standard partial pressure of hydrogen gas ($pH_2$) or a concentration of two atoms of hydrogen with the amount of hydrogen in the oxidized form $[H^+]^2$:

$$E_h = E_o + \frac{RT}{2F}\ln\frac{[H^+]^2}{[H_2]} \tag{4}$$

where $E_h$ = potential referred to the normal hydrogen electrode
$E_0$ = standard potential of the system at 25°C when the activities of reactants are unity
$F$ = Faraday constant
$R$ = gas constant
$T$ = absolute temperature

The term $[H^+]^2/[H_2]$ can represent any compound which has the capability to exist in two states, reduced or oxidized respectively, in aqueous solution.

*(1) Extracellular Measurement of ORP.* The above characteristics represent both advantages and disadvantages for extracellular ORP measurement. According to experimental reports, the hydrogen ion concentration (pH) (Clark and Cohen, 1923), the presence of any inorganic or organic compound with oxidizing and/or reducing capability (including oxygen or metals), as well as temperature, will influence the output potential of a direct sensing ORP probe. This makes the interpretation of extracellular ORP data difficult. Correlations of practical significance were found between ORP and various physiological and biochemical events in microbial cultures (Hewitt, 1950; Rabotnova, 1963). Also, the dependency of mammalian cell growth on the status of redox potential of

the culture media has been demonstrated (Daniels *et al.*, 1970; Toth, 1975). The problems in interpretation of ORP data, however, have handicapped the widespread application of this measurement in fermentation technology. The most probable future use for ORP is in the detection of minute amounts of dissolved oxygen (<1 ppm) in anaerobic processes (between $-150$ and $-450$ mV) which are of importance for toxin production and in anaerobic biosynthesis of several organic solvents (acetone, *n*-butanol) where renewed commerical interest has recently occurred due to the energy crisis.

Steam sterilizable ORP sensors are manufactured by various companies, including Ingold, where combined platinum calomel as well as combined pH–redox probes are available. The ORP analyzer is a simple potentiometer with a measuring range of $\pm 500$ mV or $\pm 1000$ mV and is available from most pH analyzer manufacturers. Redox sensor probes are available from most suppliers of pH electrodes, such as Ingold and Leeds & Northrup. Based on reported field experiences, accuracy of ORP measurement is $\pm 5$ mV (FS).

*(2) Intracellular Measurement of ORP.* An interesting approach was developed by Harrison and Chance (1970) to determine intercellular redox potential by means of measuring $NAD–NADH_2$ conditions. The method involved sending and receiving light through the culture broth in the wavelength range of 360–460 nm. Although the device does have selectivity (Harrison and Harmes, 1972), the scattering of light caused by gas bubbles produces background noise which can only be resolved by the use of laser light. Cost of the equipment, however, has handicapped the commerical application of such a device.

**d. Carbohydrate Measurement.** Solomons (1969b) discussed a direct reading glucose sensor whose operating principle was based on glucose oxidase enzyme activity (Updike and Hicks, 1967). The reaction results in the formation of hydrogen peroxide which is monitored by electrochemical, amperometric detectors.

Several reactions of this technique have been developed both for off-line and on-line measurement purposes. The significance of these sensors should be emphasized in view of the reportedly developing techniques for biochemical conversion of carbohydrates (e.g., cellulose–glucose, starch–glucose conversions) for a wide range of applications.

Introduction by Leeds & Northrup of an on-line operating glucose analyzer with a sterilizable probe is considered to be a significant development and will be discussed later. Based on current trends (Zabriskie and Humphrey, 1978) it is anticipated that dialysis membrane-based continuous culture broth sampling will be one of the solutions to overcoming the difficulty of sample removal and treatment prior to wet chemical analyses in fermentation technology (cf Section III, D, 2, c).

**e. Cell Mass Measurement.** Automatic determination of fermentation broth cell mass is still among the most difficult of problems to be resolved. The most common technique is the measurement of light transmittance through the broth. The basis for the method used is the assumption that the intensity of the transmitted light is proportional to the turbidity, which in turn is related to the number of cells present. Devices developed during the past 10 years (e.g., Nephelostat, Peraino and Eisler, 1973; Turbidostat, New Brunswick Scientific Co.) all have similar limitations which can be enumerated as follows: (1) Only discrete "spherical" objects can be measured; i.e., devices are applicable for certain bacteria, yeast, or animal cell cultures, but not for filamentous fungi. (2) The technique is only applicable with clear solutions of soluble culture media, i.e., not applicable for industrial culture media which may contain insoluble and opaque materials, such as starch, corn steep liquor, molasses, or calcium carbonate. (3) The relationship between the cell concentration and light transmittance usually becomes nonlinear beyond 5 gm/liter cell concentration. These problems limit the applicability of such equipment for general scientific investigation. In those cases where application is possible, separate analyses must be made to determine the correlation between the transmitted light intensity and the actual cell population or mass present. The literature generally gives data either in cell numbers (e.g., $10^3$ cells/ml) or dry weight (DW) (e.g., 10 gm/liter cells). From an engineering point of view this latter dimension is preferred since on this basis a material balance and conversion yield can be calculated. Automatic and successful direct determination of dry cell mass for general application has not been reported in the literature to date.

## 2. Off-Line Automatic Analyzers

Difficulties in the development of steam sterilizable on-line reading sensors for detection of various metabolites have resulted in the use of off-line analytical devices utilizing repetitive and/or continuous sampling of exhaust gases to monitor specific parameters. Sampling devices usually consist of a pipeline (sterilizable) through which gases released by the culture are connected to the particular analyzer. The following paragraphs discuss several approaches which have passed the rigor of preliminary field experience and will probably be used in the future for quasi on-line chemical analysis of compounds of interest in biochemical processes.

**a. Gas Chromatography.**   Off-line determination of various volatile compounds (e.g., methane, acetaldehyde, ethanol) using gas chromatographs was successfully tested in biochemical processes by several researchers (Hamer, 1967; George and Gaudy, 1973; Abbott et al., 1973).

One of the most elegant applications was reported by O'Brien and Cecchini (1970), who analyzed the metabolic patterns of bacteria by determining their end products (alcohols, di- and tricarboxylic acids) and using this "finger print" for taxonomic identification. It is interesting to note that this approach might also have practical significance for the detection of certain types of contamination in pure culture, though such application has not yet been reported in the literature.

Recent developments in gas chromatograph (GC) technology have made it possible to connect them to the exit gas lines of fermentors. As an example, there is Hewlett-Packard's Series 5830A, Reporting Gas Chromatograph. This instrument can be connected to a maximum of 16 sample lines with an automatic sampler. It can be programmed to control oven temperature, has a dual thermal conductivity detector, and automatically integrates the data and reports them in alphanumeric and/or graphical formats. Signals can also be obtained in binary coded decimal (BCD) form for computer interfacing. An HP5830A GC has been used in the verification of the effects of C/N feed ratios on protein and ethanol biosynthesis for yeast grown on molasses. Figure 6 (G. M. Toth and L. K. Nyiri, unpublished observation), shows the GC profile of a C. utilis culture sampled at 15-minute intervals for $N_2$, $CO_2$, $H_2O$, acetaldehyde, and ethanol. Note that the total sample retention time was about 3 minutes for molecular weights up to ethanol.

The advantage of gas chromatography is the ability for automatic determination of a relatively wide range of compounds. Its disadvantages are (1) the discontinuous signal produced and the time length between signals, since the reporting frequency is controlled by the retention time and the number of compounds being analyzed; (2) the relatively high price which, however, can be partially offset if several fermentors are to be monitored; and (3) the practical problem which requires investigation for each scale of use—to what extent do the volatile compound concentrations in the exit gas reflect the actual concentration in the culture broth? For the example shown (acetaldehyde and ethanol) the liquid–gas distribution ratio could be determined for a given gas flow rate.

**b. Mass Spectrometer.**   In recent years reports have appeared about the use of mass spectrometry for monitoring low-molecular-weight compounds in physiological fluids, e.g., ethyl alcohol and dimethylsul-

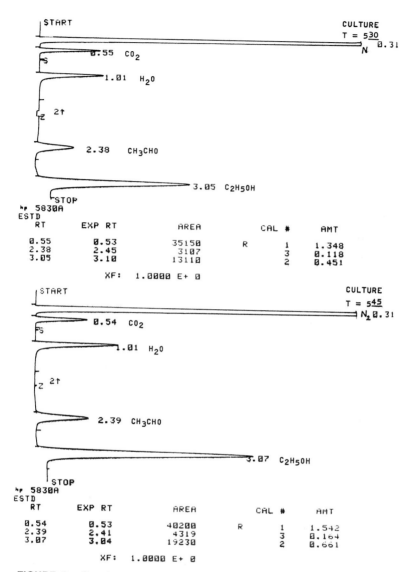

**FIGURE 6.**    Gas chromatographic profile on *Candida utilis* culture. (From Nyiri and Toth 1975).

foxide and $O_2/CO_2$ in blood (Green, 1970; Wald et al. 1971). This practice was currently extended to the field of biochemical engineering.

Weaver and co-workers (1976) reported the mass spectrometric measurement of volatile organic chemicals in aqueous solutions whereas Reuss et al. (1975, 1976) applied mass spectrometer for determination of dissolved oxygen, carbon dioxide, and methanol in fermentation broth.

The advantages of application of mass spectrometer (MS) to monitor dissolved volatile compounds in biological fluids can be listed as follows:

1. Response time is rapid (<1 minute).
2. Sensitivity (detection limit) is in the range of $10^{-5}$ $M$.
3. There is a linear response to compound concentration over wide range (e.g., between $10^{-5}$ $M–10^{-1}$ $M$).
4. It is possible to determine the compound concentration both in liquid and gaseous phases.

In the case reported by Reuss et al. (1975, 1976) a quadruple type mass spectrometer (QMS) was connected to a bacterial culture broth for the analysis of dissolved oxygen, dissolved carbon dioxide, and methanol concentrations. The physical connection between the culture liquid and the QMS sensing element was established by a sensor covered by a silicon rubber membrane. The thickness of the membrane made it possible to uncouple the pressure conditions existing inside the fermentor and in the high vacuum of the ion source in the mass spectrometer.

Application of mechanically supported semipermeable membranes as separators makes it possible to transfer the dissolved, low-molecular-weight (LMW) volatile compounds into the MS, which at the same time can also be connected to the exit gas phase of the fermentor. This technical approach makes it possible to determine the concentration difference of compounds existing between the dissolved and gaseous phase (e.g., $HCO_3^-/CO_2$ proportion).

The application of mass spectrometry to monitor concentration changes of important compounds in fermentation is in its early stage. Certain technical problems are anticipated (e.g., proper resolution of compounds to be measured, limitation of measurement speed by the diffusion rates of compounds through the membranes, preparation of sample to volatilize the compounds); however, these problems can be solved by technical improvements (e.g., selection of suitable membranes, application of selected ion-peak monitoring systems for resolution, and microprocessors for calibration—peak holding as well as tuning).

In view of the fact that more than one fermentor can be connected to

one MS, the relatively high unit cost of the mass spectrometer is eliminated. In addition, it is anticipated that with the development of the
methodology less expensive and more versatile mass spectrometers will
appear on the market, bringing this analytical technique into the realm of
widespread applicability.

**c. Dialysis Systems.** Development of plastic membranes with various molecular weight cutoff values has opened up the possibility for
selective separation of molecules from the culture broth (on-line) and
analysis off-line in readily available equipment. This same principle has
been applied in mass spectrometers. In a recent study Zabriske and
Humphrey (1978) described a dialysis system which is attached to a
modified baffle in the fermentor. Compounds with low molecular weight
are continuously dialyzed from the culture fluid into the water–buffer
solution circulating in the dialysis tube. Glucose was measured by a
glucose analyzer with a response time of 4–7 minutes to step changes in
the culture broth's glucose content. The system is a modified version of a
device developed by Glaxo Ltd. (England) for continuous determination
of penicillin.

Advantages of the dialysis approach are (1) selectivity for the compound of interest, (2) continuous signal production (analog signal), and
(3) relative simplicity of the sampling system. The potential disadvantage of the approach is the relatively long response time which is a
function of the distance between the dialysis system and the analyzer
and the assay procedure selected.

**d. Automatic Wet Chemical Analyzers.** A general purpose and
sophisticated device which can automatically and semicontinuously
determine a variety of organic and inorganic constituents of a culture
broth is the Technicon AutoAnalyzer. This instrument consists of a train of
modules (Fig. 7) which can pretreat, dilute, sample, react, and analyze
the sample broth by a specific quantitative detection device (colorime-
ter, flame photometer, flowmeter, potentiometer, etc.) for the compound
in question. Results of the analysis can be either recorded and/or sent to
an interface in the form of a digital signal for further data processing.
Typical of the assay procedures for fermentation processes which have
been reported for the AutoAnalyzer are mono- and disaccharides, inorganic and organic nitrogens, phosphorus, amino acids, vitamins, and
certain antibiotics. Although the AutoAnalyzer is considered an off-line
device because of the relatively long elapsed time (typically 10–30
minutes) between the sampling and availability of results, it will find
increased "on-line" application because of its accuracy, reliability, and
versatility of available measurements. Process information obtained by

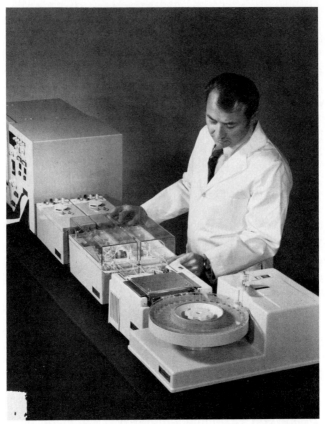

**FIGURE 7.** Technicon dual channel AutoAnalyzer II Continuous-flow modular analytical system for wet-chemical analyses at rates of up to 60 samples per hour.

this method has the potential for process monitoring and possible control response as well as postexperimental comparison of data obtained with on-line operating sensors and other wet-chemical analyzers.


## IV. ELECTRONIC ANALYSIS OF PROCESS DATA

The instruments previously described produce information on process status, equipment performance, substrate uptake, and product formation. Sensors, which reflect the process status, generally detect an end result of several metabolic activities and not the metabolic activity itself (e.g., pH is the result of the accumulation of many compounds of acidic or alkaline characteristics as a result of composite metabolic activity).

Furthermore, the specific broth concentration measurements reflect a condition which can be meaningful only if it is placed in the proper time dimension to describe the trend of the process. Consequently, concentration changes in time dimensions (rates) are more useful for experimental and process-control purposes than an absolute concentration value. For instance, oxygen uptake rate gives more information about the biological process conditions than the percentage of $O_2$ in the exit gas. Finally, as the number of instrumentation signals increases, the sheer management and timely reporting of process data in a usable fashion becomes more difficult.

Based on these considerations, starting in the early 1970s, attempts were made to (1) combine individual process sensor signals to obtain otherwise inaccessible process information, (2) place the direct and combined data outputs in a real-time dimension, and (3) store this data for later retrieval and data reduction (Nyiri, 1971, 1972; Humphrey, 1971; Nyiri and Humphrey, 1972). For these purposes real-time operating digital computers were interfaced with the fermentor's instrumentation. The computer was programmed to acquire the data and combine them to produce new process information. Signals from the sensors (called "gateway sensors," Aiba et al., 1973) are mathematically combined according to the theory of graphs. Nodes of the graph represent a series of computations resulting in a new, previously inaccessible process status indicator. As an example, determination of the respiratory quotient [$RQ$ = $CO_2$ produced (moles)/$O_2$ consumed (moles)] is used. $O_2$ and $CO_2$ gas concentrations are measured in percentage of the effluent gas. In order to convert these raw signals into mole concentrations, the gas flow rate, temperature, and pressure data of the fermentor as well as the gas constants are required (Fig. 8).

Based on a program containing all these inputs, a computer acquires the sensor data from the instruments and performs the computations resulting in this new information, respiratory quotient. Problems with respect to the use of $RQ$ information and its applicability for identification of the physiological status of a C. utilis culture were discussed by Nyiri et al. (1975a). Computer determinations of $RQ$ were made by several investigators with promising results using on-line optimization and process control (Nyiri and Krishnaswami, 1974; Aiba et al., 1976). In addition, both material and heat balances have also been successfully computed by this technique (Cooney et al., 1975).

Fermentation data processing hardware requirements are shown in a typical schematic, Fig. 9. The hardware requirements will change in the future as further developments in the electronic state of the art occur. Essentially, however, the following system elements will be required:

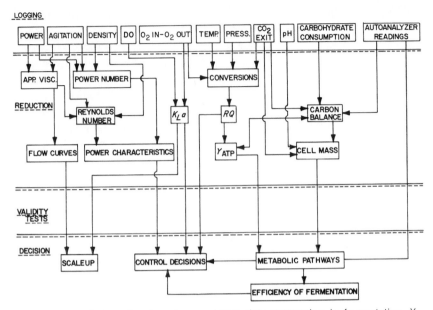

**FIGURE 8.** Schematic of on-line, real-time data processing in fermentation. $Y_{ATP}$, organic energy yield factor.

1. Sensor
2. Amplifier
3. Signal conditioning
4. Analog-to-digital converter (for production of BCD signal generally 16-bit word length)
5. Electronic interface
6. Central processing unit (CPU)
7. Real-time clock
8. Data and/or program storage
9. Communication link [teletype (TTY)], video terminal with possible hard-copy capability
10. Data plotter (graphical display of data)

Experience has shown that total execution time for a data acquisition and analysis cycle for a fermentor with 42 separate signal pickups and 61 information outputs required about 1 second, while printing the information on a high-speed printer required about 16 seconds (Nyiri et al., 1975b). These processing times make it possible to acquire data in 1-minute intervals which, for most biological processes, can be considered "continuous" signal pickup and data processing. Metabolic responses to external frequency pulses usually exhibit themselves be-

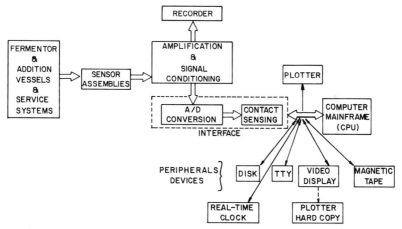

**FIGURE 9.** Schematics of a computer-aided fermentation system (data acquisition, data processing).

tween 2 and 5 after the step change and (depending on the environmental conditions) can last more than 20 minutes (Nyiri and Toth, 1975). These observations are of importance when process control is to be implemented based on physiological analysis of the culture.

Rapid development in microelectronics has resulted in drastic reduction in the cost per bit of computer memory (Noyce, 1977) and has made it possible to develop microprocessors and microelectronic memories of high density. As a result, low-cost programmable data processors with all of the capabilities needed to perform data acquisition and analysis in real-time and store this information will soon be available for biological processes.

## V. EVALUATION OF CONTROL CONCEPTS IN BIOLOGICAL PROCESSES

As was pointed out previously the objective of process control in biological processes is to create a suitable environment for the culture to express the maximum or desired metabolic activity. This primary objective is accomplished by a secondary series of control objectives which maintain a series of set-point conditions in individual control loops. The control problem is complicated further if one considers the fact that the set points may be changed during the process due to changes in physiological conditions of the culture.

Over the past years several control strategies were developed for fermentation processes. These have been discussed by Aiba (1966).

Current fermentation (cell culture) techniques utilize primarily feedback control which can be implemented manually or by automatic (electronic) means (Fig. 10A and B). This method is a closed loop configuration, where the control (environmental) variable is measured (sensed) and is compared with a reference set point. If a difference or error exists between the actual and desired levels of the controlled variable, a control "action" takes place. In its simplest form (Fig. 10A) the measuring device has an analog output (e.g., a pressure gauge) which is "sensed" by an operator who knows the desired value he or she wishes to maintain and who manually adjusts the final control element to achieve the corrective action. In the automatic mode (Fig. 10B), the operator's function is replaced by (1) a transmitter (analogous to the operator's eye), and (2) a controller which produces an output control

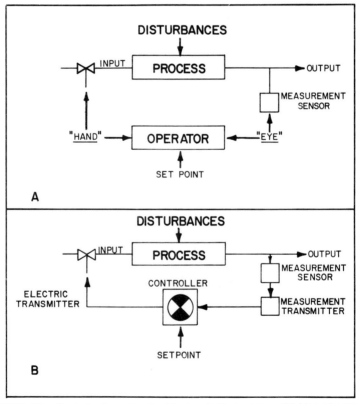

**FIGURE 10.** Control schematics for biochemical processes. (a) Manual control of a process variable; (b) automatic control of a process variable (feedback control).

signal to the final control element which is a function of the input error signal into the transmitter.

With respect to control variables normally found in most fermentation processes it can usually be stated that (1) the optimum value of the controlled variable (e.g., temperature) was chosen by trial and error over a protracted period of time; (2) the set-point value is generally held constant over the course of the process independent of the physiological stage of the culture (this is mainly due to the lack of information about correlations between the environment and cell physiology); and (3) the sensor accuracy usually exceeds the accuracy of the controller's performance, and to date no evidence is yet available to define how accurate the environmental control must be to maximize cell metabolic activity.

Recent investigations indicate that the optimum environmental conditions change over the course of the process and that certain environmental variable "profiles" need to be developed in an attempt to adjust the environment to cope with the demands of changing cell physiology and enzyme activity (Constantinides et al., 1970 a,b; Rai and Constantinides, 1973; E. B. Lynch and W. F. Ramirez, personal communication, 1975). Though publication of such information has not been made, the authors are aware of several commerical processes which have used this approach to optimize a particular facet of a culture's performance. In addition, changes in one environmental variable may require readjustment of other controlled variables due to the complex interactions between the cells and the environment (Nyiri et al., 1973; Unden and Heden, 1973). Because of the complexity of the biological process, the use of computers to search out optima and apply this knowledge to establish proper environmental conditions for cells of changing physiology is an inevitable step in the future development of both instrumentation and process analysis and control techniques in the fermentation industry (Fuld, 1960; Humphrey, 1971; Nyiri, 1971, 1972).

The first known application of computers to control fermentation process variables was Dista's effort in a penicillin fermentation plant (Grayson, 1969). In this instance, direct digital control was employed where the computer replaced the controller's comparative function in feedback control loops by determining the signal error and computing the necessary final control element action for the manipulated variable. The advantage of this approach was a more accurate set-point control of the individual process variables. However, one potential disadvantage was the loss of process control in the event of computer failure. The overall strategy of control was based on individual set-point control and did not take into account the interactions between the culture and its environment.

Recognition of interactions between the environment and the process culture made it necessary to attempt to develop control modes where more than one process variable was changed simultaneously and where the degree of change was based on (a) the condition of one process status indicator, and (b) the condition of other controlled variables which might interactively influence the process status. The first example (on a bench-scale system) was the interactive control of dissolved oxygen on the basis of computed values of $k_La$ using agitation speed and/or air-flow rate for control purposes (Nyiri et al., 1974) (Fig. 11). Computer linearization of this control system resulted in more stable control performance (Jefferis et al., 1973).

Based on experience and current developments in analog control techniques a modified version of cascade (supervisory) control strategy was recently proposed for biological processes (Nyiri and Jefferis, 1974; Wilson et al., 1975; Nyiri et al., 1975b). Figure 12 illustrates the concept. Environmental variables (e.g., temperature, pressure, and pH) are controlled by individual analog controllers. The control loop consists of conventional elements including sensor, transmitters, controller, and final control element. The set point of every controller, however, is generated by a digital computer. Signals originating from the sensors that indicate the process controller's performance (e.g., temperature reading) as well as signals reflecting biological process conditions, e.g., $O_2\%/CO_2\%$ in the exit gas, are introduced into the computer which (a) reduces the individual signals to process status indicators [e.g., rate of

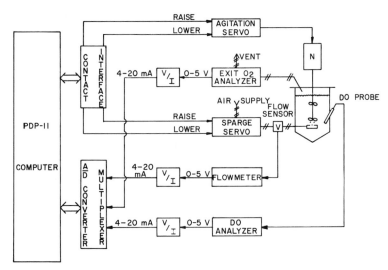

**FIGURE 11.** Computer control of dissolved oxygen. N, agitation; V/I, voltage–current signal conditioning; and V, volumetric gas flow rate.

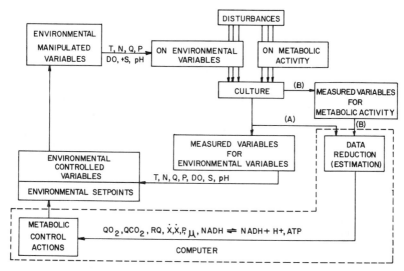

**FIGURE 12.** Cascade control of fermentation process variables. *DO,* dissolved oxygen concentration; *N,* agitation speed; *P,* pressure, product; *Q,* gas flow rate; *S,* substrate; *+S,* substrate addition rate; *T,* temperature; $\dot{X}$, cell mass formation rate; and $\mu$, specific growth rate.

oxygen uptake $(Q_{O_2})$ and respiratory quotient $(RQ)$] and (b), based on experimentally defined interactions between the environment and the cells, alters the set point of one or more individual controllers in order to keep the environmental conditions at the desired optimum.

The individual controllers can be substituted by a digital computer, thereby creating a direct digital control structure (Jefferis, 1973). The decision to rely fully on computers or to use analog controllers interfaced to the computer is at the discretion of the user.

In another reported version, the control of individual environmental variables is performed by microprocessors and the definition of the biological process status (data acquisition, data reduction) is implemented by a real-time digital computer with large memory capacity (Jefferis, 1973). Development of such a structure, however, requires a wealth of information on the culture–environment interactions which, at this moment, is still very much in the exploratory state.

## VI. AUXILIARY INSTRUMENTATION

In this category we list those devices which are necessary to complete an instrumentation package, which include digital panel meters and recorders.

## A. Digital Panel Meters

Digital panel meters (DPMs) are extremely useful and valuable tools for making precise electrical measurements. They provide accurate and unambiguous readings which satisfy a wide variety of application requirements for laboratory, pilot, and production electronic and electrical measuring equipment. Digital panel meters can be used with any transducer to provide digital readout of such process parameters as temperature, pressure, air flow, agitation, pH, dissolved oxygen, or any other physical parameter. Though their cost is usually greater than standard analog meters, their accuracy and readability is more than worth the added expenditure. For example, in the standardization of instruments (zeroing, span, calibration) or controller set-point establishment, their accuracy is far superior to analog devices.

Selection of a suitable DPM for an application requires establishment of the following criteria: (1) resolution, (2) accuracy, and (3) stability. In DPM nomenclature the number of digits readable on a DPM are given as shown in the tabulation below (excluding decimal point).

| DPM digits | Full-scale maximum count |
|------------|--------------------------|
| $2\frac{1}{2}$ | 199 |
| 3 | 999 |
| $3\frac{1}{2}$ | 1999 |
| $3\frac{3}{4}$ | 3999 |
| 4 | 9999 |
| $4\frac{1}{2}$ | 19999 |
| $4\frac{3}{4}$ | 3999 |

Sources of digital panel meters are Data Technology, Analog Devices, Canyon Electronics, Datel Systems, Analogic Corp., and Fairchild Instrumentation.

From our own experience, despite the fact that DPMs have output signals directly acceptable by some computer interfaces, the natural oscillation of most DPM input signals will cause "floating" of the last output digit, thereby creating latching problems at the moment of acquisition by the computer. It is therefore preferred to separate the signals (from the DPM) for interfacing. Hence, one of the benefits of DPMs, the fact that they provide their own analog-to-digital (A/D) conversion, is not considered by us to be a general plus if computer coupling is the objective in the electronic system design.

## B. Recorders

For permanent record keeping of acquired signals recorders are usually used with most of the various instrumentation systems employed for process monitoring and control. Recorders can be purchased starting with single-point recording units up to recorders which will accept 30 points. Printing can be either by traditional ink–pen systems or newer inkless pressure or thermal recording styli. Essentially all instrument manufacturers are a source of recorders (e.g., Honeywell, L&N, Foxboro, Taylor).

The rationale for recording any variable which is monitored depends to a large extent on the organization doing the work, the need for permanent records (such as in production operations to meet government requirements for product process history), and the nature and type of process work being performed. For several reasons, one of which is the difficulty of simple storage of printed data in standard files, strip-type chart recorders are preferred over circular ones. Also, multipoint recording devices provide a convenient method of recording several relatable variables on the same time axis, making visual data analyses much easier. However, the number of simultaneous recordings on a single chart must be limited to between three and six, depending on the variables involved, if one wishes to utilize full recorder scale width for each variable range. Field experience has shown that beyond six variables on a single recorder chart, regardless of recording chart width, there tends to be added confusion of analyses and not the improvement which is desirable.

Based on the foregoing, it is recommended that multipoint strip chart-type recorders be considered for process data recording. Newer designs in recording equipment have resulted in smaller units which can be easily panel mounted. In addition, several manufacturers are beginning to include microprocessor devices in their recorder which allows the selection and frequency of recording points, as well as variable expansion of recording ranges.

## VII. SUMMARY

We have attempted to provide a very broad overview of currently available instrumentation systems based on a generic description of the areas of their application in fermentation. Unfortunately, in a review such as this, it is always possible to overlook both current research work and possibly newer vendor developments. We have also tended to "take a stand" on those classes of instruments which we feel are better suited for use in fermentation work. Some readers will accept this, and others will certainly have their own "best solutions" to champion.

Because of the very rapid developments occurring in the electronics industry, newer techniques and equipment methods are probably already coming to market even during publication of this review. If fermentation processing is to advance, newer instrumentation and, more specifically, newer sensors must be pursued and developed to meet the critical needs of the biological processes with which we must deal. Until the time comes when such equipment is readily available to all, we must rely on the ingenuity and determined efforts of process development pioneers to continue to adapt what is available to the "advancement of the art."

## VIII. APPENDIX: LIST OF EQUIPMENT MANUFACTURERS

Activion Glass, Ltd., Halstead, Essex, England
Analog Devices, Norwood, Massachusetts
Analogic Corp., Wakefield, Massachusetts
Anarad, Inc., Santa Barbara, California
Automation Products, Inc., Houston, Texas
Beckman Instruments, Fullerton, California
BLH Electronics, Waltham, Massachusetts
Brooks Instruments Div., Hatfield, Pennsylvania
Canyon Electronics, Phoenix, Arizona
Chemcon, Inc., Medfield, Massachusetts
Data Technology, Santa Ana, California
Datel Systems, Canton, Massachusetts
Electro Corp., Sarasota, Florida
Fairchild Instrumentation, San Jose, California
Fisher and Porter Co., Warminster, Pennsylvania
Foxboro Co., Foxboro, Massachusetts
Gulton Industries, Costa Mesa, California
Hewlett Packard Corp., Santa Clara, California
Honeywell, Process Control Div., Ft. Washington, Pennsylvania
Infrared Industries, Santa Barbara, California
Ingold Electrodes, Inc., Lexington, Massachusetts
Instrumentation Laboratory, Inc., Lexington, Massachusetts
Interpace Corp. (Lapp), Rochester, New York
Leeds and Northrop, North Wales, Pennsylvania
LFE Corp., Waltham, Massachusetts
Meriam Instrument Div. (Scott & Fetzer), Cleveland, Ohio
Micro-strain, Inc., Spring City, Pennsylvania
Milton Roy Co., St. Petersburg, Florida
Mine Safety Appliance Co., Pittsburgh, Pennsylvania
New Brunswick Scientific Co., Edison, New Jersey
Ohio Semitronics Co., Columbus, Ohio
Orion Research, Inc., Cambridge, Massachusetts
Radiometer, Copenhagen, Denmark
Ranco Controls, Columbus, Ohio
Robertshaw Controls Co., Richmond, Virginia
Rosemount, Inc., Minneapolis, Minnesota
Schaevitz Engineering, Pennsauken, New Jersey
Schutte and Koerting, Cornwells Heights, Pennsylvania
A. O. Smith Corp., Erie, Pennsylvania
Taylor Instrument Co., Rochester, New York

Technicon Instrument Corp., Tarrytown, New York
Transducers, Inc., Whittier, California
Vibrac Corporation, Chelmsford, Massachusetts
Yellow Springs Instrument Co., Yellow Springs, Ohio

## REFERENCES

Abbot, B. J., Laskin, A. I., and McCoy, C. J. (1973). *Appl. Microbiol.* **25,** 787.
Aiba, S. (1966). *Proc. Int. Cong. Microbiol., 9th, 1966.*
Aiba, S., Humphrey, A. E., and Millis, N. F. (1973). "Biochemical Engineering," 2nd ed. Academic Press, New York.
Aiba, S., Nagai, S., and Nishizawa, Y. (1977). *Biotechnol. Bioeng.* **18,** 1001.
Bandyopadhyay, B., Humphrey, A. E., and Taguchi, H. (1967). *Biotechnol. Bioeng.* **3,** 533.
Charles, M., and Toth, G. M. (1974). *Ann. Meeting Am. Chem. Soc. 168th.*
Clark, W. M., and Cohen, B. (1923). *Public Health Rep.* **38,** 666.
Considine, D. M., ed. (1957). "Process Instruments and Controls Handbook." McGraw-Hill, New York.
Constantinides, A., Spencer, J. L., and Gaden, G. L., Jr. (1970a). *Biotechnol. Bioeng.* **12,** 803.
Constantinides, A., Spencer, J. L., and Gaden, G. L., Jr. (1970b). *Biotechnol. Bioeng.* **12,** 1081.
Cooney, C. L., Wang, H., and Wang, D. I. C. (1975). *U.S./U.S.S.R. Joint Symp. Data Acquisition Process.*
Daniels, W. F., Garcia, L. M., and Rosenstahl, J. F. (1970). *Biotechnol. Bioeng.* **12,** 403.
Fiechter, A., and von Meyenburg, K. (1963). *Biotechnol. Bioeng.* **10,** 535.
Fuld, G. J. (1960). *Adv. Appl. Microbiol.* **2,** 351.
Gandhi, A. P., and Kjaergaard, L. (1975). *Biotechnol. Bioeng.* **17,** 1103.
George, T. K., and Gaudy, A. F. (1973). *Appl. Microbiol.* **26,** 736.
Grayson, P. (1963). *Process Biochem.* **4,** 43.
Green, D. W. (1970). *Inter-Science Chem. Reports* **4,** 211.
Hamer, G. (1967). *Biotechnol. Bioeng.* **8,** 433.
Harmes, C. S., III (1972). *Dev. Ind. Microbiol.* **13,** 146.
Harrison, D. E. F., and Chance, B. (1970). *J. Appl. Microbiol.* **19,** 446.
Harrison, D. E. F., and Harmes, C. S., III (1972). *Process Biochem.* **13,** 1.
Hewitt, L. V. (1950). "Oxidation—Reduction Potentials in Bacteriology and Biochemistry." Livingstone, Edinburgh.
Humphrey, A. E. (1971). *Proc. Labex Symp. Comput. Control Ferment. Process.*
Ishizaki, A., and Hirose, Y. (1973). *Agric. Biol. Chem.* **37,** 1235.
Ishizaki, A., Shibai, H., Hirose, Y., and Shiro, T. (1973). *Agric. Biol. Chem.* **37,** 33.
Jefferis, R. P., III (1973). *Proc. Eur. Conf. Comput. Process Control Ferment., 1st.*
Jefferis, R. P., III, Biela, J. A., and Nyiri, L. K. (1973). *75th Natl. AIChE Meet.*
Kjaergaard, L. (1977). *Adv. in Biochem. Eng.* **7,** 131.
Kruse, P. F., Jr., and Patterson, M. K., eds. (1973). "Tissue Culture: Methods and Applications." Academic Press, New York.
Lengyel, Z. L., and Nyiri, L. K. (1965a). *Biotechnol. Bioeng.* **7,** 31.
Lengyel, Z. L., and Nyiri, L. K. (1965b). *Biotechnol. Bioeng.* **7,** 91.
Lengyel, Z. L., and Nyiri, L. K. (1966). *Biotechnol. Bioeng.* **8,** 337.
Linek, V., Sobotka, M., and Prokop, A. (1973). *Biotechnol. Bioeng.* **4,** 423.
Mackereth, F. J. H. (1964). *J. Sev. Instrum.* **41,** 38.
Metzner, A. B., and Taylor, J. S. (1962). *AIChE J.* **6,** 103.

Noyce, R. N. (1977). *Sci. Am.* **237,** 62.

Nyiri, L. K. (1971). *Proc. Labex. Symp. Comput. Control Ferment. Process.*

Nyiri, L. K. (1972). *Adv. Biochem. Eng.* **2,** 49.

Nyiri, L. K. (1978). *Adv. Biochem. Eng.* **8** (in press)

Nyiri, L. K., and Humphrey, A. E. (1972). *Ferment. Technol. Today, Proc. Ferment. Int. Symp. 4th, 1972.*

Nyiri, L. K., and Jefferis, R. P., III (1974). *In* "Single-Cell Protein II" (S. R. Tannenbaum and D. I. C. Wang, eds.), p. 105, MIT Press, Cambridge, Massachusetts.

Nyiri, L. K., and Krishnaswami, C. S. (1974). *75th Meet. Am. Soc. Microbiol.*

Nyiri, L. K., and Lengyel, Z. L. (1968). *Biotechnol. Bioeng.* **10,** 133.

Nyiri, L. K., and Toth, G. M. (1975). *Annu. SIM Meet. 1975.*

Nyiri, L. K., Jefferis, R. P., III, and Humphrey, A. E. (1973). *Abstr. Pap., 166th Meet., Am. Chem. Soc.*

Nyiri, L. K., Jefferis, R. P., III, and Humphrey, A. E. (1974). *Biotechnol. Bioeng. Symp.* **4,** 613.

Nyiri, L. K., Wilson, J. D., Humphrey, A. E., and Harmes, C. S., III (1975a). U.S. Patent 3,326,738.

Nyiri, L. K., Toth, G. M., and Charles, M. (1975b). *Biotechnol. Bioeng.* **17,** 1663.

O'Brien, R. T., and Cecchini, G. L. (1970). *Dev. Ind. Microbiol.* **11,** 99.

Peraino, C., and Eisler, W. J. Jr. (1973). *In* "Tissue Culture. Methods and Applications" (P. F. Kruse, Jr. and M. K. Patterson, eds.), p. 351. Academic Press, New York.

Rabotnowa, I. L. (1963). "Die Bedeutung physikalisch-chemisher Faktoren (pH und rH$_2$) für die Lebenstätigkeit der Mikroorganismen." Fischer, Jena.

Rai, V. R., and Constantinides, A. (1973). *AIChE Symp. Ser.* **23,** 114.

Reuss, M., Piehl, H., and Wagner, F. (1975). *Europ. J. Appl. Microbiol.* **1,** 323.

Reuss, M., Piehl, H., and Wagner, F. (1976). *Abstr. Int. Ferment. Symp., 5th, 1976,* p. 24.

Schroder, K. D., and Weide, H. (1974). *Biotechnol. Bioeng. Symp.* **4,** 713.

Severinghaus, J. W., and Bradley, A. F. (1958). *J. Appl. Physiol.* **13,** 515.

Shibai, H., Ishizaki, A., Mizuno, H., and Hirose, Y. (1973). *Agric. Biol. Chem.* **37,** 91.

Siegel, S. O., and Gaden, E. I., Jr. (1962). *Biotechnol. Bioeng.* **4,** 345.

Solomons, G. L. (1969a). "Materials and Methods in Fermentation." Academic Press, New York.

Solomons, G. L. (1969b) *Chem. Eng. (London)* **215,** p. 55.

Taguchi, H., and Humphrey, A. E. (1966). *J. Ferment. Technol.* **44,** 881.

Taylor, G. W., Kondig, J. P., Nagle, S. C., and Higuchi, K. (1971). *Appl. Microbiol.* **21,** 928.

Toth, G. M. (1975). *Proc. Cell Cult. Congr.*

Tuffile, C. M., and Pinho, F. (1970). *Biotechnol. Bioeng.* **12,** 843.

Unden, G. A., and Hedén, C. G. (1973). *Proc. Eur. Conf. Comput. Process Control. Ferment., 1st.*

Umbreit, W. W., Stauffer, R. H., and Stauffer, J. R. (1957). "Manometric Techniques." Burgess, Minneapolis, Minnesota.

Updike, S. J., and Hicks, G. P. (1967). *Nature (London)* **214,** 986.

Vincent, W. A., and Priestley, G. (1975). "Handbook of Enzyme Biotechnology." Academic Press, New York.

Wald, A., Hass, W. K., and Ransohoff, J. (1971). *J. Assoc. Adv. Med. Inst.* **5,** 325.

Weaver, J. C., Mason, M. K., Jarrel, J. A., and Peterson, J. W. (1976). *Biochim. Biophys. Acta.* **438,** 296.

Wilson, D. J., Nyiri, L. K., Humphrey, A. E., and Harmes, C. S., III (1975). U.S. Patent 3,326,737.

Wimpenny, J. W. T. (1968). *Symp. Soc. Gen. Microbiol.* **19.**

Zabriskie, D. W., and Humphrey, A. E. (1978). *Biotechnol. Bioeng.* **20** (in press).

Zajic, J. E., and Liu, F. S. (1970). *Dev. Ind. Microbiol.* **11,** 350.

Chapter 15

# Computer Applications in Fermentation Technology

WILLIAM B. ARMIGER
ARTHUR E. HUMPHREY

## I. INTRODUCTION

Recent advances in biochemistry and microbiology have resulted in a more complete understanding of fermentations, particularly with respect to substrate metabolism, growth requirements, and enzyme regulation. These developments are enabling the fermentation technologist to describe the dynamic behavior of fermentation processes in terms of kinetic models.

**375**

MICROBIAL TECHNOLOGY, 2nd ed., VOL.II
Copyright © 1979 by Academic Press, Inc.
All rights of reproduction in any form reserved. ISBN 0-12-551502-2

With this understanding, it is now possible to expand the control of fermentation processes beyond independent closed-loop feedback control of culture conditions, such as temperature, pH, and dissolved oxygen, into the more sophisticated strategies of adaptive or interactive control. Eventually, on-line optimization of a fermentation based upon model reference control of a process will be possible. This accomplishment will mean that a feed-forward, interactive control strategy will be utilized. In order to reach this capability, it is necessary to develop a system which encompasses both hardware and software needed to generate the raw data, analyze the resulting information, determine an optimal solution, and implement a control decision. A highly instrumented computer-coupled fermentor is capable of meeting this task.

The accomplishment of on-line optimization of a fermentation is an evolutionary process. The process began with the achievement of fermentation data acquisition by a computer system. Most fermentation companies now routinely utilize computers in data acquisition. This development involved on-line process monitoring and scanning, data logging, and generation of alarm signals. In many cases, this information was collected in a form that was transferred to a computer system for off-line analysis. With the advent of inexpensive and reliable minicomputers, it has become economical to utilize a computer system on-line. With this capability, it has been possible to move rapidly to a computer applications level which has involved partial process control based upon using predetermined set points for controlling primary fermentation parameters, such as temperature, pH, and dissolved oxygen. It has also been possible to utilize the information generated on-line to study the process and to elaborate the process kinetics. Today, on-line computer systems are being utilized for reducing the time required for determining the dynamics of fermentation processes, for testing control strategies, and for modeling the process.

With the development of more powerful minicomputer operating system software, it is possible to operate the computer system using both foreground/background tasks, i.e., simultaneously performing the on-line data acquisition and analysis tasks and simultaneously doing on-line process simulation. This allows the fermentation technologist to quickly develop and refine mathematical models and to identify the model parameters so that methods can be developed for predicting the interactive relationships between primary fermentation parameters. Once significant operating confidence is obtained with process kinetic models, computer systems can advance to the final stage of automatic process control, i.e., on-line optimization based upon model reference control.

## II. HISTORY OF COMPUTER APPLICATIONS

Computer applications in the chemical process industry began in the late 1950s. By the late 1960s, publications on computer applications in batch process control were fairly common. However, it was the widespread availability of minicomputers in the 1970s that led to an abundance of process-control applications.

Yamashita and Murao (1967), and Koga et al. (1967) published the first papers on the use of digital computers specifically for modeling fermentation processes. Yamashita et al. (1969), Grayson (1969), and Harrison et al. (1971) published the first real description of a computer-coupled fermentation system. Since that time, numerous articles dealing with system design and software utilization have been published. (These are listed in the Supplementary Reading Section.)

Due to corporate secrecy policies, it is not clear which companies were the first to utilize computers in fermentation processes. However, there are some publications dealing with industrial applications. In 1966, The Ajinomoto Company in Japan utilized a YODIC-500 computer for direct digital control of their glutamic acid fermentation, and Dista Products utilized an ARCH 102 computer for direct digital control of 114 loops in a new fermentation plant. In 1969, Glaxo Laboratories, Ltd., converted an existing fermentation plant from analog to direct digital control. Since the early 1970s, many research-oriented pilot-plant systems have been developed for on-line data acquisition, analysis, and digital set-point control. The University of Pennsylvania (Armiger and Humphrey 1974; Armiger et al., 1975, 1976; Harrison et al., 1971; Harrison and Harmes, 1972), E. R. Squibb (Moes et al., 1971; Young and Koplove, 1972), Lord Rank Research Center (Flynn, 1974), Station de-Gene Microbiologique (Blachere 1973; Corrieu et al., 1974; Lane, 1973), Karolinska Institute (Unden and Heden, 1973), Gesselschaft Fur Biotechnologische Forschung MBH (Jefferis, 1975), and many others (Cooney et al., 1975; Jefferis, 1973; Nyiri et al., 1975; Nagai et al., 1976) have interfaced computers with their fermentation systems.

Computer control and computer applications in fermentation cover a broad spectrum of uses depending upon the needs of a particular process. The requirements for a pilot plant are quite different from those for a production plant. The degree of sophistication of the computer software depends upon the use of the system. Moving from data logging to data analysis, process modeling, process control, and finally to process optimization requires an increasingly more complex system in terms of both the hardware and the software. All of these considerations along with future trends for computer applications will be discussed in the following sections.

## III. GENERAL APPLICATIONS

The application of computers to fermentation technology is very dependent upon the end use of the computer system. For example, requirements for a pilot-plant system are quite different from those for a production plant. In a pilot plant, the computer is used primarily as a research tool. It is necessary that the system be designed to be as flexible as possible. It should be particularly useful in obtaining kinetic data necessary for modeling the process and for developing control strategies aimed at process optimization. Pilot-plant computer-coupled systems should be structured to accommodate software development and software testing.

There are two levels in which a computer system can be utilized in a pilot plant. The first level involves operator-oriented programs which can be utilized by a technician. To run these programs, it is necessary for a technician to know only the computer command necessary to start them. The programs are designed to prompt the operator with the responses necessary for implementation. These programs include such functions as data logging, data analysis, and process control. By using operator-oriented routines, the technician can generate the necessary documentation and procedures for system operation and measurements. It is also helpful if the system contains programs for generating standard reports and graphical output.

The second level of computer utilization requires an engineer with programming ability to develop and insert into the operator-oriented programs new routines to be run by the technician. Also, on this level the engineer interacts with the computer to develop control strategies and process optimization routines. The engineer can interact with the system to quickly access large volumes of data. The easy access to this information frees him or her to develop and test control strategies and process optimization routines. Thus, the computer-coupled system is a research tool which allows the engineer to spend most of his or her time interpreting data rather than merely collecting and plotting it.

For a full-scale production plant, the requirements for a computer-coupled system are quite different. The primary objective is to produce a product in an optimal manner. As a result, the computer is used for process control and optimization aimed at maintaining quality control and product uniformity. Here again, it is necessary to interact with the computer system on two levels. The first level involves an operator-oriented command language allowing easy access to the entire process. The command language will contain routines for generating process status information, calculating process variables, and scheduling process operations.

Plant status routines are used to send alarm messages to the operator and to supply him or her with information on the overall operation of the process. Raw material inputs, flow rates, and utility requirements in terms of steam, cooling water, and electric power demands are just a few of the process parameters and variables that are routinely monitored. Scheduling routines provide the operator with information for sequencing operations. This is particularly valuable for a batch process where there are strict protocols for cleaning, media sterilization, vessels charging, and product harvesting. The sequencing routines provide information to step the operator through the process in order to start up or to shut down the plant. With a sequencing routine, it is possible for the computer to verify a command by the operator and to prevent him or her from moving to the next sequence of events until the previous request has been satisfied. In this way, the computer system can be utilized to ensure that the same procedures are followed for each batch run which helps to assure product quality and product uniformity. These routines also monitor raw material inventories in order to maintain proper deliveries, and they assist the operator in planning maintenance schedules for major equipment. Finally, the command language generates specific process information regarding the status of each vessel and automatically produces reports and documentation at the end of each run.

The second level on which the plant-scale computer system can be utilized is similar to that in the pilot plant. At this level, an engineer is required for making changes in the process-control or monitoring routines. These changes are required as process improvements are developed in the pilot plant and transferred to plant-scale operations.

## IV. SPECIFIC APPLICATIONS

### A. Data Logging

Table I lists variables that are commonly monitored during a fermentation. In data logging applications the raw data are fed into a computer system where they are compared with calibration curves and correction factors in order to obtain values in standard engineering units. The data are usually printed out on a hard-copy device and written onto a mass storage device for a more detailed analysis by a larger computer system at a later time.

**TABLE I.**  Variables Logged and Controlled

Measured parameters
  Inlet gas flow rate
  Inlet gas pressure
  Inlet gas temperature
  Inlet gas carbon dioxide
  Inlet gas oxygen
  Cooling jacket flow rate
  Cooling jacket temperature differences
  Agitator speed
  Power to shaft
  Vessel pressure
  Glucose concentration
  Culture turbidity
  Culture fluorescence
  Exit gas carbon dioxide
  Exit gas oxygen
Controlled variables
  Dissolved oxygen
  Vessel temperature
  pH
  Foam
  Substrate addition rate
  Acid addition rate
  Base addition rate

## B. Data Analysis

Numerous papers have been published dealing with the analysis of fermentation data (Yamashita and Murao, 1967; Yamashita et al., 1969; Humphrey, 1971; Nyiri, 1971; Jefferis et al., 1972; Wilson, 1975; Wilson et al., 1975; Wang, 1976; Zabriskie, 1976). Many of the analytical techniques depend upon the particular process for which they were developed. However, there are certain calculations that are common for most fermentations, These include computations for the oxygen uptake rate, carbon dioxide evolution rate, respiratory quotient, overall mass transfer coefficient, heat balance, power requirements, cell mass, and cell growth rate. Table II is a summary of some of these calculations.

These calculations require combining inputs from a number of measured variables. As a result, a computer system is a valuable tool for obtaining information on-line concerning the status of the fermentation process. Often, it is not practical to obtain these parameters by any other means. A good example of this is the measurement of biomass concentration and growth rate. Several papers have appeared in the literature concerning the indirect measurement of cell mass concentration, cell growth rate, and substrate concentration (Jefferis et al., 1972; Zabriskie, 1976; Wang, 1976; Armiger, 1977; Harima, 1977). Utilization of com-

**TABLE II.** Summary of Calculations

Oxygen uptake rate
Carbon dioxide evolution rate
Respiratory quotient
Total oxygen utilized
Total carbon dioxide evolved
Overall oxygen mass transfer rate
Heat balance
Metabolic heat evolution rate
Cell mass concentration
Cell mass growth rate
Specific growth rate
Cell yield
Nitrogen utilization rate
Glucose utilization rate
Carbon balance
Nitrogen balance
Oxygen balance
Apparent viscosity

puters for indirect measurement to give on-line biomass and growth rate data will have a major impact on the fermentation industry.

Two of these techniques have been derived by performing material balances around the fermentor. The first approach selects a specific molecular component and uses a kinetic model relating cell growth with the material balance. Figure 1 shows a balance around a fermentor for component X. Zabriskie (1976) estimated biomass concentration by a material balance for oxygen. He corrected the balance for secondary metabolite formation by also monitoring the carbon dioxide evaluation rate. The basic relationship is shown as Eq. (1) (see Fig. 2), where the

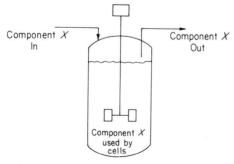

Component X In

Component X Out

Component X used by cells

Balance: in - out = used by cells for growth and maintenance

**FIGURE 1.** Material balance around a fermentor.

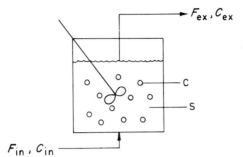

**FIGURE 2.** Oxygen balance around a fermentor. (From Zabriskie, 1976.)

rate of change of component $X$ with time is equal to the difference in input and output calculations plus an accumulation term.

Gas phase balance:
$$\frac{d}{dt}(V_g C)\, dt = F_{in}C_{in} - F_{ex}C_{ex} - J_{g \to a} \qquad (1)$$

Utilizing molecular oxygen for the material balance has several advantages over other components. First, there is a close relationship between oxygen and the energetics of growth. Second, oxygen concentration can be measured continuously in both the gaseous and the aqueous phase. Finally the solubility of oxygen in the aqueous phase is so low that a quasi-steady state may be assumed for the concentration of oxygen in the liquid phase. This simplifies the material balance equation. Thus, for oxygen, the uptake rate can be simply expressed as the difference between the inlet and outlet streams [See Eq. (2)].

$$F_{in}C_{in} - F_{ex}C_{ex} - R_a \approx 0 \qquad (2)$$

If one assumes the fermentor is essentially in steady state conditions with respect to oxygen utilization, then the oxygen uptake rate is equal to the oxygen consumption rate. A kinetic model for oxygen consumption, first proposed by Marr, *et al.* and by Luedeking and Piret (1959) is shown in Eq. (3).

Model:
$$R_a = (M_{i/x}X + \frac{1}{Y_{X/i}}\frac{dX}{dt})V_a \qquad (3)$$

This is a two-parameter model which assumes that the oxygen is utilized for two basic purposes within the organism. A portion of the oxygen is consumed for cellular maintenance functions which are necessary to sustain cell viability and must be satisfied at all times. The balance of the oxygen is utilized for cell growth, Thus, the two parameters in the model are termed the growth yield and cell maintenance constants.

By combining Eqs. (1) and (3), a first-order linear differential equation is obtained [Eqs. (4a) and (4b)].

$$\frac{F_{in}C_{in} - F_{ex}C_{ex}}{V_a} = M_{i/x}X + \frac{1}{Y_{X/i}}\frac{dX}{dt} \tag{4a}$$

In the case of oxygen:     $$OUR = M_{O_2/x}X + \frac{1}{Y_{X/O_2}}\frac{dX}{dt} \tag{4b}$$

Equations (5) and (6) are analytical solutions of this equation for the cell mass concentration and growth rate, respectively:

$$X(t) = \exp\left(-M_{O_2/x}Y_{X/O_2}t\right)\left[\int_0^t Y_{X/O_2}\exp\left(M_{O_2/x}Y_{X/O_2}t\right)\right. \tag{5}$$

$$\left. OUR\ (t)\ dt + X_o\right]$$

$$\frac{dX}{dt} = Y_{X/O_2}(OUR) - M_{O_2/x}Y_{X/O_2}X \tag{6}$$

To solve this equation only a knowledge of the yield and maintenance parameters plus some way of estimating the initial biomass concentration are needed. Computer results for batch culture of *Thermoactinomyces* sp. and *Saccharomyces cerevisiae* using these correlations based upon real-time oxygen uptake rate data are given in Fig. 3 and 4. For the *Thermoactinomyces* sp. there is good agreement between the on-line estimates of biomass and the laboratory data, especially before the beginning of the stationary phase.

The results are not as good for the experiment involving *S. cerevisiae* due to the diauxic growth of the organism. This observation points to one of the limitations of the first model which assumed that the growth yield and cell maintenance parameters are constant. Another limitation of the basic model is that it does not account for the formation of a secondary product. For example, in order to apply this technique to an antibiotic fermentation, a third term involving a product yield parameter must be included.

In order to deal with the problem of diauxic growth, Zabriskie (1976) introduced a metabolic correction factor ($\beta$) into Eq. (5) which is a function of the respiratory quotient. Figure 5 shows the estimates of the biomass concentration for the yeast incorporating the metabolic correction function into the model. It is obvious that a significant improvement in the estimation technique is obtained.

A biomass estimation procedure analogous to the oxygen balancing may be followed using, carbon dioxide. This method was also examined

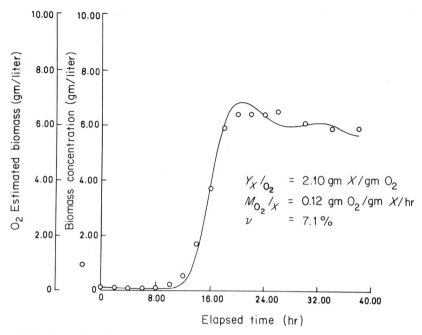

**FIGURE 3.** *Thermoactinomyces* sp. biomass concentration data and correlation results. (From Zabriskie, 1976.)

by Zabriskie (1976). However, the results show that this technque was much less accurate than oxygen balancing. This is probably due to the fact that carbon dioxide is not directly linked to oxidative phosphorylation so that the relationship between the metabolic correction factor and the respiratory quotient is not as simple.

The second method involving material balancing is based upon the principle of conservation of mass. It does not rely upon determining conversion yields or metabolic rates of activity. However, it does assume that the elemental compositions of the cells and extracellular products are known. The fundamental relationship states that biomass and extra-cellular products are produced in stoichiometric ratios. The substrate, along with oxygen and ammonia, is converted to biomass, product, carbon dioxide, and water, i.e.,

$$\text{Substrate} + \text{oxygen} + \text{ammonia}$$
$$= \text{biomass} + \text{product} + \text{carbon dioxide}$$
$$+ \text{water (carbon source)}$$

Once the molecular composition of the biomass and the product have been determined, it is possible to utilize the stoichiometric relationships along with monitoring the oxygen uptake rate, and the substrate or ammonia consumption rate, to continuously monitor the biomass con-

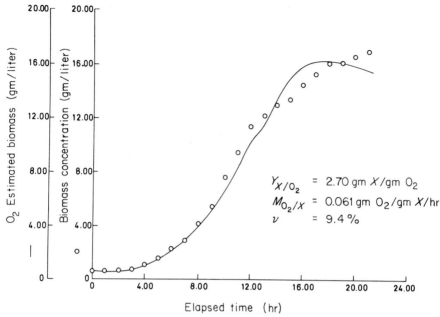

**FIGURE 4.** *Saccharomyces cerevisiae* biomass concentration data and correlation results. (From Zabriskie, 1976.)

centration and growth rate. The results of Wang (1976) are shown in Fig. 6.

The method of Zabriskie for determining biomass by oxygen uptake has been combined with Wang's technique for material balancing to derive a method of estimating the substrate concentration in a fermentation process that is impossible to measure on-line, i.e., cellulose (Armiger 1977).

This method assumes that there are two stoichiometric equations for the mass balance. One equation is for the growth of cells and production of product:

$$\text{Ammonia} + \text{carbohydrate} + \text{oxygen}$$
$$= \text{cells} + \text{product} + \text{carbon dioxide} + \text{water}$$

The other equation is used for cell maintenance:

$$\text{Carbohydrate} + \text{oxygen} = \text{carbon dioxide} + \text{water}$$

This technique was demonstrated by growing *Thermoactinomyces* sp. on cellulose where the product formed was extracellular protein. To balance these equations, compositions for the cell mass and protein were determined along with a product yield constant. The stoichiometric coefficients for oxygen and for carbon dioxide were measured experimentally and then the remainder of the coefficients were determined by

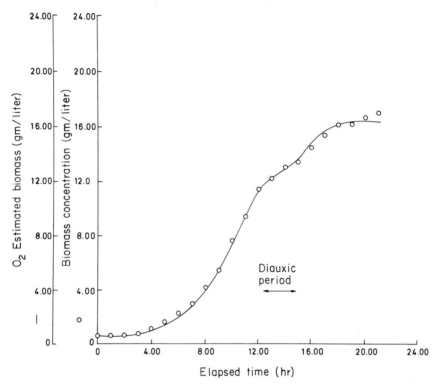

**FIGURE 5.** *Saccharomyces cerevisiae* biomass concentration data and correlation results with correction function.

balancing the equation. The cellulose concentration was calculated in the following manner: From a given inoculum size and oxygen uptake rate at any particular time, the biomass concentration was estimated. This estimate was then used in the stoichiometric equation to estimate the carbohydrate utilization and hence the concentration of cellulose.

Figure 7 illustrates the success of this technique by comparing laboratory analyses for cellulose and biomass with those values determined on-line. For checking the consistency of the stoichiometry, the on-line analyses for carbon dioxide and nitrogen were compared with the predicted results (Figs. 8 and 9).

Three previous examples all illustrate the value of an on-line computer system for the indirect estimate of important fermentation process parameters. Without the development of these and similar techniques, it would not be possible to evolve more sophisticated control strategies or to realize the on-line optimization of a fermentation.

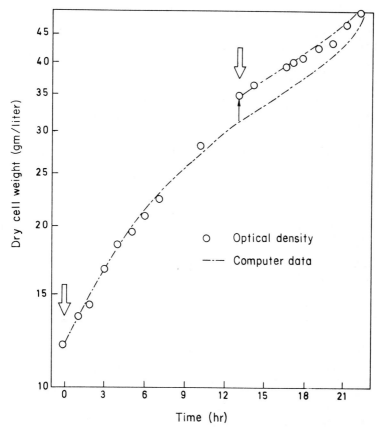

**FIGURE 6.**   Computer results of the cell measurement compared with the actual cell densities reinitialization ( ↓ ) at the thirteenth hour. (From Wang, 1976.)

## C. Process Modeling

Describing microbial processes by mathematical models is of increasing interest to microbiologists and biochemical engineers. Models are valuable for understanding processes and interactions among variables. In addition, models are necessary to provide the control engineer with insight into possible control strategies. Further, if the model can predict the dynamic behavior of a process, it can be useful for process optimization. Dynamic models are valuable for simulating experiments and testing control strategies. They can be utilized to eliminate many expensive and time-consuming experiments and to design key experiments to test a hypothesis. A large number of models for microbial processes have been proposed in the literature. These range from very

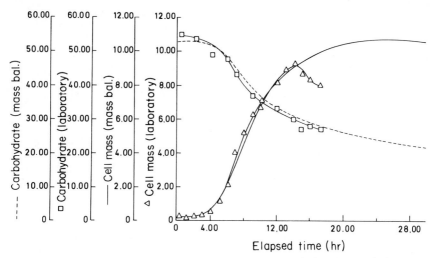

**FIGURE 7.** Cell mass and carbohydrate concentrations. Laboratory analysis versus mass balance estimation (From Armiger, 1977).

simple to very complex systems. Most of the models involve multivariable, nonlinear differential equations which must be solved by numerical techniques. Consequently, the utilization of a computer is absolutely essential in the solution of these fermentation models.

The objective of most fermentation models is to describe the formation of a primary product, such as biomass, or a secondary product, such as

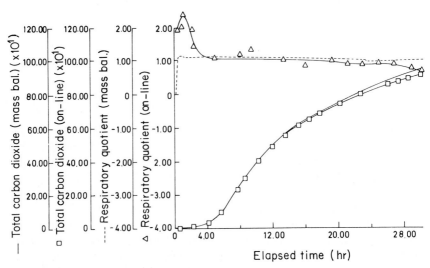

**FIGURE 8.** Total carbon dioxide (grams) and respiratory quotient on-line analysis versus mass balance estimate. (From Armiger, 1977.)

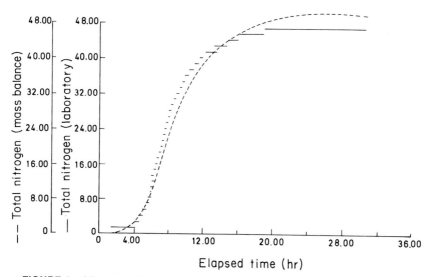

**FIGURE 9.** Nitrogen utilized (grams). Laboratory analysis versus mass balance estimation. (From Armiger, 1977.)

an antibiotic. These two situations have been modeled for batch, fed-batch, and continuous fermentations. There are two fundamental approaches to modeling systems. These involve the use of structured and unstructured models. Most fermentation models described unstructured systems. They do not account for individual differences between cells but examine variables that are characteristics of the macroscopic environment. Structured models do describe differences between cells and cell components; for example, one might describe the gross cell biomass, an enzyme component of that biomass, and the genetic mechanism activity. Unfortunately, the components of the structured model are difficult to measure; hence, they have not had the utility of the unstructured model. Tsuchiya et al. (1966), Fredrickson et al. (1970), and Yoshida and Taguchi (1977) all provide reviews of these approaches.

## D. Process Control and Optimization

The application of computer control to the fermentation industry has lagged behind similar advances in the chemical process industry by about 10 years. When the complexity of the kinetics of microbial systems is compared to the kinetics of chemical systems, the reasons for this lag become obvious. In spite of these difficulties, process control of fermentations can be implemented. The ways in which control has been implemented include regulation, sequencing, and optimization methods.

Regulation is concerned with maintaining individual process loops at preset values. Sequencing simply involves stepping through a predetermined program of events. Optimization is responsible for determining the best overall operating conditions for the process as a function of time. (All three of these functions are being utilized to some extent in fermentation production plants today.)

There are two basic approaches for using computers in process regulation—direct digital control (DDC) and digital set-point control (DSC). In DDC, as the name implies, the computer system interacts directly with the manipulated variables of the process. It has the advantage that the control algorithm can be totally specified by the control engineer. However, it has the disadvantage of requiring backup controllers to cover for computer system failures. With DSC, the computer simply changes the set points on a slave analog controller. This means that a computer system failure does not result in a total loss of control of the process. However, the modes of control are limited to those available on an analog device, i.e., proportional, integral, and differential.

In selecting a control system for a fermentation plant, the number of control loops required will determine which is the most economical approach. From a functional standpoint, DSC offers some distinct advantages over DDC. For DSC, the computer system only has to calculate set-point changes which are implemented at low frequency so the system is available for more meaningful or complex process calculations. A good example would involve running programs for calculation of operating parameter profiles associated with process optimization.

The regulation of individual process variables and the sequencing of plant operations is not unique to fermentation process control. However, the strategy utilized for manipulating and sequencing variables based upon process models is the area where the biochemical engineer can have the greatest impact. Yamashita et al. (1969) discuss these control functions as they relate to the production of glutamic acid. Several papers applying both discrete and continuous maximum principles for optimizing fermentation process have appeared (Constantinides and Vishva, 1974). These studies determined optimum profiles of operating conditions, such as temperature, pH, and dissolved oxygen. However, the optimization of the fermentation step exclusive of the entire process is of limited value. The real objective function of an optimization must involve the entire process from raw materials through product recovery. Blanch and Rogers (1972) approached this problem for the continuous production of gramicidin S. Okabe and Aiba (1973) have approached the problem in a more general manner for antibiotic production. Okabe and Aiba (1974a,b, 1975a,b) have split apart the process into a series of steps that are optimized individually. These parts were recombined to arrive at a global optimization for the process. This more fundamental approach to process optimization is the direction in which the field must move if it is going to have an important impact on fermentation processes. However, these advances will not be realized until more accurate dynamic models of systems are developed.

## V. SYSTEM CONFIGURATION

A highly instrumented fermentation system coupled to a digital computer is capable of generating on-line information for conceptualizing kinetic models and implementing strategies for process control and optimization. The process of utilizing the signal produced by an instrument and integrating it into a computer-controlled system involves both hardware and software development. With the advent of inexpensive but very powerful minicomputers, the problem of hardware development is basically one of selecting electronically compatible modules. The development of techniques and software for data analysis and process control is the area in which further research is needed. Configuration of system hardware and software must be structured to operate as an integrated unit. It must also be flexible enough to accommodate growth and expansion of both hardware and software as new products and techniques become available.

### A. System Hardware Configuration

The system hardware consists of three main segments: (1) sensors and instrumentation, (2) computer interface, and (3) computer system. In general, the weakest segment in any system involves the sensors. As shown in Table I the number of measurements is very limited. Most of the sensors are measuring variables in the macroscopic environment of the fermentation broth, while the key factors controlling growth and product formation involve the microscopic environment internal to the cell. The development of sensors and techniques for monitoring levels and activities of cellular constituents is only beginning. The physical problem of measuring fermentation parameters is further complicated by the sterilization requirements which severely restrict the choice of materials of construction of the sensors and often limit their sensitivity and accuracy.

The hardware for the computer interface is readily available. With sufficient attention to detail, it will reliably function. However, the architecture of the interface depends upon the application. Analog scanning is the most economical method of interfacing if there is a large number (over 25) of channels involved. However, digital multiplexing, which is cost competitive on a small system (under 25 channels), offers many advantages. First, it allows each channel to function as a separate unit, and it provides the operator with a visual display of all the channels at one time. Second, it allows for all of the data to be logged simultaneously rather than sequentially so that calculations can be made on data from a specific instant in time. For large systems, the advantages of digital multiplexing are difficult to justify because of cost. Therefore, in most plant or pilot-plant applications involving many vessels, an analog scanning interface is selected.

Computer interfacing involves a choice between increased flexibility and increased software requirements. This includes trade-offs such as random access scanning and increased hardware costs with less flexibility but lower software cost. Again, the number of channels involved in the interface will determine if the cost savings are to be gained by using more sophisticated hardware or software.

The selection of a computer system usually starts with the choice between a dedicated computer and a time-sharing system. With the availability of inexpensive but very powerful minicomputers, a dedicated system can usually be justified very easily on the basis of lower operating costs and increased system reliability. The details of selecting the size of computer memory and the peripheral devices are dependent upon the specific application. A disk-based system with sufficient memory to allow for foreground/background operation and some graphics capability is a good starting point. In many cases, communications with a large computer installation can be very beneficial for accessing special software or performing complex calculations.

Figure 10 is a hardware schematic and signal flow diagram for a typical installation. The analog signals from the various sensors are fed to the interface where the analog voltage is converted to a binary coded decimal (BCD) word. The BCD output of the interface is transmitted to the computer under control of the computer software which specifies the scanning rate and sequence. Once the data reach the computer, they are stored on the system disk and analyzed by the system programs. If the results of the analysis are to be used for process control, signals from the computer are then transmitted back to the fermentor to manipulate set points on analog controllers (DSC) or to manipulate variables directly (DDC). The results generated can also be output on a hard-copy device, stored on the disk or tapes, plotted on a graphical device, or transmitted to another computer system.

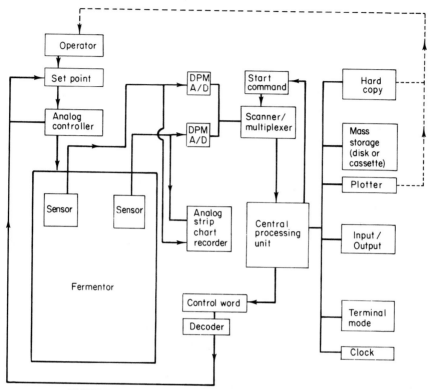

**FIGURE 10.** Schematic of physical configuration of system.

## B. System Software Configuration

The first step in system software configuration is the selection of the proper computer operating system. An operating system with foreground/background, multiprogramming, and multitasking capabilities and a high-level language (i.e., FORTRAN) is preferred. The man–machine interface should be as simple as possible in order to make the computer operator as effective and reliable as possible.

An overview of the system software organization is illustrated in Fig. 11. The computer system can be conceptualized as two parts, the central processor, which contains the core memory; and the system disk. All other computer equipment are really peripheral devices to support these two systems. The central processor contains the computer's instructional logic and a relatively small region of core memory which is allocated into three sections. The first part contains the operating system resident monitor, which is the software enabling the operator to issue commands from the terminal to load programs into core, execute them, and control

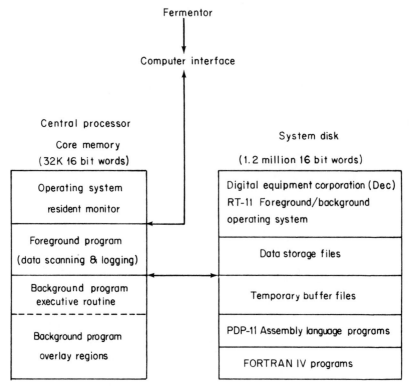

**FIGURE 11.**  Overview of system software.

and monitor information transmitted to or from all system devices. The second portion contains the foreground program, which is responsible for sending the start command at the proper time to the scanner and logging the data on the disk. The third section is for the background program, which is divided into many smaller regions. The first region contains the executive routine for managing the sequence in which different program modules enter, execute their function, and leave the overlay regions.

Since computer core is limited in minicomputers, only data and program modules required at a particular point in time reside in an overlay region. The bulk of the software remains on the disk. As time progresses and certain tasks are completed, data are transferred to a file on the disk and a new program module is read from the disk and overlaid on top of a module whose function is complete. The system disk has no computational power; it is a storage device only. It contains vast amounts of information consisting of the operating system modules required by the resident monitor, data storage files, temporary buffer files, FORTRAN programs, and assembly language routines.

In a foreground/background operating system, tasks in the foreground have system priority. When the foreground has nothing to perform, control is automatically switched to the background to execute its task. Therefore, software must be divided into foreground and background tasks. The foreground functions include starting the scanner, logging data on the disk, and generating alarm messages. The benefits to this arrangement are manifest. The most time-dependent functions are performed first.

Clearly, it would be unacceptable if data acquisition had to be delayed because the computer was occupied by producing a report which could take as long as an hour. Typically, the tasks performed in the foreground can be executed very rapidly (seconds). Thus, without foreground/background capability, all of the time between the end of the foreground function and the generation of a successive data set by the scanner (1–10 minutes) would be wasted. However, with this arrangement, the system can be utilized automatically between scans for many different functions from analyzing laboratory data, simulating the fermentation, producing graphs, or developing new programs.

## VI. FUTURE TRENDS

Computer applications in fermentation technology will increase. The availability of microprocessors and software to support these systems will change the hardware configuration of a system and greatly reduce the costs. Soon the computer associated costs will be much less than those related to the sensor systems. However, the real future lies in developing improved mathematical models and strategies for process control and optimization. At this time, the capabilities of computer hardware greatly exceed the capabilities of the software that has been developed. Therefore, the greatest challenges for future application of computers to fermentation are involved with interpreting results rather than physically implementing control decisions. The time has arrived for the biochemical and control engineers to utilize their skills to bring computer applications in fermentation technology up to the level of the available computer technology.

## VII. SUMMARY

Computers have a variety of applications in the fermentation industry ranging from data logging and on-line analysis to process modeling, control, and optimization. Only a small fraction of the potential has been realized. This is mainly in the areas of data logging and on-line analysis. Many groups are working on modeling, control, and optimization, but there has not yet been significant developments in terms of interactive control. One of the major problems in developing mathematical models and control strategies lies in the inability to measure on-line many of the important process parameters. Significant improvements in existing sensors and the development of new sensors is needed. Finally, fermentation-related software development needs to be brought to the high level that will fully utilize computer hardware developments, particularly in the microprocessor area.

## REFERENCES

Armiger, W. B. (1977). Ph.D. Thesis, University of Pennsylvania, Philadelphia.
Armiger W. B., and Humphrey, A. E. (1974). *Proc. Intersect. Congr. IAMS, 1975 1st,* **5,** 99–119.
Armiger, W. B., Zabriskie, D. W., Humphrey, A. E., Lee S. E., and Jolly, G. (1975). *AIChE Symp. Ser. No. 158* **72,** 77–85.
Armiger, W. B., Zabriskie, D. W., and Humphrey, A. E. (1976). *Proc. Int. Ferment. Symp., 5th, 1976* p. 33.
Blachere, H. T. (1973). *Proc. Eur. Conf. Comput. Process Control Ferment., 1st, 1973,* Session 1, Paper 2.
Blanch, H. W., and Rogers, P. L. (1972). *Biotechnol. Bioeng.* **14,** 151–171.
Constantinides, A., and Vishva, R. R. (1974). *Biotechnol. Bioeng. Symp.* **4,** 663–680.
Cooney, C. L., Wang, H., and Wang, D. I. C. (1975). *Pap., US/U.S.S.R. Semin. Meas. Ferment. Processes.*
Corrieu, G., Blachere, H., and Geranton, A. (1974). *Biotechnol. Bioeng. Symp.* **4,** 607.
Flynn, D. S. (1974). *Biotechnol. Bioeng. Symp.* **4,** 597–605.
Fredrickson, A. G. Megee R.D., and Tsuchiya, H. M. (1970). *Adv. Appl. Microbiol.* **13,** 15–30.
Grayson, P. (1969). *Process Biochem.* **4,** 43–44, 61.
Harima, T. (1977). Ph.D. Thesis, University of Pennsylvania, Philadelphia.
Harrison, D. E. F., and Harmes, C. S. (1972). *Process Biochem.* **7**(4), 13–16.
Harrison, D. E. F. (1971). *Proc. Int. Congr. Microbiol., 10th, 1970.*
Humphrey, A. E. (1971). *Proc. LABEX Symp. Comput. Control Ferment. Processes, 1971* p. 1.
Jefferis, R. (1973). *Proc. Eur. Conf. Comput. Process Control Ferment. 1st, 1973,* Session 1, Paper 2.
Jefferis, R. (1975). *Process Biochem.* **10,** 15.
Jefferis, R., Humphrey, A. E., and Nyiri, L. K. (1972). *164th Annu. Meet., Am. Chem. Soc.*
Koga, S., Berg, C. R., and Humphrey, A. E. *Appl. Microbiol.* **15,** 683.
Lane, A. (1973). *Proc. Eur. Conf. Comput. Process Control Ferment. 1st, 1973,* Session 2, Paper 2.
Luedeking, R., and Pirt, E. L. (1959). *J. Biochem. Microbiol. Technol. Eng.* **1,** 393.
Marr, A. G., Nilsson, E. H., and Clark, D. J. (1962). *Ann. N.Y. Acad. Sci.* **102,** Art. 3, 536.
Moes, H., Ryu, D. Y., Brannick, L., and Cohan, M. (1971). *162nd Natl. Meet. Am. Chem. Soc.*
Nagai, S., Nishizawa, Y., and Yamagata, T. (1976). *Proc. Int. Ferment. Symp., 5th, 1976,* p. 30.
Nyiri, L. K. (1971). *Proc. LABEX Symp. Comput. Control Ferment. Process, 1972* **13,** 13.

Nyiri, L. K., Toth G. M., and Charles, M. (1975). *Biotechnol. Bioeng.* **17,** 1663
Okabe, M., Aiba, S., Okado, M. (1973). *J. Ferment. Technol.* **51,** 594–605.
Okabe, M., and Aiba, S. (1974a). *J. Ferment. Technol.* **52,** 279–292.
Okabe, M., and Aiba, S. (1974b). *J. Ferment. Technol.* **52,** 759–777.
Okabe, M., and Aiba, S. (1975a). *J. Ferment. Technol.* **53,** 230–240.
Okabe, M., and Aiba, S. (1975b). *J. Ferment. Technol.* **53,** 730–743.
Tsuchiya, H. M., Fredrickson A. G., and Aris, R. (1966). *Adv. Chem. Eng.* **6,** 125–206.
Unden, A. G., and Heden, C. G. (1973). *Proc. Eur. Conf. Comput. Process Control Ferment. 1st, 1973,* Session 2, Paper 1.
Wang, H. Y. (1976). Ph.D. Thesis, MIT, Cambridge. Massachusetts.
Wilson, J. D. (1975). U.S. Patent 3,926,738.
Wilson, J. D., Nyiri, L. K., Humphrey, A. E., and Harmes, C. S. (1975). U. S. Patent 3,926,737.
Yamashita, S. and Murao, C. (1967). *J. Soc. Instrum. Control Eng.* **6,** 735.
Yamashita, S., Hoshi, H. and Inagaki, T. (1969). *Ferment. Adv. Pap. Int. Ferment. Symp., 1968* pp. 441–463.
Yoshida, T., and Taguchi, H. (1977). *In* "Workshop on Computer Applications in Fermentation Technology 1976" (R. P. Jefferson ed.); GBF Monogr. Ser. No. 3, pp. 93–106. Verlag Chemie, Weinheim.
Young, T. B., and Koplove, H. M. (1972). *Ferment. Technol. Today, Proc. Int. Ferment. Symp., 4th, 1972.*
Zabriskie, D. W. (1976). Ph.D. Thesis, University of Pennsylvania, Philadelphia.

## SUPPLEMENTARY READING

### General Papers and Reviews

Armiger, W. B., ed. (1979). "Computer Applications in Fermentation Technology," *Vol. 9. Biotechnol. Bioeng. Symp. Ser.,* Wiley, New York.
Dolby, D. D. and Jost, J. L. (1977). Computer applications to fermentation operations. *In* "Annual Reports on Fermentation Processes" (D. Perlman, ed.), pp. 114. Academic Press, New York. Vol 1.
Fuld, G. J. (1960). Control applications in fermentation processes. *Adv. Appl. Microbiol.* **2,** 351–360.
Humphrey, A. E. (1971). Present limitations to the control and understanding of a fermentation process. *Proc. LABEX Symp. Comput. Control Ferment. Processes, 1971* p. 1.
Humphrey, A. E. (1973). Rationale and economics of computer process control *Proc. Eur. Conf. Comput. Process Control Ferment. 1st, 1973* Session 1, paper 1.
Humphrey. A. E. (1974). Current developments in fermentation *Chem. Eng.* p. 98.
Humphrey, A. E. (1975). Rationale and problems in the use of computer-coupled fermentation systems. *Pap., US/USSR Semin. Meas. Ferment. Processes.*
Humphrey, A. E., and Jefferis R. P. (1973). Optimization of batch fermentation processes. *Pap. GIAM Meet. 4th 1973.*
Humphrey, A. E., and Jefferis R. P. (1977). Computer-assisted fermentation developments. *Dev. Ind. Microbiol.* **18,** 58–70.
Khosrovi, B., and Topiwala, H. H. (1974). Mathematical modeling and computer control in fermentation processes. *Annu. Rep. Prog. Appl. Chem.* **59,** 357–365.

Mueller, H., and Mueller, F. (1973). Examples and studies of continuous fermentation optimization by computer. *Pap., GIAM Meet., 4th, 1973.*

Nyiri, L. K. (1972). A philosophy of data acquisition, analysis, and computer control of fermentation processes. *Dev. Ind. Microbiol.* **13,** 136–145.

Nyiri, L. K. (1973). Application of computers in biochemical engineering. *Adv. Biochem. Eng.* **2,** 49–93.

Nyiri, L. K. (1974). Strategies and problems related to computer coupled fermentation systems. *Proc. Intersect. Congr. IAMS 1st. 1975* Vol. 5; pp. 120–136.

Nyiri. L. K., and Jefferis, R. P. (1973). Process control aspects of single-cell production in submerged cultures. *Proc. Int. Conf. Single-Cell Protein, 2nd, 1973,* p. 105.

Nyiri, L. K., Jefferis, R. P., and Humphrey, A. E. (1974). Applications of computers to the analysis and control of microbiological processes. *Biotechnol. Bioeng. Symp.* **4,** 613.

Ryu, D. D. V., and Humphrey, A. E. (1973). Examples of computer-aided fermentation systems. *J. Appl. Chem. Biotechnol.* **25,** 283–295. 1973.

Trambouze, P., and Mueller, H. (1973). Computers in fermentation control. *Process Biochem.* **8,** 7–10.

## System Design and Application

Geranton, A. (1973). Process control in fermentation—design and construction of software *Proc. Eur. Conf. Comput. Process Control Ferment., 1st, 1973, Session 3, Paper 2.*

Huang, I., and Sonn, M. (1972). Computer control of batch processes. *Br. Chem. Eng. & Process Technol.* **17,** 507–512.

Jefferis, R. P. (1977). Software and file structures for computer-coupled pilot fermentation plants. *In* "Workshop on Computer Applications in Fermentation Technology 1976", (R. P. Jefferis, ed.), GBF Mono. Ser. No. 3, pp. 21–36. Verlag Chemie, Weinheim.

Lundell, R., and Laiho, P. (1976). Engineering of fermentation plants. *Process Biochem.* **11,** 13–17.

Meskanen, A., Lundell R., and Laiho, P. (1976). Engineering of fermentation plants. Part 3. *Process Biochem.* **11,** 31–36.

Shave, J. D. (1976). The computer in the brewhouse. *Process Biochem.* **11,** 10–12.

Spruytenburg, R., Dang, A. D. P., Dunn, I. J., Mart, J. R., Einsele, A., Fiechter, A., and Bourne, J. R. (1976). Experience with a computer-coupled bioreactor. *Chem. Eng. (London)* **310,** 447.

Wilson, J. (1975). Planning a computer-coupled fermentation production plant. Pap., US/USSR Semin. Meas. Ferment. Processes, 1975.

Wilson, J. D., Humphrey, A. E., Harmes, C. S., and Nyiri, L. K. (1974). Procède et appariel pour controler les réactions biochimiques. French Patent 2,184,035,

## Data Analysis and Modeling

Armiger, W. B., Moreira, A. R., Phillips, J. A., and Humphery, A. E. (1977). Modeling cellulose digestion for single cell protein *PACHEC, 2nd 1977* pp. 196–205.

Blanch, H. W., and Rogers, P. L. (1977). Production of Gramicidin S. in batch and continuous culture. *Biotechnol. Bioeng.* **13,** 843–864.

Blanch, H. W., and Dunn, I. J. (1974). Modeling and simulation in biochemical engineering. Adv. Biochem. Eng. **3,** 127–165.

Brown, D. E., and Vass, R. C. (1973). Maturity and product formation in cultures of microorganisms. *Biotechnol. Bioeng.* **15,** 321–330.

Cooney, C. L., Wang, H., and Wang, D. I. C. (1975). Computer-aided fermentation monitoring and diagnostics. *Pap., US/USSR Semin. Meas. Ferment. Processes, 1975.*

Cooney, C. L., Wang, H., and Wang, D. I. C. (1977).Computer-aided material balancing for prediction of fermentation processes. *Biotechnol. Bioeng.* **19,** 55–69.

Dang, N. D. P. (1976). The building and use of dynamic models for biological studies. *J. Ferment. Technol.* **54,** 396–405.

Dang, N. D. P., Dunn, I. J., and Mor, J-R. (1975). Modeling of dynamic biological processes with empirical transfer functions. *J. Ferment. Technol.* **53,** 885–894.

Edwards, V. H., and Wilke, C. R., (1968) Mathematical representation of batch culture data. *Biotechnol. Bioeng.* **10,** 205–232.

Erickson, L. E., Nakahara, T., Gutierrez, J. R., MacLean, G. T., and Fan, L. T. (1975). Modeling and characterization of hydrocarbon fermentations. *Pap., US/USSR Joint Semin. Univ. Pennsylvania, 1975,* p. 75.

Fiechter, A., and von Meyenburg, K. (1968). Automatic analysis of gas exchange in microbial systems. *Biotechnol. Bioeng.* **10,** 535–549.

Fishman, V. M., and Biryukov, V. V. (1974). Kinetic model of secondary metabolite production and its use in computation of optimal conditions. *Biotechnol. Bioeng. Symp.* **4,** 647–662.

Imanaka, T., Kaiedo T., Soto K., and Taguchi H. (1972). Optimization of alpha-galactosidase production by mold. *J. Ferment. Technol.* **50,** 633–646.

Imanaka, T., Kaiedo, T., and Taguchi H. (1973). Unsteady-state analysis of a kinetic model for cell growth and alpha-galactosidase production in mold. *J. Ferment. Technol.* **51,** 423–430.

Johnson, D. B., and Berthouex, P. M. (1975). Using multiresponse data to estimate biokinetic parameters. *Biotechnol. Bioeng.* **17,** 571–583.

Khan, A. H., and Ghose T. K. (1973). Kinetics of citric acid fermentation by *Aspergillus Niger. J. Ferment. Technol.* **51,** 734–741.

Kishimoto, M., Yamane, T., and Yoshida, F. (1976). Sensitivity analysis for exponential fed-batch culture. *J. Ferment. Technol.* **54,** 891–901.

Kono, T., and Asai, T. (1969a). Kinetics of continuous cultivation. *Biotechnol. Bioeng.* **11,** 19–36.

Kono, T., and Asai, T. (1969b). Kinetics of fermentation processes. *Biotechnol. Bioeng.* **11,** 293–321.

Kremen, A. (1974). Possible control mechanisms in tetracycline Production: A kinetic model. *Biotechnol. Bioeng. Symp.* **4,** 105–113.

Matsumura, M., Muhotaka, S., Yoshetomo, H., and Kobayashi J. (1973). Analysis of cell behavior in transient state of continuous culture with an apparent age model. *J. Ferment. Technol.* **51,** 904–916.

Minkevich, I. C., and Eroshin, V. K. (1974). Theoretical calculations of mass balance during the cultivation of microorganisms. *Biotechnol. Bioeng. Symp.* **4,** 21.

Moreira, A. R. (1977) Model cellulose degradation by *Thermoactinomyces 168th Annu. Meet. Am. Chem. Soc.*

Nagatani, M., Shoda, M. and Aiba S. (1968). Kinetics of product inhibition in alcohol fermentation. *J. Ferment. Technol.* **46,** 241–248.

Nyiri, L. K., Toth, G. M., Krishnaswami, C. S., and Parmenter, D. V. (1977). On-line analysis and control of fermentation processes. *In* "Workshop on Computer Applications in Fermentation Technology 1976" (R. P. Jefferis, ed.), GBF Monogr. Ser. No. 3, pp. 37–46. Verlag Chemie, Weinheim.

Paynter, M. J. B., and Bungay, H. R. (1970). Responses in continuous culture of lysogenic. *Escherichia coli* following induction. *Biotechnol. Bioeng.* **12,** 347–351. 1970.

Peringer, P. Blachere, H., Borrieu, G., and Lane, A. G. (1973). Mathematical model of the kinetics of growth of *Saccharomyces cerevisiae*. Biotech, Bioeng. Symp. **4**, 27–42. 1973.

Perringer, P., Blachere, H., Borrieu, G., and Lane, A. G. (1974). A generalized mathematical model for the growth kinetics of *Saccharomyces cerevisiae* with experimental determination of parameters. *Biotechnol. Bioeng.* **16**, 431–454. 1974.

Ribot, D. (1977). Polynomial identification methods *In* "Workshop on computer applications in Fermentation Technology 1976" (R. P. Jefferis, ed.), GBF Monogr. Ser. No. 3, pp. 125–140. Verlag Chemie, Weinheim.

Sundstrom, H. E., Klei, H. E., and Brookman, G. T. (1976). Response of biological reactors to sinusoidal variations of substrate concentration. *Biotechnol. Bioeng.* **18**, 1–14.

Svrcek, W. Y., Elliott R. F., and Zajic J. E. (1974). The extended Kalman filter applied to a continuous culture model. *Biotechnol. Bioeng.* **16**, 827–846. 1974.

Swartz, J. R., Wang, H., Cooney, C. L., and Wang, D. I. C. (1976). Computer-aided yeast fermentation. *Proc. Int. Ferment. Symp., 5th, 1976*, pp. 29–63.

Van Dedem, G., and Moo-Young, M. (1975). A model for diauxic growth. *Biotechnol. Bioeng.* **17**, 1301.

Wang, D. I. C. (1975). Modeling and control on the utilization of mixed substrates by mixed cultures. *Pap., US/USSR Joint Semin. Data Acquisition Process. Lab. Ind. Meas. Ferment. Processes, 1975.*

Wang, H., Cooney, C. L., and Wang, D. I. C. (1977). Computer-aided baker's yeast fermentation. *Biotechnol. Bioeng.* **19**, 69–86.

Yamane, T., and Hirano, S. (1977). Semi-batch culture of microorganisms with constant feed of substrate. *J. Ferment. Technol.* **55**, 156–165. 1977.

Yoshida, T., and Taguchi, H. (1977). The use of models in fermentation control. *In* "Workshop on Computer Applications in Fermentation Technology 1976" (R. P. Jefferis, ed.), GBF Monogr. Ser. No. 3, pp. 93–106. Verlag Chemie, Weinheim.

Young, T. B., and Koplove, H. M. (1972). A systems approach to design and control of antibiotic fermentations. *Ferment. Technol. Today, Proc. Int. Ferment. Symp., 4th, 1972.*

Zabriskie, D. W. (1977). Real-time estimation of fermentation biomass concentration using component balancing techniques. *Proc. US/USSR Joint Semin. 1975.*

Zabriskie, D. W., Armiger, W. B., and Humphrey, A. E. (1977). Applications of computers to the indirect measurement of biomass concentration and growth rate by component balancing. *In* "Workshop on Computer Applications in Fermentation Technology 1976." (R. P. Jefferis, ed.) GBF Monogr. Ser. No. 3, pp. 59–72. Verlag Chemie, Weinheim.

Zines, D. O., and Rogers, P. L. 1971). A chemostat study of ethanol inhibition. *Biotechnol. Bioeng.* **13**, 293–308. 1971.

## Process Control and Optimization

Aiba, S., Nagai, S., and Nishizawa, Y. (1976). Fed batch culture of *Saccharomyces cerevisiae*—a perspective of computer control to enhance the productivity in baker's yeast cultivation. *Biotechnol. Bioeng.* **18**, 1001–1016.

Blachere, H. T., Peringer P., and Corrieu, G. V. (1977). Optimization of fermentation plants. *In* "Workshop on Computer Applications in Fermentation Technology 1976." (R. P. Jefferis, ed.), GBF Monogr. Ser. No. 3, pp. 1–10. Verlag Chemie, Weinheim.

Bourdaud, D., and Foulard, C. (1973). Identification and optimization of batch culture fermentation processes. *Proc. Eur. Conf. Comput. Process Control 1st, 1973* Session 4, Paper 1.

Chen, H. T., and Fan, L. T. (1975), Principles of on-line dynamic optimization in fermentation systems. *Pap., US/USSR Joint Semin. Data Acquisition Process. Lab. Ind. Meas. Ferment. Processes, 1975.*

Constantinides, A., Spencer, J. L., and Gaden, E. L. (1970a). Optimization of batch fermentation processes. I. Development of mathematical models for batch penicillin fermentations. *Biotechnol. Bioeng.* **12,** 803.

Constantinides, A., Spencer, J. L., and Gaden, E. L. (1970b). Optimization of batch fermentation processes. II. Optimum temperature profiles for batch penicillin fermentations. *Biotechnol. Bioeng.* **12,** 1081.

Gaudy, A. F. (1975). The transient response to pH and temperature shock loading of fermentation systems. *Biotechnol. Bioeng.* **17,** 1051–1064.

Humphrey, A. E., and Jefferis, R. P. (1973). Optimization of batch fermentation processes *Proc. GIAM Meet. 4th, 1973.*

Imanaka, T., Kaieda, Takeji and Taguchi, Hesahaw (1973). Optimization of alpha-galactosidase production in multi-stage continuous culture of mold. *J. Ferment. Technol.* **51,** 431–439.

Lane, A. C., and Blachere, H. (1973). Computer control and optimization in the cultivation of yeast. *Proc. GIAM Meet. 4th, 1973.*

Muzychenko, L. A., Mascheva, L. A., and Yakovleva, G. V. (1974). Algorithm for optimal control of microbiological synthesis. *Biotechnol. Bioeng. Symp.* **4,** 629–645.

Paschold, H., and Verstoep, N. D. L. (1977). Computer based control of fermentation processes. *Soc. Chem. Ind. Meet.*

Perringer, P., Blum, J., and Blachere, H. (1974). Optimal single-cell production from yeast in a continuous fermentation process. *Proc. Intersect. Meet. IAMS,* **5,** pp. 82–98.

Ryu, D. Y., and Humphrey, A. E. (1972). A reassessment of oxygen-transfer rates in antibiotic fermentations. *J. Ferment. Technol.* **50,** 424–431.

Spitzer, D. W. (1976). Maximization of steady-state bacterial production in a chemostat with pH and substrate control. *Biotechnol. Bioeng.* **18,** 167–178.

Wang, D. I. C., Wilcox, R. P., and Evans, L. B. (1975). Modeling and control on the utilization of mixed substrates by mixed cultures *Pap., Joint US/USSR Semin., Univ. Pennsylvania, 1975.*

Chapter 16

# General Procedures for Isolation of Fermentation Products

PAUL A. BELTER

## I. INTRODUCTION

The subject, recovery and isolation of products produced by fermentations, is a branch of a much larger topic, the separation of natural

**403**

MICROBIAL TECHNOLOGY, 2nd ed., VOL. II
Copyright © 1979 by Academic Press, Inc.
All rights of reproduction in any form reserved. ISBN 0-12-551502-2

products, which in itself is a subdivision of the overall subject of separation science. Books have been written on the subject of separation, and the techniques evolved to obtain "pure" compounds. Scientists and technologists have spent an appreciable part of their technical lives devising methods to separate the complex mixtures that nature produces and, according to the second law of thermodynamics, prefers. This effort requires a great deal of knowledge, ingenuity, and creativity to devise better separation tools and, subsequently, integrated optimal isolation procedures. In the fermentation industry the isolation operations often require more energy, equipment, and labor than the fermentation per se. The recovery plant usually represents a major investment and the isolation costs are a substantial fraction of the cost of the product. It is, therefore, important that the isolation processes be well conceived and the plants well designed.

The isolation of products produced by fermentation is a specialized branch of separation science because of the unique character of the materials being handled rather than any separation science peculiarity. Many of the techniques useful for carbohydrate purification, oil extraction, and protein fractionation are applicable. The wide diversity of product types, and the fact that often living dynamic organisms and/or their enzyme systems are involved, complicate the situation. In spite of this, general principles, theories, and models for separation operations have been developed that are applicable to the isolation of fermentation products (Karger et al., 1973). Many of these fundamentals have been covered in detail in the literature, especially the chemical engineering literature (Holland, 1975), but usually are illustrated by examples that are not related to the fermentation industry. For example, the fermentation of various carbohydrate materials to make ethyl alcohol and the subsequent recovery of "spirits" by distillation was practiced long before fractional distillation theory was developed. In fact, Sorel's (1893) original work was conducted to understand the empirically practiced distillations of the time. Today, after much improvement and refinement, distillation theory forms the backbone of the petroleum industry and is equally applicable in certain areas of the fermentation industry.

King (1971) divided separation processes into three categories depending on their fundamental controlling phenomenon: mechanical, equilibrium and rate-controlled processes. Thirty of the forty-two processes listed have been or are being used to isolate fermentation products. The spectrum of product types discussed in this book covers widely varying categories ranging from relatively simple amino and organic acids to complex mixtures of antibiotics and enzyme preparations. In practice, operational scales range from the laboratory to commercial plants. Batch and continuous operations, both concurrent and countercurrent, are used. In addition, various modes of operation, such as single-stage, multiple discrete-stage, and differential contacting techniques, are employed. Consequently, both steady state and transient state considerations are involved. For the application of these techniques knowledge of the pertinent material and energy balances, equilibrium relationships, and rate phenomena is essential. Unfortunately, both the physical and chemical information for fermentation products are rarely available and must be determined before separation theory and models can be used. This has led in the past to a high degree of empiricism that it is hoped will be minimized in the future. It is obvious from the foregoing that the proposed subject is an extremely diverse one which can only be covered superficially in this chapter. Consequently, the format of this chapter is the discussion of segments of recovery processes, their underlying theory, and general utility for all products rather than to give specific processing details for individual products. In addition, we concentrate on developments that have achieved commercial importance, while laboratory and small-scale techniques are mentioned only briefly.

The basic problems that confront the separation technologists, whether they be chemists or engineers, are the complexity of the mixture and the relatively low concentration of the desired product in the starting material or feed to the isolation process, the harvested fermentation broth or "beer." This heterogeneous mixture contains the intact organisms or mycelium, or fragments thereof, plus other insoluble fractions of medium ingredients, residual substrates, and, in some cases, product. The soluble portions of residual substrates, metabolic pathway intermediates, and usually the desired product are present in the liquid part of this mixture. Although not always, the product usually is present in a relatively low concentration compared to the concentration of other materials present.

Questions that must be answered to devise an isolation process for a given product are listed below.

1. What is the value of this product?
2. What is an acceptable product quality for the proposed end use?
3. Where is the product in the complex mixture?
4. Where are the impurities in the complex mixture?
5. What are the physical and chemical properties of the product and the impurities?
6. What are the economics of various isolation possibilities?

Careful consideration of these questions will provide answers that will enable the isolation technologist to reach the goals of adequate product quality and high recovery with minimum effort. Obviously, the last two items affect the final cost of the product appreciably.

## II. GENERAL PROCESSING

Although isolation processes are usually similar for a given class of compound, they vary appreciably between classes. Obviously, one would not use the same isolation techniques for volatile organic solvents that one would use to isolate sensitive and unstable enzymes or antibiotics.

Fortunately, an overall similarity exists in the processing sequence of all of the recovery schemes. In general, conventional isolation processes can be divided into the following four segments.

1. *Removal of insolubles.* Filtration, centrifugation, and decantation are the usual unit operations used in this segment. Relatively little concentration of product occurs during these steps.

2. *Primary isolation of product.* These processes are relatively nonselective in performance and only separate materials of widely divergent polarities. Appreciable concentration and quality increases usually occur during these sequences. Typical operations are sorption (physical and/or ion exchange), solvent extraction, precipitation, and ultrafiltration.

3. *Purification.* Processing techniques which are highly selective for the product and remove impurities of similar chemical functionality and physical properties are used in this segment. Typical approaches are fractional precipitation, chromatography of various types (adsorption, partition, ion exchange, and affinity), chemical derivatization, and decolorization.

4. *Final product isolation.* The desired characteristics of the product for its ultimate use dictate the final sequence utilized at this point. Crystallization followed by centrifugation or filtration and drying are typical steps used for high-quality products. Other products may be drum or spray dried directly from solution if the solvent is water, or desolventized if an organic solvent has been used.

A hypothetical though reasonably typical processing profile for a product is given in Table I.

Several isolation processes have been developed in recent years that simplify the older conventional sequence by combining the first two steps. These developments, whole-broth extraction and fluidized ion-exchange processes, are discussed in Section III, D.

**TABLE I.** Typical Processing Profile

| Step | Product Conc. (gm/liter) | Product Quality (%) |
|---|---|---|
| Harvest broth | 0.1–5 | 0.1–1.0 |
| Filtration | 0.1–5 | 0.1–2.0 |
| Primary isolation | 5–10 | 1–10 |
| Purification | 50–200 | 50–80 |
| Crystallization | | 90–100 |

## III. SOLIDS REMOVAL

The conventional first step in the isolation of a fermentation product is the separation of the heterogeneous fermentation broth into two fractions, "solubles" and "insolubles." This division usually simplifies or makes possible the application of other techniques to the stream that contains the desired material that would not be possible otherwise; i.e., the use of sorption or ion exchange columns in the fixed-bed mode that would be inoperative unless the feed solution was devoid of particulate matter, the selective precipitation of crude product with salt forms or flocculants, and/or the prevention of excessive emulsification during extraction with an organic solvent. In the cases where the product resides in the insoluble fraction, an appreciable reduction in the volume of material to be processed occurs simultaneously with the removal of most of the soluble impurities.

### A. Filtration

In general, fermentation broths are difficult to filter for one or more of a variety of reasons. The nature, sliminess, morphology, and size of the microorganisms; pH and viscosity of the broth; temperature history of the broth; contamination by undesired microorganisms; and the nature of the insoluble portion of residual substrates are a few of the factors that contribute to the filtration difficulties.

Special techniques have been developed to study the filtration characteristics of these broths in the laboratory and to simulate in the laboratory the operation of continuous rotary vacuum filters that are used for filtering these intractable mixtures on a production scale.

The rotary vacuum precoat filter is probably the most widely used filter in the fermentation industry. The drum, which rotates in a trough of fermentation broth, is covered with a layer of precoat, usually diatomaceous earth, prior to filtration. A small quantity of admix is added and thoroughly mixed with the broth before it is pumped into the filter trough. As the drum under vacuum rotates, a thin layer of admix and mycelium adhere to the drum. The layer of solids is dewatered during its passage on the drum to the discharge point, where a knife blade cuts off a thin layer of filter aid precoat and fermentation solids. Usually the cake is washed with water several times during its passage over the drum. These washes serve to displace the broth solution and solubles that are held in the cake voids.

A second type of continuous vacuum filter having a string discharge has been useful for mold fermentations, such as penicillin, where the mold mycelium interlaces upon itself on the drum to form a mat which is subsequently lifted from the drum by the filter strings.

It should be evident that the nature of the insolubles affects the type of filter that can be used for a given application.

### *Theory*

The important variables for a constant pressure filtration can be correlated in a simple fashion by modification of the Poiseuille equation. Integration of that relationship gives Eq. (1).

$$\Theta/V = \mu\alpha CV/2\Delta PA^2 + \mu R_m/\Delta PA \tag{1}$$

where $\Theta$ is time, $V$ is the volume of filtrate, $\mu$ is the viscosity of the filtrate, $C$ is the volume of cake deposited per volume of filtrate, $A$ is the filtration area, $R_m$ is the initial resistance of filter medium, $\Delta P$ is the pressure drop across the bed, and $\alpha$ is the average specific cake resistance.

A plot of $\Theta/V$ versus $V$ results in a straight line having a slope of $\mu\alpha C/2\Delta PA^2$, for a given set of filtration conditions if the particles are noncompressible ($\alpha$ is also constant). Cakes from fermentation broths are seldom noncompressible and experimental measurements of filtration rates are necessary for design. Since filtration theories deal only with attainable rates, no answers are given for important questions, such as filtrate clarity, tenacity of cake to adhere to the drum, ease and efficiency of washing, and dryness of the resultant cake. However, much insight can be gained by the proper interpretation of filter leaf data obtained in the laboratory. Dahlstrom and Pruchas (1957) have described correlating methods for the various rates involved in the application of continuous filters. In addition, the selection of the proper filter aid for a given application requires that a compromise be made between filtration rate and filtrate clarity.

The ability of aminoglycoside antibiotics to bind to diatomaceous earths and cellulosic filter aids has been reported by Wagman *et al.*

(1975). This phenomenon results in an additional source of loss and must also be considered in the selection of a filter aid.

## B. Centrifugation

Although numerous types of centrifuges have been developed over the years that can separate liquid phases and minor amounts of suspended solids, the design of relatively high-gravity machines that will handle appreciable volumes of particulate matter is relatively recent. Only in special cases will these units give the clarity of effluent or the "dryness" of cake that can be obtained with a precoat filter. Consequently, these factors plus the existence of plants based on filtration technology have slowed down their acceptance as a tool to remove the insolubles portion of a fermentation broth. As pointed out by Gray *et al.* (1972), the exception appears to be in the continuous processing of enzymes, where centrifugation plays a significant role. Recent operating data for the recovery of microorganisms by centrifugation is given in *Chemical Processing* (Anonymous 1977). The theoretical basis for the scale-up of centrifugation operations have been covered by Aiba *et al.* (1973).

## C. Coagulation and Flocculation

Coagulation, defined as the conversion of dispersed colloids into small floc using simple electrolytes, and flocculation, defined as the agglomeration of these small flocs into larger settleable particles using polyelectrolytes, are useful approaches that can improve filtration and centrifugation rates and effluent clarity. Bacterial cells and proteinaceous material in the broth are hydrophilic colloids that usually have negatively charged surfaces. Hydrophobic colloids, such as oil droplets, that lack colloidial stability in the presence of simple electrolytes, are present also.

The simple electrolytes, such as acids and bases, salts, and multivalent ions, are relatively cheap but in general less effective than the polyelectrolytes. Certain finely divided solids, such as clays, activated silica, or carbon, which serve as absorbents and nucleation sites, can also serve as coagulant aids. Polyelectrolytes are high molecular weight, water-soluble organic compounds that are available in forms that give either nonionic, anionic, or cationic functionality.

Each fermentation broth must be studied empirically to establish the usefulness of this approach. Parameters that need to be investigated are the concentrations of colloid, coagulant, and/or flocculant; the effect of severity of agitation; and the kinetics of the binding and settling phenomena. Consequently, subjective responses are flocculation rate, floc size, sedimentation rate, and clarity of the supernatant liquor.

Design approaches based on Stokes' and Newton's laws have been used to predict sedimentation rates for flocced particles. However, complications relating to the presence of nonspherical rather than spherical particles, varying surface charges of polyelectrolytes, and the physical considerations of convection currents, particle–particle interactions (hindered settling), and particle flotation leave much to be desired. These techniques have been developed for waste treatment applications (Shuster and Wang, 1977) but are slowly being introduced into processes for fermentation broths (Treweek and Morgan, 1977).

### D. Whole-Broth Processing

In the late 1950s technology was reported (Bartels *et al.*, 1958) for the sorption of a basic antibiotic, streptomycin, on a cationic ion-exchange resin directly from the unfiltered fermentation broth. This approach circumvented the problems and expense of conventional techniques for removing the insolubles. Various investigators recognized the advantages of this type of processing for suspensions of nonfermentation materials as well and have published methods for predicting performance (Prout and Fernandez, 1961; Marchello and Davis, 1963). Belter *et al.* (1973) have described a periodic countercurrent whole-broth process for the recovery of novobiocin. The whole broth, direct from the fermentor, is passed over a high-capacity vibrating screen to remove large particulate matter that would accumulate in the resin bed if not removed. The resulting screened broth is then contacted with an anionic-exchange resin in a series of specially designed, well-mixed columns. Each of the columns is equipped with a special head screen which retains the resin particles but permits passage of the mycelia and other insolubles. After a predetermined time of operation the resin in the lead column is loaded and the column is removed from the train, washed free of insolubles, and eluted with a methanolic ammonium chloride solution. This operation simultaneously removes the novobiocin and restores the resin to the chloride form. The resulting eluates are processed subsequently to crystals by conventional methods. Periodic countercurrent operation is obtained by advancing each of the partially loaded columns one position and adding a freshly eluted and regenerated column in the trail position of the train. At this point the sorption cycle is repeated. It is important to recognize that the operation described is transient in nature during an individual sorption cycle but approaches a pseudo-steady state process when the column positions are changed in the sequence outlined.

A general mathematical model was developed and used to predict the performance of such a system. The model consists of a suitable continuity equation for each stage coupled with the prevailing mass transfer and equilibrium relationships in each stage.

Assuming complete mixing in the column the continuity equation is given for the $n$th column in the train by Eq. (2).

$$V_c \frac{dC_n}{dt} = FC_{n-1} - FC_n - V_R \frac{dq_n}{dt} \tag{2}$$

where $V_c$ is the liquid volume in column $n$, $C_n$ is the antibiotic concentration in the bulk effluent from the column, $C_{n-1}$ is the antibiotic concentration in the bulk feed to column $n$, $F$ is the volumetric flow rate, $V_R$ is the volume of resin in the column, $q_n$ is the antibiotic concentration in the resin phase, and $t$ is time.

The antibiotic transfer rate is expressed in Eq. (3) as the product of an overall mass transfer coefficient and a concentration difference or driving force.

$$\frac{dq_n}{dt} = K_a(C_n - C_n{}^*) \tag{3}$$

Equilibrium is expressed by Eq. (4) as a Freundlich-type relationship.

$$C_n{}^* = b(q_n)^a \tag{4}$$

where $C^*$ is the antibiotic concentration in the bulk liquid that is in equilibrium with the resin phase and $a$ and $b$ are empirical constants.

The mass transfer coefficient, $K_a$, can be expressed as a function of resin loading by the empirical correlation given in Eq. (5).

$$K_a = Ae^{-B} \frac{q}{q_\infty} + Ce^{-D} \frac{q}{\eta_\infty} \tag{5}$$

where $q_\infty$ is the ultimate resin loading and constants $A$, $B$, $C$, and $D$ determined by laboratory sorption data.

This set of relationships, Eqs. (2)–(5), describes mathematically the time-dependent performance of each column in the sorption train. Modeling of the sorption sequence consists of solving this set of simultaneous equations for each of the columns in the system. Obviously, a computer is required to solve these nonlinear equations for a multiple-stage system. Such calculations permit the determination of effluent concentrations as a function of time (broth loss curves) and concentration on the resin as a function of time (resin loading curves). Subsequent integration of the broth loss profiles can establish total or percentage loss for process evaluation studies.

**FIGURE 1.** Production-scale columns. (Courtesy of The Upjohn Co.)

It should be realized that these equations apply to all constituents of the broth as well as to the antibiotic; and with additional assay information, resin loading and loss curves could be established for other components and, if present, interactions between competing components evaluated.

Production-scale columns for this type of processing are shown in Fig. 1. This method of primary recovery has general utility for materials that have ionic functionality.

## IV. PRIMARY SEPARATIONS

After separation of the fermentation broth into soluble and insoluble fractions, the location of the desired product governs the next step in the sequence. In the case of certain conversions, such as steroid dehydrogenations, intracellular enzymes, and several antibiotics, the desired product remains with the insoluble fraction and the clarified broth can be discarded. In cases where the product is merely insoluble, it can be leached from the mycelia. Standard engineering approaches (Brian, 1972) are available for evaluation and design of this operation. In the cases of intracellular enzymes and bound or associated antibiotics some method of disrupting the cell matrix or membranes and releasing the desired product is needed prior to leaching. Heat, pH adjustment, chemical and/or mechanical disintegration, and osmotic shock are methods that have been used successfully in the laboratory to release the bound product.

However, according to Dunnill (1972) the disruption method that appears to have the greatest commercial potential is the mechanical approach (homogenization). Hetterington *et al.* (1971) and Gray *et al.* (1972) have studied the disintegration of bakers' yeast and *E. coli* cells, respectively. These authors have developed and used Eq. (6), which correlates the release of enzymes with the number of passes through the machine that the cell suspension was given.

$$\log \frac{R_m}{R_m - R} = KNP^{2.9} \tag{6}$$

where $R_m$ is the maximum soluble protein available for release, $R$ is the amount of soluble protein released after $N$ cycles, $K$ is a temperature-dependent rate constant, and $P$ is the homogenizer operating pressure.

After the cell has been lysed and the product released into the liquid, one or more of the approaches discussed in Section III are suitable to remove the cell debris. In most cases the remaining steps in the sequence will be similar to those described for products which are originally extracellular.

However, in the case of enzyme separation, the clear extract contains numerous heat-labile components and the relatively harsh techniques used for extracellular products may not be applicable. Milder fractionating techniques that have been used for these proteinaceous products are ultracentrifugation, ultrafiltration, and various types of chromatography. These techniques are typical ones used for the fractionation of proteins and peptides regardless of their source.

Most materials, antibiotics, extracellular enzymes, organic acids, amino acids, and solvents, are present originally in the liquid portion of the broth. The remainder of this section covers the relatively nonselective separation approaches that are in general use for the primary separation steps.

## A. Liquid Extraction

When exposed to a two-phase immiscible liquid system, an organic compound will be distributed in a definite ratio between these two phases. This ratio has been given the name of the distribution coefficient and is defined in Eq. (7).

$$K \equiv C_L/C_H \tag{7}$$

where $C_L$ and $C_H$ are the concentration of the organic compound in the light and heavy phase solvents, respectively. It is a characteristic property of the compound as are melting point, optical rotation, or light absorption properties. It is this property that is exploited in the extraction of clarified fermentation broths with an organic solvent. Choice of the extracting solvent depends on a number of factors, such as distribution properties, volatility, ease of recovery, and cost. The selectivity of a solvent system is characterized by the ratio of distribution coefficients of components under consideration and is designated as the $\beta$ value, a parameter synonymous to the $\alpha$ value used for distillation calculations.

For the cases where the distribution coefficient for a given compound is linear, independent of concentration, material balance considerations used to predict and evaluate the performance of extraction systems are relatively simple. The definition of an extraction factor, $E$, is given in Eq. (8).

$$E \equiv LK/H \tag{8}$$

where $L$ is the volumetric flow rate of the phase with the lesser density and $H$ is the volumetric flow rate of the more dense phase. It can be derived that for a single equilibrium contact the fraction unextracted, $X/X_0$, is equal to $1/1 + E$ where $X$ is the final concentration and $X_0$ is the initial concentration. Also, this type of reasoning can be extended to predict the performance of multiple-stage contacting. If a heavy phase liquid containing solute is subjected to multiple countercurrent contacts with a light immiscible solvent, each contact attaining equilibrium, it can be derived by a series of material balances that the rejection ratio, $R$, is given by Eq. (9).

$$R = \frac{E(E^n - 1)}{E - 1} \tag{9}$$

where $R$ is the ratio of the weight of solute leaving in the light phase to the weight of solute leaving in the heavy phase and $n$ is the number of equilibrium stages.

It follows from the definition of $R$ that the fraction of solute extracted into the light phase is $R/(R + 1)$ or $R/(R + 1) \times 100$ to give the yield of desired product. The plot in Fig. 2 shows graphically the relationship between extracted solute, extraction factor, and number of stages used in a countercurrent extraction unit.

When the situation is reversed and the solute enters the system in the light phase, the rejection ratio is that given by Eq. (10).

$$R = \frac{E^m(E - 1)}{E^m - 1} \tag{10}$$

where $m$ equals the number of stages in the unit.

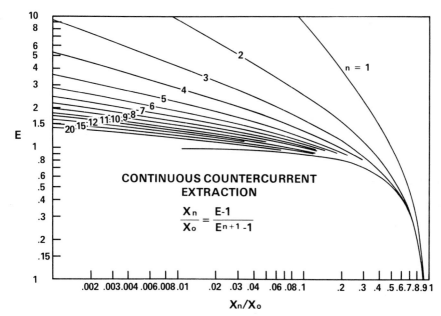

**FIGURE 2.** Relationship between unextracted solute, extraction factor, and number of stages in continuous countercurrent extraction. (Craig and Craig, 1956; reproduced by permission of Interscience Publishers, Inc.)

It should be remembered that the applicable rejection ratio relationship holds for all components present in the filtered broth and the success of this approach depends on choosing a sufficiently selective solvent to recover the desired product relatively devoid of impurities. Unfortunately, the few solvents that have the proper handling characteristics are not sufficiently selective to permit appreciable purification during the extraction step.

Historically, Craig and Craig (1956) developed this technique and demonstrated its usefulness in the laboratory. The countercurrent distribution system developed by Craig and Post consisted of a series of specially designed tubes, which are equivalent to a series of separatory funnels, connected so that the upper phase solvent can be transferred to successive tubes in the train. The operation of the series of tubes is automated so that a complete cycle consists of agitation to mix the two phases and simultaneous distribution of the solute between the two immiscible phases, a time delay period to allow the equilibrated phases to settle, and subsequent transfer of the upper phase into the next tube leaving the lower phase in the same tube. Simultaneously, a solute-free portion of the upper phase is added to the first tube of the extraction train. If the extraction factor is constant throughout the operation (no change of intrinsic distribution coefficient or solvent volumes), the distribution of solute in the extraction system, after a given number of transfers, can be determined by successive terms of the binomial expansion, Eq. (11).

$$F_{n,r} = \frac{n!}{r!(n-r)!} \left(\frac{1}{1+E}\right)^n E^r \tag{11}$$

where $F_{n,r}$ = the fraction of the feed material in tube $r$ after $n$ transfers.

Although this relationship gives the profile of product within the unit after a given number of transfers or a function of time, it is an important one that can also form the basis for the analysis of discrete stage models for sorption and chromatographic separation. These are discussed in Sections IV, C and V, C.

Although many types of liquid extraction equipment have been developed and described in the literature (King, 1971), the Podbielniak centrifugal extractor has played the greatest role in the recovery of antibiotics. Its chief advantages that led originally to its development and use for penicillin processing were small holdup of liquid, high capacity, and the development of high centrifugal forces which enabled it to handle intractible emulsions. Other centrifugal units have been developed and are used extensively to extract clarified fermentation broths.

## B. Dissociation Extraction

A number of products to be recovered from fermentation broths are either weak acids or weak bases. In general, if a compound is to be extracted by an organic solvent, pH conditions are selected so that the material is largely in the un-ionized form. Thereby, weak bases are extracted at relatively high pH values and weak acids under low pH situations. These conditions result in rapid and complete extraction but little separation of impurities of a similar nature. Usually the $\beta$ values, the ratio of the distribution coefficients, remain relatively constant at extreme pH levels. As discussed in the previous section on liquid extraction the ease of separation depends on this ratio. In some cases the $\beta$ value will change appreciably with a pH shift and, consequently, the ease of separation will be improved. A typical situation of this nature is shown in Fig. 3, where the distribution properties of penicillins F, G, and K are compared with other major acidic impurities of unknown composition. Unfortunately, the solvent system is not given but it serves to illustrate the phenomenon under discussion.

If the assumption is made that at extreme pH values the ionized species is not soluble in the organic phase and the un-ionized species is soluble in either phase, the mathematical concepts covered in Section

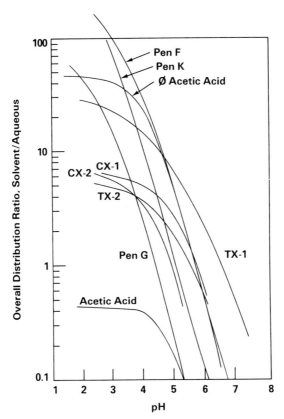

**FIGURE 3.** Distribution ratios for various penicillins and impurities. (Sounders *et al.*, 1970; reprinted by permission of AIChE.)

IV, A can be used to define an intrinsic distribution coefficient and an extraction factor and, subsequently, an extract purity based on the solvent selectivity or $\beta$ value. The performance of the system can be evaluated in the usual way. However, when the extraction is conducted at intermediate pH values where the materials are partially ionized, evaluation and design of the extraction process is considerably more complicated. Sounders *et al.* (1970) have given the following relationship shown as Eq. (12) for weak acids.

$$K = \frac{K^0[H^+]}{[H^+] + k_1} \tag{12}$$

where K is the apparent distribution ratio for the weak acid, $K^0$ is the intrinsic distribution ratio for the un-ionized species, and $k_1$ is the ionization equilibrium constant for the weak acid.

An analogous relationship for weak bases can be developed as given by Eq. (13).

$$K = \frac{K^0 k_1}{1 + [H^+]}$$
(13)

Equations (14) and (15) present Eqs. (12) and (13) in logarithmic form.

$$pK - pH = \log (K^0/K - 1) \quad \text{for weak bases} \quad (14)$$

$$pH - pK = \log (K^0/K - 1) \quad \text{for weak acids} \quad (15)$$

## C. Sorption Processing

Although this section emphasizes adsorption operations, it must be realized that the fundamental concepts of equilibrium between the liquid and solid phases, the mass transfer relationship, and the prevailing hydrodynamics apply to any type of transfer between the liquid and solid phases. The difference occurs in the mechanisms by which the solute is held in the solid phase. In the case of ion exchange relatively strong chemical bonds are involved, while weaker forces are utilized in the case of physical adsorption. The manufacture of activated carbons, the usual sorption solid for fermentation processing, is conducted in such a way that tremendous surface area per weight of carbon is formed. As the carbonaceous substance (coke, char, charcoal) is activated a network of large internal spaces, cracks, and crevices (pores) are formed. A good adsorbent has a combination of large surface area with openings sufficiently sized to allow the sorbate passage to that area. The nature of the adsorbate influences the capacity for a specific carbon. Since the dual influences of the nature of the carbon surface and the adsorbate are little understood, equilibrium data must be determined experimentally for a given application and correlated empirically. The adsorption data can usually be represented by a linear relationship, such as $C = mq$, over short ranges of concentrations or by the familiar Freundlich-type equation, $C = mq^n$, over extended concentration ranges. In these relationships, $C$ and $q$ refer to the concentration of the material in the liquid and solid phases, respectively. The $m$ and $n$ symbols are empirically determined constants.

The methods of contacting the liquid and solid phases vary from simple batch contact and multiple-batch contacts followed by filtration, to the usual fixed-bed approach. Prediction of the results of batch operation is relatively simple, involving only material balance and equilibrium considerations. However, prediction of the performance of a fixed bed, which is the most efficient way to use a given amount of carbon, is more difficult because of its inherent transient state nature and the resulting mathematical complexity of two independent variables, time and position.

Fixed-bed contacting has been modeled adequately by establishing the continuity equation across a differential height of solid adsorbant during an increment of time, $dt$. The resulting partial differential equation is given by Eq. (16).

$$\frac{U}{\epsilon}\frac{dC}{dZ} + \frac{dC}{dt} = \frac{-\rho}{\epsilon}\frac{dq}{dt} \tag{16}$$

where $C$ is the concentration of sorbed material in the solution surrounding the sorbent, $Z$ is the height in the column, $t$ is time, $\epsilon$ is the porosity of the bed, $U$ is the superficial velocity of the solution, $\rho$ is the bulk density of the sorbent, and $q$ is the concentration of sorbed material in the solid phase.

If it is also assumed that only mass transfer through the liquid film surrounding the particle and the subsequent diffusion of solute into the solid particle form the major resistances to that transfer, it is possible to write a relationship that describes the transfer rate in terms of an overall mass transfer coefficient. Such a relationship is given by Eq. (17).

$$dq/dt = K_a(C - C^*) \tag{17}$$

where $C^*$ is the concentration of solute in equilibrium with $q$ and $K_a$ is the overall mass transfer coefficient.

In general, $K_a$ can be expressed as a function of $q$ for a given particle size and contacting conditions. The actual relationship must be determined experimentally for each application since it is dependent on the composition of the feed solution, the inherent properties of the solute and sorbent, and the liquid and solid phase diffusion mechanisms.

The continuity equation (a partial differential equation), the rate equation for sorption, the mass transfer relationship, and the equilibrium relationship in combination constitute the mathematical model which can be used to describe quantitatively the performance of a fixed-bed sorption process.

Chen et al. (1968) have used the approach described above with the help of large-scale computers to study and design fixed-bed sorption processes. This attack had been used by others to predict the performance of fixed-bed columns and processes in other industries.

However, the approach described previously makes the basic assumption that no back-mixing occurs during the contacting operation. In practice, additional zone spreading occurs for a variety of reasons, such as, nonequilibrium contact, axial and radial diffusion which results from molecular and eddy disturbances, irregular flow profiles, thermal effects, and channeling due to nonuniform packing characteristics. Historically, the overall result of these nonidealities has been tackled by introducing a "dispersion" coefficient into the analysis. Tracer techniques and correlations have been established to measure and predict the magnitude of the additional dispersion by a number of authors (Chung and Wen 1968; Hashimoto et al. 1977). Since it is difficult to separate and measure the effects of the individual nonidealities, Chen et al. (1972) circumvented the problem by simulating an actual sorption train as a series of tanks operating under ideal conditions of back-mixing. Use of the set of equations given in Section III, D for ion-exchange operations and the required number of tanks permit an approximation for quantitative evaluation and/or prediction of fixed-bed sorptions.

A third method of approximation for this problem is that of assuming a number of discrete stages operating with fixed distribution parameters. The eluant and solids profile can be determined by summation of the fundamental distribution for each of the feed increments and its subsequent distributions as given by Eq. (18).

$$\sum_{n=1}^{n=f(\Theta)} \sum_{r=1}^{r=N} \frac{n!}{r!\,(n-r)!} \left(\frac{1}{1+E}\right)^n E^r \tag{18}$$

where $n$ is the number of feed increments and, consequently, the number of equilibrations; $r$ is the stage number under consideration; $E$ is the sorption parameter for the liquid and solid distribution; and $N$ is the total number of stages.

Use of Eq. (18) permits the calculation of solute profiles throughout the fixed bed in both liquid and solid phases as well as solute concentration in the emerging liquid as a function of feed increments (or time).

Physical sorption is a processing approach that has been used for the recovery of numerous fermentation products, particularly antibiotics.

## D. Precipitation

The inherent lack of solubility for salts of various products can be used advantageously to recover selectively and/or purify the desired product. Numerous examples, such as the lactate and oxalate salts of erythromycin, calcium salt of citric acid, and sodium dodecal sulfonate salt of gentamicin, are given in the antibiotic literature. However, its greatest use is in the elimination of proteinaceous impurities and/or the purification of enzymes. Typically, this is accomplished by adding inorganic salts to give relatively high ionic strengths, by reducing the solubility with the addition of organic solvents to change the dielectric strength of the solvent system, or by a combination of both. The last technique is used to precipitate and fractionate dextrans into useful materials of narrow molecular weight ranges.

The underlying principles for this separation approach have been covered by Aiba *et al.* (1973).


## V. PURIFICATION OPERATIONS

The unit operations described previously produce mixtures of materials from the fermentation broth that are similar in physicochemical properties to the desired product and therefore require the use of additional separation techniques that are more powerful than those presented, to obtain high-quality products. These techniques gain their separation power because they consist of multiple stages assembled in a cascade so that both product quality and yield can be obtained.

### A. Fractional Liquid–Liquid Extraction

The simple concept and basic definitions for this unit operation have been covered in Section IV, A.

A mixture can be separated into only two streams during continuous operation with a single column. A cascade of contacting stages can be assembled as shown in Fig. 4 with feed introduced at or near the center of the column. The section above the feed stage functions as an enriching section, removing impurities from the product solute. The lower section below the feed stage acts to strip the desired product from the impurities. Consequently, the goals of high-quality and improved yields can be met.

If no solvent is added to the column with the feed mixture, the rejection ratio can be shown to be that given in Eq. (19).

$$R = \frac{E^m(E^n - 1)}{E^m - 1} \tag{19}$$

where $R$ is the rejection ratio, $E$ is the extraction factor, $m$ is the number of enriching stages plus 1, and $n$ is the number of stripping stages plus 1. Simplification of this relationship can be made in special cases as shown in Table II. More complex relationships, which deal with the case where the solute mixture is added to the column dissolved in either the upper or the lower phase, may be found in the literature (Brian, 1972). Belter (1967) has published a method by which the separation perfor-

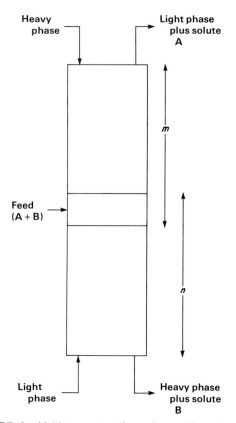

**FIGURE 4.** Multistage extraction column with center feed.

mance of a continuous column can be simulated in a normal Craig distribution.

Numerous mechanical devices have been conceived for multiple-stage contacting of two liquids. Most of these units have been described and discussed by King (1971). Bartels and Kleiman (1949) described and presented a simplified mathematical prediction approach for the fractionation of streptomycins using a cascade of disk- and bowl-type centrifuges.

## B. Chromatography

Chromatography is defined as a method of separation in which the flow of a fluid promotes the fractionation of mixtures by differential migration in a porous sorptive medium (usually a solid) from a narrow initial feed zone. The operation is a time-dependent one but is equilib-

**TABLE II.** Simplified Forms of Equation (19)

| Extraction factor $E$ | Rejection ratio |
|:---:|:---:|
| 1 | $n/m$ |
| $>> 1$ | $E^n$ |
| $<< 1$ | $E^m$ |
| 1 (and center feed with $n = m$) | $E^n$ |

rium controlled. A schematic illustration of the separation of these components by a differential migration is shown in Fig. 5. Three components are added to the leading edge of the chromatography system prior to elution as shown in Fig. 5a. As the flow of fluid passes through the system, the components separate. Figures 5b–d show the position and separation of the three components at increasing time intervals of fluid flow. Figure 5e shows the concentration of each of the components in the system effluent versus time. A similar plot, which could have the order of components reversed, can be obtained for concentrations of components prior to migration from the system, such as shown in Fig. 5d.

Useful types of chromatography for a separation and purification of fermentation products are adsorption, partition, ion exchange, and gel filtration. Also, dry column chromatography, affinity, and ion exclusion have been advocated. These operations are similar in their mathematical developments but differ considerably in the predominant mechanism by which the differential migration occurs. Separation occurs in adsorption chromatography because of the different strengths of van der Waal's forces binding the solutes to the solid phase. It is accomplished in partition chromatography by the different distribution coefficients of solutes between the mobile and stagnant adsorbed liquid phase. The governing mechanism for ion-exchange chromatography is the varying strengths of the chemical bonds formed with the ionized solutes and the ion-exchange functionality. Gel filtration depends on the ability of molecules of different size and shape to penetrate the matrix of the column material. These mechanisms overlap to some extent and it should not be construed that any chromatographic separation is dependent on only one mechanism.

The basic theory for chromatographic separations parallels those described previously for fixed-bed operations with the mathematical exception that different boundary conditions are employed in the solution of the partial differential equations. Since the concentration of an individual solute is dependent on location within the column and time, it is apparent that the continuity equation will be a partial differential equation (as given in Section IV, C for sorption processes). The transfer rate

# DIRECTION OF FLUID FLOW

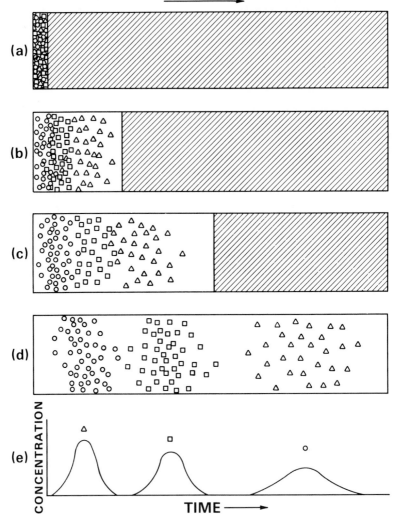

**FIGURE 5.** Schematic illustration of separation by the differential migration mode. (Karger *et al.*, 1973; reproduced by permission of Wiley, New York.)

and equilibrium relationships can be established experimentally as before.

The fundamental theory for determining concentration profiles during chromatographic separations has been discussed and evaluated by Lightfoot *et al.* (1962). Examples are given for cases of increasing complexity ranging from the simple situation of equilibrium, negligible dispersion and linear sorption isotherms to those cases where none of these ideal conditions exists.

## C. Carbon Decolorization

The presence of color bodies and their precursors in the final solutions can seriously affect the quality of the ultimate product and is of concern in the preparation of high-quality products. The usual approach to their removal is by adsorption on activated carbon, either powdered or granular, using techniques and modes of operation previously described.

## D. Crystallization

Usually crystallization is the final purification step in the processes that must produce high-quality products. It is an effective and efficient method of purification and possesses some advantages over competitive methods. Relatively low-temperature operations minimize undesirable degradation of heat-sensitive materials. Operations are usually conducted at high concentrations and, consequently, unit costs are low. Separation factors are usually high, which makes this operation an effective fractionation tool.

Various aspects of crystallization phenomena, such as nucleation of different types and growth kinetics, for simple systems have been explored and reported in the literature (Mullins, 1972). Separation or fractionation by crystallization is of special concern but, unfortunately, the prerequisites for prediction—knowledge of phase diagrams for a given mixture, knowledge of true equilibrium states between the crystals and the surrounding solution, and knowledge of the pertinent kinetics—are rarely available. Consequently, determination of optimal crystallization conditions for a given problem becomes a matter of empirical experimentation. Obviously, crystallization is not necessary for all products and in several isolated instances, such as neomycin and certain enzymes, crystallization of the product has never been accomplished.

## VI. PRODUCT ISOLATION

Recovery of the product in its final form depends on the prior processing and its ultimate use. Basically, the operations can be divided into two areas; first, processing of crystalline products and, second, drying of products directly from solution.

## A. Crystal Processing

High-quality crystals are recovered in conventional batch Nutsche-type filters or centrifugal filters. In the first of these techniques the same theory that was covered in Section III, A for constant pressure filtrations applies to the filtration of crystalline materials, and considerable simplification can be made. If it is assumed that the crystal slurry is uniform and, therefore, the filtrate volume is proportional to the crystal weight, the crystals are noncompressible and that the resistance of filter medium is negligible, Eq. (1) can be simplified to give Eq. (20).

$$dV/d\theta = K(\Delta PA/T) \tag{20}$$

where $T$ is the thickness of crystal layer and the other symbols are as defined for Eq. (1). Upon substitution of $W \cong V$ and integration one obtains Eq. (21).

$$\Theta = K \, (W^2/\Delta PA^2) \tag{21}$$

This relationship can be used to predict filtration and washing rates after the permability factor, $K$, has been determined from laboratory tests.

A typical centrifugal filter is shown in Fig. 6. The following relationship, Eq. (22), for the centrifugation rate has been developed by Valleroy and Maloney (1960).

$$\frac{dV}{d\theta} = \frac{(2\pi N)^2 \rho L (r_0^2 - r_L^2)}{2\mu(\alpha W/A_m^2 + R_M/A_0)} \tag{22}$$

where $N$ is the speed, $\mu$ is the viscosity, $\alpha$ is the specific resistance of the cake, $W$ is the solids concentration in the cake, $A_m$ is the log mean area, and $R_M$ is the medium resistance.

It is readily apparent that this relationship is equivalent to that used previously with the necessary modifications to account for the different geometry and gravitational driving force.

Since the crystals deposited in the basket are subsequently washed to remove adhering mother liquor, the nature and size of the crystals affect the centrifugation and washing rates. Consequently, the crystallization and recovery steps are interrelated and these factors should be considered in the optimization of the crystallization sequence.

After washing the crystals are spun dry and discharged from the basket to trays for subsequent drying or desolventizing, depending on the nature of the solvent employed for crystallization and washing.

## B. Drying

The removal of liquid (water or an organic solvent) from the purified product stream, either as liquid-wet crystals or from a solution, usually

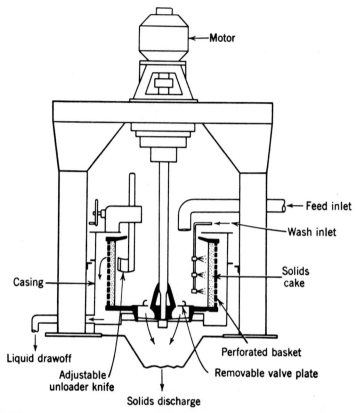

**FIGURE 6.** Top suspended basket-type centrifugal filter. (Jordan, 1968a; reproduced by permission of Interscience Publishers.)

constitutes the final step in the manufacture of a given product. The underlying principles of drying rates, which are governed by the physicochemical properties of the material to be processed, the psychrometric relationships of the drying medium, and the heat and mass transfer characteristics of both the material and the drying equipment used, have been covered elsewhere (Nonhebel and Moss, 1971). The physical nature of the product, its heat sensitivity, and, again, the physical characteristics of the final product that are acceptable for its ultimate use dictate the drying approach selected. Thirty-four parameters affecting drying have been listed by Nonhebel and Moss (1971). These parameters were divided into four classes according to whether they affect heat transfer, properties of the drying environment, physical properties of the solid–liquid system, or the inherent properties of the solid. Equilibrium relationships between solvent and solid, energy, and mass balances must be considered in the design of optimum drying processes and systems.

Because of the wide variety of drying problems numerous types of drying equipment are available. The vacuum tray dryer, consisting of heated shelves within a single chamber that may be evacuated, is used extensively for pharmaceutical products. Although labor intensive, it is particularly useful for small batches of expensive materials where loss must be avoided and for heat-sensitive products.

Lyophilization or freeze-drying is a highly specialized method of removing liquid. The liquid, usually water, is removed by sublimation from the frozen solution. The initial freezing of the solution can be accomplished either outside or inside of the vacuum chamber prior to drying. The equipment for freeze-drying illustrated in Fig. 7 is similar to that used for vacuum tray drying. Special auxiliaries, such as high-vacuum pumps, low-temperature condensers, and means for removing ice from the condenser surfaces, must be provided. This drying technique is useful for antibiotic solutions, enzyme preparations, and bacterial suspensions.

Equipment for other drying techniques is illustrated in Fig. 8, for drum dryers, and Fig. 9, for spray drying. These approaches are used for situations where crystallization either is not important or has not been accomplished. In the case of drum dryers, the water is removed by heat conduction through a thin film of solution adhering to the rotating steam

**FIGURE 7.**  Freeze–drying system. (Courtesy of The Upjohn Co.)

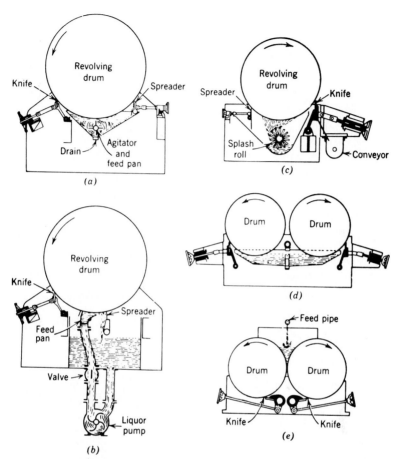

**FIGURE 8.**   Drum dryers. Arrangements for feed input and product take-off. (a) Single drum, dip-feed; (b) single drum, pan-feed; (c) single drum, splash-feed; (d) double drum, dip-feed; (e) double drum, top-feed. (Jordan, 1968b, reproduced by permission of Interscience Publishers.)

heated drum. As the drum rotates, the product is dried and scraped from the drum at the discharge point.

In spray drying the feed solution or slurry is atomized into a chamber through which heated air passes. The sensible heat of the hot gas supplies the latent heat requirements for vaporization of the liquid, leaving small particles of dry powder. Subsequently, the dry product must be separated from the exit air. The solution is dispersed by one of three methods, pressure nozzles, two fluid nozzles, and rotating disks. The product usually dries as small spheres and consequently has a relatively low bulk density. Although relatively expensive because of

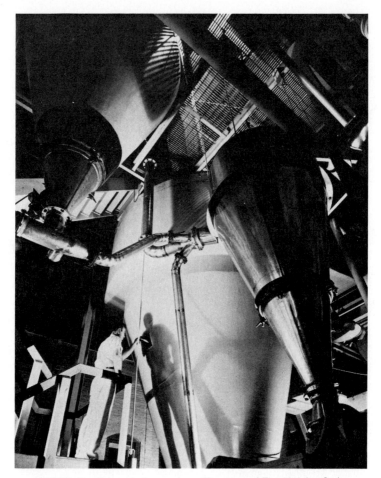

**FIGURE 9.** Spray-drying system. (Courtesy of The Upjohn Co.)

poor thermal efficiency, the technique is preferred for heat-sensitive materials.

## VII. ANCILLARY OPERATIONS

Several operations exist that are essential to an integrated process for the manufacture of a fermentation product but serve only supporting roles. Typical of such operations are the recovery of residual organic solvents from extracted spent broths and the subsequent treatment of the spent broths.

The nature and concentration of materials in the residual fermentation broths depend on their source but, in general, these waste streams contain extremely high levels of biodegradable compounds, resulting in high biological oxygen demand (BOD) requirements and nitrification demands when discharged. The need to maintain and improve the quality of our water resources and our environment has led to extensive studies on ways of handling the solids in these streams.

For economic reasons and since residual organic solvents would contribute appreciably to the BOD load, residual extraction solvents are removed from the liquid streams prior to treatment or disposal. The solvent recovery procedures are reasonably routine stripping operations and can be evaluated and designed by distillation techniques covered in the literature (King, 1971; Brown, 1950).

Numerous methods of waste treatment and disposal for industrial waste treatment have been studied and reported. Approaches that have been studied for spent broths include processes based on activated sludge and trickling filter aerobic digestion, anaerobic filters, spray irrigation, oxidation ponds, deep well injection, and incineration. Comprehensive descriptions of these techniques and examples from the pharmaceutical industry and for other fermentation products are given by Struzeski (1975) and Nemerow (1963).

Use of concentrated spent broths for liquid supplement feeding of ruminants and for the recovery of solids to be used in animal feeds for their nutritive value and the presence of unidentified growth factors constitute another approach to the problem of disposal.

## REFERENCES

Aiba, S., Humphrey, A. E., and Millis, N. F. (1973). "Biochemical Engineering," 2nd ed. Academic Press, New York.

Anonymous (1977). *Chem. Process.* **40**, 10, 154.

Bartels, C. R., and Kleiman, G. (1949). *Chem. Eng. Prog.* **45**, 589–594.

Bartels, C. R., Kleiman, G., Korzun, J. N., and Irish, D. B. (1958). *Chem. Eng. Prog.* **54**, 49.

Belter, P. A. (1967). *Ind. Eng. Chem.* **59**, 3.

Belter, P. A., Cunningham, F. L., and Chen, J. W. (1973). *Biotechnol. Bioeng.* **15**, 533–549.

Brian, P. L. T. (1972). "Staged Cascades in Chemical Processing," Prentice-Hall, Englewood Cliffs, New Jersey.

Brown, G. G. (1950). "Unit Operations," Wiley, New York.

Chen, J. W., Buege, J. A., Cunningham, F. L., and Northam, J. I. (1968). *Ind. Eng. Chem., Process Des. Dev.* **7**, 26.

Chen. J. W., Cunningham, F. L., and Buege, J. A. (1972). *Ind. Eng. Chem., Process Des. Dev.* **11**, 430.

Chung, S. F., and Wen, C. Y. (1968). *AIChE J.* **14**, 857.

Craig, L. C., and Craig, D. (1956). *In* "Technique of Organic Chemistry" (A. Weissberger, ed.), 2nd ed., Vol. 3, Part 1. Wiley (Interscience), New York.

Dahlstrom, D. A., and Pruchas, D. B. (1957). "Scaleup Methods for Continuous Filtration Equipment," Inst. Chem. Eng., London.

Dunnill, P. (1972). *Fermen. Technol. Today, Proc. Int. Fermen. Symp. 4th, 1972* pp. 186–194.

Gray, P. P., Dunnill, P., and Lilly, M. D. (1972). *Fermen. Technol. Today, Proc. Int. Ferment. Symp. 4th, 1972* pp. 347–351.

Hashimoto, K., Minura, K., and Tsukano, M. (1977). *J. Chem. Eng. Jpn.* **10**, 1.

Hetherington, P. J., Follows, M., Dunnill, P., and Lilly, M. D. (1971). *Trans. Inst. Chem. Eng.* **49**, 142.

Holland, C. D. (1975). "Fundamentals and Modeling of Separation Processes: Absorption, Distillation, Evaporation and Extraction," Prentice-Hall, Englewood Cliffs, New Jersey.

Jordan, D. G. (1968a). *Chem. Process Dev.* Part 2, p. 846.

Jordan, D. G. (1968b). *Chem. Process Dev.* Part 2, p. 910.

Karger, B. L., Snyder, L. R., and Horvath, C. (1973). "An Introduction to Separation Science," Wiley, New York.

King, C. J. (1971). "Separation Processes," McGraw-Hill, New York.

Lightfoot, E. N., Sanchez-Palma, R. J., and Edwards, D. O. (1962). *In* "Chromatography and Fixed-Bed Separations" (H. M. Schoen, ed.), pp. 99–181. Wiley, New York.

Marchello, J. M., and Davis, M. W., Jr. (1963). *Ind. Eng. Chem.* **2**, 27.

Mullins, J. W. (1972). "Crystallization," CRC Press, Cleveland, Ohio.

Nemerow, N. L. (1963). "Theories and Practices of Industrial Waste Treatment," Addison-Wesley, Reading, Massachusetts.

Nonhebel, G., and Moss, A. A. H. (1971). "Drying of Solids in the Chemical Industry." CRC Press, Cleveland, Ohio.

Prout, W. E., and Fernandez, L. P. (1961). *Ind. Eng. Chem.* **53**, 449.

Shuster, W. W., and Wang, L. K. (1977). *Sep. Purif. Methods* **6**, 153–187.

Sorel, M. (1893). "La rectification de l'alcool." Paris.

Sounders, M., Pierotti, G. J., and Dunn, C. L. (1970). *Chem. Eng. Prog. Symp. Ser.* **66**, 40–42.

Struzeski, E. J., Jr. (1975). "Waste Treatment and Disposal Methods for the Pharmaceutical Industry," EPA 330/1-75-001. EPA Office of Enforcement, Washington, D.C.

Treweek, G. P., and Morgan, J. J. (1977). *J. Colloid Interface Sci.* **60**, 2.

Valleroy, V. V., and Maloney, J. O. (1960). *AIChE J.* **6**, 382.

Wagman, G. H., Bailey, J. V., and Weinstein, M. J. (1975). *Antimicrob. Agents Chemother.* pp. 316–319.

Chapter 17

# Use of Immobilized Cell Systems to Prepare Fine Chemicals

ICHIRO CHIBATA
TETSUYA TOSA
TADASHI SATO

MICROBIAL TECHNOLOGY, 2nd ed., VOL. II

## I. INTRODUCTION

Since 1960, a number of papers have been published on preparations, properties, and potential utilizations of immobilized enzymes. In 1969, the authors (Tosa *et al.*, 1966, 1967; Chibata *et al.*, 1972) first succeeded in the industrial application of immobilized enzyme, i.e., immobilized aminoacylase, for continuous production of L-amino acids from acyl-DL-amino acids. Production of 6-aminopenicillanic acid and fructose by immobilized penicillin amidase and by immobilized glucose isomerase, respectively, soon followed in the United States, Europe, and Japan.

At present, enzymes derived from microorganisms are used advantageously for industrial purposes. Microbial enzymes are classified into two forms. One is the extracellular enzyme excreted from microbial cells during culture, while the other is the intracellular enzyme which remains in the cells. For the immobilization of the latter, the enzymes must be extracted from cells by some technique, and in some cases further purification is necessary. Extracted intracellular enzymes are, in many cases, unstable. If whole microbial cells can be immobilized directly without extracting enzyme, the immobilized cells can be used as a solid catalyst. In such immobilized microbial cells, the procedure for extraction of enzyme can be omitted, and loss of enzyme activity is expected to be kept to a minimum.

In addition, if the microbial cells having multienzyme systems are immobilized and can be used as a solid catalyst, the possibility exists that fermentative methods involving multienzyme reactions may be replaced by the continuous enzyme reaction using immobilized cells.

More recently, on the basis of these ideas, the studies on immobilization of whole microbial cells have been the subject of increased interest. However, there are fewer reports on immobilized cells than on immobilized enzymes. Chibata *et al.* (1974e) and Tosa *et al.* (1974) studied the immobilization of whole microbial cells, i.e., *Escherichia coli,* having high aspartase activity and in 1973 succeeded in the industrial application of immobilized cells to the continuous production of L-aspartic acid from ammonium fumarate. This was the first international industrial application of immobilized microbial cells.

In 1974 Chibata *et al.* (1975c) and Yamamoto *et al.* (1976, 1977) succeeded in applying industrial production of L-malic acid from fumarate using immobilized *Brevibacterium ammoniagenes* having high fumarase activity.

In this chapter the examples mentioned above and some other applications of immobilized cell systems for preparation of fine chemicals are reviewed.

## II. IMMOBILIZATION OF MICROBIAL CELLS

At present, many papers on immobilization of microbial cells have been published (Table I).

Methods for immobilization of microbial cells can be classified into three categories analogous to the immobilization of enzymes, that is, carrier-binding, cross-linking, and entrapping methods. Of these methods, the entrapping method has been most extensively investigated.

### A. Carrier-Binding Method

The carrier-binding method is based on direct binding of microbial cells to water-insoluble carriers. Microbial cells are ionically bound to water-insoluble carriers containing ion-exchange residues. Hattori and Furusaka (1960, 1961) bound *E. coli* cells and *Azotobacter agile* cells to Dowex-1 (Cl form). They observed that these immobilized cells showed oxidation activity for glucose and succinic acid.

However, this method is considered not to be advantageous because enzymes and cells may leak out from the carrier due to autolysis during the enzyme reaction.

### B. Cross-Linking Method

Microbial cells can be immobilized by cross-linking with bi- or multifunctional reagents such as glutaraldehyde or toluenediisocyanate. Chibata *et al.* (1974e) investigated immobilization of *E. coli* cells having aspartase activity by cross-linking with glutaraldehyde as the bifunctional reagent. The immobilized cells showed aspartase activity corresponding to 34.2% of that of intact cells.

So far very few papers have appeared on this method. However, there is a possibility that suitable cross-linking reagents for immobilization of cells will be found in the future.

### C. Entrapping Method

So far the method of directly entrapping microbial cells into polymer matrices has been most extensively investigated for immobilization of cells. For this method, the following matrices are employed: (1) collagen, (2) gelatin, (3) agar, (4) alginate, (5) carrageenan, (6) cellulose triacetate, (7) polyacrylamide, and (8) polystyrene. Among these matrices, polyacrylamide gel has been extensively used for immobilization of many kinds of microbial cells, as shown in Table I. This polyacrylamide

**TABLE I.**  Preparations and Applications of Immobilized Microbial Cells

| Enzyme reaction and application[a] | Microorganism | Immobilization method | Reference |
|---|---|---|---|
| **[I] Oxidoreductase reaction** | | | |
| L-Sorbose $\xrightarrow{\text{L-sorbose dehydrogenase}}$ L-sorbosone | Gluconobacter melanotenus | Polyacrylamide | Martin and Perlman (1976a) |
| Cortisone $\xrightarrow{\text{steroid 11}\beta\text{-hydroxylase}}$ cortisol | Curvularia lunata | Polyacrylamide | Mosbach and Larsson (1970) |
| **[II] Transferase reaction** | | | |
| NAD + ATP $\xrightarrow{\text{NAD kinase}}$ NADP + ADP | Achromobacter aceris | Polyacrylamide | Chibata et al. (1975b); Uchida et al. (1977) |
| 2 Glucose + 2 p-nitrophenyl phosphate $\xrightarrow{\text{acid phosphatase}}$ glucose 6-phosphate + glucose 1-phosphate + 2 p-nitrophenol | Escherichia freundii | Polyacrylamide | Saif et al. (1975) |
| **[III] Hydrolase reaction** | | | |
| Triacetin or tributyrin + $H_2O$ $\xrightarrow{\text{lipase}}$ glycerol + acetic acid or butyric acid | Pseudomonas mephitica var. lipolytica | Heat treatment | Kosugi and Suzuki (1973) |
| dl-1-Menthol ester + $H_2O$ $\xrightarrow{\text{l-menthol ester hydrolase}}$ l-menthol + d-menthol ester + organic acid | Alginomonas nonfermentas | Polyacrylamide | Nonomura et al. (1976) |

| Reaction | Organism | Support | Reference |
|---|---|---|---|
| Lactose + H₂O $\xrightarrow{\beta\text{-galactosidase}}$ D-glucose + D-galactose | Lactobacillus bulgaricus | Agar | Miyata and Kikuchi (1976) |
| | | Polyacrylamide | Ohmiya et al. (1977) |
| | Saccharomyces lactis | Cellulose triacetate | Dinelli (1972) |
| | Escherichia coli | Polyacrylamide | Ohmiya et al. (1977) |
| | | Polyacrylamide | Ohmiya et al. (1977) |
| Sucrose + H₂O $\xrightarrow{\text{invertase}}$ D-glucose + D-fructose | Fungal spores | Ion-exchange cellulose | Johnson and Ciegler (1969) |
| | Saccharomyces pastorianus | Agar | Toda and Shoda (1975) |
| | Yeast | Freeze-dried | Nanba and Matsuo (1970) |
| Raffinose + H₂O $\xrightarrow{\text{raffinase}}$ D-glucose + D-. . . . galactose + D-fructose | Mold | Glutaraldehyde | Nishimaru et al. (1975) |
| Penicillin G + H₂O $\xrightarrow{\text{penicillin amidase}}$ 6-APA + phenylacetic acid | Escherichia coli | Polyacrylamide | Sato et al. (1976) |
| | Bacillus megaterium | Cellulose triacetate DEAE–cellulose | Dinelli (1972) Fujii et al. (1973) |
| 6-APA + D-phenylglycine methyl ester $\xrightarrow{\text{penicillin amidase}}$ ampicillin + methanol | Achrombacter sp. | DEAE–cellulose | Fujii et al. (1973) |
| | Escherichia coli | Cellulose triacetate | Marconi et al. (1975) |

(continued)

**TABLE I** (Continued)

| Enzyme reaction and application[a] | Microorganism | Immobilization method | Reference |
|---|---|---|---|
| 7-ADCA + D-phenylglycine methyl ester $\xrightarrow{\text{penicillin amidase}}$ cephalexin + methanol | Achromobacter sp. | DEAE–cellulose | Fujii et al. (1974) |
| Acetyl-DL-methionine + $H_2O$ $\xrightarrow{\text{aminoacylase}}$ L-methionine + acetyl-D-methionine + acetic acid | Aspergillus ochraceus | Egg albumin and glutaraldehyde | Hirano et al. (1977) |
| ε-Aminocaproic acid cyclic dimer + $H_2O$ $\xrightarrow{\text{cyclic dimer hydrolase}}$ ε-aminocaproic acid ε-aminocaproic acid | Achromobacter guttatus | Polyacrylamide | Kinoshita et al. (1975) |
| L-Arginine + $H_2O$ $\xrightarrow{\text{L-Arginine deiminase}}$ L-citrulline + $NH_3$ <br> [IV] Lyase reaction | Pseudomonas putida | Polyacrylamide | Yamamoto et al. (1974a) |
| Diaminopimelic acid $\xrightarrow{\text{diaminopimelic acid decarboxylase}}$ L-lysine + $CO_2$ | Microbacterium ammoniaphilum | Polyacrylamide | Kanemitsu (1975) |
| L-Aspartic acid $\xrightarrow{\text{aspartate 4-decarboxylase}}$ L-alanine + $CO_2$ | Pseudomonas dacunhae | Polyacrylamide | Chibata et al. (1975a) |
| Pyruvate + phenol + $NH_3$ $\xrightarrow{\beta\text{-tyrosinase}}$ L-tyrosine | Erwinia herbicola | Collagen and dialdehyde starch | Yamada et al. (1975) |
| Pyruvate + pyrocatechol + $NH_3$ $\xrightarrow{\beta\text{-tyrosinase}}$ L-DOPA | Erwinia herbicola | Collagen and dialdehyde starch | Kumagai et al. (1976) |

| Reaction | Microorganism | Method | Reference |
|---|---|---|---|
| Fumaric acid + $H_2O$ $\xrightarrow{\text{fumarase}}$ L-malic acid | Brevibacterium ammoniagenes | Polyacrylamide | Yamamoto et al. (1976) |
| | Brevibacterium flavum | Carrageenan | Tosa et al. (1977) |
| Fumaric acid + $NH_3$ $\xrightarrow{\text{aspartase}}$ L-aspartic acid | Escherichia coli | Carrageenan | Tosa et al. (1977) |
| | | Glutaraldehyde | Chibata et al. (1974e) |
| | | Polyacrylamide | Chibata et al. (1974e) |
| Indole + L-serine $\xrightarrow{\text{tryptophan synthetase}}$ L-... tryptophan | Escherichia coli | Polyacrylamide | Chibata et al. (1974c) |
| 5-Hydroxyindole + L-serine $\xrightarrow{\text{tryptophan synthetase}}$ 5-hydroxy-L-tryptophan | Escherichia coli | Polyacrylamide | Chibata et al. (1974b) |
| L-Histidine $\xrightarrow{\text{L-histidine ammonia-lyase}}$ urocanic acid + $NH_3$ | Achromobacter liquidum | Polyacrylamide | Yamamoto et al. (1974b) |
| [V] Isomerase reaction | | | |
| D-Glucose $\xrightarrow{\text{glucose isomerase}}$ D-fructose | Actinomyces | Anion-exchange resin | Ishimatsu et al. (1976) |
| | | Drying after dipping in organic acid | Tsumura and Kasumi (1977) |
| | | Gelatin and glutaraldehyde | Yuta et al. (1975) |
| | | Heat treatment | Takasaki and Kanbayashi (1969) |
| | | β-Ray irradiation | Tsumura (1969) |

*(continued)*

**TABLE I** (Continued)

| Enzyme reaction and application[a] | Microorganism | Immobilization method | Reference |
| --- | --- | --- | --- |
| | *Bacillus coagulans* | Glutaraldehyde | Novo Industri (1976a,b) |
| | *Streptomyces griseus* | Polyacrylamide | Chibata et al. (1974d) |
| | *Streptomyces phaeochromogenes* | Carrageenan | Tosa et al. (1977) |
| | | Cellulose triacetate | Kolarik et al. (1974) |
| | | Chitosan | Tsumura et al. (1976) |
| | | Collagen and glutaraldehyde | Vieth et al. (1973) |
| | | | Wang et al. (1973) |
| | *Streptomyces venezuelae* | | Venkatasubramanian et al. (1974) |
| [VI] Multienzyme reaction | | | |
| Oxidation of glucose and succinic acid | *Azotobacter agile* | Dowex-1 | Hattori and Furusaka (1961) |
| | *Escherichia coli* | Dowex-1 | Hattori and Furusaka (1960) |
| Survival of microorganism in polymer lattice | *Tetrahymena pyriformis* | Polyacrylamide | Updike et al. (1969) |
| Reduction of nitrate and nitrite $NO_3^- \longrightarrow NO_2^- \longrightarrow NH_4^+$ | *Micrococcus denitrificans* | Liquid membrane | Mohan and Li (1975) |
| Production of hydrogen from glucose (hydrogen battery) | *Clostridium butyricum* | Polyacrylamide | Karube et al. (1976a) |

| Process | Organism | Support | Reference |
|---|---|---|---|
| Photosynthetic regeneration of ATP | *Rhodospirillum rubrum* chromatophores | Polyacrylamide | Yang et al. (1976) |
| Degradation of phenol<br>Phenol $\longrightarrow$ $CO_2 + H_2O$ | *Candida tropicalis* | Aluminium alginate<br>Polystyrene | Hackel et al. (1975)<br>Hackel et al. (1975) |
| | | Polyacrylamide | Hackel et al. (1975) |
| Conversion of L-sorbose to 2-keto-L-gulonic acid<br>L-Sorbose $\xrightarrow{\text{L-sorbose dehydrogenase}}$ L-sorbosone<br>$\xrightarrow{\text{L-sorbosone oxidase}}$ 2-keto-L-gulonic acid | *Gluconobacter melanogenus* and *Pseudomonas syringae* | Polyacrylamide | Martin and Perlman (1976b) |
| Preparation of orcinol and orcinol monomethyl ester from depside evernic acid by the esterase and decarboxylase activities of the cells | *Umbilicaria pastulate* | Polyacrylamide | Mosbach and Mosbach (1966) |
| Preparation of coenzyme A from pantothenic acid, L-cysteine, and ATP | *Brevibacterium ammoniagenes* | Polyacrylamide | Shimizu et al. (1975) |
| Catabolism of L-arginine | *Streptococcus faecalis* | Polyacrylamide | Franks (1971) |
| Production of lactic acid from 1,2-propanediol | *Arthrobacter oxydans* | Polyacrylamide | Yagi et al. (1976) |
| Production of L-glutamic acid from glucose | *Corynebacterium glutamicum* | Collagen | Brownstein et al. (1974) |
| | | Polyacrylamide | Slowinski and Charm (1973) |
| Production of ethanol from glucose | *Saccharomyces cerevisiae* | Calcium alginate | Kierstan and Bucke (1977) |
| Production of ethanol from inulin | *Kluyveromyces marxianus* | Calcium alginate | Kierstan and Bucke (1977) |

*(continued)*

**TABLE I** *(Continued)*

| Enzyme reaction and application[a] | Microorganism | Immobilization Method | Reference |
|---|---|---|---|
| Production of beer from wort | Yeast | Polyvinyl chloride and porous bricks | Corrieu et al. (1976) |
| Bread making | Bakers' yeast | Alkaline metal salts of alginic acid | Chibata et al. (1974a) |
| BOD sensor | Soil bacteria | Collagen and glutaraldehyde Polyacrylamide | Karube et al. (1976b) Karube et al. (1977) |

[a] Key to abbreviations: 7-ADCA, 7-aminodesacetoxycephalosporanic acid; 6-APA, 6-aminopenicillanic acid; ATP, adenosine triphosphate; BOD, biochemical oxygen demand; L-DOPA, 3,4-dihydroxy-L-phenylalanine; NAD, nicotinamide adenine dinucleotide.

gel method was first used for immobilization of enzymes by Bernfeld and Wan (1963). This technique was later applied to immobilization of lichen cells, *Umbilicaria pustulata,* by Mosbach and Mosbach (1966). Chibata *et al.* (1974e) then investigated the conditions of immobilizing bacterial cells, such as *E. coli,* with aspartase activity, by the same method for industrial use.

In order to most efficiently prepare immobilized *E. coli* cells, the type and the concentration of bifunctional reagents for cross-linking and the concentration of acrylamide monomer were investigated. The following bifunctional reagents were used for lattice formation: *N,N'*-methylene bisacrylamide, *N,N'*-propylene bisacrylamide, diacrylamide dimethyl ether, 1, 2-diacrylamide ethyleneglycol, *N,N'*-diallyl tartardiamide, ethyl-eneurea bisacrylamide, and 1,3,5-triacryloyl hexahydro-s-triazine. The activities of the immobilized *E. coli* obtained were almost the same except for those with ethyleneurea bisacrylamide and 1,3,5-triacryloyl hexahydro-s-triazine. *N,N'*-Methylene bisacrylamide was chosen since it was commercially available at low cost.

Further, the concentrations of acrylamide monomer and *N,N'*-methylene bisacrylamide and the amount of cells to be entrapped were investigated. As a result, the optimum conditions were established for immobilization of *E. coli* cells.

*Escherichia coli* cells (10 gm, wet weight) collected from cultured broth were suspended in 40 ml of physiological saline. To this suspension were added 7.5 gm of acrylamide monomer and 0.4 gm of *N,N'*-methylene bisacrylamide at 8°C, and the total was mixed. To the mixture, 5 ml of 5% $\beta$-dimethylaminopropionitrile, an accelerator of polymerization, and 5 ml of 2.5% potassium persulfate, an initiator of polymerization, were added. The reaction mixture was allowed to stand at below 40°C for 10–15 minutes to give a stiff gel. The resultant stiff gel was granulated to a suitable size and was thoroughly washed with physiological saline.

The authors succeeded in immobilizing many kinds of microbial cells using this polyacrylamide gel method. The authors also found (Tosa *et al.,* 1977) that carrageenan, a polysaccharide widely used as a food additive, is a more advantageous matrix for immobilization of microbial cells because the procedure is easy and can be carried out under relatively mild conditions. Consequently, the obtained immobilized cells have higher activity than those prepared by the polyacrylamide gel method and, moreover, the immobilized cells are more stable. Microbial cells are suspended in physiological saline and carrageenan is dissolved in the same saline. After mixing both solutions, gelation occurs easily and rapidly by cooling or contacting with an aqueous solution containing cations, such as $K^+$, $NH_4^+$, $Ca^{2+}$, $Cu^{2+}$, $Mg^{2+}$, $Mn^{2+}$, $Fe^{3+}$, or amines. For example, 5 gm of whole microbial cells are suspended in 5 ml of physiological saline at 45°–50°C, and 1.7 gm of carrageenan are dissolved in 34 ml of the same saline at 45°–60°C. Both are mixed, and the mixture is cooled at around 10°C. In order to increase the gel strength, the resultant gel is soaked in cold 0.3 $M$ potassium chloride solution. After this treatment, the stiff gel which is obtained is granulated to a suitable particle size. If the operational stability of immobilized cells is not satisfied, the immobilized cells are treated with hardening agents such as tannin, glutaraldehyde, or glutaraldehyde and hexamethylenediamine. As a result, more stable immobilized cells can be obtained.

By this method, the authors successfully immobilized whole cells of *E. coli* having aspartase activity, *Brevibacterium flavum* having fumarase activity and *Streptomyces phaeochromogenes* having glucose isomerase activity, respectively. As shown in Table II, the enzyme activities of these immobilized cells were higher than those of the preparations obtained by the polyacrylamide gel method. Moreover, these immobilized cells were stable, and a column packed with them could be used for continuous reaction for long periods. Operational stabilities of immobilized *E. coli* and immobilized *S. phaeochromogenes* were markedly increased by hardening with glutaraldehyde and hexamethylenediamine.

**TABLE II.** Comparison of Enzyme Activities and Stabilities of Immobilized Microbial Cells Prepared by Carrageenan and Polyacrylamide Gel Methods

| Microorganism | Enzyme | Enzyme activity ($\mu$mole/hour/gm cells) | | Half-life (days) | |
|---|---|---|---|---|---|
| | | Polyacryl-amide | Carra-geenan | Polyacryl-amide | Carra-geenan |
| *Eschericha coli* | Aspartase | 19,000 (29.2%)[b] | 30,400 (46.8%) 21,400[a] (32.9%)[b] | 120 | 50 686[a] |
| *Streptomyces phaeochromogenes* | Glucose isomerase | 3,880 (49.1%)[b] | 4,280 (54.1%) 4,310[a] (54.5%)[b] | 53 | 53 289[a] |
| *Brevibacterium flavum* | Fumarase | 6,680 (34.0%)[b] | 9,920 (50.5%) 9,120[a] (46.4%)[b] | 72 | 70 120[a] |

[a] Value after hardening with glutaraldehyde and hexamethylenediamine.
[b] Yield of activity after immobilization or after hardening shown in parentheses.

The carrageenan method is applicable for immobilization of many kinds of microbial cells and is considered to be more advantageous for industrial purposes than the polyacrylamide gel method.

Immobilization of microbial cells by collagen was carried out using the same technique for immobilization of enzymes developed by Vieth *et al.* (1973). This method is suitable for preparation of a membrane type of immobilized cells. After immobilization, various hardening treatments are carried out, since, in this case, the strength of collagen membrane is weak and enzymes easily leak out from the matrix.

Immobilization of microbial cells into cellulose fiber was developed by Dinelli (1972), and the enzyme activity of the immobilized cells prepared by this method is said to be very stable.

Besides entrapping methods described above, agar (Miyata and Kikuchi, 1976), alginate, and polystyrene (Hackel *et al.*, 1975) have been reported as matrices for immobilization of microbial cells. However, the detailed conditions for immobilization of microbial cells and the characteristics of resulting immobilized cells have not been clarified.

More recently, immobilization of cells by liquid membrane has been reported (Mohan and Li, 1975). In this method, a buffer solution containing cells is dispersed and emulsified in an oil phase composed of surfactants, various additives, and hydrocarbon solvent to form encapsulated cells with a liquid membrane.

## D. Miscellaneous Immobilization Methods

Fixation of enzyme within microbial cells has been carried out by heat treatment at a temperature which does not inactivate the desired enzyme. This method, however, does not always result in immobilization of microbial cells (Takasaki and Kanbayashi, 1969). When whole cells of *Streptomyces* sp. having glucose isomerase activity were heated at 60°–85°C for 10 minutes, the enzyme was fixed inside the cells and did not leak out from the cells when the cells were incubated under the conditions for enzyme reaction. These immobilized cells were used for industrial production of high-fructose syrup from D-glucose. Glucose isomerase was also fixed inside of *S. phaeochromogenes* cells by $\beta$-ray irradiation (Tsumura, 1969).

A method for immobilization of *S. phaeochromogenes* cells by complex formation with chitosan was reported (Tsumura *et al.*, 1976). When the cells were added to chitosan solution dissolved in acetic acid and the mixture adjusted to pH 6.0, the complex of cells and chitosan was formed and precipitated by aggregation.

## III. CHARACTERISTICS OF IMMOBILIZED MICROBIAL CELLS

Some enzymatic properties of immobilized microbial cells have been investigated, as in the case of immobilized enzymes, and have been compared with those of intact cells or native enzymes.

### A. Optimum pH

A number of papers investigating the effect of pH on enzyme reaction by immobilized microbial cells have been published, and their optimum pHs have been compared with those of intact cells or native enzymes.

Among 15 cases listed in Table III, the shift of optimum pH was observed in 5 cases, and 10 cases did not show the shift of optimum pH.

In *E. coli* and *A. agile,* immobilized using Dowex-1 (C1-form) (Hattori and Furusaka, 1960, 1961) or *B. ammoniagenes* immobilized by entrapping with polyacrylamide gel (Shimizu *et al.*, 1975), the optimum pH for oxidation of glucose or succinate or for synthesis of coenzyme A shifted 1–2 pH units to the alkaline range in comparison with those of intact cells. Further, the optimum pH of β-galactosidase activity of *E. coli* immobilized by entrapping with polyacrylamide gel also shifted 0.5 pH units to the alkaline side in comparison with that of native enzyme (Ohmiya *et al.*, 1977). The shift of optimum pH to the acid range was observed in aspartase activity of *E. coli* by the same entrapping method (Chibata *et al.*, 1974e).

### B. Optimum Temperature

Effects of temperature on enzyme reaction using immobilized microbial cells have been investigated, and the optimum temperature of immobilized cells has been compared with that of intact cells or native enzymes.

As shown in Table III, the optimum temperature of the immobilized microbial cells was similar to that of intact cells or native enzymes except for L-arginine deiminase activity of *Pseudomonas putida* immobilized by entrapping with polyacrylamide gel (Yamamoto *et al.*, 1974a).

### C. Stability

The most important factor in the large-scale use of immobilized microbial cells is their stability. A number of papers on thermal stability of immobilized cells have been published as in the case of immobilized enzymes. In many cases so far reported, thermal stability of microbial

**TABLE III.** Comparison of Optimum pH and Temperature of Intact and Immobilized Microbial Cells on Enzyme Reaction

| Microorganism | Enzyme | Optimum pH | | Optimum temp. (°C) | | Reference |
|---|---|---|---|---|---|---|
| | | Intact | Immobilized | Intact | Immobilized | |
| Achromobacter aceris | NAD kinase | 7.0 | 7.0 | 45 | 45 | Uchida et al. (1977) |
| Achromobacter liquidum | L-Histidine ammonia-lyase | 9.0[a] | 9.0[a] | 60[a] | 60[a] | Yamamoto et al. (1974b) |
| Aspergillus ochraceus | Aminoacylase | [8.0][b] | 8.0 | — | — | Hirano et al. (1977) |
| Azotobacter agile | Oxidation of glucose or succinate | 6.0 | 7.0–8.0 | — | — | Hattori and Furusaka (1961) |
| Brevibacterium ammoniagenes | Fumarase | [7.0–7.5][b] | 7.0–7.5 | [60][b] | 60 | Yamamoto et al. (1976) |
| | Synthesis of coenzyme A | 6.5 | 7.5 | — | — | Shimizu et al. (1975) |
| Escherichia coli | Oxidation of glucose or succinate | 6.0–7.0 | 7.0–8.0 | — | — | Hattori and Furusaka (1960) |
| | Aspartase | 10.5 | 8.5 | 50 | 50 | Chibata et al. (1974e) |
| | β-Galactosidase | [7.5][b] | 8.0 | [55][b] | 55 | Ohmiya et al. (1977) |
| Gluconobacter melanogenus | Penicillin amidase | 8.5 | 8.5 | 40 | 40 | Sato et al. (1976) |
| | L-Sorbose dehydrogenase | 6.3 | 6.3 | 50 | 50 | Martin and Perlman (1976a) |
| Kluyveromyces lactis | β-Galactosidase | [6.3][b] | 6.3 | [37][b] | 37 | Ohmiya et al. (1977) |
| Lactobacillus bulgaricus | β-Galactosidase | [5.5][b] | 5.5 | [55][b] | 55 | Ohmiya et al. (1977) |
| Pseudomonas putida | L-Arginine deiminase | 5.5–6.0[a] | 5.5–6.0[a] | 37[a] | 55[a] | Yamamoto et al. (1974a) |
| Rhodospirillum rubrum chromatophores | Photophosphorylating activity | 7.5–8.0 | 7.5–8.0 | 43 | 43 | Yang et al. (1976) |

[a] In the presence of cetylpyridinium chloride.
[b] Data for native enzyme in brackets.

cells was increased by immobilization. A significant reduction in thermal stability was observed only for L-sorbose dehydrogenase of *Gluconobacter melanogenus* immobilized by entrapping with polyacrylamide gel (Martin and Perlman, 1976a). When immobilized *G. melanogenus* cells were incubated in saline for 30 minutes at temperatures between 45° and 60°C, the immobilized cells lost activity more rapidly than the intact cells. However, when L-sorbose was added during this heat treatment, the activity of immobilized cells remained stable. At 50° and 55°C an increase in activity was observed.

In the presence of certain bivalent cations, several immobilized cells showed improved resistance to thermal inactivation. Chibata *et al.* (1974e) revealed that bivalent metal ions, such as $Ba^{2+}$, $Ca^{2+}$, $Mg^{2+}$, $Mn^{2+}$, and $Sr^{2+}$, helped to prevent thermal inactivation of aspartase activity of immobilized *E. coli* cells. It was also found that these metal ions enhanced operational stability of aspartase activity of immobilized *E. coli* cells during the continuous enzyme reaction.

Besides metal ions, it has been reported that the addition of antibiotics, such as neomycin, ampicillin, chloramphenicol, and tetracycline, increased the stability of L-sorbose dehydrogenase activity of immobilized *G. melanogenus* cells (Martin and Perlman, 1976a).

### D. Activation of Immobilized Microbial Cells

It has been observed that the enzyme activities of immobilized microbial cells increase when autolysis of the cells within matrices occurs or when cells are treated with surfactants.

The authors (Chibata *et al.*, 1974e) at first found that when the immobilized *E. coli* cells prepared by the polyacrylamide gel method were incubated at 37°C for 24–48 hours in a substrate solution, i.e., 1 $M$ ammonium fumarate solution (pH 8.5) containing 1m$M$ $Mg^{2+}$, the aspartase activity increased nine to ten times. This phenomenon was considered to be either an adaptive formation of aspartase–protein in the presence of substrate or an increase of membrane permeability for substrate and/or product due to autolysis of *E. coli* cells in the gel lattice.

In order to investigate the adaptive formation of the enzyme, immobilized cells were incubated in 1 $M$ ammonium fumarate solution for 48 hours at 37°C in the absence or presence of chloramphenicol, an inhibitor of protein synthesis, at concentrations that completely inhibited protein synthesis. As a result, the enzyme activity increased, even in the presence of chloramphenicol. Therefore, this activation was not considered to be the result of protein synthesis of *E. coli* cells in the gel lattice. This was also confirmed by electron micrographs of immobilized *E. coli* cells. Electron micrographs taken after activation indicated that lysis of cells had occurred. Martin and Perlman (1976a) also reported lysis of immobilized G. *melanogenus* leading to an increase of L-sorbose dehydrogenase activity during early stages of incubation. Whole cells of *Streptococcus faecalis* immobilized by entrapping with polyacrylamide catalyze the degradative conversion of L-arginine to putrescine, and the degradative activity of the immobilized cells was reported to be increased by treating with lysozyme (Franks, 1971).

Enzyme activities of immobilized cells were increased by treatment with surfactants. Yamamoto *et al.* (1976) revealed that fumarase activity of B. *ammoniagenes* immobilized by entrapping with polyacrylamide gel was markedly increased by treating with surfactants. These immobilized cells form L-malic acid as a main product with succinic acid as a by-product. Separation of succinic acid from L-malic acid is very difficult. The formation of succinic acid was found to be suppressed by the treatment with surfactants, such as deoxycholate, bile acid, and bile extract. When immobilized B. *ammoniagenes* cells prepared by polyacrylamide gel method were treated with sodium laurylsulfate, the formation of coenzyme A from pantothenic acid, cysteine, and ATP was also reported to be increased (Shimizu *et al.*, 1975).

Without the treatments described above, Yamamoto *et al.* (1974a) revealed that, in some cases, the permeability of substrate and/or product through the cell membrane is increased by immobilization of microbial cells with polyacrylamide gel. That is, formation of L-citrulline from L-arginine by L-arginine deiminase activity is rarely observed in the case of intact cells of P. *putida* in the absence of surfactant (cetyltrimethylammonium bromide), while the formation occurs without the surfactant by immobilized cells. This phenomenon was also observed in formation of urocanic acid from L-histidine by using intact and immobilized cells of *Achromobacter liquidum* having L-histidine ammonialyase activity (Yamamoto *et al.*, 1974b).

## IV. PRODUCTION OF CHEMICALS BY IMMOBILIZED MICROBIAL CELLS

Studies so far reported on chemical processes using immobilized microbial cells are listed in Table I. Enzyme reactions are classified according to enzyme reactions recommended in 1972 by the enzyme commission of the International Union of Pure and Applied Chemistry and the International Union of Biochemistry.

In this section we will describe utilization of immobilized microbial cells for the production of useful chemicals.

## A. Oxidoreductase Reaction

As shown in Table I, two examples have been reported on immobilization of microbial cells having oxidoreductase activity and its application for the formation of organic compounds.

For example, whole cells of G. *melanogenus* having L-sorbose dehydrogenase were immobilized by the entrapping method using polyacrylamide gel, and the immobilized cells were used for preparation of L-sorbosone from L-sorbose (Martin and Perlman, 1976a). L-Sorbosone is an intermediate for the synthesis of vitamin C. In this study, a very interesting phenomenon was found; namely, addition of electron acceptors, such as phenazine methosulfate or methylene blue, or addition of antibiotics, such as neomycin, ampicillin, chloramphenicol, or tetracycline, increases the stability of the enzyme activity.

Immobilized cells of Curvularia lunata entrapped in polyacrylamide gel are used for production of cortisol by 11$\beta$-hydroxylation of Reichstein compound S (Mosbach and Larsson, 1970). The immobilized cells lose part of the 11$\beta$-hydroxylase activity during storage, but the lost activity is reactivated by incubating the immobilized cells in a nutrient solution containing cortisone for 16 hours. The reactivation of activity was considered to be probably due to growth of the cells and/or enzyme formation by induction with cortisone during the incubation.

## B. Transferase Reaction

As shown in Table I, two papers on the use of immobilized microbial cells for transferase reaction have been published. One is on production of NADP from NAD and ATP by the action of NAD kinase (Chibata *et al.*, 1975b; Uchida *et al.*, 1977). That is, *Achromobacter aceris* having higher NAD kinase activity was immobilized into a polyacrylamide gel lattice, and the immobilized cells were used for continuous production of NADP. Although this microorganism had ATP degradation activity, acid treatment of the immobilized cells at pH 4.0 completely removed the undesired activity and at the same time activated the desired NAD kinase. When the acid-treated immobilized cells were packed into a column and the substrate solution (pH 7.0) containing 2.7 m$M$ NAD, 2.7 m$M$ ATP, and 3.0 m$M$ MnCl$_2$ was passed through the column at a flow rate of space velocity (SV, hour$^{-1}$) 0.1 at 37°C, 76% of the NAD in the substrate solution was converted to NADP. The enzyme activity of the immobilized cells was more stable in comparison with that of intact cells. Their half-lives were calculated to be around 20 days for immobilized cells and 6 days for intact cells at 37°C.

Continuous production of glucose 6-phosphate and glucose 1-phosphate from glucose and *p*-nitrophenyl phosphate by the action of acid phosphatase was achieved by Saif et al. (1975). The cells of Escherichia freundii having acid phosphatase activity were immobilized by the entrapping method using polyacrylamide gel. When the immobilized cells were packed into a column and the mixture of equal volumes of 100 mM glucose and 25 mM p-nitrophenyl phosphate was passed through the column at a flow rate of SV = 0.128 at 37°C, glucose 6-phosphate and glucose 1-phosphate were isolated from the effluent in high yield. The enzyme activity of the immobilized cells was very stable, and the column retained 50% of the initial activity after 120 days of operation.

## C. Hydrolase Reaction

There are many examples of the application of immobilized microbial cells for formation of useful organic compounds by hydrolase reaction. Yamamoto et al. (1974a) studied continuous production of L-citrulline, which is used medicinally, from L-arginine by using immobilized P. putida having high L-arginine deiminase activity. The cells, after entrapment in a polyacrylamide gel lattice, exhibited 56% of the arginine deiminase activity of that of intact cells. In the case of intact cells, the addition of surfactant, such as cetyltrimethylammonium bromide, was essential for a high rate of L-citrulline formation, but in the case of immobilized cells, the surfactant had no effect on L-citrulline formation.

Recently, many studies have been carried out on the continuous production of 6-aminopenicillanic acid (6-APA) from penicillin by using a column packed with immobilized penicillin amidase. 6-APA is used as an intermediate in the production of synthetic penicillin. Some of the immobilized systems are said to be used in industry in the United States and Europe.

Sato et al. (1976) studied an efficient continuous method for the production of 6-APA from penicillin G by using immobilized E. coli having high penicillin amidase activity. Whole cells of E. coli were immobilized by the entrapping method using polyacrylamide gel. These microbial cells have penicillinase activity, decomposing both penicillin and 6-APA. Specific inactivation of penicillinase activity was very difficult. However, penicillinase activity was much lower than penicillin amidase activity. Therefore, optimum conditions for the continuous production of 6-APA without removing the penicillinase activity were investigated. By choosing suitable flow rate, that is, when 50 mM penicillin solution (pH 8.5) was passed through a column packed with the immobilized cells at a flow rate of SV = 0.24 at 40°C, 6-APA was found to be efficiently produced in 78% yield.

Immobilized microbial cells having penicillin amidase activity were also used to synthesize ampicillin from 6-APA and D-phenylglycine methyl ester (Fujii *et al.*, 1973). Whole cells of *Achromobacter* sp. or *Bacillus megaterium* were immobilized by ionic binding to DEAE–cellulose. The immobilized cells were used not only to synthesize penicillins but also to acylate the cephalosporin nucleus. However, commercial development of this technique is retarded by low production yield because of product or substrate inhibition and reversibility of the reaction.

A number of papers have been published on the immobilization of microbial cells having invertase activity and their applications for production of invert sugar. This system will compete with immobilized enzyme system.

For the production of L-amino acids, Tosa *et al.* (1966, 1967) and Chibata *et al.* (1972) succeeded in the industrial application of immobilized aminoacylase for continuous production of L-amino acids from acyl-DL-amino acids. Instead of this system, *Aspergillus ochraceus* having aminoacylase activity was immobilized by cross-linking with egg albumin and glutaraldehyde, and the continuous optical resolution of acetyl-DL-methionine was investigated by using these immobilized cells (Hirano *et al.*, 1977).

Further, *Alginomonas nonfermentas* having *l*-menthol ester hydrolase activity was immobilized by the entrapping method using polyacrylamide gel, and preparation of *l*-menthol from synthetic *dl*-menthol ester was investigated by Nonomura *et al.* (1976).

Besides the processes described above, continuous hydrolysis of ε-aminocaproic acid cyclic dimer contained in waste water from nylon factories was investigated by immobilizing *Achromobacter guttatus* by the entrapping method using polyacrylamide gel (Kinoshita *et al.*, 1975).

In the food industry, immobilized microbial cells having β-galactosidase and raffinase activities were used for hydrolysis of lactose in milk and of raffinose interfering with crystallization of sugar from beet, as shown in Table I. These systems will also compete with immobilized enzyme systems in the future.

## D. Lyase Reaction

Lyase is an enzyme cleaving C—C, C—O, C—N, or other bonds by means other than hydrolysis or oxidation. A number of studies of the immobilization of lyases and their applications have been published. Many kinds of microbial cells having lyase activity can be immobilized, as shown in Table I, and the immobilized cells have been used for production of useful organic compounds.

## 1. Carbon–Carbon Lyase Reaction

In the carbon–carbon lyase reaction, carboxyl groups are eliminated from organic compounds by the action of decarboxylase of immobilized microbial cells.

Chibata *et al.* (1975a) immobilized whole cells of *Pseudomonas dacunhae* having high aspartate 4-decarboxylase activity and investigated the conditions for continuous production of L-alanine from L-aspartic acid by using immobilized cells. In this system, additions of pyridoxal phosphate and $Co^{2+}$ to the substrate solution were effective as a cofactor for enzyme reaction and as an enzyme stabilizer, respectively.

Further, *Microbacterium ammoniaphilum* having diaminopimelic acid decarboxylase activity was immobilized into a polyacrylamide gel lattice, and the immobilized cells were used for production of L-lysine from LL-2,6-diaminopimelic acid (Kanemitsu, 1975). When the immobilized cells (amounts corresponding to 200 mg dry cells) were incubated with 100 ml of 0.2 $M$ diaminopimelic acid solution (pH 8.0) at 40°C for 8 hours, 280 mg of L-lysine was obtained from the reaction mixture.

In addition to the decarboxylation reactions described above, *Erwinia herbicola* having $\beta$-tyrosinase activity was immobilized by cross-linking with collagen and dialdehyde starch, and the immobilized cells were used for production of L-tyrosine from phenol, pyruvate, and ammonia (Yamada *et al.*, 1975), and production of L-DOPA from pyrocatechol, pyruvate, and ammonia (Kumagai *et al.*, 1976), respectively.

## 2. Carbon–Oxygen Lyase Reaction

The carbon–oxygen lyase reaction has been studied for use in the production of L-malic acid from fumaric acid (Chibata *et al.*, 1975c; Yamamoto *et al.*, 1976), L-tryptophan from indole and DL-serine (Chibata *et al.*, 1974c) and 5-hydroxy-L-tryptophan from 5-hydroxyindole and DL-serine (Chibata *et al.*, 1974b).

L-Malic acid is used in the pharmaceutical field as an antidote for hyperammoniemia and a component of amino acid infusion. This acid is industrially produced from fumaric acid by the action of fumarase in batch process using cultured broth of microbial cells.

For the continuous production of L-malic acid, Chibata *et al.* (1975c) studied the immobilization of whole cells of *B. ammoniagenes* having high fumarase activity by the polyacrylamide gel method. However, such immobilized cells formed succinic acid as a by-product, the separation of which from L-malic acid is very difficult. Therefore, the successful utilization of this system for industrial production of pure L-malic acid is to prevent succinic acid formation during the enzyme reaction. Various treatments of immobilized cells were carried out, as described in previous sections. The treatment with detergent, such as deoxycholic acid, bile acid, and bile extract, was found to be very effective for suppression of succinic acid formation. These detergents also markedly enhanced the formation of L-malic acid by the immobilized cells. For example, when the immobilized cells were allowed to stand in 1 *M* sodium fumarate (pH 7.5) containing 0.3% bile extract at 37°C for 20 hours, succinic acid formation by the immobilized cells was completely suppressed and its fumarase activity was enhanced at about 10 times in comparison with that of untreated cells.

When 1 *M* sodium fumarate solution (pH 7.0) was passed through the column packed with the bile extract-treated immobilized cells at 37°C at an SV = 0.23, the reaction reached equilibrium with about 80% conversion of fumaric acid to L-malic acid. Fumaric acid was removed by acidification from the effluent and pure L-malic acid was separated in about 70% yield from fumaric acid. The fumarase activity of the immobilized cells was very stable and its half-life was calculated to be 53.5 days at 37°C.

Under the conditions described above, 15.4 metric tons of theoretical yield of L-malic acid could be produced per month, using a 1000-liter column. This production system has been utilized industrially since 1974 by Tanabe Seiyaku Co. Ltd., Japan.

For production of L-tryptophan from indole and DL-serine, the authors immobilized whole cells of *E. coli* having tryptophan synthetase activity into a polyacrylamide gel lattice. When the immobilized cells (30 gm) were incubated with 100 ml of substrate solution containing 1 gm of indole and 2 gm of DL-serine at 30°C for 24 hours, 1.5 gm of L-tryptophan was isolated from the reaction mixture. From a similar reaction mixture containing 1 gm of 5-hydroxyindole in place of indole, 1.1 gm of crystalline 5-hydroxytryptophan was obtained.

### 3. Carbon–Nitrogen Lyase Reaction

Two studies by the authors on production of L-aspartic acid from fumaric acid and ammonia, and production of urocanic acid from L-histidine are examples of the carbon–nitrogen lyase reaction.

L-Aspartic acid is widely used as a medicine and food additive. The acid had been industrially produced by fermentative and enzymatic methods from fumaric acid and ammonia by the action of aspartase.

Since this reaction had been carried out in batch process, there were some disadvantages for industrial purposes. To overcome the disadvantages, Tosa *et al.* (1973) investigated the immobilization of aspartase. The active immobilized aspartase was obtained by entrapment of extracted aspartase in a polyacrylamide gel lattice. However, this immobilized enzyme was relatively unstable; i.e., its half-life was 27 days at 37°C. Therefore, this immobilized aspartase was not considered satisfactory for the industrial production of L-aspartic acid. The authors thus considered that if the microbial cells could be immobilized directly, these disadvantages might be overcome. The immobilization of whole microbial cells was studied from these points of view, and this technique was incorporated into industrial use in 1973 (Chibata et al., 1973, 1974e; Tosa et al., 1974; Sato et al., 1975). This was considered the first industrial application of immobilized microbial cells in the world.

Whole cells of E. coli having high aspartase activity were immobilized into a polyacrylamide gel lattice, and the immobilized cells were activated by soaking in 1 M ammonium fumarate (pH 8.5) containing 1 mM $Mg^{2+}$ at 37°C for 24–48 hours as described in Section III, D.

By using a column packed with the activated immobilized E. coli cells, the conditions for the continuous production of L-aspartic acid were investigated. When an aqueous solution of 1 M ammonium fumarate (pH 8.5) containing 1 mM $Mg^{2+}$ was passed through the column at a flow rate of SV = 0.6 at 37°C, the reaction proceeded to completion. The effluent, 2400 liters, was adjusted to pH 2.8 with 60% $H_2SO_4$ at 90°C and then cooled at 15°C for 2 hours. The L-aspartic acid that crystallized out was collected by centrifugation and washed with water. This product was pure without the need for recrystallization, and the yield was 304.8 kg (98% of theoretical). The immobilized cell column was very stable and its half-life was estimated to be 120 days at 37°C. The overall production cost of this system was reduced to about 60% of the conventional batch process using intact cells due to marked reduction of the cost of preparing the catalysts and to reduction of labor cost by automation.

Whole cells of Achromobacter liquidum having higher L-histidine ammonia-lyase activity were immobilized into a polyacrylamide gel lattice, and the immobilized cells were used for production of urocanic acid from L-histidine (Yamamoto et al., 1974b). Urocanic acid is used as a sun-screening agent in the pharmaceutical and cosmetic fields. Although the microorganism has urocanase activity which decomposes urocanic acid to imidazolone propionic acid, this activity was removed by simple heat treatment (70°C, 30 minutes) of the cells before immobilization.

By using a column packed with the immobilized heat-treated cells, the conditions for the continuous production of urocanic acid were investigated. When an aqueous solution of 0.25 M L-histidine (pH 9.0) containing 1 mM $Mg^{2+}$ was passed through the column at flow rates of SV below 0.06, L-histidine was completely converted to urocanic acid. Pure urocanic acid was crystallized in a high yield from the effluent by merely adjusting the pH of the effluent to 4.7. The enzyme activity of the column was very stable in the presence of Mg²⁺, and its half-life was estimated to be about 180 days at 37°C.

### E. Isomerase Reaction

As shown in Table I, many papers have been published on the isomerization of D-glucose to D-fructose by immobilized microbial cells having glucose isomerase activity.

Takasaki and Kanbayashi (1969) investigated the use of heat-treated *Streptomyces* sp. having higher glucose isomerase activity for conversion of D-glucose to D-fructose. Into a column packed with the heat-treated cells, 40% glucose solution (pH 8.0) containing 5 m$M$ Mg²⁺ and 1 m$M$ Co²⁺ was fed by the method of upward flow at 70°C. The continuous reaction could be carried out for 15 days at the average isomerization rate of 40%. However, there are a number of problems yet to be solved for industrial application of this column technique. The industrial production of fructose from glucose is presently accomplished by a batch reaction process, reusing the heat-treated cells two or three times. The preparation of improved stabilized immobilized cells and the development of a continuous reactor system suitable for industrial uses are expected in the future.

Vieth *et al.* (1973) immobilized *S. phaeochromogenes* cells having glucose isomerase activity by the collagen membrane method after heat treatment of the cells. The cell–collagen–membrane was tanned by dipping in alkaline formaldehyde or glutaraldehyde solution, and the tanned cell–collagen–membrane was cut into small chips. When 1 $M$ D-glucose solution (pH 7.0) containing 10 m$M$ Mg²⁺ was passed through a column packed with the chips at flow rate of 11–12 ml/hour at 70°C for 40 days, the continuous reaction could be carried out at the isomerization rate of 40% for 15 days. After that the activity of the column gradually decreased.

The authors (Chibata *et al.*, 1974d) immobilized whole cells of *Streptomyces griseus* having glucose isomerase activity into a polyacrylamide gel lattice and investigated the use of this preparation for continuous conversion of D-glucose to D-fructose. In this system, Mg²⁺ was necessary for the activation of enzyme reaction and Co²⁺ was useful for the stabilization of enzyme activity.

The industrial isomerization of D-glucose to D-fructose by immobilized glucose isomerase has been carried out in Japan, the United States, and in Europe, and this immobilized microbial cell system is expected to compete with the immobilized enzyme system.

## F. Multienzyme Reaction

Applications for enzyme reactions using immobilized microbial cells are carried out mainly by the action of a single enzyme. However, many useful chemicals can be produced by multistep reactions with the action of several kinds of enzymes, especially in fermentative methods. The immobilization of microbial cells for multienzyme reactions has been attempted, as shown in Table I.

Slowinski and Charm (1973) immobilized whole cells of *Corynebacterium glutamicum* (a glutamic acid-producing bacteria) by the entrapping method using polyacrylamide gel and investigated the formation of L-glutamic acid from glucose by using these cells. The immobilized *C. glutamicum* cells corresponding to 2.8 mg (dry weight) of the cells were suspended in 40 ml of fermentation medium containing glucose as the carbon source, inorganic ammonium salts as the nitrogen source, and several kinds of metal salts. The suspension was incubated at 30°C under stirring. After 144 hours of processing, 15 mg/ml of L-glutamic acid were accumulated in the reaction mixture. In this case, productivity of L-glutamic acid by immobilized cells was said to be superior in comparison with that of intact cells. From this work they estimated that the immobilized cells probably can be used in a column with continuous processing.

In addition, continuous production of coenzyme A from pantothenic acid, L-cysteine, and ATP was investigated by immobilized *B. ammoniagenes* cells (Shimizu *et al.*, 1975). The reaction consisted of the following five steps: panthothenic acid → phosphopantothenic acid → phosphopanthothenoylcysteine → phosphopantetheine → dephosphocoenzyme A → coenzyme A. When a substrate solution (pH 6.5) containing sodium pantothenate, cysteine, ATP, $MgSO_4$, and potassium phosphate was passed through a column (0.9 × 30 cm) packed with the immobilized cells at a flow rate of SV = 0.1–0.2 at 37°C, 500 $\mu$g/ml of coenzyme A were produced in the effluent. The half-life of the immobilized cell column was 5 days at 37°C.

Recently whole cells of yeast bound to fragments of polyvinyl chloride and porous bricks were used for continuous production of beer from wort (Corrieu *et al.*, 1976). The immobilization of yeast cells was carried out by directly circulating cultured broth into a column packed with polyvinyl chloride and porous bricks. For continuous production of beer, two immobilized cell columns were prepared. One column was used to change wort to beer and another column was used for the maturation of beer. The quality of the beer produced is said to be no different from ordinary beer.

In addition, more recently continuous ethanol production from glucose by *Saccharomyces cerevisiae* cells immobilized in calcium alginate gel was reported (Kierstan and Bucke, 1977). When a solution of 10% (w/v) glucose containing $0.05 M$ $CaCl_2$ was passed through a column ($0.9 \times 8$ cm) packed with the immobilized cells at a flow rate of 50 ml/24 hr, the maximum efficiency of glucose to ethanol conversion was 90% of the theoretical maximum yield. This system has a half-life of about 10 days and has been used for the efficient continuous production of ethanol over a total period of 24 days.

As described above, the examples pose the possibilities that fermentative methods involving multienzyme reactions may be replaced by the continuous enzyme reaction using a column packed with immobilized microbial cells. However, there are a number of difficulties to be overcome, especially the supply of air or oxygen into the column or reactor. If these difficulties can be overcome and energy regeneration can be efficiently carried out, immobilized microbial cell systems are expected to become advantageous bioreactors or catalysts for industrial production of many useful chemicals in the future.

## V. CONCLUSION

As described above, many kinds of microbial cells having an enzyme of higher activity can be easily immobilized. Continuous enzyme processing by the immobilized microbial cells will be advantageous in the following cases: (1) when the enzymes are intracellular, (2) when the enzymes extracted from microbial cells are unstable, (3) when the enzymes are unstable during and after immobilization, (4) when the microorganism contains no other enzymes that catalyze interfering side reactions or when those interfering enzymes can be readily inactivated or removed, and (5) when the substrates and products are not high molecular compounds.

Another aspect to be considered is the volume of liquid to be produced. For the unit production of a desired compound, the volume of fermentation broth is much smaller in the case of the continuous method using immobilized cells than in the case of conventional batch fermentative methods. The indication is that the former method is very advantageous from the point of view of water pollution control in plants. Moreover, in the case of the continuous reaction using immobilized cells, it requires less space—mass production becomes possible by using much smaller enzyme reactors than conventional batch fermenters.

In addition, taking into consideration the problems concerning new treatment of fermentative waste for prevention of water pollution, cost of energy, and others, this technique is very promising and is expected to develop along with immobilized enzyme systems in the future.

## REFERENCES

Bernfeld, P., and Wan, J. (1963). *Science* **142,** 678–679.

Brownstein, A. M., Vieth, W. R., and Constantinides, A. (1974). *168th Meet., Am. Chem. Soc. Meet.*

Chibata, I., Tosa, T., Sato, T., Mori, T., and Matuo, Y. (1972). *Ferment. Technol. Today, Proc. Int. Ferment. Tec. Symp., 4th, 1972* pp. 383–389.

Chibata, I., Tosa, T., Sato, T., Mori, T., and Yamamoto, K. (1973). *Enzyme Eng.* **2,** 303–313.

Chibata, I., Tosa, T., and Mori, T. (1974a). Japanese Patent Kokai 74/30,582.

Chibata, I., Kakimoto, T., and Nabe, K. (1974b). Japanese Patent Kokai 74/81,590.

Chibata, I., Kakimoto, T., and Nabe, K. (1974c). Japanese Patent Kokai 74/81,591.

Chibata, I., Tosa, T., and Sato, T. (1974d). Japanese Patent Kokai 74/132,290.

Chibata, I., Tosa, T., and Sato, T. (1974e). *Appl. Microbiol.* **27,** 878–885.

Chibata, I., Tosa, T., Sato, T., and Yamamoto, K. (1975a). Japanese Patent Kokai 75/100,289.

Chibata, I., Kato, J., Watanabe, T., and Uchida, T. (1975b). Japanese Patent Kokai 75/135,290.

Chibata, I., Tosa, T., and Yamamoto, K. (1975c). *Enzyme Eng.* **3,** 463–468.

Corrieu, G., Blachere, H., Ramirez, A., Navarro, J. M., Durand, G., Toulouse, I. N. S. A., Duteurtre, B., and Moll, M. (1976). *Proc. Int. Ferment. Symp. 5th, 1976* p. 294.

Dinelli, D. (1972). *Process Biochem.* **7**(8), 9–12.

Franks, N. E. (1971). *Biochim. Biophys. Acta* **252,** 246–254.

Fujii, T., Hanamitsu, K., Izumi, R., Yamada, T., and Watanabe, T. (1973). Japanese Patent Kokai 73/99,393.

Fujii, T., Matsumoto, K., Shibuya, Y., Hanamitsu, K., Yamaguchi, T., Watanabe, T., and Abe, S. (1974). British Patent 1,347,665.

Hackel, U., Klein, J., Megnet, R., and Wagner, F. (1975). *Eur. J. Appl. Microbiol.* **1,** 291–293.

Hattori, T., and Furusaka, C. (1960). *J. Biochem. (Tokyo)* **48,** 831–837.

Hattori, T., and Furusaka, C. (1961). *J. Biochem. (Tokyo)* **50,** 312–315.

Hirano, K., Karube, I., and Suzuki, S. (1977). *Biotechnol. Bioeng.* **19,** 311–321.

Ishimatsu, Y., Shigesada, S., and Kimura, S. (1976). Japanese Patent Kokai 76/44,688.

Johnson, D. E., and Ciegler, A. (1969). *Arch. Biochem. Biophys.* **130,** 384–388.

Kanemitsu, O. (1975). Japanese Patent Kokai 75/132,181.

Karube, I., Matsunaga, T., Tsuru, S., and Suzuki, S. (1976a). *Biochim. Biophys. Acta* **144,** 338–343.

Karube, I., Mitsuda, S., and Suzuki, S. (1976b). *Annu. Meet. Soc. Ferment. Technol. Jpn, 28th, 1976* p. 127.

Karube, I., Mitsuda, S., Matsunaga, T., and Suzuki, S. (1977). *J. Ferment. Technol.* **55,** 243–248.

Kierstan, M., and Bucke, C. (1977). *Biotechnol. Bioeng.* **19,** 387–397.

Kinoshita, S., Muranaka, M., and Okada, H. (1975). *J. Ferment. Technol.* **53,** 223–229.

Kolarik, M. J., Chen, B. J., Emergy, A. H., Jr., and Lim, H. C. (1974). *In* "Immobilized Enzymes in Food and Microbial Processes" (A.C. Olson and C.L. Cooney, eds.), pp. 71–83. Plenum, New York.

Kosugi, Y., and Suzuki, H. (1973). *J. Ferment. Technol.* **51,** 895–903.

Kumagai, H., Sezima, S., Yamada, H., Hino, T., and Okamura, S. (1976). *Annu. Meet. Agric. Chem. Soc. Jpn., 52nd, 1976* p. 233.

Marconi, W., Bartoli, F., Cerere, F., Galli, G., and Morisi, F. (1975). *Agric. Biol. Chem.* **39,** 277–279.

Martin, C. K. A., and Perlman, D. (1976a). *Biotechnol. Bioeng.* **18,** 217–237.

Martin, C. K. A., and Perlman, D. (1976b). *Eur. J. Appl. Microbiol.* **3,** 91–95.

Miyata, N., and Kikuchi, T. (1976). Japanese Patent Kokai 76/133,484.

Mohan, R. R., and Li, N. N. (1975). *Biotechnol. Bioeng.* **17,** 1137–1156.

Mosbach, K., and Larsson, P. O. (1970). *Biotechnol. Bioeng.* **12,** 19–27.

Mosbach, K., and Mosbach, R. (1966). *Acta Chem. Scand.* **20,** 2807–2810.

Nanba, A., and Matuo, Y. (1970). *Annu. Meet. Agric. Chem. Soc. Jpn, 45th, 1970* p. 251.

Nishimaru, H., Izumi, C., Narita, S., and Yamada, K. (1975). Japanese Patent Kokai 75/140,680.

Nonomura, S., Watanabe, M., and Inagaki, T. (1976). Japanese Patent Kokai 76/48,488.

Novo Industri (1976a). Japanese Patent Kokai 76/51,580.

Novo Industri (1976b). U.S. Patent Application No. 501,292.

Ohmiya, K., Ohashi, H., Kobayashi, T., and Shimizu, S. (1977). *Appl. Environ. Microbiol.* **33,** 137–146.

Saif, S. R., Tani, Y., and Ogata, K. (1975). *J. Ferment. Technol.* **53,** 380–385.

Sato, T., Mori, T., Tosa, T., Chibata, I., Furui, M., Yamashita, K., and Sumi, A. (1975). *Biotechnol. Bioeng.* **17,** 1797–1804.

Sato, T., Tosa, T., and Chibata, I. (1976). *Eur. J. Appl. Microbiol.* **2,** 153–160.

Shimizu, S., Morioka, H., Tani, Y., and Ogata, K. (1975). *J. Ferment. Technol.* **53,** 77–83.

Slowinski, W., and Charm, S. E. (1973). *Biotechnol. Bioeng.* **15,** 973–979.

Takasaki, Y., and Kanbayashi, A. (1969). *Kogyo Gijutsuin Biseibutsu Kogyo Gijutsu Kenkyusho Kenkyu Kokoku* **37,** 31–37; *Chem. Abstr.* **74,** 139,538 (1970).

Toda, K., and Shoda, M. (1975). *Biotechnol. Bioeng.* **17,** 481–497.

Tosa, T., Mori, T., Fuse, N., and Chibata, I. (1966). *Enzymologia* **31,** 214–224.

Tosa, T., Mori, T., Fuse, N., and Chibata, I. (1967). *Biotechnol. Bioeng.* **9,** 603–615.

Tosa, T., Sato, T., Mori, T., Matuo, Y., and Chibata, I. (1973). *Biotechnol. Bioeng.* **15,** 69–84.

Tosa, T., Sato, T., Mori, T., and Chibata, I. (1974). *Appl. Microbiol.* **27,** 886–889.

Tosa, T., Sato, T., Yamamoto, K., Takata, I., Nishida, Y., and Chibata, I. (1977). *Int. Cong. Pure Appl. Chem.* **26,** 267.

Tsumura, N. (1969). *Annu. Meet. Soc. Ferment. Technol., Jpn, 21st, 1969* p. 81.

Tsumura, N., and Kasumi, T. (1977). Japanese Patent Kokai 77/44,285.

Tsumura, N., Kasumi, T., and Ishikawa, M. (1976). *Rep. Natl. Food. Res. Inst. (Tokyo)* **31,** 71–75 (in Japanese).

Uchida, T., Watanabe, T., Kato, J., and Chibata, I. (1978). *Biotechnol. Bioeng.* **20,** 255–266.

Updike, S. J., Harris, D. R., and Shrago, E. (1969). *Nature (London)* **224,** 1122–1123.

Venkatasubramanian, K., Saini, R., and Vieth, W. R. (1974). *J. Ferment. Technol.* **52,** 268–278.

Vieth, W. R., Wang, S. S., and Saini, R. (1973). *Biotechnol. Bioeng.* **15,** 565–569.

Wang, S. S., Vieth, W. R., and Constantinides, A. (1973). *Enzyme Eng.* **2,** 123–129.

Yagi, S., Toda, Y., and Minoda, T. (1976). *Annu. Meet. Agric. Chem. Soc. Jpn, 51st, 1976* p. 414.

Yamada, H., Yamada, M., Nakazawa, E., Kumagai, H., Hino, T., and Okamura, S. (1975). *Annu. Meet. Agric. Chem. Soc. Jpn, 50th, 1975* p. 336.

Yamamoto, K., Sato, T., Tosa, T., and Chibata, I. (1974a). *Biotechnol. Bioeng.* **16,** 1589–1599.

Yamamoto, K., Sato, T., Tosa, T., and Chibata, I. (1974b). *Biotechnol. Bioeng.* **16,** 1601–1610.

Yamamoto, K., Tosa, T., Yamashita, K., and Chibata, I. (1976). *Eur. J. Appl. Microbiol.* **3,** 169–183.

Yamamoto, K., Tosa, T., Yamashita, K., and Chibata, I. (1977). *Biotechnol. Bioeng.* **19,** 1101–1114.

Yang, H. S., Leung, K.-H., and Archer, M. C, (1976). *Biotechnol. Bioeng.* **18,** 1425–1432.

Yuta, S., Bhatt, R. R., Yoshida, T., and Taguchi, K. (1975). *Annu. Meet. Soc. Ferment. Technol., Jpn, 27th, 1975* p. 250.

Chapter 18

# Economics of Fermentation Processes

WILLIAM H. BARTHOLOMEW
HAROLD B. REISMAN

## I. INTRODUCTION

Fermentation is one of the oldest known chemical processes. Fermented foods and alcoholic beverages have been produced for thousands of years, but it is only in the last three to four decades that

MICROBIAL TECHNOLOGY, 2nd ed., VOL.II
Copyright © 1979 by Academic Press, Inc.
All rights of reproduction in any form reserved. ISBN 0-12-551502-2

biotechnology and chemical engineering have been applied on a large scale to bring to the world fermentation products such as antibiotics, flavoring agents, amino and organic acids, and vitamins.

Fermentation processes, and hence the economics of fermentation, depend upon specificity of enzymatic reactions to produce complex products too costly to produce by direct synthesis. Relatively high efficiency, coupled with availability and cost of agricultural raw materials, allows use of biological systems to produce needed chemical products. In recent years, production of biocatalysts themselves has become an important industry. Large-scale conversion of glucose to high-fructose corn syrup requires equally large-scale production of the needed enzyme—glucose isomerase.

Prior to commercialization, extensive economic evaluation is necessary. Unique characteristics of fermentation processes, including use of agricultural commodities, high-energy use per unit of product, and need for aseptic operation, require involvement of life scientists, chemical engineers, and other engineering disciplines needed for plant construction and startup. If it is assumed that a totally new facility is to be built, site selection is an early requirement. The usual criteria for chemical processes are used to narrow site selection, but there are certain unique requirements which should be considered in fermentation plant design. Process design follows both conventional engineering needs as well as those derived from unique biochemical inputs from laboratory and pilot plant data. Instrumentation and mechanical design follow, again referenced to certain unique control and cleanliness requirements. It is best to consider immediately from the outset that the plant will most likely fall under jurisdiction of one or another governmental regulatory body since a food or pharmaceutical product may be manufactured. Moreover the increasingly complex environmental and regulatory requirements are often recognized late in the design and construction periods, or worse, they sometimes are recognized in the course of operation, particularly those resulting from increasing environmental pressures. Costly correction or modification may then be necessary.

The capital and gross operating cost of a "conventional" fermentation plant is roughly independent of the product produced. At current prices, as a first approximation for a grass-roots fermentation and finishing plant cost, one can use a figure within the range of $20 to $50 per liter of installed fermentation capacity. The value within this range depends upon scale and complexity. The economics of scale-up of capital costs will likely have an exponent of 0.75 for the fermentation section of the plant relative to the 0.6 exponent commonly used for a first approximation in processing plants. The fermentation operating costs per unit volume per unit time will vary somewhat but will hold generally within a reasonably narrow range. That is, barring use of an exotic raw material, the cost of labor, utilities, and materials would not vary greatly if put on a unit volume and unit cycle time basis. Also, processing or extraction can vary in complexity but, by and large, they too will hold to a reasonably narrow cost range. The capital and operating costs of all auxiliary operations for fermentation and extraction per se also exert a fly wheel effect relative to total installation cost and cost of operation.

Thus, the unit cost of the bulk product is much a function of fermentation yield and fermentation cycle time. This is illustrated in Table I, presenting certain yield, price, and sales parameters over the 20-year period 1950–1970 for various products. During this period the price of penicillin dropped to one twenty-fifth of the 1950 value, reflecting the substantial increase in yield through strain selection and mutation as well as sales volume increase without substantial new investment. In the overall period from the early 1940s to today this product has enjoyed approximately a 1000-fold improvement in titer. In products like antibiotics and vitamin $B_{12}$, the organism converts a substantial part of the carbohydrate to cells and $CO_2$, with the product synthesized only as an incidental by-product of cell growth. In contrast, for production of such commodities as glutamic, gluconic, and citric acids, the organism converts 60–95% of the carbohydrate to the end product on a weight basis. Process improvement by strain selection, engineering, and medium changes, as with antibiotics, has resulted in rate and yield gain. However, the total genetic improvement such as evidenced in antibiotics has necessarily been limited by the nature of these processes which yield such high conversions from the start.

Monosodium glutamate was not produced in any quantity by fermentation until after 1960. Upon changeover from the natural product extraction, the cost of production was reduced substantially, reflected in a selling price reduction to the range $1.75–$2.00 per kilogram in 1961. Over the ensuing years most of the process improvements and increased volume of operation have barely held the line against inflation. Citric acid as a mature product has not been able to hold this line, despite significant process changes, and it has increased in cost.

In the discussion which follows, general principles in both cost estimation and fermentation design will be reviewed and specific details elaborated. Certain tradeoffs will be reviewed and alternatives given to achieve a specific result. Although both cost estimation and biotechnology have advanced to become more or less rigorous sciences, economics has not yet reached that same level of exactitude. In the realm of fermentation economics, therefore, a certain amount of intuition and perhaps even emotion can do a great deal to reach the optimum least cost design.

**TABLE I.** Cycle Time, Yield, and Sales Volume Parameters

| | Bulk price $/kg | | U.S. sales kg/year | | Approximate cycle time and yield, 1970 | |
|---|---|---|---|---|---|---|
| | 1950 | 1970 | 1950 | 1970 | Time (days) | gm/liter |
| Vitamin $B_{12}$ | 450,000 | 8,000 | 20 | 1,000 | 5 | 0.03–0.08 |
| Penicillin G | 580 | 23 | 160,000 | 1,250,000 | 7 | 10–15 |
| Monosodium glutamate | 4[a] | 1 | 4,300,000 | 22,000,000 | 1.5 | 80–100 |
| Citric acid | 0.54 | 0.76 | 19,000,000 | 57,000,000 | 4–5 | 120–150 |

[a] Extracted from natural products; dropped to below $2.00 per kg in 1961 with startup of fermentation processes.

## II. PLANT DESIGN

### A. Fermentors

Fermentors for submerged, aerobic fermentations are generally stirred, gas–liquid contactors; i.e., a closed cylindrical vessel with a diameter/height ratio between 1:1 and 4:1, having concave heads, an agitator, internal baffles, internal heat-exchange coils, and/or a jacket. Steam sparging, automatic temperature, air flow, pressure, pH, and foam controls, internal bearings, and entry methods are almost always included. The main design criteria concern sterility or, perhaps more exactly, asepsis. Design criteria must ensure that foreign microorganisms cannot enter to disrupt normal transformations within the fermentor. Contamination is due to one or more breakdowns in the following steps or precautions: (1) the vessel itself or charged materials are insufficiently sterilized; (2) feed streams (including defoamer) are not properly sterilized; (3) foreign organisms enter through seals or any entry connecting to nonsterile environment; (4) process upsets or mechanical failures.

Fermentor volume may be any convenient value but is normally limited by construction/transportation constraints and potential loss due to contamination of a large broth volume. In recent years, vessel volume has increased for aerobic processes, and there is a report of over 800-m³ vessel size. When construction information is given, 230- to 380-m³ vessels are often noted. Internal finish is very important and the relatively small saving achieved by avoiding internal polishing is usually lost many times over by greater effort on cleaning and contamination loss.

The fermentor shaft seal very often creates sterility problems. A stuffing box, while less costly, allows leakage and creates sanitation problems. In recent years, double mechanical seals have become common. Sufficiently high temperatures must be maintained, using pressurized steam, at or near the seal face to prevent entry of foreign organisms.

As to construction specifics, field construction is usually more expensive and has more problems than fabrication in a vendor's shop. Thus, off-site fabrication is a normal choice where possible. A 4- to 4.5-m diameter tank is the usual limiting range for rail or road transport. At the earlier day design of 50- .to 75-m³ volume fermentation vessels, this resulted in a configuration ranging 1–1.5:1 height/diameter ratio for liquid volume. As the requirement for larger vessels developed, they were increased in height and in some cases shipped in sections for final field connection. This resulted in vessels on the order of 3 or 4:1 height diameter configuration. Such design calls for multiple agitators and usually the air sparging has been limited to the bottom impeller. Thus a constraint on fermentor field erection caused an evolution to tall, multiple-impeller vessels. Such vertical configuration creates a requirement for higher pressure air compressors which are more costly. In certain instances this has resulted in lengthened vessels designed to lie on their side so lower cost air blowers can be used with multiple agitators and air spargers. This configuration can be satisfactory if the system is not extremely aerobic and has no inherent foaming problem. Surface superificial air velocity will increase rapidly with operating volumes above 50% in such configuration. Certain unique designs have been presented (Heden, 1976).

The tall vertical fermentor can be satisfactory in comparatively lengthy fermentations lasting several days. In such cases, where usually less than 5–10% of the oxygen is removed, the air volume can be maintained at a comparatively low practical level to keep foaming under control and still ensure adequate molal supply of oxygen. Generally one is dealing with a viscous non-Newtonian broth, where the controlling factor is getting dissolved oxygen from the bulk liquid to the cell and not from the gas into the liquid. In this system a relatively greater amount of the energy can be put into mixing and pumping to enhance dissolved oxygen distribution within cell clumps and less into the air introduction and transfer of oxygen to the liquid system.

In a highly aerobic, low-viscosity fermentation, lasting on the order of a day or so, as much as 50% of the oxygen can be removed from comparatively large volumes of air passing through the fermentor. Such a system can require as much energy for supplying the air as is used by the agitator for mixing and pumping. Thus a squat 1–1.5:1 height/diameter ratio is more satisfactory for easier foam control and good oxygen transfer from gas to liquid. One can concentrate the energy at a single sparged impeller to obtain the maximum oxygen transfer into the liquid for the energy expended in balance with the requirements for sufficient turnover to ensure uniformity. If one desires a large vessel for the fermentation, consideration must be given to the added problems and cost of field erection versus the long-term advantages of such a squat fermentor over a tall one.

Another consideration in overall plant design must be the number of fermentors. Assuming requirements are for 750 m³ of installed capacity, one normally would not erect one 750-m³ fermentor. Either three 250- or four 190-m³ vessels would be more rational. Each design will have to consider a balance of support facilities, such as utilities, the holding tank size, and size of the initial separation steps to handle the volume from the whole culture storage vessel within a time constraint. Tailored into the consideration is the optimum size of the operating crew needed to run the fermentor system.

Materials of construction are normally rather limited. In some instances a case may possibly be made on paper for utilization of carbon steel fermentors. However, the greater maintenance cost, shorter life, and greater problems in maintaining sterile operation render such construction completely undesirable for pharmaceutical and food operations in particular. In a choice between 304 and 316 stainless steel, consideration for an even longer life favors 316 stainless steel at least for the coils, since 316 does not pit to the extent of 304. Life of coils can be affected particularly in high chloride water when the fermentor is sterilized empty, concentrating chloride ion in residual water on the inside of the coil.

A basic well-designed fermentor system will survive many process and product changes through the years. Even though the plant may be justified for a specific process and product, it is well for the long-term economics of the plant to provide flexible and enduring fermentation equipment.

Engineers in particular like to think in terms of continuous operation to maximize production rates. Such a system can be practical and economic when one is dealing with a product associated with growth of the organism cells themselves or an intracellular product. Also low sensitivity to contamination, such as with a thermophile or production at a lowered pH, is another criterion. It is impractical to consider anything but batch fermentation when the product is an extracellular metabolite that is sensitive to culture and medium balance. Similarly batch fermentation is preferred when the culture can undergo mutation at extended operating time and is sensitive to contamination. Other design criteria are discussed by Lundell and Laiho (1976), Meskanen et al. (1976), and Ovaskainen et al. (1976).

## B. Utilities

Utilities for a fermentation and extraction process represent a major capital cost and operating expense. Steam is required for sterilization of vessels and medium, maintenance of sterility (steam seals), water removal by evaporation, and drying of product. Fermentation processes are most often highly exothermic and considerable amounts of cooling water are needed. Further, medium and feed streams must be cooled after sterilization. Even with recuperation, there is need for large volumes of cooling water either from wells or produced via towers or mechanical means. Process water needs are also high (media preparation, crystallization, cake washing, dilution, equipment washing). Electricity demand is high not only for fermentor drives but for a profusion of rotating machinery in extraction, such as pumps, vacuum systems, agitators, filter, and dryers. With energy costs rising so rapidly it is essential that cost estimates be updated frequently. Various estimates are given below (Table II) for early 1977 but these values must be used with caution even as a base for escalation.

Agitation–aeration requirements are critical for most fermentation processes and both capital and operating costs are appreciable. An article (Ryu and Oldshue, 1977) discusses impact of changes in agitator design and its appreciable effect on overall costs. Capital cost for mixers and blowers, including installation, is given at approximately $900/kW.

Alternatives for consideration in sterile air delivery will affect initial investment and long-term operating costs. An air system can be designed utilizing high-pressure (about 4.5 atm) oil-free compressors (Bartholomew et al., 1974) with exhaust temperature in excess of 220°C. This procedure, combined with a pressure holding tank to maintain adequate retention time, provides sterile air without the need for intermittent inspection and changeout of packed fibrous or membrane filters that must function in the micron interception range. Impingement filters are susceptible to bypass or breakage which would permit entry of nonsterile air. In general, the hot air system requires lower initial investment and reduced maintenance. The choice of systems will vary with the individual plant. One must balance the extra energy cost of somewhat less efficient compression which results in an adequate sterilization temperature, against continued higher labor and maintenace cost for the

**TABLE II.**  Utilities Operating Costs and Range of Energy Prices

| | | |
|---|---|---|
| Cooling Water | | |
|   Well | 0.3–1.5 cents/m³ | |
|   Cooling tower | 0.7–2.2 cents/m³ | |
|   Mechanical refrigeration | 20–25 cents/m³ | |
| Steam | | |
|   At $2.40/therm condensate recovery cost is approximately $6.59–$9.00/1000 kg | | |
| Electricity | | |
|   Range is 25–35 mils/kWh | | |
| | Electricity (mils/kWh) | No. 2 fuel oil ($/therm) |
| Eastern United States | 30–35 | 2.40[a] |
| Europe | 25–30 | 2.60–2.80 |

[a] Coal cost is estimated at $1.50–$2/per therm with minimal transportation.

more complex filtration system which has the increased probability of failure—resulting in contamination.

### C. Pumps and Piping

Standard pumps and piping may be utilized for utilities and services when nonsterile service is concerned. Since construction material is very often stainless steel or clad steel, pipe connections, bolts, and flanges must be correctly chosen to prevent corrosion or galling. Where sterile service is involved, lines must be arranged for (1) complete drainage, (2) sterilization, and (3) ease of cleaning. Pockets and "dead spots" are to be avoided since solids deposition occurs, heat transfer is affected adversely, and an inoculum source for undesirable organisms is created. Valves must be full-flow and closely coupled; need for cleanliness and temperature service (sterile side) means that expensive valves must be used. Materials of construction are usually stainless steel, Teflon, and high-temperature resistant rubber or silicone materials. Normal piping factors must be increased for appropriate costing. All-welded construction is frequently used with further internal polishing of pipe and fittings. Welds should be tested ultrasonically and radiographically.

Sterile transfers are often made by use of sterile air or steam pressure above the material to be transferred. However, special pumps may be used for a low rate feeding of speciality materials (precursors, defoamers, sterility control agents). Diaphragm pumps, which may be steam sterilized, can be used as can various peristaltic pumps. Design criteria are the same as detailed above.

It is vital to decrease turnaround times when dealing with a short-time fermentation. There is a need for rapid transfer of the completed batch to the harvest tank. Sizing of the bottom outlet and the transfer pump combined with a positive pressure in the vessel must be such that the fermentor can be emptied within 15–20 minutes. This may seem axiomatic but it is surprising how such a simple item can be overlooked in the initial design.

### D. Heat Exchange

Heat exchange in the normally exothermic biochemical reactions occurring within a fermentor can occur within the vessel or in external exchangers with broth recycle. If heat exchange occurs within a vessel using coils, a balance (or tradeoff) must be maintained between process result versus heat removal. There are two general configurations shown in the tabulation below:

| A Best process result | B Best heat transfer |
|---|---|
| Vessel wall | Vessel wall |
| Cooling coils | Baffles |
| (1.0–1.5 coil diam. from wall) | Cooling coils |
| | Internal volume |
| Baffles | |
| Internal volume | |

In the authors' experience, a design with configuration A will yield a superior process result. Coils close to the wall leave less annular space where liquor is "hidden" from the impeller. With highly viscous non-Newtonian broths, as much as twice the power input is required by a vessel with configuration B to reach the optimum process plateau compared to configuration A. However, configuration B will provide overall heat transfer coefficients about 40% greater than the former.

Use of broth cycling and external heat exchange offers some advantages but oxygen deprivation may be a problem as well as added contamination potential. Coupling external heat exchange with external aeration (as in a continuous loop design) has been reported for specific applications.

Sterilization is another critical unit operation in fermentation processes affecting economics. Nature of medium, length of cycle, and process conditions will determine whether batch or continuous sterilization is used.

Some processes require a certain minimum cooking time for best results and some may be nearly independent of sterilization up to an hour or more at 121°C. Batch sterilization is perfectly satisfactory in such instances, particularly if the process runs for several days so an 8–10 hour turnaround is not severely limiting.

Short fermentations, 1–1.5 days, almost demand a continuous sterilization system to shorten the turnaround time to the order of 3 hours to maximize fermentor productivity.

Regardless of fermentation time, high concentration of carbohydrate utilized in conjunction with soluble proteins frequently results in a requirement for short interaction. This can be accomplished batchwise by separate sterilization of components. Such a process leads to a cumbersome system of satellite vessels and may not always be economical. Because of the well-known fact that protein denaturation in cellular material occurs at a high activation energy level compared to usual chemical reactions, the high temperature–short time operation in a continuous unit is the more economic and practical solution for sterilization of systems sensitive to chemical changes. If an economizer heat exchanger is used to heat the raw medium or makeup water while cooling the sterilized medium, the steam requirement for sterilization can be reduced at least 50%. Thus, economics of energy utilization favor continuous sterilization.

The choice of sterilization system can affect the design and cost of accessory equipment on the fermentor. In a continuous system, the fermentor is sterilized separately. A single-speed motor and agitator is satisfactory. The design can be made to utilize full power at the air flow selected since the agitator need never be used except in a flowing air system. Batch sterilization calls for a slightly more expensive alternate. Optimum agitator selection is achieved by use of a two-speed motor or a variable-speed drive. Thus one can sterilize the batch at a lower speed in the ungassed condition without overloading the motor, and the higher speed can be used in the gassed condition to draw maximum power. Response times to changes during cooling are vastly different in batch versus continuous sterilization. The vacuum breaker system to avoid tank collapse must be considerably larger for a tank sterilized empty than for one full of liquid. The initial cost and maintenance of this breaker system is thus greater for the continuous sterilization operation.

The cooling system to remove the heat of fermentation and, in the case of batch sterilization, to cool the batch after sterilization can vary considerably in design and economics. Long-duration fermentations present relatively little problem for heat removal. Although more costly to fabricate, jacketed fermentors or utilization of plate coils as baffles can provide adequate surface area. In a low-temperature fermentation, it may be necessary to provide mechanical cooling if adequate chilled well or lake water is not available. This availability of cool water can have considerable influence on overall economics and be a prime consideration in site selection.

When one is not so fortunate as to have access to cold water and must utilize cooling towers, the approach temperature may not allow too much leeway. In such situations, considerable energy can be utilized for pumping large volumes of cooling water. Sufficient volumes often cannot be introduced without excessive pressure drop in the cooling coils. To accommodate the pressure drop and approach temperature problems, it may be necessary to have multiple segments of coils connected in parallel. This all increases the installation cost and influences the economics. It adds another point bearing on overall selection of site as well as the research approach to the process. As noted before, the microbiological approach adds many constraints not required for a strictly chemical process, but there are some advantages. The engineer can only go so far, but the microbial geneticist can extend the degrees of freedom by organism manipulation. Thus when the preliminary design indicates a real problem, such as stated above for the cooling water, a program to develop the fermentation to operate at a somewhat higher temperature allows a desirable alternate not always available in ordinary chemical processing.

Heat exchangers in fermentation applications can be of any type, but preferred configurations are the spiral type and plate type. In these designs, high turbulence is maintained (there may be solids present in suspension), even heating/cooling is achieved, and ease of cleaning is assured. For reasons of cleanliness and corrosion control, stainless steel is a preferred construction material. Direct steam injection, followed by a retention section, is another convenient mode for continuous sterilization. Sterilization, as well as other design criteria, is well covered by Blakebrough (1967, 1969) and Yamada et al. (1972).

### E. Instrumentation and Control

Fermentation processes are highly instrumented. Instrumentation costs—initial purchase, installation, and maintenance—are therefore relatively high for such processes. Complexity is also high since a centralized control room is desirable (one for all fermentors; one or more for extraction and purification) and local readouts are usually essential. Costs of micro- and minicomputers and even more complex systems have fallen to levels which permit computer control of many variables. Either individual readout or cyclic sensing of control variables is in use. All systems require careful review for degree of interlocking design (example is agitator power interlocked with positive air flow), number and types of alarms, panel arrangement (local and centralized), type of cabling, wiring design, and need for permanent record. Programmable systems must be considered for certain repetitive or sequential functions, such as batching and sterilization.

Table III lists estimates and ranges for instrumentation costs (early 1977). Electronic systems (as pH) include pneumatic converters. A pre-piped and prewired panel would cost an additional $600 to $1000 per running foot.

Process computers have been applied to fermentation plants. Requirements and some applications effort are detailed in a number of reports (Humphrey, 1977; Jefferis, 1977; Nyiri, 1972; Nyiri and Charles, 1977). The advantages of computer control are: (1) increased throughput, more consistent quality; (2) increased operating consistency, reduced contamination losses; (3) decreased material and energy re-

**TABLE III.** Cost of Instrumentation (Pneumatic Systems)[a]

| Instrument | Cost ($) |
|---|---|
| Controllers | |
| Air-flow control | 1,600–1,700 (includes valve) |
| Pressure control | 1,700–1,850 (includes valve) |
| pH control | 4,000–4,500 (includes valve) |
| Temperature control (dual range) | 2,800–3,000 (valves excluded) |
| Dissolved $O_2$ control | 3,000 (valve excluded) |
| Defoamer probe | 2,600–3,000(includes valve) |
| Level (diaphragm) | 1,000–1,100 (no valve) |
| Flow (magmeter) | 4,500–5,000 (for 2-inch unit including valve) |
| Analyzers | |
| Oxygen analyzer | 2,000 |
| Carbon dioxide analyzer | 3,500 |
| AutoAnalyzer single channel | 25,000 |
| pH meter | 3,000 |
| Indicators | |
| Rpm | 300 |
| Alarms | 50 per point (plus $25/point per pressure switch) |
| Utility monitor | 50 –100 (plus $400 for transmitter) |
| Integrator | 400 |

[a] Add $300 for recorder function and $200 for second pen. Installation cost factor: 30–40% (exclude laboratory equipment installation for bench equipment as pH meter and Auto Analyzer) Costs as of early 1977.

quirement, greater ease in process development; (4) decreased man-power requirements; and (5) improved data storage.

Interactive control sequences have given higher yields and significantly improved process control. Costs of such systems (including developmental time) are quite high and a clear understanding of potential benefits must be developed. While new computer systems have been developed for pilot-plant use in fermentation research, plant-scale systems have usually been installed in an existing facility rather than in a grass-roots plant. Computer control concepts have appeared in the patent literature (Wilson, 1975; Wilson et al., 1975; and this volume, Chapter 15).

## F. Foam Control

Foam generation is a normal accompaniment to highly aerated aqueous suspensions containing proteins and other surface-active materials. Foam formation results in product loss, lowered aeration capacity (reduced broth density), potential for contamination, and need for lowered working volume. Chemical means of foam control are common but problems of toxicity and reduction in oxygen transfer rate must be overcome or at least controlled. Commonly used defoamers include higher carbon alcohols, fatty acids and derivatives, polyglycols, and silicones. Defoamers may be diluted prior to use and are normally added via sterile overpressure (air or steam) or with a controlled flow pump; addition may be on a timed basis, on demand, or by use of a combination of methods. Alarm systems to prevent overaddition are usually included and add to the cost. Defoamer must be sterilized prior to use. Batch methods are common but circulating aseptic systems (common headers) are also utilized.

Mechanical foam breakers are used in conjunction with chemical systems to reduce cost and associated problems. Most such mechanical systems are internal (paddles, ultrasonic devices, high-speed disk or centrifuge bowl) but some recirculating devices are known.

## G. Design Checklist

At some point in the design phase, a completely detailed design and construction checklist is needed. The following list is needed not only for project engineering but for investment analysis as well (Weaver and Bauman, 1973). The design and construction detail will bring together process parameters and site construction information which will permit rational meshing of schedules. All details will receive attention commensurate with the degree of sophistication required by design needs.

*Design and Construction Checklist*

1. *Process*
   Plant and process streams (on process flow diagram)
   Material and energy balance
   Physical/biochemical data
   Equipment performance
   Operating conditions (average, peak)
   Materials of construction
   Relief/blowdown systems
   Raw material availability
   Raw material impurity levels
   Impurity buildup (recycle)
   Risk analysis
   Relate capacity and sales forecast
   Useful life (major elements)
   Temperature, pressure, pH
   Yield, rate, fermentation cycle time
   Batch turnaround time

Data on specific gravity, viscosity, heat of vaporization
Sterilization requirements
Site availability and requirements
Emergency control plan

2. *Design*
   Annual operating factor
   Expansion capability
   Process flexibility (alternate processes)
   Instrumentation—need and complexity
   In-process storage, spare parts, warehousing
   a. *Buildings*
      Floors
      Utilities distribution
      Pressurization
      Type of construction
      Sanitation
      Fire protection and safety
      Corrosion resistance
      Access (stairways, platforms, elevators)
      Cranes or monorails
      Foundations
      Ground water and soil
      Communications
      Regulatory and local codes
   b. *Equipment*
      Design conditions (including access and safety)
      Need for stress relieving
      Davits or monorails
      Inspection/cleaning openings
      Pressure/vacuum design
      Internal specifications (baffling, demisters, piping, agitators, etc.)
      Skirt/support leg designs
      Paint and insulation
      Code requirements
      Process review
      Heat exchangers (code and TEMA class, type of bundle, expansion joints)
      Need for special gas purges (or blanketing)
      Heat recovery
   c. *Compressors, blowers*
      Type and specification
      Driver
      Space for auxiliaries and rotor removal
      Local/central control panels
      Auxilliary lubrication/seal oil pumps
      Relief and safety devices
      Filters, separators, knockout drum design
      Location of air intake
      Ambient condition variation
      Degree of overdesign
      Mode of air sterilization
      Heat recovery

d. *Pumps, piping*
  Type and number
  Driver and horsepower
  Seal types
  Seal fluid requirements
  Temperature, pressure, material, schedule
  Proper arrangement (access and safety)
  Underground piping (fire lines, water, sewer, drains)
  Special fabrication (rubber lines, plastic, glass)
  Ambient temperatures
  Tracing requirements
  Off-sites (pipe bridges, supports, buried or overhead with clearances)
  Shielding, jacketed lines, extreme conditions
e. *Fired heaters*
  Source, availability, price of primary and secondary fuel sources
  Standby fuel storage
  Design pressure and temperature
  Tube and fitting material
  Spare burners, tubes, instrumentation
  Space for erection and tube pulling
  Waste heat generator
  Stack height and need for support
  Fireproofing of main support
  Need for water pretreatment (fresh or recycle)
  Emergency alarms and control plan
f. *Electricity and instrumentation*
  Electrical classification by area
  Motor list
  Voltage variation
  Extremes of moisture, temperature, vibration
  Grounding requirements
  Emergency power requirements
  Type of instrumentation
  Redundancy and type of alarms
g. *Utilities*
  Need for new facilities
  Physical conditions (flow, pressure, temperature, losses)

|                     |                  |
| ------------------- | ---------------- |
| Steam               | Potable water    |
| Condensate          | Utility air      |
| Cooling water       | Instrument air   |
| Chilled water       | Inert gas        |
| Deionized water     | Refrigeration    |

  Define all tie-ins and recycle lines
h. *Nonprocess requirements*
  Climatic and geological conditions
  Roads, pavings, sprinklers
  Sewer systems (storm, sanitary, chemical)
  Walks, parking, rail spurs, fencing
  Special buildings
    Change houses for both sexes

| Maintenance | Laboratories (include inoculum, in-process, product testing, |
| Guard house | technical support, and quality control) |
| Office | Garage |
| Control room | Stores |
| Lunch room | Utilities |

Warehouse with associated equipment
*Waste treatment and pollution control*
Regulatory requirements and plans
Treatment methods
Need for sludge disposal
Need for effluent holding
Plant emissions and compliance
Plant compliance with OSHA

| Dust | Noise |
| Fumes | Failsafe design |
| Vapors | Biological hazards |
| Heat | |

Laboratory (testing) requirements
j. *Tank Farm*
Location
Dikes
Truck access
Utility requirements
Venting
Lighting
Loading and unloading
Cleaning/sanitation
Inventory monitoring
k. *Loading and unloading*
Material receipts and shipments
Frequency of tank truck and tank car movement
Spotting
Number of shifts
Weather protection
Office space
Tie-ins (drainage, fire protection, sanitation, communications)
Process and utility requirements
Need for weighing
Turning clearance
Tank car holding area
Railroad standards and responsibilities
Docks and wharves

| Pilings | Telephone |
| Dredging | Gangplanks |
| Derricks | Fire fighting |
| Offices | Lighting |
| Utilities | Drinking water |

Car puller
Special warehouse equipment
Warehousing and loading must fulfill regulatory requirements
Need for retention samples and coding

### III. PROCESS DESIGN

The assessment of profitability of a process or product is an on-going exercise. It should start in the research stage when technical and commercial data are limited. Further evaluation is made at intervals when justified by new information. Decisions after such evaluations can range from increased effort to abandonment of the project. Fermentation processes often can be improved vastly by microbial manipulation, more so than usually possible in chemical conversions. Therefore, it behooves the microbial scientist or engineer closely involved with a project to make frequent reassessments for evaluation of alternatives during the course of a project. The evaluation will help identify where the data are imprecise, what is important in terms of profitability, and where to concentrate the effort. It will help also to clarify objectives of the necessary program and reveal the supplementary commercial information and market research requirements. Such process evaluation is frequently complex and difficult. A key factor in industry is profitability regardless of technical achievement. Objective procedures to aid such assessment are based on the return on investment (ROI) as the criterion.

This need for objectivity is paramount. A potentially good project or process can be terminated too soon by pessimistic or overly conservative estimates; in contrast, optimistic estimates based on "pie in the sky" thinking can lead to unwarranted project expenditures and, if continued through to a final plant, can lead to economic disaster.

The accuracy of an estimate and the resultant basis for a project decision are no better than the least reliable data. The probability for commercial success lies in achieving projected costs, the target sales and marketing goals, and in overcoming all the other problems, such as FDA and USDA needs and the other regulatory requirements. As the probability for success for any one of these categories approaches zero, the project has little chance for economic success even if the others approach 100%.

The development of a profitability assessment at the product research and early development stages must necessarily be made without a firm cost estimate prepared from detailed design. Nevertheless, an indication of the construction cost is one of the first requirements. This predesign or order of magnitude estimate is usually prepared by a scientist and/or engineer closely concerned with the process. It can vary widely in dependability and in amount of time spent to develop the necessary data.

Preparation of a reasonable estimate may be impeded by lack of typical cost data for the installation of the process and related auxiliary facilities. Chilton (1949, 1960), Aries and Newton (1955), Bauman

(1964), Guthrie (1969), Pikulik and Diaz (1977), and Zimmerman and Lavine (1962) have published cost estimating data for the process industries. Throughout the past quarter century such journals as *Chemical Engineering* and *Cost Engineering* have had a series of articles updating various facets of equipment and process costing. A particularly valuable summation of the entire procedure for cost and profitability estimation is that of Weaver and Bauman (1973). Their chapter in the "Chemical Engineers Handbook" also contains many references.

Chilton (1949) was one of the first to organize data for a systematic estimation of costs for new plants. It is based on developing ratios to installed major process equipment costs as means for arriving at a method for rapid estimation of construction costs. His equipment cost data are keyed to the *Engineering News-Record* (ENR) Construction Cost Index at 400 to facilitate correction for inflation. Since 1949, this ratio has advanced to about 2600 in 1977! Correcting his equipment cost information by such a ratio could be misleading, but his principles and the procedure for selecting ratio factors as percentages of installed major process equipment plus others expressed as percentages of total physical cost can still be applied. Utilizing this procedure provides a means for rapid and comparatively inexpensive preliminary capital cost estimation.

The introductory phases involve preparation of a flow diagram and the selection of processing steps in recovery operations. A generalized flow diagram is given in Fig. 1 and possible unit operations are given for various processes in Table IV. It should be understood that not all steps

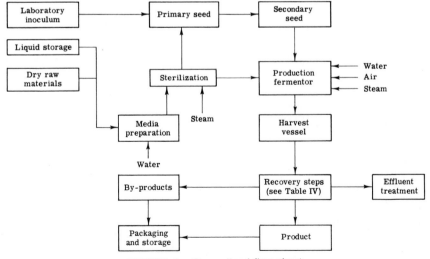

**FIGURE 1.** Generalized flow sheet.

**TABLE IV.** Typical Process Showing Various Unit Operations

*Submerged fermentation*

| | | | | |
|---|---|---|---|---|
| Storage tanks | Media preparation | Sterilization | Seed fermentor | Harvest tank |
| Warehouse | Weigh tanks | Batch | Production fermentor | |
| (raw materials) | Metering | Continuous | Agitation | |
| | Dilution | Air sterilization | Aeration | |
| | Auxiliaries (both fermentation and recovery) | | | Laboratories |
| | Air compressor | Cooling water | Steam | |
| | Process water | Refrigeration | Electricity | |

*Recovery operation*

| | | | | |
|---|---|---|---|---|
| | Cell separation | Precipitation | Separation | Reslurry |
| | Centrifugation | pH/salt adjustment | Ion exchange | Wash |
| | Filtration | Heat treatment | Rotary filter | Concentration |
| | Coagulation | | Centrifugation | Evaporation |
| | | | Solvent extraction | |
| | | | Ion exclusion | |
| | | | Membrane processes | |

*Crude separation*

| | | | |
|---|---|---|---|
| | Carbon treatment | *Purification* | Product separation |
| | Adsorption | Electrodialysis | Filter |
| | Solvent extraction | Crystallization | Centrifuge |
| | Membrane/osmotic separations | Ion exchange | |
| | Storage | *Packaging* | *Warehouse and shipping* |

*Drying*
Freeze
Fluid bed
Spray
Vacuum
Rotary
Drum

need be included nor all possible operations run; selection will be based upon laboratory and pilot-plant data for the individual process.

As an example of considerations in fermentation process design, an outline of information needed is given in Table V. Selection is based upon economics, reaction type, laboratory data, and information developed in the course of scaleup. Once the information is selected and

**TABLE V.**  Process Design Criteria for Fermentation

1. Substrate (energy, maintenance, conversion)
   | Carbohydrate | Fats and oils |
   | Molasses | Waste liquors |
   | Grains | Hydrocarbons |
   | Spent grains | Starch (and hydrolysates) |

2. Other raw materials
   Nitrogen sources (ammonia, ammonium salts, urea, proteins)
   Cations (Fe, Mg, Mn, Zn, Cu, Na, K, Ca, as examples)
   Anionic constituents (phosphates, for example)
   Vitamins and cofactors
   Growth factors (usually from natural products)

3. Type of reaction (including rate and yield)
   | Oxidation | Hydrolysis |
   | Reduction | Hydroxylation |
   | Decarboxylation | Esterification |
   | Deamination | Amination |

4. Principal products and by-products

5. Organism utilized (including inoculum development)
   | Bacteria | Animal, plant cells |
   | Mold | Algae |
   | Actinomycete | Yeast |

6. Agitation–aeration requirements
   | Dissolved oxygen level | Air flow |
   | Oxygen transfer rate | Power input |
   | Viscosity effect | Transfer resistances |

7. Special operations
   | Semibatch | Feeding methods |
   | Continuous | Defoamer type and addition procedure |

8. Special equipment or process specification
   | Membrane | Corrosion (material of construction) |
   | Recycle | Horizontal fermentor |
   | Autolysis | Air lift |
   | Cleaning | Sterilization requirements |

9. Control requirements
   | Temperature | Media concentration |
   | pH | $pCO_2$ |
   | Pressure | Redox potential |

combined with a flow sheet, plant design and operating cost estimation can begin.

Two detailed and complete reports covering capital and operating cost calculations for numerous fermentation processes have been completed by Fong (1975a,b). Other methods of optimization and cost calculations in fermentation processes can be found in two articles by Richards (1968), in procedure by Whitaker and Walker (1973) for classroom instruction, and in a design report by Schepers (1974).

## IV. CASE STUDY—PROJECT EVALUATION

With the preceding information in mind, it is the intention of the authors to develop a project model and to present some data and procedures we have found useful for such preliminary process evaluations. There is no intent to get into the more sophisticated corporate procedures that apply and are necessary as a project develops into the final stages of a capital appropriation request.

### A. Selected Equipment Costs

Table VI presents approximate costs for some equipment items frequently used in the fermentation industry. These are derived from recent quotations corrected to current cost indices where necessary and by current contact with some of the manufacturers. It is by no means complete but indicates a sampling of typical equipment. Information of this nature should be kept current by those interested in rapid evaluation techniques. Likely sources are recent quotes, recent company capital appropriation requests, and the corporate engineering department files in addition to the continuing stream of cost update articles in the journals.

### B. Production Cost Factors

Table VII contains various approximations for rapid production cost estimation. Percentage ranges of the order indicated have been expressed in many articles, including those referred to above, and in general are confirmed by the authors' experiences.

### C. Assumptions and Materials Usage

As an example for our case study, we will assume a scenario that research has discovered, partially developed, and conducted prelimi-

**TABLE VI.** Approximate Cost of Equipment Frequently Used in the Fermentation Industry

*Indices*
*Engineering and News-Record* (ENR) Construction Cost Index at 2600 (1913 = 100)
Marshall & Swift Equipment Cost Index at 500 (1926 = 100)
*Agitators and drives*
Standard (304 stainless steel unit with stuffing box)
  Range:                        7.5–375 kW
  Scale factor exponent:     0.75
  75 kW unit cost:          $41,000
Deluxe (316 stainless steel unit with mechanical seal)
  7.5 kW unit cost:        $13,000
  75 kW unit cost:         $58,000
(Data courtesy of J. A. Mason, Mixing Equipment Company)
*Centrifuges*
Centrifugal desludgers (316 stainless steel)
  100–200 liter/minute unit with accessories:        $125,000
  50–100 liter/minute unit with accessories:        $ 75,000
(Rates can vary considerably, depending on feed material properties and clarity required)
Basket centrifuge (304 stainless steel automated batch)

                                       $ 75,000
  76 × 122 cm basket:             (installed $200,000)

| *Evaporators* (stainless steel) | *Per kg of water evaporated per hour* |
|---|---|
| Double effect: | $18–$26 |
| Double effect with steam recompression: | $20–$29 |
| Triple effect: | $26–$40 |
| Quadruple effect: | $31–$44 |

Between 10,000 and 35,000 kg/hour range
  Boiling range:         93°C − 50°C
  Final viscosity:       Up to 200 cP
(Data courtesy of D. F. Dinnage, APV Company, Inc.)
*Fermentors*
    15 m³:           $ 35,000
    113 m³:         $ 95,000
    225 m³:        $165,000
(304 stainless steel with coils, baffles, 2.75 atm rating,
standard openings and connections)

*Holding tank* (304 stainless steel)
  190-m³ silo:              $37,000
  Installed with pvf, insts. on pad    $60,000
*Spray dryers* (304 stainless steel)
  Nominal water removal

| kg/hour | *Installed cost* |
|---|---|
| 1800 | $ 600,000 |
| 3200 | $1,250,000 |
| 4500 | $1,600,000 |

(Complete with fans, ducts, air filters, power lines, steam preheater, gas burner, dryer chamber, cyclones, wet collection system, bagging system, buildings, etc.)

**TABLE VII.**   Approximations for Production Cost Estimation

*Production costs*

Depreciation: 6.67–10% of capital (10- to 15-year straight line depreciation)

Taxes and Insurance: 2–6% of capital (varies with location)

All salaries except maintenance supervision: 70–100% of direct labor (administration, supervisory, clerical, technical, engineering)

Indirect Labor: 30–50% of direct labor (includes control and quality control labs, yard, materials handling, cleanup, etc.)

Associated payroll costs: 20–40% of payroll

Supplies:                           20–30% of direct labor

Other plant overhead:      50–70% of direct labor

   (telephone, outside services, rentals, tools, travel expenses, demurrage, freight, dues, donations, legal, medical and other professional services, sewer charges, etc.)

Maintenance: 4–8% of capital

     (salaries, labor, and associated payroll costs are approximately 50% of total)

*General cost factors*

Sales and Administration plus Research

and Development expense:      10–15% of net sales

Provision for taxes:                 50% of profit from operations

Startup expenses:                   5–10% of capital

Working capital:                     20–30% of net sales

---

nary field tests on a new bacterial control agent for a major United States crop of 5,000,000 ha planted annually. Furthermore, there are indications of favorable patent coverage. Field trials with standardized units of activity in 1 kg of product indicate treatment of 3 kg/ha is required. Requirements for control may range from four to eight treatments per crop season with six treatments being the weighted average over the entire country.

Preliminary market research, based on these results, indicates a possible market penetration at a level of 15–25% of the crop and at a $2.50/kg of product cost to the grower.

This results in a potential market for between 4.5 and 7.5 million ha-treatments which reduces to 13.5–22.5 million kg of product per year. Thus a preliminary evaluation near the midrange level is in order. Assuming a fermentor with an easily manageable capacity of 225 m³, the bases determined by pilot data and further assumptions are listed in Table VIII. Raw material costs are also calculated for the indicated production level of 17 million kg, a rate fairly near the average of the projected market range.

## D. Process Flow Diagram, Material Balance, and Utilities

Simultaneous with development of these numbers is the requirement for a process flow diagram (Fig. 2) and a material balance flow sheet

**TABLE VIII.**  Materials Usage and Cost

*Basis*
  Volume of production fermentors: 225 m³
  Operating volume: 180 m³
  Number of fermentors: 4
  Stream factor: 97%
  Cycle time: 40 hours (includes 4-hour turnaround)
  Production/batch: 20,000 kg (at a standardized activity per kg)
  Batches/year: 850
  Production/year: 17,000 kg

| Raw materials | kg/batch | 1000 kg/year | Unit cost ($/kg) | Cost ($1000) |
|---|---|---|---|---|
| Fermentation | | | | |
| Molasses | 11475 | 9754 | 0.07 | 683 |
| Adjuvant solids | 3668 | 3118 | 0.30 | 935 |
| Misc. salts, etc. | 922 | 784 | 0.70 (avg.) | 549 |
| Subtotal | | | | 2167 |
| Formulation | | | | |
| Diluent | 9100 | 7735 | 0.30 | 2320 |
| Stabilizers | 100 | 85 | 2.00 | 170 |
| Sticker | 400 | 340 | 0.50 | 170 |
| Anticaking agent | 400 | 340 | 0.24 | 82 |
| Subtotal | | | | 2742 |
| Packaging | 25 kg/bag | 680,000 bags | 0.40/bag | 272 |
| | | | | 5181 |

(Fig. 3). These augment the assumptions and process requirements as well. From these, the nature of the equipment can be resolved and sizing determined. A utilities usage (Table IX) can be calculated based on the equipment selected and the process requirements.

## E. Capital Investment, Direct Labor Requirements, and Production Cost

Next a capital investment can be calculated in the manner outlined in Table X, followed by an estimation of the operating labor (Table XI), and finally by the production cost estimate (Table XII). There are many ways to assemble the information and calculations merely by individual preference, but the scheme presented in Figs. 2 and 3 and Tables VIII–XII is fairly straightforward and presents the assumptions conveniently. It outlines these assumptions, the thought process, and calculations in sufficient detail to permit ready cross-checking or easy substitutions of alternatives. The presentation, done in this manner, requires little verbal explanation for easy understanding.

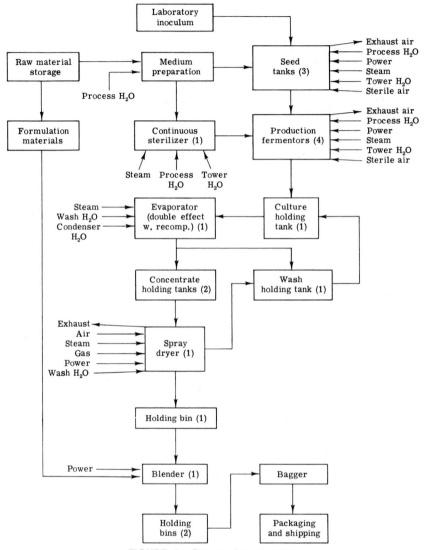

**FIGURE 2.** Process flow diagram.

This model outlines a simple process of fermentation, evaporation, drying, and blending. If the activity is contained as an intracellular product, then there is an alternative which would substitute centrifugation to separate cells followed by cell drying rather than the concentration step. A calculation exercise for the alternate process could explore the advantages of reduced energy cost versus the inherent problem of waste disposal created by the centrifugation step. If the anticaking agent

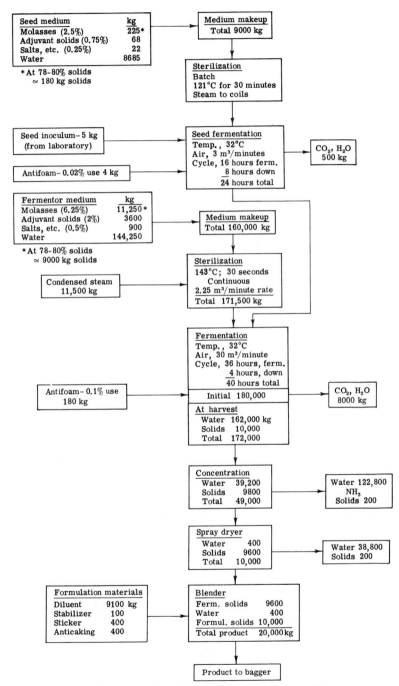

**FIGURE 3.** Material balance flow sheet (per batch).

**TABLE IX.**   Utilities

| | Steam (kg) | Process water (m³) | Tower water (m³) | Electrical (kWh) | Gas (m³) |
|---|---|---|---|---|---|
| Seed | 7,000/batch | 9 | 300 | 300 | — |
| Fermentor | 25,000 | 147 | 6,000 | 5,700 | — |
| Concentration | 53,000 | — | 1,200 | 650 | — |
| Dryer | 120,000 | — | — | 5,400 | 1,800 |
| Compressed air | — | — | — | 9,250 | — |
| Blender | — | — | — | 900 | — |
| Subtotal | 205,000 | 156 | 7,500 | 22,200 | 1,800 |
| Misc. usage and contingency at 33% | 68,000 | 52 | 2,500 | 7,325 | 600 |
| Total/batch | 273,000 | 208 | 10,000 | 29,525 | 2,400 |
| Per day | 636,000 | 484 | 23,300 | 68,800 | 5,600 |
| Installed | 1,000,000 | — | 27,000 | 100,000 | — |
| Installed cost[a] ($1000) | 600 | 100 | 450 | 375 | 25 |

[a] The sum of the installed costs of utilities estimated from rule of thumb factors adds up to $1,550,000. This is a cross-check and fortunately in close agreement with the $1,590,000 figure used in Table X as obtained by arbitrary ratio of installed major equipment costs.

effectively overcomes hygroscopicity of the molasses-based medium, environmental considerations might outweigh other factors to favor the evaporation process. In the process selected for our model, waste problems are minimized to the point there would be no need for additional disposal facilities. This is not the usual case for fermentation processes. Waste disposal can be a major factor in both capital and operating costs.

### F. Return on Investment: Sensitivity Analysis

Returning to the mechanics of the process evaluation, the next step is in evaluation of the ROI based on the assumptions and standardized factors. The summary of various cases is contained in Table XIII. The base case A results in an ROI of 21.2%. Normally, a 20% ROI would be considered satisfactory for a venture of this nature, depending on the risk relative to alternative utilization of capital within a company. In cases B, C, and D, the effect of 20% reduction in selling price and sales/ production volume or a 20% capital overrun are considered. The indication of greatest sensitivity is in a selling price reduction, although ROIs for all cases are substantially reduced. Thus, the effect of wrong market judgment considerably outweighs the effect of a poor capital projection. If there were concern about selling price holding at the design level but more apt to hold at the lowest volume projected, then case E can be calculated readily to show the effect of the conservative basis of a smaller plant. Quantifiable conceptual information of this nature, developed at reasonably early stages of the project, supplies management with a satisfactory basis of judgment for further action. This is about the simplest form for presenting profitability data. It must be recognized that it is only for a broad look at profitability and a useful tool for consideration of alternatives.

If the project progresses to a capital request stage, by then the process and the plant design should be firm. All major equipment should be specified and quotes obtained with plant capacity firm and process flow sheets, energy and material balances optimized, plus process and instrumentation diagrams nearly complete. Such an estimate for major capital funding can take a 2- to 4-month period to prepare. The financial information is presented in year to year breakdown for the construction year(s) plus 10 years of the project (or whatever is standard for a given company). In such an analysis, the effect of inflation escalators and other capital additions through the years are considered item by item in the cost development. The ROI, cash flows, and payout times are calculated, and usually a present value return analysis is made at the standardized discount rate (or rates) of the company. Some companies will have other financial judgement criteria, but the items mentioned above are basic requirements for a finalized capital request.

In summation, it will be readily apparent that the project model selected is somewhat hypothetical. The authors are not aware of any microbial crop-control agents of such potential in the literature. The only reasonable success story for such a product is that of *Bacillus thuringiensis* which in a 15-year period from 1962 to 1977 has grown from sales in the low hundreds of thousands of dollars to a reported sales volume of approximately $13 million shared by three companies. Satisfactory profitability can be obtained only in multiple-use, shared fermentation facilities at such low volumes. Moreover, this type of bacterial control agent generally will be quite specific relative to chemical control agents. This creates a limited market. Since the cost of research and clearance is nearly as much for a low-dollar volume product as for a large-volume one, it is quite apparent why major fermentation and chemical companies have limited their investments in this direction, although from an ecology viewpoint, it is the desirable approach.

The key point is that such commodity items are sometimes difficult to justify from an ROI basis. For example, SCP processes from hydrocarbons—or for that matter any waste substrate—are often not

**TABLE X.** Capital Investment Estimate

| | Number | Size | Unit cost | Extended cost | Installed cost |
|---|---|---|---|---|---|
| | | | | Cost (in thousands) | |
| Fermentation | | | | | |
| Formulation tanks | 2 | 38 m³ | $ 15 | $ 30 | |
| Agitator | 2 | 3.75 kW | 4 | 8 | |
| Continuous sterilizer: 304 Stainless Steel | 1 | 2.25 m³/min | 100 | 100 | |
| Compressors | 2 | 85 m³/min | 225 | 450 | |
| Antifoam system | | | 100 | 100 | |
| Seed tanks: 304 Stainless Steel | 3 | 15 m³ | 35 | 105 | |
| Agitator | 3 | 19 kW | 14 | 42 | |
| Production fermentors: 304 Stainless Steel | 4 | 225 m³ | 165 | 660 | |
| Agitator | 4 | 150 kW | 70 | 280 | |
| Holding tanks: 304 Stainless Steel | 2 | 190 m³ | 40 | 80 | |
| Pumps | | | | 150 | |
| Separators and special equipment | | | | 100 | |
| Subtotal: Major fermentation equipment | | | | $2,105 | |
| Installed cost: 2.5 × major fermentation equipment[a] | | | | | $5,263 |
| Instrumentation: 10% (avg. to complex installation)[b] | | | | | 526 |
| Process piping: 20% (avg. installation)[b] | | | | | 1,053 |
| Buildings: Mixed outdoor and indoor const., 40%[b] | | | | | 2,105 |
| Total: Fermentation capital | | | | | $8,947 |
| Processing | | | | | |
| Evaporator: 304 Stainless Steel | 1 | 18,000 kg H₂O/hr | $400 | $ 400 | $ 600 |

| | | | | | |
|---|---|---|---|---|---|
| Holding tanks: 304 Stainless Steel | 3 | 75 m³ | 24 | 72 | 180 |
| Holding bins: 304 Stainless Steel | 3 | 75 m³ | 24 | 72 | 180 |
| Blender: 304 Stainless Steel | 1 | 15 m³ | | | 75 |
| Vibrating screen | 1 | 4 m²-3 deck | | | 50 |
| Subtotal: Process equipment | | | | | $1,085 |
| Instrumentation: 10%[b] | | | | | 109 |
| Process piping: 20%[b] | | | | | 217 |
| Buildings: Indoor construction, 75%[b] | | | | | 814 |
| Subtotal | | | | | $2,225 |
| Spray dryer: 304 Stainless Steel | 1 | 4,500 kg H$_2$O/hour | $1070 | $1070 | 1,600 |
| Total: Process capital | | | | | $3,825 |
| Utilities and Tankage | | | | | |
| Utilities: (major addition at existing site 20% of installed major equipment) | | | | | |
| 0.2 (5263 + 1085 + 1600) | — | | — | — | $ 1,590 |
| Molasses storage tanks | 2 | 2800 m³ | — | — | 270 |
| Subtotal | | | | | $ 1,860 |
| Total fixed capital | | | | | $14,632 |
| Engineering and construction: 20% straightforward[c] | | | | | 2,926 |
| Contingency; average, 20%; process subject to change[c] | | | | | 2,926 |
| Total capital investment (exclusive of land) | | | | | $20,484 |
| Use | | | | | $20,500 |

[a] Authors' experience.
[b] Chilton's factors on installed equipment costs (1949).
[c] Chilton's factors on total physical cost (1949).

**TABLE XI.**  Direct Labor Requirements

|  | Per Shift |
|---|---|
| Fermentation | 3 |
| Evaporator | 1 |
| Dryer | 1 |
| Blending | 1 |
| Packing and shipping | 2 |
| Total/shift | 8 |
| Total/4 shifts[a] | 32 |
| Wages/operator | $16,000 per year[a] |
| Total direct labor | $512,000 |

[a] 168 hours/week ÷ 4 = 42 hours/week/operator. Each operator thus has the equivalent of 43 hours straight time pay built into the schedule.

**TABLE XII.**  Production Cost Estimate

|  | Cost (in thousands) | |
|---|---|---|
| Raw materials | | |
| Fermentation | $ 2167 | |
| Formulation | 2742 | |
| Packaging | 272 | |
| Subtotal | | $ 5,181 |
| Payroll Charges | | |
| Direct labor | $ 512 | |
| Indirect labor (40% of direct labor) | 205 | |
| Salaried payroll (85% of direct labor) | 435 | |
| Associated payroll costs (30% of payroll) | 346 | |
| Maintenance salaries, labor and assoc. (3% of capital) | 615 | |
| Subtotal | | $ 2,113 |
| Supplies and expenses | | |
| Maintenance (3% of capital) | $ 615 | |
| Supplies (25% of direct labor) | 128 | |
| Subtotal | | $ 743 |
| Utilities | | |
| Steam: $8.50/1000 kg × 273 × 850 | $ 1972 | |
| Electricity: $0.035/kWH × 30,000 × 850 | 893 | |
| Water: $0.008/m³ × 10,000 × 850 | 68 | |
| Gas: 2400 m³ × $0.09/m³ × 850 | 184 | |
| Subtotal | | $ 3,117 |
| Other Costs | | |
| Depreciation (12 years) | $ 1708 | |
| Taxes and insurance (4% of capital) | 820 | |
| Other plant overhead (60% of direct labor) | 307 | |
| Subtotal | | $ 2,835 |
| Total | | $13.989 |
| | | $ 0. 82 per kg |

**TABLE XIII.** Return on Investment—10-Year Basis

| | Value (dollars in thousands) | | | | |
|---|---|---|---|---|---|
| | A<br>Base case<br>17 million kg/Year<br>at $2.50/kg | B<br>20% Reduced<br>selling price<br>at $2.00/kg | C<br>20% Reduced<br>volume at<br>13.6 kg/Year | D<br>20%<br>Capital<br>overrun | E<br>13.6 kg/Year<br>plant<br>design |
| 1. Sales to growers | 42,500 | 34,000 | 34,000 | 42,500 | 34,000 |
| 2. Dealer discounts, distribution, freight (30%) | (12,750) | (10,200) | (10,200) | (12,750) | (10,200) |
| 3. Net sales | 29,750 | 23,800 | 23,800 | 29,750 | 23,800 |
| 4. Cost of goods sold | (13,989) | (13,989) | (12,210)[b] | (14,743)[b] | (11,818)[b] |
| 5. Gross profit | 15,761 | 9,811 | 11,590 | 15,007 | 11,982 |
| 6. Sales and Administration plus Research and Development (12.5% of net sales) | (3,719) | (2,975) | (2,975) | (3,719) | (2,975) |
| 7. Profit from operations (2–10 years) | 12,042 | 6,836 | 8,615 | 11,288 | 9,007 |
| 8. Startup costs, first year (10% of capital) | (2,050) | (2,050) | (2,050) | (2,460) | (1,770) |
| 9. Profit from operations (first year) | 9,992 | 4,786 | 6,565 | 8,828 | 7,237 |
| 10. Net earnings after taxes (50%)—first year | 4,996 | 2,393 | 3,283 | 4,414 | 3,619 |
| 11. Net earnings after taxes (50%)—2–10 years | 6,021 | 3,418 | 4,308 | 5,644 | 4,504 |
| 12. Working capital (25% of net sales) | 7,438 | 5,950 | 5,950 | 7,438 | 5,950 |
| 13. Original fixed capital investment | 20,500 | 20,500 | 20,500 | 24,600 | 17,700[c] |
| 14. Return on investment (%)[a] | 21.2 | 12.5 | 15.9 | 17.2 | 18.7 |

[a] ROI = $\frac{(\text{line } 10 + 9 \times \text{line } 11)/10}{\text{line } 13 + \text{line } 12}$

[b] Variable costs of materials and utilities × 0.8. Semifixed costs of labor related items × 0.94 (direct labor to 30 men from 32). Capital related costs follow their percentage of capital investment.

[c] Calculated based on (17/13.6)^{0.67} ratio of plant size to obtain new capital cost. The 0.67 factor is a balance between the 0.75 for fermentation section scale factor and the often used 0.6 for process plant scale factor.

viable in the United States in the context of our low-cost competitive protein products from agriculture.

Thus, it is not surprising that about three-quarters of the dollar volume of aerobic microbial products resides in high-markup pharmaceuticals represented by antibiotics and conversion products.

## V. REFERENCES

Aries, R., and Newton, R. (1955). "Chemical Engineering Cost Estimation." McGraw-Hill, New York.

Bartholomew, W., Engstrom, D., Goodman, N., O'Toole, A., Shelton, J., and Tannen, L. (1974). *Biotechnol. Bioeng.* **16,** 1005.

Bauman, H. (1964). "Fundamentals of Cost Engineering in the Chemical Industry." Von Nostrand-Reinhold, Princeton, New Jersey.

Blakebrough, N., ed. (1967). "Biochemical and Biological Engineering Science," Vol. I. Academic Press, New York.

Blakebrough, N., ed. (1969). "Biochemical and Biological Engineering Science," Vol. 2. Academic Press, New York.

Chilton, C. (1949). *Chem. Eng. (N.Y.)* **56** (6), 97.

Chilton, C. (1960). "Cost Engineering in the Process Industries." McGraw-Hill, New York.

Fong, W. (1975a). "Proteins from Hydrocarbons," Rep. No. 60A Suppl. Process Econ. Program, Stanford Res. Inst., Menlo Park, California.

Fong, W. (1975b). "Fermentation Processes," Rep. No. 95. Process Econ. Program, Stanford Res. Inst., Menlo Park, California.

Guthrie, K. (1969). *Chem. Eng.* (N.Y.) **46**(6), 114.

Heden, C.-G. (1976). U.S. Patent 3,997,400.

Humphrey, A. (1977). *Process Biochem.* **12,** 19.

Jefferis, R., ed. (1977). "Workshop—Computer Applications in Fermentation Technology, 1976," Ges. Biotechnol. Forsch. mbH. Verlag Chemie, Weinheim.

Lundell, R., and Laiho, P. (1976). *Process Biochem.* **11,** 13.

Meskanen, A., Lundell, R., and Laiho, P. (1976). *Process Biochem.* **11,** 31.

Nyiri, L. (1972). "Advances in Biochemical Engineering", Vol. 2, p. 49. Springer-Verlag, New York.

Nyiri, L., and Charles, M. (1977). "Annual Reports on Fermentation Processes" (D. Perlman, ed.), Vol. 1, p. 365. Academic Press, New York.

Ovaskainen, P., Lundell, R., and Laiho, P. (1976). *Process Biochem.* **11,** 37.

Pikulik, A., and Diaz, H. (1977). *Chem. Eng.* **84** (21), 107.

Richards, J. (1968). *Process Biochem.* **3,** 56.

Ryu, D., and Oldshue, J. (1977). *Biotechnol. Bioeng.* **19,** 621.

Schepers, E. (1974). *Natl. Chem. Eng. Conf. Process Ind. Aust.—Impact Growth, 1974* p. 60.

Weaver, J. and Bauman, H. (1973). *In* "Chemical Engineers Handbook" (R. Perry and C. Chilton, eds.), 5th ed., p. 25–13. McGraw-Hill, New York.

Whitaker, A., and Walker, J. (1973). *J. Biol. Educ.* **7,** 37.

Wilson, J. (1975). U.S. Patent 3,926,738.

Wilson, J., Nyiri, L., Humphrey, A., and Harms, C. (1975). U.S. Patent 3,926,737.

Yamada, K., Kinoshita, S., Tsunoda, T., and Aida, K., eds. (1972). "The Microbial Production of Amino Acids." Kodansha, Ltd., Tokyo.

Zimmerman, O., and Lavine, I. (1962). *Cost Eng.* **4,** 7.

Chapter 19

# Fermentation Processes and Products: Problems in Patenting

IRVING MARCUS

## I. INTRODUCTION

A patent combines the mix of a scientific document with a legal document. The patent stems from our Constitutional provision:

> . . . to promote the progress of Science and Useful Arts, by securing for limited times

**497**

MICROBIAL TECHNOLOGY, 2nd ed., VOL. II
Copyright © 1979 by Academic Press, Inc.
All rights of reproduction in any form reserved. ISBN 0-12-551502-2

to Authors and Inventors the exclusive right to their respective writings and discoveries. (U.S. Constitution, Article I, Section 8)

## A. The Patent Statute: Title 35—Patents

Section 101, Inventions patentable, provides

> Whoever invents any new and useful process machine, manufacture or composition of matter on any new or useful improvement thereof may obtain a patent therefor, subject to the conditions and requirements of this title.

The basic conditions for patentability are set forth in the negative. A person is entitled to a patent *unless* the invention was not new (Sec. 102) and if not identically disclosed if it would have been obvious (Sec. 103). Thus, there are, in the positive, three main conditions for obtaining a patent. The invention must be new, must be useful, and must be unobvious.

## B. Requirements of a Patent Disclosure

In order, however, to obtain a patent there must be an application for a patent by the inventor with certain prescribed conditions, one of which is set forth in Section 112 entitled "Specification":

> The specification shall contain a written description of the invention, and of the manner and process of making and using it, in such full, clear, concise, and exact terms as to enable any person skilled in the art to which it pertains, or with which it is most nearly connected, to make and use the same, and shall set forth the best mode contemplated by the inventor of carrying out his invention.

These are basics which are required for an understanding of patents that deal with microbiological inventions. The principal microbiological invention is, of course, the fermentation process and processes involving fermentations are in many parts of the patent system.

The microorganism may be referred to as a starting material in process inventions, it may be considered a catalyst, or it may itself be an end product, the result of a man-made variation of an available microorganism.

Thus, there is a mixture of science and the law embodied in a patent. The subject matter is within the realm of science, while the claiming of the subject matter with supportive disclosure is the legal aspect. The legal problems in disclosing and claiming inventions dealing with the action of microorganisms have been in existence for almost 30 years. These problems come to light during the examination of the application and the search of the prior art (patents, domestic and foreign, and the literature).

## II. BACKGROUND

### A. Applications for Patent Involving the Action Of Microorganisms

For many years inventions dealing with microorganisms received only sporadic attention in the patent area. The cases that came to light involved esoteric legal points that were generally *sui generis*, of first impression. There was little, if any, uniformity either in the patent practice by the examiner in the United States Patent and Trademark Office (PTO) or in the decisions affecting such practice. Case by case, we struggled with the problems that arose. At this time, we are far more knowledgeable in the subject from a legal sense. Meanwhile, the subject began receiving the attention of those working in the patent field and various articles appeared covering the aspects of the application of patent laws to the technology (Marcus, 1975; Levitt, 1976). One aspect of the problem will be covered more fully below with particular attention to the recent diplomatic activity which is now in treaty form. Most recently, the subject attained notoriety with the publicity concerning genetic research and recombinant DNA.

We will concentrate on the practice in patent law dealing with inventions in microbiology from the viewpoint of that law as practiced in the United States. Some mention of foreign practices will, however, be reviewed when the aforementioned treaty is discussed, but basically it is with United States patent laws that we will be dealing. The reasons are that the scope of patent protection permitted by U.S. patent laws is of greater breadth than those of certain of the foreign countries, each of which may be part of a convention under which reciprocal rights are exchanged, all subject, however, to national laws as to the type of patent protection granted.

### B. Classification in the United States Patent and Trademark Office (PTO)

In the PTO, there is an entire classification, a single class, that deals with the chemistry of fermentation, Class 195 (Manual of Classification, 1936). The classification is ancient but still applicable, although currently under revision. Most of the inventions are in the fields of fermentation to yield useful products, and the class also provides for enzymes and compositions. However, there are other classifications that deal with microbiological processes, e.g., processes of preparing vaccines and sera using animals as an essential part of the process. Products prepared by microbiological processes are generally classified and examined in the PTO according to their chemical characteristics. Microorganisms can be used in many chemical reactions as the catalytic agent, examples of which include oxidation, reduction, and hydrolysis reactions.

My first contact with the subject in the PTO was as classification examiner charged with the duties of maintaining the classification of chemical patents. With the advent of antibacterials, mostly known as antibiotics, the field of microbiology became very active in patent filings. Each newly discovered organism was carefully cultured by researchers in their search for improvements over penicillin. Soil samples from all over the world were carried to the researchers in the field, and I am certain that many research dollars were spent on the culturing of new organisms that had interesting antibacterial properties but could not be reproduced outside of the laboratory. Although there were many that failed, there were some notorious successes, namely, tetracycline (Conover, 1955), oxytetracycline (Terramycin) (Sobin et al., 1950), chlortetracycline (Aureomycin) (Duggar, 1949), streptomycin (Waksman and Schatz, 1948), erythromycin (Bunch and McGuire, 1953), and fumagillin (Hanson and Eble, 1953).

## C. Examination in the United States Patent and Trademark Office

The patent applications filed thus covered many aspects of the inventive concepts. There were process applications for using a newly discovered microorganism to prepare a new product as well as known products. There were applications for improvements in process conditions, in media used, and in the recovery steps. In some cases the recovery was purely chemical and began not with the fermentation itself but with the broth end product, which included the undesirable components. There were also claims for the products themselves. In some cases, the products were defined not by chemical formula but only in terms of characteristics and properties. In some cases, claims were in the form of the product as prepared by the claimed process. There were patents for derivatives of the antibiotics already produced. These are all illustrated by the sample claims provided in Appendixes 1 through 4.

The Classification Division of the PTO made the decision as to which patent examining division each type of claim should be submitted. Inventions in fermentation chemistry, organic chemistry, and medicines were being examined in separate and distinct examining divisions. Where the product was medicinal, and its structure undetermined, the medicine division was charged with the examination. When the structure was elucidated, the burden of examination was shifted to the examiner in organic chemistry. The working arrangement satisfied the parties but had little effect on the patent law at that time. Although this may seem to be an unusual arrangement, it is based on actual experience. No matter how uniform an examining system, how consistent the guidelines, and how highly skilled the examiners, each is a distinct personality. The individual treatment by an examiner in organic chemistry reviewing a product claim might be different from that of an examiner in the medicine arts looking at a product claim that has no known chemical composition but has medicinal properties. Thus, differences in handling applications did occur in spite of PTO efforts to minimize if not prevent all such occurrences. The reason for this background information will become apparent as we discuss how the practice evolved to its current status. Some of this early treatment, however, is discussed by Levy and Wendt (1955). Their review covered deposit of cultures of microorganisms as well as how to claim the products of the fermentation, particularly when the structure is unknown and use is made of infrared absorption spectrum, which is incorporated into the application by a formal drawing in lieu of the structure. These infrared absorption spectra are characterized as "finger prints" for identification purposes insofar as the product is concerned, although they are combined with other known characteristics, including elemental analysis, melting point, molecular weight, X-ray diffraction patterns and optical rotation as well as biological activity against microorganisms. Many other characteristics are used as can be seen by the examples of claims in the appendixes to this chapter.

A real problem occurred in examination when two or more applicants claimed the same product but each used different identifying data and often named the microorganism by their own "catchword" (Reynolds, 1967). The examiners were astute enough to discover this and did not issue patents to two applicants for the same product. In some cases one application defined the product in terms of properties and the other by an elucidated formula. Since the inventions were the same, and only one patent could be granted, it was the examiner's burden to make the discovery of identity and set in motion an interference, which is a contest to determine who was the first inventor. This interference is in the PTO and the final decisions there are appealable.

Thus during the trying years of handling patent applications involving the use of microorganisms which could not be identified, there was no "official" Patent Office policy that instructed the examiners as to the handling of the application and in effect all decisions were those of the individual examiners and their immediate supervisors in their respective "art" areas. So while there was a rule as to which type of application would be examined in the respective examining areas there was no policy to provide uniformity. As case law developed, decisions were made. This meant that examination techniques were based on case law by courts unfamiliar with the technology about microorganisms.

## D. Statutory Proposals

It was not until the patent reform legislation was being proposed that statutory provision was suggested. In 1971, S.643 provided:

Section 112—Specification

(d) When the invention relates to a process involving the action of a microorganism not already known and available to the public or to a product of such a process, the written description required by subsection (a) of this section shall be sufficient as to said microorganism, if—

(1) not later than the date that the United States application is filed, an approved deposit of a culture of the microorganism is made by or on behalf of the applicant or his predecessor in title, and

(2) the written description includes the name of the depository and its designation of the approved deposit and, taken as a whole, is in such descriptive terms as to enable any person skilled in the art to which the invention pertains to make and use the same.

An approved deposit shall be a deposit which—

(1) is made in any public depository in the United States which shall be designated for such deposits by the Commissioner of Patents by publication, and

(2) is available, except as otherwise prohibited by law, in accordance with such regulations as may be prescribed,

(a) to the public upon issuance of a United States patent to the applicant or his predecessor or successor in title which refers to such deposit, and

(b) prior to issuance of said patent, under the conditions specified in Section 122.

It is to be noted, however, that S.643 required that the approved deposit be made in "any public depository in the United States." In some cases, this would be impossible for foreign inventors to do since certain pathogenic organisms are subject to import control by the Department of Agriculture as well as other government agencies.

Meanwhile, patent reform legislation was having a stormy passage through Congress and none has yet been approved. However, in the 94th Congress, S.2255 was passed by the Senate on February 26, 1976. Section 112(d) became Section 112(f) and under this was provided:

(g) For the purpose of subsection (f) (1) of this section. an approved deposit shall be a deposit which—

(1) is made in any public depository which shall have been designated for such deposits by the Commissioner by publication, and

(2) is available, except as otherwise prohibited by law, in accordance with such regulations as the Commissioner may prescribe.

Thus the deposit can be made anywhere in the world subject to the guidelines issued by the PTO now in effect. As discussed below the Budapest Treaty provides for the recognition of an International Depository Authority where a deposit of culture of a microorganism may be made in connection with a patent application filed in the patent offices of the respective contracting countries or in designated regional patent offices. The treaty refers to these as Industrial Property Offices. The European Patent Office is a regional patent office.

This discussion is necessitated by the fact that in most of these applications, the subject matter deals with a microorganism which is new, with no known prior description, but which is useful to produce a valuable result, be it in a process or to produce a new chemical product.

How does one meet the statutory provisions of 35 U.S.C 112 by adequately describing such a microorganism? The deposit of a culture in a public depository became the answer. This is a major variation from a conventional patent disclosure and, of course, led to many contentious problems until it was finally resolved in 1971. Since then, the proposed legislation and treaty have been suggested to officially sanction the practice.

It is apparent from this discussion that a microorganism cannot be prepared readily since in most cases it is a naturally occurring living material, and even in those cases where it is prepared or altered by mutation, it is questionable whether it is readily and uniformily reproducible.

## E. Patent and Trademark Office Guidelines

The problems in this field leading to the Guidelines for Deposit of Microorganisms (1971) were set forth in an earlier report (Marcus, 1975). Most important was the fact that the idea was conceived by applicants for their own benefit and, for a long time, the contract was under no control by anyone outside of the depositor–depository relationship. As a result, there was confusion about statutory basis and it was not until judicial decisions were made that the PTO took steps to resolve the problems, as is discussed later.

## III. PATENT PRACTICE AND PROBLEMS

### A. Review of Case Law

To set the stage for a discussion of the patent practice, I will start with a review of some important decisions involving microbiological inventions and then consider current practices based on these decisions keeping in mind the chronology of these decisions. *Guaranty Trust Company of New York et al. v. Union Solvents Corporation* (1931) involved a suit on an invention of Weizmann for "Production of Acetone and Alcohol by Bacteriological Processes" (1919). The 1931 decision of the Delaware District Court found Claims 1 and 3 of the patent valid and infringed. These are reproduced:

> 1. The process of producing acetone and butyl alcohol by the fermentation of liquids containing natural substances rich in starch by means of the herein described bacteria which are capable unaided by converting sterile fermentable grain starch substantially into acetone and butyl alcohol, and also liquefying gelatin.
> 3. The process of producing acetone and butyl alcohol by the inoculation of a cereal composition with the herein described bacteria which are capable unaided of converting sterile fermentable cereals substantially into acetone and butyl alcohol.

The question of patent disclosure of the bacteria was considered and effectively handled by the court by acceptance of the plaintiff's Exhibit 12 by plaintiff's expert and by evidence that the selection of the amount of inoculum was left to the judgement and experience of the operator to determine. The specification had referred to the well-defined sources from which Weizmann obtained his bacteria and the precise procedure. The court, therefore, concluded that anyone following this procedure could obtain the bacteria and would have considerable success.

The contention of the defendant that the invention in the Weizmann patent was for the life cycle of a living organism was overcome by pointing out that it is not the bacteria per se that is being claimed but a fermentation process using such bacteria, which the inventor discovered, under the disclosed operating conditions. Thus the inventive concept for Weizmann's discovery was for the use of such bacteria in a process to obtain a very useful known product. The Court of Appeals sustained this decision in 1932.

The Court of Customs and Patent Appeals (CCPA) held the following claim unpatentable in *In re Arzberger* (1940):

> 1. Bacteria herein described and designated as clostridium saccharobutyl-acetonicum-liquefaciens.

The holding was based in the proposition that this bacterium was not statutory subject matter under the plant patent section of the patent laws (Patents for Plants, 1954):

> Sec. 161. Whoever invents or discovers and asexually reproduces any distinct and new variety of plant, including cultivated sports, mutants, hybrids, and newly found seedlings, other than a tuber propagated plant or a plant found in an uncultivated state, may obtain a patent therefor, subject to the conditions and requirements of title. (Amended September 3, 1954, 68 Stat. 1190.)

The provisions of this title relating to patents for inventions shall apply to patents for plants, except as otherwise provided.

In the common language of the people, "plant" as used by Congress did not encompass bacteria even though its reproduction is asexual, by binary fission. The CCPA never decided whether it was encompassed by the statutes as a composition of matter (as will be later discussed).

Next in time was *Ex parte Kropp* (1949), a decision of the PTO Board of Appeals in 1959. The central issue involved the deposit of the culture of the microorganism used in the claimed process. The case held that the hitherto unidentified species of microorganism set forth in the claimed process:

> 4. A process for producing jm-57h which comprises cultivating the organism *Streptomyces* sp. JM-57h in an aqueous nutrient medium containing a source of protein and a source of carbohydrate under aerobic conditions until antibacterial activity is produced in said medium

was insufficiently disclosed if the microorganism had not been deposited in a recognized depository *prior* to filing the application. An offer to deposit was made but the decision held that the deposit, which would have been sufficient to overcome the rejection, had not been made by the time of filing of the application and the subsequent deposit or offer to deposit could not overcome this defect in disclosure.

An interesting variation of this concept was subsequently decided by the CCPA in 1971. The case *In re Argoudelis et al.* (1970) held that the deposit of the microorganism, made no later than the date of filing, could be restricted so that no disclosure of it would be available until the patent would issue. The PTO had held that the Statute (Sec. 112) required a written description, and in the absence of a written description reference to something outside the specification must be a reference to something in the public domain, hence no availability of the "starting material." Since the depositor restricted access to the microorganism only to nominees of the applicant or the PTO, the organism was not adequately described. This PTO position was reversed by the CCPA. Thus a restricted deposit was an enabling disclosure, under 35 U.S.C 112 during the pendency of the application, with the deposit becoming available to the public after grant of the patent as the quid for the patent grant. This decision caused the PTO to publish "Guidelines for Deposit of Microorganisms" (1971), which are still in effect.

Since the publication of these guidelines, problems concerning deposit of microorganisms have been minimized. Without any legislation to the contrary, the PTO will accept a deposit anywhere in the world, if proper access is given to that culture deposit when the United States patent issues.

This is the current practice in the United States and, of course, the practice approved by the Budapest Treaty. The Treaty also provides for access to the organism when there is a publication of the application. This precise point was decided by the CCPA in *Feldman v. Aunstrup* (1975). The respective applicants were contesting an interference as to a fermentation process. The sole count reads:

> A process for the preparation of a milk-coagulating enzyme which comprises cultivating a milk-coagulating enzyme producing strain of *Mucor miehei* Cooney et Emerson or a natural or artificial variant or mutant thereof in a suitable nutrient medium, and thereafter recovering the milk-coagulating enzyme from the medium.

The microorganism is the essence of the invention. Feldman deposited his culture with the Northern Regional Laboratory of the ARS, and Aunstrup with the Centraalbureau voor Schimmel cultures (CBS) (Central Bureau for Mold Cultures) in Baarn, Netherlands (Table I). The latter deposit was timely, prior to his convention foreign application which was granted priority for his corresponding United States application. The CBS, at the time a private institution, is now part of the Royal Netherlands Academy of Sciences and Arts, a governmental organization. The deposit, at one time restricted, was removed from restriction in 1969. The CBS is a qualified culture collection and distributes samples. The CCPA found the deposit in CBS to satisfy the conditions set forth in *In re Argoudelis et al.* (1970) and in the Patent Office Guidelines. Not only were samples available to the public at the proper time, but the PTO could also obtain access to the CBS sample at any time during pendency of the patent application. The Budapest Treaty carries forward this principle by the contracting states. Thus Aunstrup's application was held to find adequate support in his foreign deposit in accordance with 35 U.S.C. 119 and he won the interference contest. Feldman carried his appeal of this decision to the Supreme Court but certiorari was denied.

This principle was also approved in a decision of the German Supreme Court in 1975 in the so-called bakers' yeast case (Bäckerhefe,

---

**TABLE I.** Culture Collection Reviews

A. Culture Collections of Microorganisms (from "Proceedings of the International Conference on Culture Collections." Tokyo, Japan, 1968).
   1. The American Type Culture Collection (W. A. Clark, pp. 3–5).
   2. The Centraalbureau voor Schimmel cultures (J. A. von Ark, pp. 7–9).
   3. History, Policy and Significance of the ARS Culture Collection (C. W. Hesseltine, W. C. Haynes, L. J. Wickersham, and J. J. Ellis).
B. Culture Collections and Patent Depositions [T. G. Pridham and C. W. Hesseltine, *Adv. Appl. Microbiol.* **19,** 1–23 (1975)].

1975). The German Court held that a deposit of a microorganism in a culture collection of an approved depository in a foreign country would satisfy the German Patent Laws if the deposit was made prior to the actual filing date of the application and if based on a convention application, prior to the home country filing. The applicant must also make an irrevocable undertaking to permit access to that deposit by the German Patent Office for examination procedure and to the interested public 18 months after the date of actual filing or the date of the Paris Convention application.

The court noted that the applicant may require third parties obtaining the sample to not only identify themselves but agree not to pass samples to others or to export the samples. This protection is due to the fact that at the time of this publication, enforceable patent rights are not yet in existence. The Budapest Treaty also provides for this notification to the depositor by the depository of those who request samples of microorganisms.

In another important decision in the field, the U.S. Supreme Court (1948) in *Funk Bros. Seed Co. v. Kalo Inoculant Co.* held the following claim invalid:

> An inoculant for leguminous plants comprising a plurality of selected mutually non-inhibitive strains of different species of bacteria of the genus *Rhizobium,* said strains being unaffected by each other in respect to their ability to fix nitrogen in the leguminous plant for which they are specific.

The patent also contains process claims.

The six well-recognized species of bacteria and the corresponding groups (cross-inoculation groups) of leguminous plants are shown in the tabulation below:

| | |
|---|---|
| *Rhizobium trifolii* | Red clover, crimson clover, mammoth clover, alsike clover |
| *Rhizobium meliloti* | Alfalfa, white or yellow sweet clovers |
| *Rhizobium phaseoli* | Garden beans |
| *Rhizobium leguminosarum* | Garden peas and vetch |
| *Rhizobium lupini* | Lupines |
| *Rhizobium japonicum* | Soy beans |

In spite of the ease of use provided by the invention, the Supreme Court found a lack of invention in the discovery because no new use of each individual bacteria was discovered:

> Discovery of the fact that certain strains of each species of these bacteria can be mixed without harmful effect to the properties of either is a discovery of their qualities of non-inhibition. It is no more than the discovery of some of the handiwork of nature and hence is not patentable. The aggregation of select strains of the several species

into one product is an application of that newly-discovered natural principle. But however ingenious the discovery of that natural principle may have been, the application of it is hardly more than an advance in the packaging of the inoculants. Each of the species of root-nodule bacteria contained in the package infects the same group of leguminous plants which it always infected. No species acquires a different use. The combination of species produces no new bacteria, no change in the six species of bacteria, and no enlargement of the range of their utility. Each species has the same effect it always had. The bacteria perform in their natural way. Their use in combination does not improve in any way their natural functioning. They serve the ends nature orginally provided and act quite independently of any effort to the patentee.

This case is of importance today because there was an appeal pending to the CCPA from a Decision of the Board of Appeals of the PTO holding unpatentable as nonstatutory. The claim:

A biologically pure culture of the microorganism *Streptomyces vellosus* having the identifying characteristics of NRRL 8037, said culture being capable of producing the antibiotic lincomycin in a recoverable quantity upon fermentation in an aqueous nutrient medium containing assimilable sources of carbon, nitrogen and inorganic substances.

This case, known as *In re Bergy et al.* (1977), was argued in March and decided on October 6, 1977; but the question as to whether microorganisms are patentable subject matter may still be in question, in view of a further case (*In re Chakrabarty\**) also pending in the same court and, at the same time, a request by the PTO for a rehearing of *In re Bergy et al.\**

As noted in *In re Arzberger,* bacteria are not "plants" within the meaning of the plant statute 35 U.S.C. 161: Patents for Plants:

Whoever invents or discovers and asexually reproduces any distinct and new variety of plant, other than a tuber propagated plant, may obtain a patent therefor, subject to the conditions of this title.

For the most part, microorganisms are products of nature and thus are not novel. The question in the *Bergy et al.* case argued by the applicants was that their culture is modified and not the microorganism as found in nature. This is supported by affidavits of well-known microbiologists including Alma Dietz who deposed that impure cultures of *S. vellosus* gave taxonomic results different from those of the claimed culture. In essence they argue the claimed culture is the product of a microbiologist. The applicants also point to many patents in which a living organism is part of the claimed invention.

On the other hand, the Patent Office Solicitor argued that the present patent laws do not encompass living organisms as statutory subject

---

\* Reargued November 1978 in the CCPA after grant of certiorari by the U.S. Supreme Court.

matter even under the statutory provisions of "manufacture" or "a composition of matter." Several writers disagree and have strongly set forth their views. One, Harold C. Wegner (1974), poses the query "Must something be dead to be patentable?"

Judge Rich in *In re Mancy et al.* (1974), reversing the rejection of

1. Process for the production of daunorubicin which comprises aerobically cultivating *Streptomyces bifurcus* strain DS 23,219 (NRRL 3539), of [sic, or] a daunorubicin-producing mutant thereof, using an aqueous nutrient medium containing assimilable sources of carbon, nitrogen and inorganic substances, and separating daunorubicin formed during the culture.

The four dependent claims specify certain further conditions of aerobic cultivation such as pH, temperature, and aeration rate, although apparently the conditions specified are conventional.

included the following statement:

The Nature of This Invention: A Process of Producing Known Antibiotic Using a New Strain of Microorganism

. . . Here appellants not only have no allowed claim to the novel strain of *Streptomyces* used in their process but would, we presume (without deciding), be unable to obtain such a claim because the strain, while new in the sense that it is not shown by any art of record, is, as we understand it, a 'product of nature.' However, it is not required for unobviousness of the method-of-use claims that the new starting material be patentable, *In re Schneider, supra;* 35 U.S.C. 103.

[4] The process of using the new *Streptomyces* strain which appellants have discovered is clearly within 35 U.S.C. 101, and we note the ready willingness of the board here, as in *Ex parte Arzberger* and *Ex parte Kropp*, to allow appellants to claim it if they show unexpected results from the use of this new strain. Indeed, the public interest appears to be well served by encouraging the patenting of such inventions. While the patent will grant appellants a limited right to exclude others from producing daunorubicin by the use of *Streptomyces bifurcus,* the public receives not only the knowledge of appellants' discovery but also access to *Streptomyces bifurcus* through its deposit with the Department of Agriculture. See *In re Argoudelis, supra.*

The importance of the question of the patentability of microorganisms per se was also set forth in the brief of the Patent Office solicitor in *In re Bergy et al.* (1977):

The age of "genetic engineering" . . . creation of new living organisms by combining genetic material from different life forms . . . is now upon us.

This concept will be further discussed but is mentioned here in its relation to the patentability of living organisms.

The decision in the *Bergy* case was by a divided court. The majority opinion was supported by a judge from the court of claims sitting by designation who agreed with the majority opinion but indicated he felt it set forth "an extremely limited holding." Only the future can tell whether this is true.

Judge Rich, speaking for the majority, thinks it is in the public interest to include microorganisms within the terms "manufacture" and "composition of matter" as set forth in 35 U.S.C. 101, quoted above. Judge Rich also points out that his views quoted above from *In re Mancy* (1974), even if considered dictum, were ill-considered since it was apparent that no claim for the microorganism could have been obtained in the Mancy case since it lacked novelty. In the *Bergy* case, however, the microorganism is found to be in its purified form and is found to meet the statutory criteria even if it is a living thing.

At this point we cannot speculate how this decision will affect cases arising from genetic engineering discussed below. Judge Miller, however, speaking for the minority, feels that microorganisms are distinct from inanimate chemical compositions and that such subject matter was never intended to be included within the present patent statute. He also refers to the Plant Variety Protection Act (1972) which he thinks further supports this conclusion that Congress did not intend organisms to be included within the scope of "manufacture" or "composition of matter." The fact that the claimed culture of the microorganism is an industrial product ignores such congressional intent. In effect the dissenting judges feel that the CCPA usurped the congressional role in reversing the rejection and that the only clear way that claims such as claim 5 of *Bergy et al.* could be found patentable would be by express statutory provision.*

---

*On March 2, 1978 in Appeal No. 77–535 the CCPA reversed the decision of the PTO Board of Appeals in *In re Ananda M. Chakrabarty* for an invention of which the following is an illustrative claim:

> 7. A bacterium from the genus *Pseudomonas* containing therein at least two stable energy-generating plasmids, each of said plasmids providing a separate hydrocarbon degradative pathway.

The Court states: "Appellant's invention involves the creation of a new strain of bacteria by the incorporation of a *single* cell, by transmission thereinto of a plurality of compatible plasmids, of a capacity for simultaneously degrading several different components of crude oil with the result that degradation occurs more rapidly. To make this non-technical description somewhat more intelligible we quote from the specification but two of its many definitions:

> *Extrachromosomal element* . . . a hereditary unit that is physically separate from the chromosome of the cell; the terms 'extrachromosomal element' and 'plasmid' are synonymous; when physically separated from the chromosome, some plasmids can be transmitted at high frequency to other cells, the transfer being without associated chromosomal transfer.

> *Degradative pathway* . . . a sequence of enzymatic reactions (e.g., 5 to 10 enzymes are produced by the microbe) converting the primary substrate [i.e., oil] to some simple common metabolite, a normal food substance for microorganisms."

and further notes that Appellant's assignee has been granted British Patent 1,436,573 containing the above and other claims for the bacterium.

On June 26, 1978, the U.S. Supreme Court on petition of the Commissioner of Patents granted certiorari (*Parker v. Bergy,* 1978), vacated the decision of the Court of Customs and Patent Appeals, and remanded the case back to the latter court in view of their earlier decision in a case involving computer software (*Parker v. Flook,* 1978) as being nonstatutory subject matter.†

I have briefly touched some of the types of inventions in the field of microbiology. Over the past 30 years, the most significant and commercially important inventions were in the field of antibiotics. The search for antibacterials since the development of penicillin and the "sulfa" drugs became an industry pastime. Everyone brought soil samples back from his vacation trip. Cultures were made and promising microorganisms were more extensively studied. As noted earlier, many never left the laboratory and many microorganisms produced interesting products which could not be made outside of the laboratory in commercial amounts. There were, of course, some highly successful developments, the most significant of which commercially were the tetracyclines. The litigation in which the tetracycline patent has been involved for the past 20 years has cost many millions of dollars to many companies, many states, and the United States government.

The most significant factor in claiming microbiological inventions occurs when the microorganism used to produce the invention, whether it be for a new product or in the process, is new and never before described in the literature. The aforementioned Section 112 of the Patent Statute requires a written description of the invention. The PTO went through some trying years until a definite practice was established in 1971. As explained by Levy and Wendt (1955) there were two serious problems: (1) How to describe unknown microorganisms; and (2) how to describe a chemical product produced in a microbiological process where there is no recognized chemical structure available. The authors relate the PTO solutions. The first was to require a deposit of the microorganism in a depository which is recognized as providing for such deposits and, complementary, the accession number given by such depository with as much of the taxonomic description as the inventors can supply at the time.

The attendant problems were discussed by Marcus (1975). First, there must be a depository that can handle the microorganism and preserve it for at least the life of the patent. Second, the PTO is not privy to the

---

The bacteria in the instant case are modified for the purpose of controlling oil spills by what they refer to as genetic engineering as the term is in the specification.

† On March 29, 1979, the CCPA decided together both the *Bergy* case and the *Chakrabarty* case. In doing so, they reaffirmed their original decision reversing the rejection from the PTO. Once again, it was a divided court. However, Judge Baldwin, who dissented in the original opinion, now joins the majority and writes a concurring opinion.

contract between the depositor and the depository so it has no voice in the deposit, yet its description and availability is a necessary adjunct of the patent application. Finally, can the depositor impose such restriction on availability of the microorganism as to preclude its release until at least the time of patenting unless they give permission for earlier release? Of course, the Commissioner of Patents must always have access to the deposit because it is part of the patent application filed with the PTO.

It was not until the *In re Argoudelis et al.* decision that the practice in the PTO was settled. The PTO issued guidelines in 1971 which are now in full use and have really minimized problems occurring in the United States. However, even with these guidelines, problems arose which, I believe, are now fairly well settled. In drafting Patent Reform Legislation, the drafters provided for the deposit of microorganisms in a public depository as designated by the Commissioner of Patents. The original legislation, as noted *supra,* also required a deposit in the United States. This caused consternation in the profession. Even though other countries at present require such a deposit under their national patent laws in a depository in that country, the prospect of that occurring in the United States appeared to have earthshaking repercussions.

In at least one instance known to me, the microorganism already deposited in a public depository in Europe could not be imported into the United States due to customs and agriculture regulations because of its pathogenicity. The following problem then arose: Can the United States grant a patent for an invention when the public in the United States can never practice the invention?

## B. Depositories: Public and International

There was another problem. The deposit of a culture in a public depository, under the guidelines, was required to be open to the public once the United States patent issued. The question, however, was the duration of availability of that culture. The fees the depositor paid theoretically stopped when the patent issued and certainly the deposit could be maintained forever. The "quid" for a patent was the giving of the invention to the public when the patent monopoly expired. But did the inventor or his assigns worry about the status of the deposit when their patent rights expired? These were serious problems since the public would have nothing when the patent expired if the "deposit" was no longer available.

Then, too, what were the controls on depositories? Some were arms of the national government, some were in universities, some were private institutions subsidized by grants, and some were company managed.

These problems became the impetus for the convening of a meeting to consider the legal and scientific aspects of an international system for the deposit of microorganisms for patent purposes. The meeting was under the auspice of World Intellectual Property Organization (WIPO). In the 3 years since the first meeting in 1974, a treaty has been approved and is now being considered for ratification. The Budapest Treaty was approved in June 1977 and is presented in Appendix 5.

The requirements of the national laws of the various countries were considered as well as the various aspects establishing international depositories or, at the least, national depositories recognized by all the participating countries. The establishment of one or more centralized international depositories was considered less advisable by the committee of experts than the recognition of national depositories whose existence would be guaranteed by the signatory nation. These recognized depositories would accept deposits for patent purposes. A deposit in any nationally recognized depository would be recognized for establishing priority rights and the single deposit in such a recognized depository would render the disclosure of a patent application "enabling" for disclosure purposes in the patent system of any of the participating nations where the application is filed.

The treaty made provisions for the status of an International Depository Authority, with assurances by the contracting state guaranteeing its existence and that the depository has the necessary staff and facilities. Of course, there are many other conditions. The requirement of a receipt of the depositor, the maintaining of the deposit, the requirement for secrecy, the ability to supply a sample when properly requested, and the ability to maintain the deposit for a designated period beyond that of the patent grant. This period is at least 30 years after the deposit, or 5 years after the last request.

There are other conditions. The virility of the deposit is checked periodically and if necessary new samples can be requested of the depositor. The depository must supply a sample of any deposited microorganism at the request of the Industrial Property Office (Patent Office) of any contracting state where an application for patent has been filed. In addition the depository will supply samples to those authorized to receive as (1) either designated by the depositor or (2) on evidence that the patent application has been published in accordance with the laws of the country where the application was filed.

Furnishing samples is made interesting. The depositor will be notified by the depository of the furnishing of samples on request, and no charge is made for furnishing a sample to a patent office, although fees may be charged either to the depositor or the requestor as per the conditions set forth in the deposit contract.

A list of recognized depository authorities will be published by the International Bureau under the treaty and from time to time each patent office will furnish the depositories with lists of publications or patents issued in that country which refer to accession numbers in the depositories.

The signatories to the treaty constitute a union (Budapest Union) and members thereof are limited to states that are members of the Paris Union.

Thus a procedure of limiting the number of deposits an applicant for patent need make is being provided by treaty. In other words, a deposit in any signatory country will suffice for a priority claim for an application relying on that deposit for its disclosure filed in that country. And when applications are filed in other countries that deposit will suffice if the proviso that it will be made available after publication of the application is complied with (Behr, 1975). Of course, there might well be national laws that will require a subsequent deposit in their country in connection with a patent application filed under its national laws. However, this later deposit in all probability need not be made before filing of the application but at any time during its pendency so that nationals of that country need not seek to obtain the culture from a depository located abroad. At present time, this is the practice in Japan. An application filed in Japan in connection with an earlier convention application in the United States relies on the United States deposit for its disclosure support. Japan will accept the application for filing and examination but will require a deposit in Japan before the patent is granted there.

Other deposits under national law may occur, but basically the treaty provisions will serve the needs of the international patent community.

We have discussed the practice about depositing and we have shown the international recognition and control of depositories. Now, what are public depositories? Basically they are culture collections with their own curators and are many in number in the United States. Most are private or for company use. As far as I know there are only two public depositories that serve the patent community at present in the United States. There is the ARS Culture Collection of the Agricultural Research Service, U.S. Department of Agriculture, located at the Northern Regional Research Laboratory, Peoria, Illinois. As reported by Hesseltine *et al.* (Table I), the culture collection came into being with that laboratory about 1940 and began with cultures from Dr. Charles Thom's collection. The NRRL services yeasts, molds, bacteria, and actinotimycetes. The designation of deposits in that culture collection follow the letters NRRL. No charge is made for the deposits or their maintenance, or for their distribution "to bona fide scientists in agriculture, in industry, and in education in the

United States and also to those outside the United States as a gesture of international cooperation and good will."

This culture collection does not publish a catalog or list of its cultures. Of course, cultures in connection with patent applications became publicly available when the U.S. patent issues and identifies the microorganism used as being in deposit with the ARS. While the application is pending, the usual secrecy and restriction of accessibility is provided.

The collection at the American Type Culture Collection, located in Rockville, Maryland, was started in 1925 (Table I). At present there is a staff of about 100. The designation of deposits follows the letters ATCC, and the precautions of secrecy and restricted access to deposits in connection with patent applications is maintained. The ATCC publishes a catalog from time to time. Provision is made at ATCC for storage of cultures of bacteria, yeasts, fungi, plant and animal virus, Rickettsiae, protozoa, and animal cell cultures (cell lines). The ATCC charges a fee for deposits and no fee after the patent issues, but requestors of culture samples must pay a service charge.

The ATCC, while funded in part from government grants, is not a governmental arm. In 1976 the ATCC issued a new regulation that in connection with patent applications in the United States and other countries, there is a flat one-time fee of $475 for 25 years of maintenance. At this time we do not know if this will be changed in view of the treaty requirements for at least 30 years of maintenance. The ATCC also has a special fee for notification of the depositors as to who requests samples. Presumably a new fee schedule will be provided when the treaty comes into force.

As far as is now known, the treaty will not convert the "guaranteed" international depositories into governmental agencies where such relationship does not now exist. The independence of the scientific organization serving the scientific as well as the patent community remains. Insofar as that international depository must serve the worldwide patent community, however, a government guarantees their existence. The overall result of nations agreeing to sponsor such institutions not only for international good will but for the protection of industrial property by the patent grant to advance the useful arts brought this treaty to fruitation in a relatively short period of time. The core of the problem—an international depository—that will safely store the culture deposit and make it available for a period of 30 years from the date of deposit or for at least 5 years after the last request is guaranteed by the signatories to the treaty.

In so doing, each state that has an Industrial Property Office—a Patent Office—thus becomes not only a guarantor of the depository's existence but a guarantor of its patent system to see that on patenting or publication the deposit will become available to the public. True, the national laws of the respective nations may vary in the way the deposit is to be made, but overall one deposit is all that is necessary at a great saving of time and money to the users of the patent systems.

While no patent reform legislation is now pending in Congress, the last of the bills, S.2255—passed by the U.S. Senate in 1976—as mentioned, *supra,* had statutory provisions for deposit of microorganisms as providing a description of the invention required by 35 U.S.C. 112, where the microorganism is not available publicly and not otherwise capable of being described in writing [Sections (f) and (g)]. Also proposed is a companion section providing for the grant of priority to an application first applied for in a foreign country which acknowledged therein a deposit in a public depository [Section 119 (d)].

The proposed European Patent Treaty, to be in operation in 1978, by the establishment of a European Patent Office has a Rule 28 (Appendix 6) dealing with the deposit of microorganisms. The conditions are quite similar to those of the Budapest Treaty.

With respect to export of samples of the culture deposits from the United States to Europe, it is my understanding that the ATCC has obtained permission by appropriate export license from the Department of Commerce to transport samples of microorganisms to the West German Patent Office in accordance with the requirements as set forth in the aforementioned bakers' yeast decision.

## C. Patentability of Microorganisms

The "burning question" in patent law today is the status of claims for microorganisms per se.

The subject has been introduced by reference to some case law, *supra.* As a product, the microorganism per se is generally found in nature and hence is unpatentable. But what about variations of the microorganism? There may be mutations some of which occur naturally, on storage and some of which are induced. In either case, is the new microorganism a natural product?

In the discussion above, it has been pointed out that most inventions are for the use of bacteria as components or intermediates of microbiological processes and as ingredients in compositions. Many patents have issued on both uses of such microorganisms and in which microorganisms are components of compositions.

In addition, animal and plant cells illustrated by deposits as cell lines in the ATCC have been used as diagnostic tools. Since these cells reproduce themselves, they have been used in mass tissue cultures in the field of testing for diseases such as malignancies, in testing for effectiveness of medicines, and in producing biological substances such as hormones. An example is the testing for the compatibility of a donor by the combining of the white cells of the donor with the white cells of the recipient.

Genetic engineering is the use of DNA to combine with a bacteria host cell and become part of its genetic complement. The ventures in recombinant DNA have been given much notoriety in recent years and the effect on patent practice must also be discussed.

The new breed of writers favors the patentability of microorganisms. Irons and Sears (1975) point out that the "great preponderance of practically useful microorganisms, however, are not natural products." Thus the application of external forces to change their characteristics results in a microorganism different from the product as formed in nature. The writers also suggest that 35 U.S.C. 101 does not limit inventions to "inanimate subject matter." This same thought is expressed by Wegner (1974). Daus et al. (1966) are in agreement that as a composition of matter, microorganisms are statutory subject matter and that In re Arzberger et al. would probably be reversed today by the CCPA. All these conjectures, however, are no substitute for statutory authority and the thought raised by the "burning question" will be the effect of the decision in the In re Bergy et al. case.

It would appear that we have avoided taking sides on the statutory standard for this important subject, while there is statutory basis for plant patents and certificates for plant varieties. It is noted that the statutory basis for deposits of microorganisms is still lacking while we have provided, through case law and PTO Guidelines, the basis for the practice. This is a strange situation.

As I pointed out above, the refusal in the Arzberger case to find bacteria within the purview of the plant patent statute as not within the congressional intent as to the word "plant," this being used in the common language of the people and not in any more specific sense.

The refusal to allow claims to microorganisms is further noted as dictum by Judge Rich's language in the Mancy case where his position was based on the product by nature theory. But consider his own statement in the Bergy et al. case as to his "ill-considered language" in the Mancy et al. case. Further, microorganisms are known to be unstable and it raises a question as to the scope of a claim for such a product and how it would be interpreted in the case law if it is subject to test in an infringement action. Would the same mutation or change occur on each application of the process conditions? This is a real problem in an area of such technical complexity, particularly since a living organism is involved.

The most interesting discussion of the problem, however, is by

Hayhurst (1971). The author not only poses the questions as to the patentability of organisms but as to man-made processes for subjecting microorganisms to treatments which result in new mutations. If the new mutant is useful, can the process be protected? which is aside from the question as to whether the product is itself patentable subject matter from a statutory viewpoint. The further question is whether the instructions for the preparation work always, invariably, or occasionally, and if they do work, will a patent issue?

The discussion takes on added importance since many countries did not allow patents for medicines, and most countries did not allow patents for chemical products per se. The Federal Republic of Germany adopted guidelines for a product system (Richtlinien, 1968). Japan followed as of June 1, 1976. In spite of that, however, claims for microorganisms are still in an ethereal state. The importance of product claims is apparent. They dominate all methods for their preparation and all uses later discovered. Processes can be altered to avoid infringement, but if the product used is patented, the patent owner dominates its uses and need not undertake to prove that the process used by others infringes his patented process.

While the CCPA's decision in *In re Argoudelis* prodded the U.S. PTO into the guidelines, and there is statutory provision proposed for microorganisms deposits if the Patent Reform Legislation is enacted, what is the road product claims for microorganisms must take? Would it be better for the PTO to propose that the statute either provide for such subject matter or preclude it and save the unnecessary expense to applicants that such cases engender?

The concern of the PTO is not without merit. If living microorganisms are held to be statutorily acceptable, would the PTO be inundated with applications for patents for newly discovered microorganisms? Of course each would have to be found useful and there would have to be a disclosed process of preparing it—since if it were in naturally occurring form it would not be new under 35 U.S.C. 101.*

---

*In a concurring opinion in *Chakrabarty*, in a footnote. Chief Judge Markey states:

As with Fulton's steamboat "folly" and Bell's telephone "toy," new technologies have historically encountered resistance. But if our patent laws are to achieve their objective, extra-legal efforts to restrict wholly new technologies to the technological parameters of the past must be eschewed. Administrative difficulties, in finding and training Patent and Trademark Office examiners in new technologies, should not frustrate the constitutional and statutory intent of encouraging invention disclosures, whether those disclosures be in familiar arts or in areas on the forefront of science and technology.

## D. Genetic Engineering

Another concern as to patentability of microorganisms is due to the advent of "genetic engineering"—the rearrangement of the basic genetic material of living things.† Rearrangement of the atoms of DNA, which is known to control the growth and reproduction of living cells, leads to a new form of DNA, known as recombinant DNA. These new molecules can enter a host cell, such as a bacterium, and become a part of its permanent genetic structure. New living microorganisms are formed. However, gene exchange among microorganisms and viruses probably occurs spontaneously in nature. At present, attempts to control recombinant DNA research are being made in Congress. The PTO issued guidelines for accelerated examination of patent applications in this field and then withdrew the practice after a joint meeting between the Secretaries of Commerce and HEW (Appendixes 7 and 8).

The scientific community is alarmed by the potential intervention of government into research activity. Are the dangers and hazards such as to warrant government intervention into the pursuit of scientific knowledge to protect the public welfare? Is there a possibility of antibiotic-resistant genes being produced? Michael Dukakis (1977), governor of Massachusetts, spoke out in *Time* magazine:

> Genetic manipulation to create new forms of life places biologists at a threshold similar to that which physicists reached when they first split the atom. I think it is fair to say that the gene is out of the bottle.

Rifkin (1977) commented that "General Electric is already out in front with the announcement that it has applied for a patent on a tiny microorganism that can eat up oil spills." This article even mentions the companies that are involved in recombinant DNA research almost as if a big

---

†In one of the Dissents, in *Chakrabarty*, Judge Baldwin notes:

> . . . a modified natural product does not become statutory subject matter until its essential nature has been substantially altered. The issue in the present case becomes whether the modification effected by appellant altered the essential nature of the starting material.

> . . . I believe that the essential nature of the unpatentable organism with which applicant started was its animateness or life. Appellant has not changed this essential nature; he has not created a new life. Rather, he has merely genetically grafted an extra plasmid on to the organism and, thereby, made the organism better in cleaning up oil spills. While this improvement in oil digesting ability does exclude the new organism from classification as a mere product of nature, like the borax-impreganted orange which was a better commercial product because it had a longer shelf life, this improvement in the utility for which the unpatentable starting material was already suited does not change the essential nature of the starting material and does not make the modified thing statutory subject matter.

secret is let out of the bag—nobody wanted to tell—nobody wanted the government to know they are in the field.

These are the new problems. At the end, the patent system will be the recipient of the research if (1) patent protection is permitted, and (2) if patent protection is sought. Whether patents will be allowed depends on the first two caveats, since without patent applications, there can be no patents.

Then too, the ultimate decision in the *Bergy et al.* case will decide if microorganisms are patentable subject matter—at least until there is statutory provision for the subject matter. This is the status of the "burning question" stated at the start of this section.

Research in inventions involving the action of microorganisms has had a peripheral problem. Unlike chemical compounds, which are readily reproducible once disclosed, microorganisms which are on deposit in a depository until a patent issues are not readily reproducible. Thus, industrial espionage, the stealing of cultures, puts someone else in possession of the "knowhow" not ordinarily available until the patent issues. The culture can even be taken abroad and used to manufacture valuable pharmaceuticals in countries where patent protection for the original owner was not available. So activity in the field of microorganisms presents an additional factor not present in ordinary research. This problem puts a burden on research organizations, the public depositories, and those who are permitted to obtain samples of microorganism cultures during the pendency of the patent application. Thus, this appears to be a reason why the depositories are asked to notify the depositor who seeks samples of the microorganisms even after the patent is obtained.

The industrial espionage problem came to light by the complaint by *American Cyanamid Co.* (1964) against a former employee who was charged with removing cultures of microorganisms from the research facilities.

The case points up the importance of microorganism deposits and the reason for some sort of controls on their availability. They are not like chemicals, purchasable by the barrel or carload. They could impose danger if improperly handled, but their importance and value to industry and science is tremendous.

## IV. SUMMARY

The merging of legal principles and a scientific subject presents many problems. Microbiology presented such problems when it became the subject of valuable patent applications. The discussion above brings together these two disciplines and shows how an attempt was made domestically, and on an international basis, to bring order to the treatment of the subject matter.

## APPENDIX 1

### 2,449,886

### STREPTOMYCIN AND PROCESS OF PREPARATION

**Selman A. Waksman, New Brunswick, and Albert Schatz, Passaic, N. J., assignors to Rutgers Research and Endowment Foundation, a nonprofit corporation of New Jersey**

**No Drawing. Application February 9, 1945, Serial No. 577,136**

**13 Claims. (Cl. 260—236.5)**

10. A process for the production of streptomycin that comprises growing a culture of a streptomycin-producing strain of *Actinomyces griseus* in a medium containing corn steep liquor, at a suitable incubation temperature and for a suitable period of cultivation, to form streptomycin in the culture broth, separating the culture broth from the organism growth, and recovering streptomycin from the broth.

11. A process for the production of streptomycin that comprises growing a culture of a streptomycin-producing strain of *Actinomyces griseus* at a suitable incubation temperature and for a suitable period of cultivation, to form streptomycin in the culture broth, separating the culture broth from the organism growth, adsorbing streptomycin from the broth, and recovering the adsorbed streptomycin.

12. A process for the production of streptomycin that comprises growing a culture of a streptomycin-producing strain of *Actinomyces griseus* at an incubation temperature of 22–28° C. for a time of the order of 6–12 days for stationary cultivation and 2–4 days for submerged aerobic cultivation, to form streptomycin in the culture broth, separating the culture broth from the organism growth, and recovering streptomycin from the broth.

13. Streptomycin.

SELMAN A. WAKSMAN

ALBERT SCHATZ

## APPENDIX 2

### 2,482,055

### AUREOMYCIN AND PREPARATION OF SAME

### B. M. Duggar

### Application February 11, 1948, Serial No. 7,592

### 8 Claims. (167–65)

I claim:

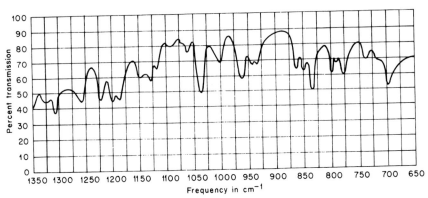

**FIGURE 1.**  Infrared absorption spectrum of HCl, salt of antibiotic.

1. Substances effective in inhibiting the growth of Gram positive and Gram negative bacteria selected from the group consisting of a substance capable of forming salts with acids, containing the elements carbon, hydrogen, nitrogen, chlorine, and oxygen, being very soluble in pyridine, soluble in methanol and in acetone and being slightly soluble in ethanol and in water, its crystals having a refractive index parallel to elongation between about 1.674 and 1.694, and exhibiting characteristic absorption bands in the infra red region of the spectrum when suspended in a hydrocarbon oil in solid form at the following frequencies expressed reciprocal centimeters: 3420, 1643, 1609, 1580, 1523, 1302, 1231, 1209, 1121, 1080, 1050, 969, 943, 867, 844, 825, 805, 794, 788, 733, 713 and the acid salts of said substance.

2. A substance effective in inhibiting the growth of Gram positive and Gram negative bacteria which is capable of forming salts with acids, containing the elements carbon, hydrogen, nitrogen, chlorine, and oxygen, being very soluble in pyridine, soluble in methanol and in acetone and being slightly soluble in ethanol and in water, its crystals having a refractive index parallel to elongation between about 1.674 and 1.694, and exhibiting characteristic absorption bands in the infra red region of the spectrum when suspended in a hydrocarbon oil in solid form at the following frequencies expressed in reciprocal centimeters: 3420, 1643, 1609, 1580, 1523, 1302, 1231, 1209, 1121, 1080, 1050, 969, 943, 867, 844, 825, 805, 794, 788, 733, and 713.

3. The monohydrochloride salt of an organic substance said salt being effective against Gram positive and Gram negative bacteria, being soluble in water and in methanol and very slightly soluble in acetone, containing the elements carbon, hydrogen, nitrogen, chlorine, and oxygen, the crystals of which have refractive indices of

$$\alpha = 1.633 \pm 0.005; \quad \beta = 1.705 \pm 0.005; \quad \text{and}$$

$\gamma = 1.730 \pm 0.005$, and exhibiting characteristic absorption bands in the infra red region of the spectrum when measured in the solid state while suspended in hydrocarbon oil at the following frequencies expressed in reciprocal centimeters: 3295, 3050, 1665, 1615, 1575, 1523, 1310, 1263, 1225, 1044, 1009, 969, 863, 851, 840, 800, 770, and 698.

4. A process for the production of aureomycin which comprises growing under aerobic conditions a culture of *Streptomyces aureofaciens* in an aqueous medium having a pH between 4 and 7 and containing a soluble carbohydrate, a source of assimilable nitrogen and essential mineral salts at temperatures within the range 20°C. to 35°C. for a period of time of about twenty-four to forty-eight hours whereby the aqueous medium is fermented and aureomycin is produced.

5. A method of producing aureomycin which comprises the step of introducing a culture of *Streptomyces aureofaciens* into an aqueous nutrient liquor having a pH between about 6 and 7 and containing fermentable carbonaceous and nitrogenous substances and mineral salts and fermenting said liquor aerobically until the pH of the liquor has dropped to below about 5.

6. A process of producing aureomycin, which comprises the steps of introducing spores of the fungus *Streptomyces aureofaciens* in an aqueous liquor containing as nutrient material a usable carbonaceous substance, a source of assimilable nitrogen, and mineral substances essential for the growth of the fungus and after said liquor has been fermented by the said fungus at a temperature within the range 20° to 35°C. at a pH between about 4 and 7 separating the insoluble mycelia from the aqueous solution.

7. A process which comprises the steps of aerobically fermenting an aqueous nutrient liquor at a temperature within the range 20° to 35°C. at a pH between about 4 and 7 with the fungus *Streptomyces aureofaciens* whereby aureomycin is produced.

8. A method which comprises the steps of growing the fungus *Streptomyces aureofaciens* in an aqueous solution containing 0.1% to 5.0% by weight of a nitrogenous substance, 0.5% to 5.0% by weight of a usable carbonaceous material, and having a pH between about 4 and 7 and a temperature of between 20°C. and 35° C., while aerating the liquor, whereby aureomycin is produced.

BENJAMIN M. DUGGAR

## APPENDIX 3

### 2,699,054

### TETRACYCLINE

### Lloyd H. Conover, Oakdale, Conn.

### No drawing. Application October 9, 1953, Serial No. 385,041

### 18 Claims. (Cl. 260–559)

What is claimed is:

1. A compound chosen from the group consisting of tetracycline, the mineral acid salts of tetracycline, the alkali metal salts of tetracycline and the alkaline earth metal salts of tetracycline.

2. Tetracycline.

3. Mineral acid salts of tetracycline.

4. Alkali metal salts of tetracycline.

5. Alkaline earth metal salts of tetracycline.

6. Tetracycline hydrochloride.

7. A process for the preparation of tetracycline hydrochloride which comprises contacting an organic solvent solution of chlortetracycline with hydrogen in the presence of a catalyst selected from the group consisting of palladium, platinum and Raney nickel until approximately one mole of hydrogen has reacted with each mole of chlortetracycline used.

8. A process for the preparation of tetracycline hydrochloride which comprises contacting a solution of chlortetracycline in an inert polar organic solvent with hydrogen in the presence of a palladium catalyst at about room temperature and under an elevated pressure up to about 200 p. s. i. until approximately one mole of hydrogen has reacted with each mole of chlortetracycline used.

9. A process for the preparation of tetracycline which comprises reacting one mole of chlortetracycline in an inert organic solvent with approximately one mole of hydrogen in the presence of palladium as a catalyst and recovering the tetracycline thus formed.

10. A process as claimed in claim 9, wherein the reaction is performed in the presence of a basic, acid binding agent.

11. A sodium salt of tetracycline.

12. A calcium salt of tetracycline.

13. A solid antibiotic preparation essentially comprising tetracycline hydrochloride.

14. Crystalline tetracycline melting at 170–175° C. with decomposition and having an optical rotation of $[\alpha]_D^{25} = 239°$, when dissolved in methanol at a concentration of 1%.

15. An antibiotic composition useful in the treatment of various infections in animals, whose antibiotic activity is due primarily to the presence of tetracycline.


## APPENDIX 4

### 2,931,798

### KANAMYCIN AND THE PROCESSES FOR THE PREPARATION THEREOF

### Hamao Umezawa, Kenji Maeda, and Masahiro Ueda, Tokyo, Japan

### Application December 16, 1957, Serial No. 703,096

### Claims priority, application Japan September 5, 1956

### 10 Claims. (Cl. 260–210)

We claim:

1. An antibiotic substance effective in inhibiting the growth of Gram-positive bacteria, Gram-negative bacteria and mycobacteria selected from the group consisting of kanamycin A and kanamycin B and acid addition salts thereof, each of said kanamycins being a substance which is soluble in water and substantially insoluble in n-butanol, ethyl acetate, butyl acetate, ether, chloroform and benzene, which forms salts with acids, which exhibits no absorpotion of untraviolet light from 220 m$\mu$ to 400 m$\mu$, which gives a positive reaction to ninhydrin reagent dissolved in pyridine and positive Molisch and Elson-Morgan reactions and negative Tollens, Sakaguchi, Fehling, maltol and Seliwanoff reactions, which contains only the elements carbon, hydrogen, oxygen and nitrogen, which contains free primary amino groups, which is dextro-rotatory in water and which exhibits characteristic absorption bands in the infra-red region of the spectrum when pelleted in the form of the free base in potassium bromide at the following wave lengths in microns: 2.96, 6.35, 6.48,

6.85, 7.25, 7.45, 7.86, 8.08, 9.65 and 10.4; further properties of said kanamycins being that kanamycin A base has the empirical formula $C_{18}H_{34-36}N_4O_{11}$, exhibits $[\alpha]_D^{24} + 146°$ (c = 1, 0.1 $NH_2SO_4$), gives a salicylidene derivative melting at 272–274° C. with decomposition, gives desoxystreptamine on strong acid hydrolysis, gives a product with an ultraviolet absorption spectrum identical to that of furfural on treatment with 40% sulfuric acid for 100 minutes at 100° C. and exhibits additional characteristic absorption bands in the infra-red region of the spectrum when pelleted in the form of the free base in potassium bromide at the following wave lengths in microns: 2.86, 2.93, 3.01, 3.03, 3.15, 3.22, 3.67, 6.23, 6.75, 6.93, 7.10, 7.33, 7.57, 7.63, 7.75, 7.82, 7.90, 7.95, 8.19, 8.43, 8.55, 8.68, 8.79, 8.90, 9.00, 9.20, 9.47, 9.85, 10.08, 10.63, 10.95, 11.25, 11.35, 11.50, 11.73, 11.95, 12.28, 12.48, 12.77 and 13.05; and that kanamycin B base exhibits $[\alpha]_D + 135°$ (c = 0.63 in water), gives a salicylidene derivative which decomposes without melting at 155–165° C., fails to give a product without melting at 255–265°C., fails to give a product with an ultraviolet absorption spectrum identical to that of furfural on treatment with 40% sulfuric acid for 100 minutes at 100° C. and exhibits additional characteristic absorption bands in the infra-red region of the spectrum when pelleted in the form of the free base in potassium bromide at the following wave lengths in microns: 3.44, 6.74, 8.28, 8.76, 9.55 and 11.15.

2. Kanamycin A free base, as defined in claim 1.

3. Kanamycin B free base, as defined in claim 1.

4. An acid addition salt of kanamycin A, as defined in claim 1.

5. An acid addition salt of kanamycin B, as defined in claim 1.

6. Kanamycin A sulfate, as defined in claim 1.

7. Kanamycin B sulfate, as defined in claim 1.

8. A process for the production of a fermentation broth containing the antibiotic described in claim 1 which comprises cultivating a strain of *Streptomyces kanamyceticus* in an aqueous carbohydrate solution containing a nitrogenous nutrient under submerged aerobic conditions until substantial antibacterial activity is imparted to said solution and then recovering said antibiotic from said solution.

9. A process according to claim 8 wherein the antibiotic is separated from the fermentation broth by adsorption on a cation exchange resin and subsequent elution therefrom.

10. A process according to claim 8 wherein the antibiotic is separated from the fermentation broth by adsorption on a cation exchange resin of the carboxylic acid type and subsequent elution therefrom by an acid.

## APPENDIX 5

### BUDAPEST MICROORGANISM TREATY

The Diplomatic Conference for the Conclusion of a Treaty on the International Recognition of the Deposit of Microorganisms for the Purposes of Patent Procedure was held in Budapest, Hungary from April 14 to 28, 1977. Some 31 States, one intergovernmental organization and 11 international non-governmental organizations were represented in the Diplomatic Conference. Following two weeks of negotiations, the Treaty was adopted by the Diplomatic Conference on April 27, 1976 and signed the following day by 13 States, including the United States. The Treaty will enter into force three months afrer ratification or accession by five States.

The ratification of the Treaty by the United States would give certain advantages to an inventor of a microbiological invention seeking patent protection in a number of countries. An inventor would be able to rely on a single deposit of the microorganism in an approved depositary to satisfy the disclosure requirements of all the member countries of the Union established by the Treaty. Each country will be able to nominate depositaries in its territory for approval, if the depositaries meet the scientific standards required by the Treaty. The public release of a deposit will be governed by the patent laws of the country or countries under which the deposit was made.

We are seeking the views of the public on the advisability of ratification by the United States. Views on the ratification of the Treaty should be submitted to this Office prior to October 14, 1977. If you have any questions or wish further information, please write or call Mr. Michael K. Kirk, Director, Office of Legislation and International Affairs, c/o The Commissioner of Patents and Trademarks, Washington, D.C. 20231 (telephone (703) 557-3065).

C. MARSHALL DANN
*Commissioner of Patents and Trademarks*

August 1, 1977

(The Treaty and Regulations below are reproduced from the May, 1977 issue of *Industrial Property,* published by the World Intellectual Property Organization, Geneva Switzerland.)

## Budapest Treaty on the International Recognition
## of the Deposit of Microorganisms
## for the Purposes of Patent Procedure*

Done at Budapest on April 28, 1977

TABLE OF CONTENTS**

*Official English title.*
    *Source:* International Bureau of WIPO.
    *Note:* This treaty was signed on April 28, 1977, at Budapest, by the following States: Bulgaria, Denmark, Finland, France, Germany (Federal Republic of), Hungary, Italy, Netherlands, Norway, Spain, Switzerland, United Kingdom, United States of America; it will remain open for signature at Budapest until December 31, 1977.
    **This Table of Contents is added for the convenience of the reader (*Editor's Note*).

## APPENDIX 6

### Rule 28

### Requirements of European patent application relating to micro-organisms

(1) If an invention concerns a microbiological process or the product thereof and involves the use of a micro-organism which is not available to the public, the European patent application and the resulting European patent shall only be regarded as disclosing the invention in a manner sufficiently clear and complete for it to be carried out by a person skilled in the art if:

(a) a culture of the micro-organism has been deposited in a culture collection not later than the date of filing of the application;

(b) the application as filed gives such relevant information as is available to the applicant on the characteristics of the micro-organism;

(c) the culture collection, the date when the culture was deposited and the file number of the deposit are given in the application.

(2) The information referred to in paragraph 1(c) may be submitted within a period of two months after the filing of the application. The communication of this information shall be considered as constituting the culture deposited being made available to the public in accordance with this Rule.

(3) The culture deposited shall be available to any person upon request from the date of publication of the application. The request shall be addressed to the culture collection and shall be deemed to have been made only if it contains:

(a) the name and adddress of the person making the request;

(b) an undertaking vis-à-vis the applicant or proprietor not to make the culture available to any other person;

(c) where the request is made before the date of publication of the mention of the grant of the patent, an undertaking vis-à-vis the applicant to use the culture for experimental purposes only.

(4) A copy of the request shall be communicated to the applicant or proprietor.

(5) The undertaking provided for in paragraph 3(b) shall cease if the application is refused or withdrawn or is demeed to be withdrawn or, if a patent is granted, on the expiry of the patent in the designated State in which it last expires.

(6) The undertaking provided for in paragraph 3(c) shall cease if the application is refused or withdrawn or is deemed to be withdrawn or, if a patent is granted, on the date of publication of the mention of the grant of the patent.

(7) The undertaking under paragraph 3(c) is not applicable insofar as the person making the request is using the culture under a compulsory licence. The term "compulsory licence" shall be construed as including ex officio licences and the right to use patented inventions in the public interest.

(8) The President of the European Patent Office shall publish in the Official Journal of the European Patent Office the culture collections which will be recognised for the purpose of this Rule and shall conclude agreements with them, in particular in respect of the deposit, storage and availability of cultures.

Article 53—Exceptions to Patentability—European patents shall not be granted in respect of: (b) plant or animal varieties or essentially biological processes for the production of plants or animals; this provision does not apply to microbiological processes or the products thereof.

## APPENDIX 7

### PATENT AND TRADEMARK OFFICE NOTICES

#### Recombinant DNA

*Accelerated Processing of Patent Applications for Inventions*

In recent years revolutionary genetic research has been conducted involving recombinant deoxyribonucleic acid ("recombinant DNA"). Recombinant DNA research appears to have extraordinary potential benefit for mankind. It has been suggested, for example, that research in this field might lead to ways of controlling or treating cancer and hereditary defects. The technology also has possible applications in agriculture and industry. It has been likened in importance to the discovery of nuclear fission and fusion. At the same time concern has been expressed over the safety of this type of research. The National Institutes of Health (NIH) has released guidelines for the conduct of research concerning recombinant DNA. "Guidelines for Research Involving Recombinant DNA Molecules," published in the Federal Register of July 7, 1976, 41 F.R. 27902–27943. NIH is sponsoring experimental work to identity possible hazards and safety practices and procedures.

In view of the exceptional importance of recombinant DNA and the desirability of prompt disclosure of developments in the field, the Assistant Secretary of Commerce for Science and Technology has requested that the Patent and Trademark Office accord "special" status to patent applications involving recombinant DNA. Upon appropriate request, the Office will make special patent applications for inventions relating to recombinant DNA, including those that contribute to safety of research in the field. Requests for special status should be written, should identify the application by serial number and filing date, and should be accompanied by affidavits or declarations under 37 CFR 1.102 by the applicant, attorney or agent explaining the relationship of the invention to recombinant DNA research. Requests also must include a statement that the NIH guidelines cited above or as amended in the future are being followed in an experimentation in this field, except that the statement may include an explanation of any deviations considered essential to avoid disclosure of proprietary information or loss of patent rights. The requests will be handled in the same manner as requests to make applications special that relate to energy or environmental quality. See Manual of Patent Examining Procedure 708.02.

Dated: Jan. 7, 1977.

C. MARSHALL DANN
*Commissioner of Patents and Trademarks*

Approved: January 10, 1977.
BETSY ANCKER-JOHNSON,
*Assistant Secretary for Science and Technology.*
[FR Doc. 77–1155; Filed 1–12–77; 8:45 am]

## APPENDIX 8

### PATENT AND TRADEMARK OFFICE NOTICES

#### Recombinant DNA

*Suspension of Accelerated Processing of Patent Applications
for Recombinant DNA Research Inventions*

On January 10, 1977, the Patent and Trademark Office issued a notice, published in the Federal Register of January 13, 1977, 42 FR 2712–2713, which provided for the accelerated processing of patent applications for inventions relating to Recombinant DNA, including those that contribute to safety of research in the field.

In order that the Federal Interagency Committee for Recombinant DNA Research may consider recommendations concerning research conducted by the private sector in this field, that part of the referenced notice dealing with accelerated processing of patent applications for Recombinant DNA research inventions is suspended until further notice.

That part of the referenced notice dealing with accelerated processing of patent applications relating to safety of research in this field will remain in force.

Dated: March 3, 1977.

RENE D. TEGTMEYER
*Acting Commissioner of Patents and Trademarks*

Approved: March 3, 1977.
BETSY ANCKER-JOHNSON,
*Assistant Secretary for Science and Technology.*
[FR Doc. 77–6908; Filed 3–8–77; 8:45 am]

## REFERENCES

*American Cyanamid Co. v. Fox* (1964). 140 U.S.P.Q. 199 New York Supreme Court, New York County.

*In re Argoudelis, De Boer, Eble and Herr* (1970). 168 U.S.P.Q. 99 (C.C.P.A.–1970).

*In re Cornelius F. Arzberger* (1940). 46 U.S.P.Q. 32 (C.C.P.A.–1940).

Bäckerhefe (1975). GRUR 1975, p. 430; German Supreme Court.

Behr, O. M. (1975). *J. Pat. Off. Soc.* **57**, 28–45.

*In re Bergy, Coats and Malik* (1977). 195 U.S.P.Q. 344 (C.C.P.A. 1977).

Bunch, R. L., and McGuire, J. M. (1953). U.S. Patent 2,653,899.

*In re Ananda M. Chakrabarty* (1978). 197 U.S.P.Q. 72. (CCPA—1978).

Conover, L. H. (1955). U.S. Patent 2,699,054.

Daus, D., Bond, R. T., and Rose, S. K. (1966). *Idea* **10**, 87–100.

Duggar, B. M. (1949). U.S. Patent 2,482,055.

Dukakis, M. (1977). *Time* April 18, p. 32.

*Feldman v. Aunstrup* (1975). 186 U.S.P.Q. 108 (C.C.P.A.—1975).

*Funk Bros. Seed Co. v. Kalo Inoculant Co.* (1948). 333 U.S. 127; 76 U.S.P.Q. 280.

*Guaranty Trust Company of New York et al. v. Union Solvents Corporation* (1931). 12 U.S.P.Q. 47 (Dist. Ct. Delaware, 1931).

Guidelines for Deposit of Microorganisms (1971). April 29; 886 O.G. Pat. Off. 638.

Hanson, F. R., and Eble, T. E. (1953). U.S. Patent 2,652,356.

Hayhurst, W. L. (1971). *Ind. Property* July, pp. 189–198.

Irons, E. S., and Sears, M. H. (1975). *Annu. Rev. Microbiol.* **29**, 319–332.

*Ex parte Eleanor J. Kropp* (1959). 143 U.S.P.Q. 148 (PTO Board of Appeals 1959).

Levitt, J. (1976). "Current Problems in Patent Law," pp. 49–95. Practising Law Institute.

Levy, D., and Wendt, L. B. (1955). *J. Pat. Off. Soc.* **37,** 855.

*In re Mancy, Florent and Preud'Homme.* (1974). 182 U.S.P.Q. 303 (C.C.P.A. 1974).

Manual of Classification (1936). U.S. Patent and Trademark Office, Class 195—Chemistry, Fermentation Original Classification.

Marcus, I. (1975). *Adv. Appl. Microbiol.* **19,** 77–83.

*Parker v. Bergy* (1978). 198 U.S.P.Q. 257 (U.S. Supreme Court—1978).

*Parker v. Flook* (1978). 198 U.S.P.Q. 193 (U.S. Supreme Court—1978).

Patents for Plants (1954). 35 U.S.C. 161; amended Sept. 3, 1954; 68 Stat. 1190.

Plant Variety Protection Act (1972). 7 U.S.C. 2321.

Reynolds, E. L. (1967). Interference Av. Bv. C; 159 U.S.P.Q. 538.

Richtlinien (1968). Guidelines German Patent Office, Jan. 2 (in Blatt für Patents—, Muster,—und Zeichenwesen).

Rifkin, J. (1977). *Mother Jones Mag.* Feb/March, pp. 23–26 and 39.

Sobin, B. A., Finlay, A. C., and Kane, J. H. (1950). U.S. Patent 2,516,080.

Umezawa, H., Maeda, K., and Ueda, M. (1960). U.S. Patent 2,931,798.

Waksman, S. A., and Schatz, A. (1948). U.S. Patent 2,449,866.

Wegner, H. C. (1974). 5 IIC 285–291.

Weizmann, C. (1919). U.S. Patent 1,315,585.

# Subject Index